FIRE SCIENCES DICTIONARY

CONTRIBUTORS AND REVIEWERS

Walter G. Berl
Applied Physics Laboratory
The Johns Hopkins University

David B. Gratz
Montgomery County, Maryland
Fire and Rescue Service

Gregory C. Helweg
Loyola College
Baltimore, Maryland

Harry E. Hickey
Fire Protection Curriculum
University of Maryland

Lester J. Holtschlag
Applied Physics Laboratory
The Johns Hopkins University

Anne B. Phillips
Smoke, Fire, and Burn Foundation
Boston, Massachusetts

Alan L. Plotkin
Loyola College
Baltimore, Maryland

Alex F. Robertson
National Bureau of
Standards
Washington, D.C.

Alan R. Taylor
U.S. Forest Service
Missoula, Montana

Gordon T. Trotter
Applied Physics Laboratory
The Johns Hopkins University

Editorial Assistants

Jeaneen B. Jernigan
Betty E. Hess
Anastacia M. Kuvshinoff

FIRE SCIENCES DICTIONARY

B. W. KUVSHINOFF
Compiler/Chief Editor
Applied Physics Laboratory
The Johns Hopkins University
Laurel, Maryland

Editors

R. M. FRISTROM
G. L. ORDWAY
R. L. TUVE

A WILEY-INTERSCIENCE PUBLICATION

JOHN WILEY & SONS, New York · London · Sydney · Toronto

To the memory of

Dr. Ralph H. Long, Jr.,

of the National Science Foundation RANN Program,
whose interest and encouragement made this work possible

Published by John Wiley & Sons, Inc.

Copyright © 1977 by Boris W. Kuvshinoff

All rights reserved. Published simultaneously in Canada.

No part of this book may be reproduced by any means,
nor transmitted, nor translated into a machine language
without the written permission of the publisher.
Except for the rights to the Work reserved by others,
the Publisher and the copyright owner hereby grant permission
to domestic persons of the United States and Canada to use this
Work in the English language in the United States and Canada after
May 31, 1984, without charge. Permission will also be granted
for foreign publication after said date. For conditions of use
and permission to use this Work for foreign publication or
publications in other than the English language apply to the Publisher.

Library of Congress Cataloging in Publication Data

Kuvshinoff, B   W
    Fire sciences dictionary.

    "A Wiley-Interscience publication."
    1. Fire prevention—Dictionaries. 2. Fire
extinction—Dictionaries. I. Title.
TH9116.K88   628.9'2'03   77–3489

ISBN 0471-51113-7

Printed in the United States of America

10 9 8 7 6 5 4 3 2 1

PREFACE

It is not difficult to realize that each profession and occupation develops a vocabulary that is often strange and even confusing to outsiders. This was the primary motivation for compiling an interdisciplinary dictionary of fire terminology: to provide a common basis of communication and understanding for the diverse members of the fire community.

This dictionary contains about 10,000 fire-related terms selected from 75 branches and fields of science and technology. The source material included many glossaries, technical dictionaries, word lists, handbooks, journal articles, books, and reports. Publications of the National Fire Prevention Association were a rich source of technical terms. Leads to many terms in emergency medical care and forest fire prevention and suppression were obtained from several excellent training manuals that were graciously provided by the Robert J. Brady Co. of Bowie, Maryland. Definitions of terms appearing in ASTM standards and in an ISO list of recommendations were studied and compared with those given in the NFPA National Fire Codes. As rich a source as any for the dictionary were terms used in the definitions that had already been prepared.

Inasmuch as a significant amount of important fire literature is available from the United Kingdom and the Commonwealth Nations, a modest effort was made to include British terms and usage. Such terms are marked by the abbreviation "Brit." To ensure the correctness of such terms, a prepublication draft of this dictionary was reviewed by various staff members of the Fire Research Station, Borehamwood, England.

In developing the definitions we found conflicts from time to time. This occurred especially between terms that were close in meaning and were sometimes used interchangeably in the literature. Wherever a choice is made, for example, to sharpen and separate the meanings of two nearly synonymous terms, a brief explanation is given in a note. In most cases, however, we chose one member of a synonym pair and cross-referenced the other to it. This was not done arbitrarily, but rather with due consideration of language, utility, usage, and often after consultation with one or more experts in the field. We also cross-referenced liberally to antonyms and related terms, especially if additional insight into meaning could be obtained under the term referred to. All senses of the terms in this dictionary are not given, just those that have some bearing on the general subject area of fire. Homonyms, words spelled identically but having distinctly different meanings, are entered individually and are distinguished by a superscripted number. Different senses of a term are introduced by a sequential number if there is only one entry for the word, or by a lowercase letter if the term is a homonym and thus appears more than once.

Extensive references to chemical compounds are available in many publications. Nevertheless, quite a few that have special significance to fire have been included. Numerous expansions of acronyms have been included as well, since they appear frequently in the literature and are puzzling to those not used to them.

Although encyclopedic treatment was not our intent, it was not categorically ruled out. Relevant comments and explanatory information is provided in the form of notes appended to many of the definitions. The definitions themselves are generally brief, and are phrased in language that would be intelligible to an educated layman. Therefore, this dictionary should be useful not only to scientists, engineers, members of the fire service, and other professionals engaged in fire research, prevention, suppression, and safety, but also to students, who in ever-increasing numbers, are taking courses in fire science and technology.

BORIS W. KUVSHINOFF

Laurel, Maryland
January 1977

ACKNOWLEDGMENTS

Work on this dictionary was initiated at the Applied Physics Laboratory of the Johns Hopkins University late in 1969. The major part of the work was done under Grants GI-12 and GI-3488X from the RANN Program (Research Applied to National Needs) of the National Science Foundation commencing in 1970. The final revisions for the present edition were done under a grant from the National Fire Prevention and Control Administration, U.S. Department of Commerce. The support received from these organizations is deeply appreciated.

The text for this dictionary was prepared and maintained on an IBM-360/91 computer, using the APL INFO-360 document-writing package of programs. Without this facility it is doubtful that this book would have reached its present state. We thank Dr. R. P. Rich and his staff, especially G. T. Trotter, for designing so much versatility into the program, making it possible to publish this book at much lower cost. Special thanks are due also to Betty E. Hess for managing the massive additions and changes to the file with smoothness and efficiency.

The principal reviewers are listed on the title page. To them we tender our sincere thanks.

Among the many individuals who commented and contributed to this book we would like to express our special appreciation to the following: Gene W. Benedict, U.S. Forest Service; A. Boyt and other staff members of the Fire Research Station, Borehamwood, U.K.; John M. Best, Montgomery County (Md.) Fire and Rescue Services; Martin M. Brown, Society of Fire Protection Engineers; Homer W. Carhart, U.S. Naval Research Laboratory; the members of the Marshalltown (Iowa) Fire Department; Robert Van Brunt, Underwriters' Laboratories, Inc.; and John P. Wagner, Gillette Research Institute.

B.W.K.

ABBREVIATIONS

a	acceleration	deg	degree(s)
A	absorptivity	Demog	demography
A	Angstrom	Dosim	dosimetry
abbrev	abbreviation	e	emissivity
Accoust	acoustics	Educ	education
Aerodyn	aerodynamics	e.g.	exempli gratia (for example)
Anat	anatomy	Electr	electricity; electronics
Ant	antonym	emf	electromotive force
Anthro	anthropology	Eng	engineering
approx	approximately	Env Prot	environmental protection
Atm Phys	atmospheric physics	esp	especially
Audio	audiometry	etc.	et cetera (and so forth)
Autom	automation	ev	electron volt(s)
Bact	bacteriology	F	Fahrenheit
Biol	biology	Fire Pr	fire practice
Build	building	First Aid	first aid
bp	boiling point	Fl Mech	fluid mechanics
Brit	British	For Serv	Forest Service
C	Celsius (centigrade)	fps	foot-pound-second; feet per second
Cartog	cartography		
cf	compare	ft	foot (feet)
cgs	centimeter-gram-second	g	gram(s)
Chem	chemistry	g	gravity
Chem Eng	chemical engineering	gal	gallon(s)
Civ Def	civil defense	Geod	geodesy; geodetics
Civ Eng	civil engineering	Geog	geography
cm	centimeter(s)	Geol	geology
colloq	colloquial	Geom	geometry
Comb	combustion	Geophys	geophysics
Commun	communications	gpm	gallons per minute
Comput	computer; computing science	Hydraul	hydraulics
cp	centipoise	Hydrol	hydrology
cps	cycles per second (see: Hz)	Hz	Hertz (cycles per second)
Cryo	cryogenics	i.e.	id est (that is)
Cyber	cybernetics	igpm	imperial gallons per minute

in.	inch(es)	Petrochem	petrochemistry
Ind Eng	industrial engineering	Pharm	pharmacy; pharmacology
Ind Hyg	industrial hygiene	Photog	photography
Info Sci	information science	Photom	photometry
Insur	insurance	Phys	physics
K	Kelvin	Physiol	physiology
l	liter(s)	psi	pounds per square inch
Law	law	psia	pounds per square inch absolute
lb	pound(s)		
m	meter(s)	psig	pounds per square inch gage
M	mass	Psych	psychology
Man Sci	management science	q.v.	quod vide (which see)
Mat Hand	materials handling	R	Rankine
Math	mathematics	rpm	revolutions per minute
Mech	mechanics	Safety	safety
Mech Eng	mechanical engineering	Salv	salvage
Med	medicine	sec	second(s)
Metall	metallurgy	Soc Sci	social science
Meteor	meteorology	Soil Mech	soil mechanics
Metrol	metrology	Spectr	spectroscopy
mev	million electron volt(s)	sp gr	specific gravity
mgpd	million gallons per day	Statist	statistics
Milit	military	Storage	storage
Mineral	mineralogy	Surg	surgery
Mining	mining	Surv	surveying
mks	meter-kilogram-second	Syn	synonym
mm	millimeter(s)	Techn	technology
mol wt	molecular weight	Testing	testing
mp	melting point	Toxicol	toxicology
mph	miles per hour	Transp	transportation
Navig	navigation	usu	usually
Nucl Phys	nuclear physics	v	velocity
Nucl Safety	nuclear safety	yd	yard (unit of length)
Opt	optics		
Packag	packaging		
Path	pathology		

FIRE SCIENCES DICTIONARY

A

A¹ (Chem) Formerly an abbreviation for *argon*.

A² (Phys) Abbreviation for *Angstrom unit*. Formerly Å.

A³ (Techn) An abbreviation, often lowercased, for absorptivity, acceleration, acre, ambient, ampere, anode, area.

AAAS = American Association for the Advancement of Science

AAIN = American Association of Industrial Nurses, Inc.

AAOO = American Academy of Ophthalmology and Otolaryngology

abandonment (First Aid; Law) Leaving an injured person before aid arrives.

abaxial Away from the axis or situated outside it. Ant: *axial*.

ABC method (First Aid) A mnemonic that stands for airway, breathe, and circulation: the sequence of operations in cardiopulmonary resuscitation.

abdomen (Anat) The part of the body between the chest and the pelvis, esp the cavity or interior part of the body containing the viscera.

abeam (Navig) At right angles to a ship's keel.

aberration¹ (Psych) Deviation from a normal course; peculiar behavior.

aberration² (Opt) Inability of an optical element, lens, or mirror to focus light rays sharply. Aberrations are classified as: astigmatism, coma, curvature of field, distortion, spherical aberration, and chromatic aberration.

ABIH = American Board of Industrial Hygiene

ability (Psych) The capacity to perform a given mental or physical act or a task.

ability grouping (Educ) Placing students according to their scholastic abilities.

abirritant (Pharm) A soothing agent that relieves an irritation. Ant: *irritant*.

ablation¹ (Phys) Removal of material by erosion, evaporation, or reaction for short-term protection against high temperatures.
NOTE: The ablating material forms a thermal barrier to protect the underlying structure. Typical materials in space reentry applications are phenolic and epoxy resins.

ablation² (Surg) Surgical removal of a part, esp by cutting, often of abnormal growths or harmful matter from the body.

ablative water (Fire Pr; For Serv) Viscous water used to wet down vegetation in fighting forest and brush fires, usu dropped from aircraft.

ablaze (Fire Pr) On fire. Burning briskly. See: *fire*.

abnormal (Psych; Statist) Diverging significantly from the average.

abortion (Med) The spontaneous or induced termination of a pregnancy before the fetus is mature enough to live independently. The removal or expulsion of a fetus, esp during the first three months of gestation. Cf: *miscarriage*.

abrasion¹ (Techn) A mark or other evidence of wear or damage left on a surface as a result of rubbing or scraping.

abrasion² (Techn) The act of wearing away by rubbing, as the wearing of moving machine parts from friction.

abrasion³ (Techn) A process in the form of grinding, lapping, or polishing used for machining and finishing of hard materials, esp when precision is required.

abrasion⁴ (Med) Surface damage to the skin or mucous membrane by rubbing or scraping.

abrasive (Techn) An extremely hard natural or synthetic material used for grinding, cutting, or polishing. Examples are emery, diamond, corundum, sand and sandstone, silicon carbide, pumice, and fused alumina. Typically, granular or gritty substances are used for grinding, while rounded particles are used for polishing and buffing.
NOTE: Abrasives are used in the form of (1) loose powders or grains, (2) paper or cloth coatings (sandpaper; emery, flint, or garnet cloth; etc), (3) grinding and polishing wheels, and (4) stones (hones, whetstones, oilstones, etc). Boron carbide (B_4C), pumice, and boron nitride, BN (borazon), approach diamond in hardness and serve as excellent abrasive materials.

abrasive blasting (Techn) A process for cleaning surfaces by means of sand, aluminum, steel grit, or other abrasives in a stream of high-pressure air. Also called: *sand blasting*.

abscissa (Math) One of the coordinates of a rectilinear (cartesian) *coordinate system*, conventionally taken parallel to the horizontal (or x) axis. Cf: *ordinate*.

absenteeism (Man Sci) The repeated failure of a worker to report to work.

absolute¹ (Abb: abs) Complete, perfect, without condition. Independent of other things.

absolute² (Metrol; Phys) Referring to measurement systems based on universal constants, primary length-mass-time-charge standards such as the *cgs*, *mks*, or *fps* systems of units.

absolute alcohol (Chem) Ethyl alcohol, $C_2H_5(OH)$, containing less than 1% by weight of water.

absolute error (Psych) The difference between the observed value and the true value or the mean.

absolute humidity (Phys) The quantity of water vapor present in a unit quantity of gas, usu expressed as the mass of water vapor per unit volume of gas including the water vapor; e.g., as grains per cubic foot or grams per cubic meter. See also: *humidity*; *relative humidity*.

absolute instrument (Metrol) An instrument that does not require calibration beyond length-time-mass measurements. Cf: *secondary instrument*.

absolute manometer See: *manometer*

absolute measurement (Math) Measurement taken without regard to algebraic sign. A measurement that is not dependent on other variables.

absolute pressure (Techn) Pressure referred to a scale on which an *absolute vacuum* is zero and expressed as pounds per square inch absolute (psia) or in pascals. By contrast, *gage pressure* indicates the difference between *atmospheric pressure* and the measured pressure. Thus, absolute pressure is the sum of gage pressure and atmospheric pressure. If the measured pressure is negative, it is commonly referred to as a *vacuum*, or partial vacuum. See: *pressure*.
NOTE: Absolute pressure is generally used in vacuum technology; gage pressure is most common in other applications.

absolute roughness (Techn) A single linear dimension used to describe the roughness characteristics of a surface.

absolute temperature (Phys) Temperature referred to *absolute zero* and expressed in degrees Kelvin or Rankine. See: *temperature scale*.
NOTE: The Kelvin scale is related to the Celsius (centigrade) scale, the Rankine to the Fahrenheit. To convert Celsius to Kelvin, add 273.15; Fahrenheit to Rankine, add 459.67.

absolute threshold (Psych) 1. The magnitude of a physical stimulus at which it can be detected reliably. 2. The minimum stimulus that can be detected at least half the time in many trials.

absolute vacuum (Phys) A theoretical concept of a space completely empty of matter (unobtainable in practice). Also called: *ideal vacuum*. See: *vacuum*.

absolute value (Math) The mathematical value of a number without regard to algebraic sign.

absolute viscosity (Phys) A coefficient of proportionality in *Newton's law of viscosity*. Also called: *viscosity coefficient*; *dynamic viscosity*.

absolute zero (Cryo) The lowest temperature (-273.16°C, 0°K, -459.67°F, or 0°R). The point on the scale at which molecular motion ceases. See: *temperature scale*.

absorbed dose (Dosim) The quantity of ionizing radiation energy, measured in *rems* or *rads*, taken up by a unit mass of a material.

absorbent[1] (Techn) Capable of taking in, blotting, or sucking up, as a sponge (liquid) or activated charcoal (gas).

absorbent[2] (Acoust; Build) A material that absorbs or dissipates sound.

absorbent cotton Dewaxed *cotton* commonly used for absorbing liquids, dressing wounds, etc.

absorber[1] (Phys) Any substance or medium that absorbs. That which is absorbed can be a gas, liquid, or any form of acoustic, mechanical, or electromagnetic energy.

absorber[2] (Nucl Phys) Generally a neutron absorber, a material in which neutrons disappear as free particles without producing other neutrons. Any material that diminishes the intensity of ionizing radiation.

absorptance; **absorbtance** (Phys; Symb: a or α) The ratio of radiant flux absorbed by a body or material to the radiant flux incident upon it. Also called: *absorption factor*.
NOTE: Absorptance measured over all wavelengths is *total absorptance*; at a given wavelength it is *spectral absorptance*; and in the visible range it is referred to as *optical absorptance*. Cf: *attenuation*.

absorption[1] (Chem; Phys) The taking up of one material into the bulk of another, as a gas by a liquid, or water by a sponge.
NOTE: The mechanism is usu physical, but chemical forces also can be involved. See also: *colloids*. Cf: *adsorption*; *sorption*.

absorption[2] (Acoust; Opt) The attenuation of electromagnetic energy, sound, or incident particles by an absorbing medium. The process in which the number or the energy of particles, photons, waves, or vibrations passing through a medium is reduced by interaction with the medium. See also: *dispersion*; *scattering*.

absorption[3] (Psych) A state of deep concentration or rapt attention to an object or activity.

absorption[4] (Med) The taking up of fluids or other substances, such as medication, by the skin, mucous surfaces, or vessels.

absorption band (Spectr) A spectral region in which absorptivity is maximum.

absorption coefficient (Spectr; Symb: a) A measure of the decrease in intensity due to absorption of light, sound, or particles passing through a substance. If the beam is also scattered, that portion is called the *scattering coefficient*. The sum of absorption and scattering coefficients is the *extinction coefficient*. See also: *absorption spectrum*.

absorption edge (Spectr) The wavelength marking an abrupt discontinuity in the intensity of an *absorption spectrum*, usu due to the onset of a process; e.g., x-ray absorption.

absorption factor See: *absorptance*

absorption line (Spectr) A narrow

characteristic electromagnetic frequency absorbed by the medium through which the energy is passing. The lines are usu associated with an atomic_spectrum.

absorption_spectroscopy (Spectr) The measurement and analysis of an absorption_spectrum. The pattern of absorption can be used to deduce atomic and molecular energy levels and structures and for chemical analyses. Ant: emission_spectroscopy.

absorption_spectrum (Spectr) The characteristic pattern of absorption vs wavelength obtained by passing continuous radiation through an absorbing medium.

absorptivity (Opt; Symb: A) The capacity of a substance to absorb radiant energy passing through it.

ABS_resin (Chem) A plastic formed by polymerization: acrylonitrile-butadiene-styrene.

abstract_ability (Psych) The capacity to use abstract ideas in problem solving and in dealing with novel situations.

abutment (Civ Eng) A supporting structure for a bridge, arch, or beam.

ac = alternating_current

accelerant (Fire Pr) A substance used to initiate or promote the spread of a fire. Flammable_liquids are the most common accelerants. Cf: accelerator[2].

acceleration (Symb: a) The time rate of increase in velocity, expressed mathematically as $a = dv/dt$, where v is velocity and t is time. Acceleration is a vector quantity, involving both magnitude (change in speed) and direction. Ant: deceleration. See also: speed.

accelerator[1] (Techn) A device for imparting or controlling acceleration.

accelerator[2] (Chem) A substance that hastens a chemical reaction; a catalyst.

accelerator[3] (Nucl Phys) A device for accelerating charged particles, such as electrons, neutrons, protons, or ions.

accelerometer (Techn) A device for measuring linear or angular acceleration. Often used to measure vibrations, shock, and other displacements. A transducer that converts acceleration into a proportional electrical signal.

acceptance_conditions (Techn) The measured or observed characteristics determining whether or not a test specimen complies by performance or specification with specified requirements; used particularly with fire endurance tests. Syn: acceptance_criteria.

acceptance_test (Fire Pr) A test conducted to determine that equipment operates satisfactorily. With fire apparatus this is commonly a pumping_test.

accepted_engineering_requirements (Build; Techn) Requirements or practices which are compatible with designs specified by a registered architect, a registered professional engineer, or other duly licensed individual or recognized authority.

access (Build) A way or means of entering. Easy reachability or freedom of passage.

access_opening (Techn) An aperture through which something can be reached, as for repair or maintenance.

accessory (Techn) 1. Anything of subordinate or secondary importance. 2. A part, device, or assembly that contributes to the effectiveness of a piece of equipment, or enables it to perform an additional function or operation.

access_panel (Build; Techn) A closure for an access opening.

accident (Safety) An unplanned event, sometimes but not necessarily injurious or damaging, which interrupts an activity. A chance occurrence arising from unknown causes; an unexpected happening due to carelessness, ignorance, and the like.

NOTE: Accidents are most often characterized as being sudden, unexpected, and upon onset beyond human control. According to researchers, accidents are not arbitrary and random; they arise from multiple factors, follow epidemiological principles, and are avoidable through multiple preventive measures.

accidental alarm (Fire Pr) A spurious signal transmitted by a fire alarm system, caused by electrical or mechanical malfunctions, heat from industrial processes, etc.

accidental death (Safety) An injury that terminates fatally and is causally related to an accident.

accidental error (Metrol) A random error occurring from unknown causes in experimental observations. Cf: systematic error.

accidental operation (Fire Pr) Transmission of an alarm by an alarm system or the release of water by a sprinkler system when no fire is present.

accident analysis (Safety) Investigation, study, and evaluation of causes and effects of accidents to determine contributing factors, identify hazards, and assess damage.

accident and sickness benefits (Insur) Periodic payments to workers who are absent from work due to off-the-job disabilities through accident or sickness.

accident cause (Safety) One or more factors associated with an accident or a potential accident.

accident costs (Insur) Monetary losses associated with an accident, consisting of: Direct costs: Monetary losses directly ensuing from an accident (e.g., costs of workmen's compensation payments and medical expenses). Indirect costs: Costs not directly associated with the accident but which are real and measurable and have been incurred because the accident happened (e.g., wages paid above compensation costs, cost of accident investigation, lost time of other workers, etc.). Insured costs: Costs covered by compensation insurance and other insurance programs (e.g., medical, property damage, etc.). Uninsured costs: Measurable costs not covered by existing insurance programs (e.g., property damage, wages paid for nonproductive hours, extra cost of overtime and wages for activities necessitated by the accident, and other miscellaneous costs such as loss of profit on contracts cancelled or orders lost, public liability claims, cost of renting replacement equipment, etc.).

accident experience (Safety) 1. One or more indexes describing accident occurrence according to various units of measurement (e.g., disabling injury frequency rate, number of lost-time accidents, disabling injury severity rate, number of first-aid cases, or dollar loss). 2. A summary statement describing accident occurrence.

accident free (Safety) A record of no accidents, sometimes of specified types, relating to an operation, activity, or worker performance during a given time period.

accident hazard (Safety) A situation present in an environment or connected with a job procedure or process that has the potential for producing an accident.

accident liability (Safety) A predisposition to have an accident. The probability of a worker becoming involved in an accident or the probability of unsafe conditions or unsafe acts producing an accident. Syn: accident probability.

accident location (Safety) The exact position of the key event that produced the accident.

accident potential (Safety) A behavior or condition likely to produce an accident.

accident prevention (Safety) The application of countermeasures designed to reduce accidents or accident potential within a system or organization. Programs directed toward accident avoidance. The reduction or elimination of behaviors and conditions having an accident potential.

accident probability (Safety) The likelihood of a worker, operation, or item of equipment becoming involved in an accident.

accident proneness (Safety) The tendency to be involved in more accidents than would be expected by chance alone. Syn: **accident susceptibility**.
NOTE: Sometimes attributed to personality factors.

accident rate (Safety) Accident experience in relation to a base unit of measure (e.g., number of disabling injuries per 1,000,000 man-hours exposure, number of accidents per 1,000,000 miles traveled, total number of accidents per 100,000 employee-days worked, or number of accidents per 100 employees).

accident records (Safety) Reports and other recorded information concerning accident experience.

accident repeater (Safety) A person who has been involved in two or more accidents in a predetermined period of time (regardless of fault).

accident reporting (Safety) Collecting information, preparing, and submitting an official report of an accident to a designated individual or agency.

accident statistics (Safety) Descriptive or inferential data that provide information about accidents.

accident susceptibility (Safety) A predisposition to a disproportionate share of accidents because of personality or behavioral characteristics. Syn: **accident proneness**.

accident type (Safety) A description of the source of injury explaining how the injury occurred; how the injured person came in contact with the object, substance, or exposure named as the source of injury; or the movement of the body that led to the injury.

acclimatization (Ind Eng; Safety) Physiological adjustment to the environment over a period of time. Adaptation to a different climate.

accommodation[1] (Phys) An adjustment to compensate for an external force.

accommodation[2] (Physiol) A change in the shape of the lens of the eye that enables it to focus on objects at different distances.
NOTE: The ciliary muscle flattens the lens for focusing on distant objects and relaxes, allowing it to become spherical for near ones.

accommodation[3] (Psych) A change in behavior, based on experience, to cope with new situations. Accurate perception and appropriate reactions to numerous new and changing cues.

accommodation coefficient (Kinetics; Symb: a) The ratio of the average actual energy transferred between a surface and gas molecules rebounding from it to the theoretical energy transfer if the rebounding molecules were in complete thermal equilibrium with the surface.

accommodation time (Physiol) The time it takes the eye to refocus when a change occurs in viewing distance.

accordion fold (Fire Pr) A method of folding back and forth, commonly used for storing salvage covers and hoses.

accordion horseshoe load (Fire Pr) A method of stowing hose on a truck bed in a nested, square-cornered, U-shape.

accordion hose (Brit) = **flaked hose**

accordion hose load (Fire Pr) A quantity of hose folded back and forth accordion style and carried in vertical stacks or horizontal layers in a hose compartment of an apparatus.

accountability (Man Sci) Responsibility for the accomplishment of or failure to perform an assigned function or task according to designated performance standards.

accreditation (Educ) The process by which an educational institution receives official recognition that it meets acceptable criteria of quality.

accumulator[1] (Phys) A device for storing energy, matter, data, etc.

Hydraulic accumulators store fluid under pressure.

accumulator² (Brit) An electrical storage battery.

accumulator³ (Comput) A device that adds numbers received and stores the sum.

accuracy (Statist) The quality of exactness or precision. A measure of the agreement between an actual value and a measured value.

acetaldehyde (Chem) A volatile, colorless, highly flammable liquid, CH_3CHO, bp 69°F (22°C), obtained by partial oxidation of ethyl alcohol. Vapor forms explosive mixtures with air over a wide range of concentrations. Used as a chemical intermediate in the manufacture of resins and other organic compounds. Also called: acetic aldehyde; ethanal. See also: formaldehyde.

acetate (Mech) A synthetic material used for fibers as cellulose acetate, commonly called rayon.
  NOTE: Cellulose acetate and rayon are now treated as different generic categories of synthetic fibers.

acetic aldehyde = acetaldehyde

acetone (Chem; Techn) A colorless, volatile, fragrant, extremely flammable liquid, $CH_3 \cdot CO \cdot CH_3$, bp 56° C, fp -95°C, widely used as a solvent.

acetone breath (Med) A sweet, fruity breath odor, often present in diabetic coma.

acetylene (Chem; Techn) A colorless, flammable gas, C^2H^2, having moderate toxicity; bp -83.6°C (sub), mp -81.8°C. Mixtures between 3% and 80% of acetylene with air and the pure gas pressurized above 2 atm are explosive. Used with oxygen in welding and extensively as a chemical intermediate. Also called: ethyne.
  NOTE: The gas is sometimes dissolved in acetone to reduce the explosion hazard.

ACGIH = American Conference of Governmental Industrial Hygienists

achieved reliability (Techn) A statistical estimate of reliability based on actual demonstrations under specified conditions.

achievement test (Educ) A test that measures how well an individual's current level of learning or experience compares to that of others.

achromatic (Opt) 1. Lacking color. 2. Producing images substantially free of color fringing, usu in reference to lenses that have been corrected at two wavelengths. See: aberration.

acid (Chem) 1. A typically sour-tasting or biting compound that turns blue litmus red and forms salts in reactions with bases. Some acids are very corrosive and attack metals, fabrics, and body tissues. The primary characteristic of an acid is that it can donate a proton or accept an unshared electron pair in a chemical reaction with a base. It is most often an electrolyte that dissociates with the formation of hydrogen ions and corresponding negative ions. This behavior depends on the solvent, which is usu water. Common acids in water are HCl (hydrochloric), H_2SO_4 (sulfuric), H_3PO_4 (phosphoric), and HNO_3 (nitric), CH_3COOH (acetic), etc. 2. An organic compound containing the -COOH group. Some compounds can be amphoteric; that is, an acid (proton donor) in one system and a base (proton acceptor) in another.

acid anhydride (Chem) A compound that reacts with water to form an acid.
  NOTE: Examples are oxides of nonmetals such as $H_2O + SO_3 \rightarrow H_2SO_4$ and organic anhydrides; e.g., $C_4H_6O_3$ (acetic anhydride) + $H_2O \rightarrow 2HCOO \cdot CH_3$ (acetic acid).

acid number (Chem) A number characterizing the amount of free acid in a substance, defined as the number of milligrams of potassium hydroxide required to neutralize the free fatty acids in one gram of material.
  NOTE: The number is often used to characterize fats, oils, etc.

acid pickling (Metall) A bath treatment, usu a sulfuric acid solution, used to remove scale and other impurities from metal surfaces prior to plating or other surface treatment.

acid salt (Chem) A salt containing replaceable hydrogen. An acid having

some of its replaceable hydrogen replaced by another positive ion.
NOTE: Often indicated by the prefix bi (as in sodium bisulfate, $NaHSO_4$).

ACIL = American Council of Independent Laboratories

Ackermann steering (Transp) The standard system of steering in which the front wheels are mounted on pivoted knuckles and are interconnected by a linkage. During a turn, the inner wheel rotates through a larger angle than does the outer wheel.

acne (Med) Oil (sebaceous) dermatitis; an inflammation of glands in the skin.

acoustic (Phys) Related to or associated with sound. Having the properties, dimensions, or physical characteristics of sound waves.

acoustical tile (Build) A sound absorbing interior finishing material, usu made of soft cellulosic matter pressed into foot-square (or larger) interlocking panels. Commonly applied to ceilings.

acoustic dispersion (Acoust) 1. The change in the speed of sound waves with a change in frequency. 2. The propagation of sound, as by a loudspeaker.

acoustic reflex (Psych) The automatic contraction of the stapedius muscle of the ear to moderately intense sounds to protect the cochlea from excessive stimulation.

acoustics (Phys; Abb: Acoust) 1. The science of sound, including its production, transmission and effects. 2. Narrowly, the study of sound transmission through various media.

acoustic trauma (Med) Hearing loss caused by a loud noise or by a blow to the head. In most cases hearing loss is temporary, although there may be some permanent loss.

acquittal (Law) A finding by a court or other legal process that a defendant is not liable for the offense with which he was charged.

acre (Metrol) An area equal to 4840 square yards. Cf: are.

acrolein (Chem) A highly irritating, toxic aldehyde, C_3H_4O, formed during the combustion of petroleum products. Also called: acrylic aldehyde.

acrylic resin (Techn) One of the thermoplastic glass-like resins of polymerized esters of acrylic or methacrylic acids. A methylmethacrylate polymer.
NOTE: Commercial trade names include Plexiglas, Lucite, and the British Perspex.

ACS = American Chemical Society

acting (Fire Pr) Descriptive of a temporary or substitute rank. A title given to an officer or firefighter who is assigned to duties of the next higher rank, or who assumes such duties in the absence of the regular firefighter or officer holding the rank. See: ranks and titles.

actinic (Opt) Pertaining to the capacity of light to induce photochemical changes in matter.

actinide series (Chem) A group of elements of increasing atomic number, starting with actinium (atomic number 89) and extending through lawrencium (the recently discovered element 103).
NOTE: Uranium-92 and plutonium-94 are included in this series.

actionable fire (For Serv) 1. A fire started illegally or allowed to burn or spread in violation of law, ordinance, or regulation. 2. Any fire that requires suppression.

activated carbon (Chem; Techn) An adsorbent charcoal made by steaming or destructive distillation of carbonaceous material under controlled conditions.
NOTE: Activated carbon or charcoal is commonly used in gas-adsorbent gas masks. Toxic substances are effectively removed when the fluid (gas) in which they are suspended is passed through a bed of granular or powdered gas-adsorbent carbon.

activated complex (Chem) A transient molecular complex formed during a reactive collision.

activation (Nucl Phys) The process of making a material radioactive by bombarding it with neutrons, protons, or other nuclear particles. Also called: **radioactivation**.

activation analysis (Chem) A sensitive method for identifying and measuring the relative abundance of chemical elements in a sample of a material. The sample is made radioactive by exposing it to bombardment by neutrons, gamma rays, or other particles. The sample then emits radiations that characterize the kinds of atoms present and their number.

activation energy (Chem; Phys) An energy barrier to a process, e.g., to chemical reactions, to the promotion of an electron to the conduction band, or to the motion of lattice defects in a crystal.

active avoidance learning (Psych) A learning task in which the subject must make a prescribed response in order to avoid an unpleasant or harmful stimulus. See: **passive avoidance learning**.

active learning (Educ) 1. Learning procedures that stress recitation and performance as opposed to simple reading of materials. 2. Learning in which the student has a strong desire and predisposition to master the subject.

active redundancy (Electr) A configuration in which circuits or other devices operate simultaneously in parallel rather than being switched on when needed. Each of the parallel units is capable of functioning satisfactorily alone. The redundancy is provided to enhance operational reliability.

active rudder (Navig) A rudder fitted with a submersible motor and propeller to provide directional thrust at low speeds.

activity coefficient (Phys; Chem) A correction factor used in fundamental calculations to account for deviations from the ideal in solutions and in the gaseous state. See: **fugacity**.

activity sampling (Ind Eng; Safety) Observations taken of the activity of an individual in a work situation to determine what he does during a given period of time.
NOTE: In occupational safety, for example, a measurement technique for evaluating potential accident-producing behavior. It involves the observation of worker behavior at random intervals and the instantaneous classification of these behaviors as safe or unsafe. Calculations are then made to determine either (a) the percentage of time the workers are involved in unsafe acts or (b) the percentage of workers involved in unsafe acts during the observation period.

act of God (Safety; Insur) An accident causal factor generally interpreted as being beyond human control (e.g., lightning, flood, tornado, earthquake, etc). See: **accident**.

actual cash value (Fire Pr; Insur; Abb: ACV) The replacement cost less depreciation of property lost by fire or other cause.

actual exposure (Ind Hyg) Employee-hours of exposure to a potential hazard taken from payroll or time-clock records, including only actual hours worked.

actuarial (Insur) Pertaining to probabilities; determined from experience or observations that serve as a basis for computing insurance rates. Syn: **probabilistic**.

actuate (Fire Pr) To make active; to energize an electrical circuit or a mechanical device. To set into operation, as a fire alarm or a fire protection device.

acuity (Physiol) The keenness of perception, esp with reference to sharpness of vision or hearing, but also to smell and touch.

acute[1] (Med) a. Having a rapid onset; severe and of short duration. b. Accompanied by symptoms of varying degrees of severity with implications of rapid onset, possibly developing into a crisis; opposed to chronic, as an acute disease.

acute[2] (Geom) Referring to an angle

of less than 90 degrees.

acute[3] (Physiol) Capable of making fine distinctions.

acute exposure (Med) Short term presence of an individual or organism in an unusual and often hazardous situation, as for example, brief presence in an atmosphere of high heat, intense cold, high bacterial or parasitic count, significant radiation, and the like.

ACV = actual cash value

adaptation (Psych) The process of adjusting to a stimulus. Becoming accustomed to something. An example is the adjustment of the iris of the eye to an increase or decrease in the level of ambient light.

adapter[1] (Techn) a. A mechanical device for connecting, joining, or fitting one thing to another. b. An accessory used to alter a device to perform a function not originally intended.

adapter[2] (Fire Pr) A fitting for connecting hose couplings of different size or thread pitch. If the coupling being attached is smaller in diameter, the adapter is called a reducer; if larger, it is called an increaser. See: coupling. See also: hose.

adaptive behavior (Psych) The capacity of an individual to respond appropriately to the environment and to cope effectively with changes occurring in it.

adaptive control system (Techn) A system that monitors the performance of another system and automatically adjusts it to produce a predetermined output. See: cybernetics.

adder (Comput) A device in a computer that performs additions.

additional alarm (Fire Pr) Any alarm subsequent to the first, calling for higher ranking officers, additional equipment, and assignment of companies to protect areas normally covered by those who have responded to the previous alarms. See: multiple alarm.
NOTE: A call for a special unit is generally not considered to be an additional alarm.

additional alarm company (Fire Pr) A company assigned or dispatched to respond to an alarm other than its first alarm assignment.

additive (Techn) A chemical substance introduced in small quantities to materials or products to improve stability, protect against deterioration, or to impart a desired property. Examples are: antioxidant; preservative; antiknock agent; antifreeze agent; etc.
NOTE: Examples in fire practice are addition of aluminum powder to increase the extinguishing effect of water, viscosity enhancers to retard water runoff, surfactants and other foam agents. See also: extinguishing agent; foam; light water; surfactant.

address (Comput) A number, name, or label identifying a specific location within a computer's memory apparatus, or identifying a device.

adhesion[1] (Phys) The tendency of matter to stick or cling to other matter. Cf: adsorption, sorption, cohesion.
NOTE: In physics, adhesion and cohesion are different aspects of intermolecular attraction. Cohesion occurs within the same body; adhesion occurs between adjacent bodies and increases with closeness of contact. Cohesion exists in gases and liquids, but is less than in solids.

adhesion[2] (Med) The growing together of previously separated tissues, particularly after surgery.

adhesive (Techn) A material used to unite two surfaces. Examples are (1) protein; (2) starch, cellulose, or gum; (3) thermoplastic synthetic resins; (4) natural resins and bitumens; (5) natural and synthetic rubbers; and (6) inorganic adhesives such as water glass (sodium silicate).
NOTE: The term cement is used frequently as a synonym for adhesive. Cf: gel; resin; epoxy.

adiabat (Thermodyn) A line representing constant enthalpy on a thermodynamic diagram.

adiabatic (Thermodyn) Occurring without gain or loss of heat during a change in condition. An example is a gas that expands without heat being transferred to it or from it. Cf: isothermal. Ant: diabatic; erroneously: nonadiabatic.

adiabatic atmosphere (Meteor) A model atmosphere in which pressure decreases uniformly with height; specifically, one in which pressure varies inversely with density to the 1.4 power. Syn: dry adiabatic atmosphere; convective atmosphere; homogeneous atmosphere. See also: barotropic atmosphere; atmosphere.

adiabatic chart (Thermodyn) A diagram used in describing an adiabatic process.

adiabatic compression (Thermodyn) Compression of a fluid without gain or loss of heat but with increase in temperature.
NOTE: Rapid compression of a flammable gas in air can raise its temperature above its ignition point.

adiabatic flame temperature See: flame temperature.

adiabatic process[1] (Thermodyn) A process in which heat neither enters nor leaves a system. Ant: diabatic process.

adiabatic process[2] (Meteor) A thermal process occurring within an air mass with no heat flowing across its boundary to or from the surrounding air.
NOTE: Air rising adiabatically in the atmosphere is cooled by expansion 9.8°C per vertical kilometer. This cooling produces clouds and precipitation in moist air. Cf: isothermal.

adiabatic recovery temperature = recovery temperature = adiabatic wall temperature. See: stagnation temperature.

adiabatic wall (Thermodyn) A wall that neither receives heat from nor gives up heat to a fluid in contact with it.

adit (Mining) A more or less horizontal entrance passageway into a mine from the outside.

adjacent box (Fire Pr) A fire alarm box in the immediate vicinity of one over which an alarm has been transmitted. Syn: vicinity box.
NOTE: Notice of an adjacent box alarm is sent to the officer in command at the working fire to investigate the call.

adjuster (Insur) An individual who determines the amount of loss suffered in an insurance claim.

adjustment (Psych) 1. A change in behavior in a problem-solving situation that cannot be dealt with in the habitual way. 2. Accommodation or conformity to the physical and social environment.

administration (Man Sci) Officers of an organization who perform executive duties. Policy is often set at a higher level, e.g., by a board of directors or commissioners.

admiralty mile See: nautical mile

admissible roughness (Hydraul) The maximum height of irregularities or roughness elements that cause no increase in drag as compared with a smooth wall.

adrenalin (Med; Pharm) A hormone produced by the adrenal gland in times of stress which constricts blood vessels (thus reducing the danger of blood loss), dilates air passages, and mobilizes sugar from the liver (thus increasing the fuel supply to muscles for flight or combat).
NOTE: Used to aid breathing in asthmatics, to stimulate arrested hearts, and in local anesthetics to reduce bleeding. The adrenal gland also produces several hormones called steroids that regulate body metabolism and sex functions.

adsorption (Chem; Phys) A process by which molecules accumulate on the surface of a solid or liquid in contact with another medium (gas, liquid, or solid). Adsorption can either be positive or negative. The forces can either be chemical (chemisorption) or physical (van der Waals adsorption). Ant: desorption. Cf: absorption; sorption.

adsorption isotherm (Chem) The curve

resulting when the amount of material absorbed per unit weight of adsorbent is plotted against the concentration at a fixed temperature.

advance a line (Fire Pr) 1. To move a hose line forward, usu in the direction of a fire; to bring a fire-fighting hose stream closer to the fire. 2. A command to move a line of hose toward a given position or area.

advection (Meteor) The horizontal transport of a property, such as heat, by the movement of air. An example is the movement of polar air toward the equator.

adverse pressure gradient (Aerodyn) A change in pressure near a wall, which together with friction causes flow separation.

adynamia = asthenia

AEC = Atomic Energy Commission; now the Energy Research and Development Administration.

aeolean tones (Acoust) Sound produced by a gas stream striking a taut wire at right angles. See: Karman vortex street.

aerate[1] (Techn) To expose a liquid or slurry (sometimes granular material) to air to release dissolved gases or to increase oxygen content. To deodorize milk, purify water, fumigate and bleach cereal grains, etc. by passing air through them.
 NOTE: Aeration has a special meaning in the generation of mechanical foam.

aerate[2] (Techn) To charge with gas, usu carbon dioxide.

aerate[3] (Physiol) To charge blood with gas, as in the oxygenation of the blood during its passage through the lung capillaries in respiration.

aerated powder (Techn) Any powdered material that has been fluidized within a container by passing air through it uniformly from below.
 NOTE: Powders are commonly fluffed with air (fluidized) to form a bed for dipping parts to be coated, in a manner similar to that used in liquid dipping. Such beds are also used as sources for powder spray operations.

aeration (Techn) Introduction of air into a material.

aerial[1] (Fire Pr) = aerial ladder

aerial[2] (Commun) = antenna

aerial[3] Pertaining to air, elevation, or aviation.

aerial attack (Fire Pr; For Serv) 1. Dropping of fire extinguishing materials in the path of forest, grass, or brush fires from aircraft. See: water bombing. 2. Dropping of equipment and supplies and laying hose from the air. See: paracargo.

aerial fuel (Fire Pr; For Serv) Combustible materials above ground, such as trees and structures. Syn: crown fuel. See: ground fuel.

aerial ladder (Fire Pr) 1. A power-operated turntable ladder attached to a ladder truck, usu in 3 or 4 sections, extending 75 to 100 feet or more. Early types were of wood (hence the slang term: big stick); current types are of aluminum or steel. Aerial ladders are raised and extended by electrical, hydraulic, or mechanical power devices. See also: ladder. 2. = aerial ladder truck.

aerial perspective (Psych) View from the air, including the fading of colors and loss of visual sharpness with increasing distance.

aerial platform = elevating platform

aerial reconnaissance (Fire Pr; For Serv) Observation of wildland fires from airplanes or helicopters for purposes of command and control of fire-fighting operations. See also: aerial attack.

Aero (Fire Pr) 1. A trade name for a fire detection system in which heat expands air in tubing, setting off an electrical alarm. 2. An alarm received from an Aero system.

aerobe (Bact) A microorganism that requires air (oxygen) for growth. See: bacteria.

aerodynamic diameter (Aerodyn) The diameter of a unit density sphere having the same settling rate as a particle in question, of whatever shape and density.

aerodynamic force (Aerodyn) The force exerted by a fluid on a body moving through it.

aerodynamics (Abb: Aerodyn) The science of gas flow; a branch of **fluid mechanics** that deals with the motion of gaseous fluids and the forces acting on bodies moving with respect to the fluids.

aeroelasticity (Fl Mech) A branch of structural and fluid mechanics dealing with the dynamic effects of aerodynamic forces on elastic bodies.

aerogel (Techn) A colloidal solution of a gas in a liquid. See: **colloids**.

aerophagia (Med) Spasmodic swallowing or gulping of air. Cf: **dyspnea**.

aerophilia (Bact) Loving or requiring air; said of certain bacteria.

aerosol[1] (Meteor; Phys) Liquid droplets or solid particles dispersed in air of fine enough particle size (0.01 to 100 micrometers in diameter) to remain dispersed for an indefinite period of time. A colloidal system in which the air is the continuous phase.
  NOTE: Common aerosols in the atmosphere are known as **fog**, **haze**, **smoke**, **dust**, or **mist**.

aerosol[2] (Techn) a. A material dispensed from its container as a mist, spray, or foam by a pressurized propellant. b. A method of packaging in which a liquid can be delivered in spray form. This is accomplished by pressurizing the container with a low boiling liquid such as **Freon** or **propane**. Uses include insecticides and room deodorants (space aerosol), and paints or other surface coatings (surface aerosol). Cf: **colloid**; **atomization**; **fallout**; **air pollution**.
  NOTE: Flammable aerosols are considered a Class 1A liquid, which falls under the requirements of the US Federal Hazardous Substances Labeling Act.

aerostatics (Fl Mech) The study of gases at rest and the equilibrium of solids immersed in them.

aerothermodynamics (Aerodyn) The study of gas flows in which heat exchange has significant effect on the flow.

affected area (Fire Pr) The zone or area in which a fire has started and smoke is being generated.

affected floor (Fire Pr) Any story, other than the fire floor, into which smoke or other fire products have penetrated. See also: **fire floor**.

AFFF = **aqueous film-forming foam**

affinity (Chem) The tendency of two **species** to react chemically.

affinity laws (Mech) A set of laws describing the performance of centrifugal pumps and compressors: (1) flow is proportional to rpm, (2) the head developed is proportional to rpm squared, (3) horsepower is proportional to rpm cubed, and (4) efficiency is approx constant.

A-frame (Build) A structure made of two independent beams fastened together at the top and separated at the bottom for stability. Derived from the shape of the letter A.

aft (Navig) Toward, at, or near the stern.

afterburner (Comb) A device for burning fuel-rich exhaust gases of turbine, rocket, or jet engines to augment thrust.

afterburning (Comb) 1. Irregular burning of fuel left in the combustion chamber of a rocket after fuel cutoff. 2. The function of an **afterburner**.

aftercooling (Techn) Cooling of a gas after compression.

afterdamp (Mining) Unbreathable gas, largely carbon monoxide, left after an explosion of **firedamp** in a mine. See: **damp**.

aftereffect (Educ; Psych) 1. The continuation of sensation after the stimulus has ceased. 2. In learning,

afterflame time (Comb) The time a specimen continues to flame after an ignition source has been removed. Syn: flame duration.

afterflaming (Comb) Continued flaming after an ignition or flame source has been removed, as of a test material under conditions specified in a particular test.

afterglow[1] (Phys) Visible radiation, usu decaying, emitted temporarily after removal of external excitation. Examples are the glow from a gas-discharge lamp or luminescent screen after excitation has ceased. Cf: phosphorescence.

afterglow[2] (Comb) Lingering visible radiation emitted by solid-phase combustion after flaming has ceased or an external flame has been removed. See: glowing combustion.

afterimage (Physiol) An apparent image remaining when a portion of the retina of the eye has been fatigued by a prolonged stimulus.
 NOTE: Afterimages can be positive or negative. A negative afterimage usu has a color complementary to that of the stimulus. A positive afterimage is similar to the original stimulus both in color and brightness.

AGC = Associated General Contractors of America

agency (Safety) The principal object, such as a tool, machine, or equipment involved in an accident. The object most directly causing the accident, or the one inflicting injury or property damage.

agency part (Safety) The specific hazardous part of the agency that contributed to an accident.

agglomeration (Phys) 1. The property of particles to cohere, thereby increasing apparent particle size. Cf: floculation. 2. An accumulation or mass of cohering particles.

aggregate (Build; Techn) A conglomerate of particles bonded together to form concrete. Common aggregates are sand, gravel, and slag. Vermiculite, perlite, and cinders are used to make lightweight concrete.

aggregation (Psych) Crowded or clustered together, sometimes descriptive of a crowd of people without a common purpose.

aggression (Psych) Hostile, forceful action directed against persons or things.

aging (Techn) A change in the properties of a metal, plastic, or other material over a period of time.

agitator (Techn) An implement for shaking or mixing a fluid or slurry.

Agrément Board (Build; Brit) A board set up by the Ministry of Public Buildings and Works to assess building materials by examination and testing.

AIA[1] = American Institute of Architects

AIA[2] = American Insurance Association

AICE = American Institute of Civil Engineers

AIChE = American Institute of Chemical Engineers

aide (Fire Pr) A personal assistant to a fire department chief. See: ranks and titles.

AIHA = American Industrial Hygiene Association

AIIE = American Institute of Industrial Engineers

AIME = American Institute of Mining, Metallurgical, and Petroleum Engineers

air[1] (Geophys) The mixture of gases surrounding the earth and constituting the normal atmosphere.

Species	% by volume
Nitrogen	78.01
Oxygen	20.95
Argon	0.93
Carbon Dioxide	0.03
Neon	0.001,8
Helium	0.000,5
Methane	0.000,2
Krypton	0.000,1

Nitrous Oxide	0.000,05
Hydrogen	0.000,05
Xenon	0.000,008
Ozone	0.000,001

NOTE: The composition varies with altitude. In addition to the listed gases, various quantities of other constituents are present, chiefly water vapor and miscellaneous pollutants. Air at sea level and 0°C (32°F) exerts a pressure of 14.7 pounds per square inch. Under the same conditions, a cubic foot of air weighs 0.08 pounds. See also: atmosphere; air pollution.

air² In archaic usage, any gas; e.g., nitrous air.

air³ (Commun; Fire Pr) Referring to radio communications. See: on the air; off the air.

air aspirator foam maker (Fire Pr) A device consisting of a large tube or nozzle in which air foam is generated by aspiration of air by venturi action and turbulence. The foam maker is supplied with a solution of foam-making concentrate. See: foam nozzle.

air-bone gap (Audiom) The difference, expressed in decibels, between the hearing levels for a particular sound frequency received by air conduction and that sensed by bone conduction.

airborne radioactivity area (Ind Hyg; Nucl Safety) Any room, enclosure, or operating area in which airborne radioactive materials exist in concentrations in excess of the specified amounts.

air brake See: brake¹

air chamber (Fire Pr) A chamber filled with air in a positive displacement pump which serves to cushion pulsations caused by the pump piston or gears. In British usage it is called an air vessel.

air circulation (Meteor) Natural or imparted motion of air.

air cleaner (Techn) A device designed to remove impurities such as dusts, gases, vapors, fumes, and smoke from the air.

air conditioning (Build) A system that washes or filters air, controls its temperature and humidity, and circulates it through a compartment or building.

air conduction (Physiol) The transmission of sound by air to the inner ear through the outer ear canal.

air contamination (Env Prot) Introduction of a foreign substance into the air. Compared to air pollution, air contamination is more specific as to impurity and is more localized.

air cooling (Techn) Reduction in air temperature by contact with a cooler medium. Cooling may be accompanied by moisture addition (evaporation), by moisture extraction (dehumidification), or by no change in moisture content.

aircraft crash-rescue vehicle (Fire Pr) A specially-designed mobile truck unit with a foam system, foam nozzles, foam concentrate, water, and ancillary equipment for controlling and extinguishing fires and rescuing persons from crashed aircraft.

aircraft rescue (Fire Pr) Activities and procedures involved in saving occupants of crashed or burning aircraft.

air curtain (Techn) A screen of bubbles released underwater, used to control oil spills. Also called: pneumatic barrier. See also: barrier.

air dose (Dosim) The measurement of x- or gamma-rays up to 3 Mev with an appropriate instrument in air at or near the surface of the body in the region of the highest dosage rate.

air drainage (Meteor) The flow of cool air down a slope due to gravity when the surrounding air is warmer and therefore less dense.

air drop (Fire Pr) 1. Delivery of firefighters, equipment, and supplies by parachute from an aircraft. 2. An instance of an aircraft dropping a load of extinguishing agent on a wildland fire.

air filter (Techn) An air-cleaning device that removes light particulate matter from normal atmospheric air before it is brought into a building or used in a process, or after such use.

air foam (Fire Pr) A type of fire-fighting foam generated mechanically and containing air mixed in a quasi-stable form with foam solution. Syn: mechanical foam. See: foam.

air-foam concentrate See: foam concentrate

air-foam nozzle (Fire Pr) A special playpipe with an air aspirator used to produce air foam.

air-foam pump (Fire Pr) A positive displacement pump that automatically proportions and mixes air, water, and foam concentrate to produce foam under pressure. See: around-the-pump proportioner.

air-foam solution (Fire Pr) A mixture of air-foam concentrate and water.

air-foam stabilizer See: foam concentrate

air freight (Transp) Materials or products transported and delivered by aircraft.

air gap¹ (Electr) A space in the iron core of a transformer or choke, provided to minimize variation in inductance.

air gap² (Build) A space between panels to improve heat insulation. Syn: interstitial space.

air-ground detection (For Serv) A fire detection network with fixed coverage of key areas by ground observers supplemented by aerial patrols.

air hammer (Techn) A percussion-type pneumatic tool, fitted with a handle at one end and a tool chuck at the other, used for a variety of impact-type operations, such as chipping or breaking concrete.

air handling system (Build) The system of fans, ducts, controls, dampers, filters, intakes, exhausts, heating and cooling equipment and the like, installed in a building to remove stale air, recondition it, and return it to occupied spaces.

air heater (Build) An indirectly fired heating appliance intended to supply heated air for space heating and other purposes, but not intended for permanent installation. Air heaters do not include kerosine stoves, oil stoves, or unit heaters.

air horn (Fire Pr) An outdoor alarm operated by compressed air for sounding coded signals.

air horsepower (Techn) The theoretical horsepower required to drive a fan if there were no losses in the fan.

air hunger = dyspnea

air lift¹ (Techn) A method of pumping water or other fluids in which compressed air is introduced at the bottom of a pipe. Water trapped between bubbles of expanding air is lifted to the surface.

air lift² (For Serv) Emergency transportation of equipment, personnel, and supplies by airplane or helicopter.

airline mask = airline respirator

airline respirator (Fire Pr) A face mask supplied with air through an air hose attached to a compressed air cylinder or a remote air compressor. Also called: airline mask.

air lock¹ (Transp) Stoppage of or interference with the flow of a liquid in a fuel or hydraulic system caused by a pocket of air or vapor. Sometimes called vapor lock or air pocket.

air lock² (Techn) A double-door entry used for access to hermetically sealed chambers. Also see: safety lock.

air mask (Fire Pr) A self-contained breathing apparatus with the mask usu connected to an air tank carried on the back. See: breathing apparatus.

air mass (Meteor) A large (500 to 5000 mile diameter) definable, more or less uniform region of air bounded by fronts at which the proper-

ties (temperature, velocity, etc.) change rapidly.
  NOTE: Air masses are classified according to their source region and the surface over which they have developed (symbol c for continental and m for maritime). Common types in the northern hemisphere are: Arctic (A), a cold mass during all seasons; Polar Continental (cP), a cold mass occurring primarily during the winter; Polar Maritime (mP), a cold mass occurring during all seasons; Tropical Continental (cT), a warm mass prominent during the summer; Equatorial (E), a warm mass occurring in all seasons; Monsoon (M), a dry mass in winter and wet mass in summer. The source region symbols are affixed with letters K and W to indicate whether the mass is colder or warmer than the surface over which it is moving. Horizontal convergence at low levels, associated with lifting, produces vertical instability of the upper air, and horizontal divergence, associated with sinking, produces vertical stability (designated u and s, respectively, on maps). Thus, for example, mPKu indicates a maritime, polar, cold, vertically unstable air mass.

air monitoring (Env Prot) The continuous sampling of pollutants in the atmosphere, both for detection and measurement.

air officer (For Serv) A staff officer, reporting to a fire boss, service chief, or camp boss, responsible for establishing and maintaining aerial support and service for firefighters at a working wildland fire.

air parcel (Meteor) A relatively small volume of homogeneous air in the atmosphere. Cf: air mass.

air pocket (Fire Pr) A bubble of air trapped in a flowing liquid. For example, a bubble that often forms when hard suction hose is elevated above the pump intake.

air pollution (Env Prot) General contamination of the atmosphere by dust, fumes, gas, mist, smog, odor, smoke, pollen, vapor, etc. in any combination and in sufficient concentration to be injurious to plant or animal life or to property. Industrial pollutants are of three basic types: photochemical (from hydrocarbons), sulfur-containing (from coal), and oxides of nitrogen (from combustion processes). Also called: atmospheric pollution. See also: water pollution.

air pressure (Fire Pr; Techn) Motive power or force provided by pressurized air, used to expel extinguishing agents from fire extinguishers, to keep water out of dry sprinkler systems, to drive pneumatic tools, and other purposes.

air quality criteria (Env Prot) The varying amounts of air pollution and lengths of exposure at which specific adverse effects to health take place.

air quality standards (Env Prot) Maximum allowable concentration levels in the atmosphere established by legislation for specific substances or combinations of substances.

air-regulating valve (Fire Pr) An adjustable valve used to regulate airflow to the facepiece, helmet, or hood of an airline respirator.

air-right structure (Build) A building or other structure that straddles a thoroughfare, railroad line, or is erected above an already existing structure.

air sampling (Env Prot) The collection and analysis of samples of air to determine quantities and types of atmospheric contaminants.
  NOTE: The most numerous environmental hazards are chemical and can be conveniently divided into (a) particulates and (b) gases or vapors. Particulates are mixtures or dispersions of solid or liquid particles in air and include dust, smoke, mist, etc. Air sampling is used also to check for radioactive fallout.

air-supply device (Fire Pr; Ind Hyg) A hand- or motor-operated blower supplying an air mask or a compressor or other source of respirable air for air and dust masks.

air-supported structure (Build) A flexible, balloon-like structure that is inflated and kept erect by a fan that pressurizes the inside air

17

sufficiently above atmospheric to prevent the structure from collapsing.

air vessel (Fire Pr; Brit) = **air chamber**

airtanker (For Serv) An aircraft equipped with special tanks or dispensing devices for dropping water or other fire extinguishing agents on wildland fires.

airway[1] (Anat) An air passage in the body, usu referring to the pharynx, trachea, bronchi, and bronchioles.

airway[2] (Fire Pr) A closed tube supplying air or oxygen to a breathing mask used during operations in noxious atmospheres.

aisle (Build; Storage) Any passageway within a storage, work, or other area. Also the spacing between rows of seats, as well as the space between blocks of seats, as in an auditorium.
 NOTE: **Cross aisle**: A passageway at right angles to main aisles used for the movement of supplies, equipment, and personnel. **Fire aisle**: A passageway established to aid in fighting fire or preventing its spread or for access to fire-fighting equipment. **Main aisle**: A passageway wide enough to permit the easy movement of equipment, supplies, and personnel; generally running the length of the building.

alarm (Fire Pr) 1. An audible or visual signal indicating the existence of a fire or other emergency condition. 2. The device used to send and receive emergency signals.

alarm circuit (Fire Pr) The electrical network connecting the central station and the individual fire department companies and civil agencies that require notification of fires.

alarm reaction (Psych) The initial response in general adaptation to an emergency situation in which the adrenalin glands begin to mobilize the body's resources.

alarm system (Fire Pr) A system for producing a warning signal to indicate a dangerous condition. Depending on its design capabilities, the system generates a signal when it detects an unusually rapid temperature rise, the presence of abnormal smoke or ionized particles, infrared or ultraviolet radiation. The alarm may be a local audible or visual signal, or an electrical signal at a remote central station.
 NOTE: In a class A alarm system, alarm boxes are connected to a **central station**, where operators relay signals and messages to the proper fire stations and units. A class B system automatically routes alarms to appropriate fire stations.

alarm valve (Fire Pr; Brit) A valve in a wet or dry sprinkler system that opens automatically when the pressure in the system drops due to the opening of a sprinkler head. A visual or audible alarm is actuated through a pressure switch or a water motor gong.

albedo (Meteor) 1. The reflected brightness of a nonluminous body, esp a planet or satellite. 2. The ratio of radiation lost by reflection to the total radiation.
 NOTE: In the case of a planet, such as the earth, the albedo controls the solar constant and indirectly affects the weather.

albumin (Physiol) A protein material found in nearly all animal and many vegetable tissues.

albuminuria (Med) The presence of albumin in the urine, usu indicating a diseased state.

ALC = **approximate lethal concentration**

alcohol (Chem) 1. Commonly **ethyl alcohol** (C_2H_5OH). 2. Any one of the organic compounds with the general formula R•OH, where R is a hydrocarbon radical.
 NOTE: Polyalcohols, such as **glycerol**, possess more than one hydroxyl group. Each OH group must be bonded to a separate carbon atom. Low molecular weight alcohols are moderately toxic flammable liquids; higher alcohols are waxy solids. Other common alcohols are **methanol**, CH_3OH (wood alcohol); and **isopropyl alcohol**, C_3H_7OH (rubbing alcohol).

alcohol foam (Fire Pr) 1. A mixture of powdered chemicals or liquids used in foam generators to produce foam to extinguish burning water-soluble flammable liquids (such as alcohol) that break down foam from ordinary foam chemicals. See: foam. 2. A special type of mechanical foam that does not require a generator.

alcoholic (Med; Psych) An individual who is physically or psychologically dependent upon alcoholic beverages. One compulsively and chronically given to drinking in excess. See: blood alcohol.

ALD = approximate lethal dose

aldehyde (Chem) One of a class of hydrocarbon derivatives obtained by partial oxidation of an alcohol. The general formula is

$$R - \overset{\overset{H}{|}}{C} = O$$

where R is a hydrocarbon radical. Examples are formaldehyde from methyl alcohol and acetaldehyde from ethyl alcohol. Aldehydes are important intermediates for the manufacture of resins, plasticizers, solvents, and dyes. See also: acetaldehyde.
NOTE: Aldehydes are obtained when primary alcohols are dehydrogenated; hence, the term. Aldehydes may irritate the respiratory tract when inhaled.

alertness (Psych) The state of being watchful or vigilantly ready. A state of keen awareness and readiness to respond to stimuli.

algebra (Math) A branch of mathematics treating relations between symbols. Elementary algebra is concerned with numerical symbols and the mathematical operations of addition, subtraction, division, multiplication, exponentiation, and root extraction. Other types exist: e.g., the algebra of logic. The algebra of sets or classes is called Boolean algebra.

algorithm (Math) A step-by-step method for solving a problem, usu involving repetition of certain operations.

alidade (Surv) A surveying instrument with a straightedge and sights, used by forest fire watchers to locate the position of a fire from a lookout station. Syn: fire finder. See: cross shot.

alienation (Psych; Soc Sci) Loss of interrelatedness with others, such as family, associates, or society in general; more rarely, insanity or mental derangement.

aliphatic (Chem) Pertaining to an open-chain carbon compound, as contrasted with the closed-ring structures characteristic of aromatic compounds.
NOTE: Often applied to petroleum products derived from a paraffin base and having a straight or branched chain, saturated or unsaturated molecular structure, but actually including open-chain hydrocarbons, alcohols, aldehydes, esters, etc.

aliquot[1] (Math) An exact proper divisor. Each of the numbers 2, 3, 4, and 6, for example, is an aliquot part of 12.

aliquot[2] (Chem) A measured fraction of a sample taken for chemical analysis.

alkali (Chem) A base, such as sodium hydroxide (NaOH), or other substance that gives a basic reaction, as for example, sodium carbonate (Na_2CO_3).
NOTE: Strong alkalis are corrosive and dangerous; they can burn the skin. Sodium hydroxide, also called caustic soda or lye, is used in soap manufacture and many other applications. An alkali turns litmus paper blue. See: base.

alkali metal (Chem) An element in the first group of the periodic table: lithium, sodium, potassium, rubidium, cesium, and francium. These metals react easily with air and violently with water.

alkaline earths (Chem) Oxides of barium, calcium, strontium, beryllium, and radium, sometimes including magnesium oxide.

alkaloid (Chem; Med; Pharm) One of a group of bitter, colorless, nitrogen-containing organic bases derived from plants. Many alkaloids, including atropine, morphine, nicotine, cocaine, strychnine, and quinine,

are physiologically active, and in quantity are toxic to humans and animals.
NOTE: Alkaloids are used extensively in medicine. For example, atropine, obtained from belladonna, is used to dilate the pupil of the eye, to dry secretions in the airway prior to surgery, and to relieve spasms. Morphine, derived from opium, is used in hydrochloride or sulfate form as an analgesic and sedative. Quinine, obtained from cinchona bark, is used as an antipyretic and antimalarial drug.

alkalosis (Med) Deficiency of hydrogen ions in the blood, or excessive alkalinity of the blood and tissues, whether due to metabolic disturbance or hyperventilation (breathing too deeply or too rapidly).

alkane (Chem) One of a series of saturated hydrocarbons with the general formula $C_{(n)}H_{(2n+2)}$, commonly called paraffin. In the formula, C is carbon and n is a whole number.
NOTE: Alkanes with fewer than five carbon atoms are gases at room temperature (e.g., methane, ethane, propane, and butane); from pentane through hexadecane they are mostly liquid, and above that, waxy solids. The compounds are flammable and are common fuels, such as natural gas (CH_4) and LP gas (mixed C_3H_8 and C_4H_{10}). Gasoline contains substantial quantities of alkanes from C_5 through C_{10}; kerosene contains alkanes in the C_{20} range.

alkene (Chem) An organic compound containing a carbon=carbon double bond as the principal functional group.

alkyd resin (Techn) A polymer of diallyl phthalate used widely in paints and enamels.
NOTE: The term alkyd is a mixed blend modification of alcohol and acid.

alkylation (Chem) The process of introducing one or more alkyl radicals ($C_{(n)}H_{(2n+1)}$) by addition or substitution into an organic compound.

allergy (Med) An abnormal response of a sensitive person to chemical and physical stimuli. Commonly, a sensitivity to dust, pollen, and to certain proteins. Symptoms are skin rashes, asthmatic conditions, intestinal disturbances, etc.
NOTE: Serious allergic manifestations occur in about 10% of the population.

all hands (Fire Pr) A condition in which all units of the first alarm assignment are actively engaged in fighting a fire.

allied lines (Insur) Insurance lines related to property insurance.
NOTE: Coverages include protection against perils, usu underwritten by fire insurance companies, such as sprinkler leakage, water damage, and earthquake.

allocated space (Storage) An area or volume of a storage space formally set aside for a particular use.

allograft (Surg) A graft of tissue from another of the same species. In man, a graft obtained from another person. See: homograft.

all-or-none law (Physiol; Psych) The discovery that the heart muscle, if stimulated, will contract to the fullest extent or not at all. The statement that a neural impulse, if it occurs at all, occurs at maximum amplitude for the neuron, regardless of the intensity of the stimulus.

allotropy (Phys) The occurrence of a chemical element, such as carbon, sulfur, phosphorus, in two or more crystalline forms.
NOTE: Charcoal, diamond, and graphite are allotropes of carbon, each form having distinctly different physical properties.

all out (Fire Pr) A signal indicating that a fire is under control or extinguished and that the units engaged are preparing to resume their normal assignments. Syn: tapped out.

allowable burned area (For Serv) The maximum average acreage burned over during a given period of years that is considered tolerable under organized fire management for a given area.
NOTE: A measure used in connection with erosion, watershed, and land management problems.

allowance¹ (Mech Eng) a. A designed difference in the sizes of mating moving mechanical parts to allow space for a film of lubricant between them. b. For nonrunning parts, allowance is the excess in size of the inner part with respect to the space into which it is forced or shrink-fitted so that pressure holds the two parts together. Cf: tolerance.

allowance² (Build; Brit) A tax deduction for fire protection, such as sprinklers.

alloy (Metall) A combination of two or more metals either in solution, as an intimate mixture, or as a definite compound. An example of a mixture is brass, which contains copper and zinc. In some instances a mixture of a metal and a nonmetal is called an alloy.

Allplas See: ball blanket

all-purpose extinguisher (Fire Pr) A chemical fire extinguisher suitable for fighting fires involving ordinary combustibles, flammable liquids, and electrical equipment. An all-class (A, B, and C) fire extinguisher.
NOTE: No single extinguisher or extinguishing agent is ideal for all classes of fire. All-purpose, therefore, means that it is relatively safe to use, although its effectiveness may be doubtful for the particular fire that is burning.

All-Service Mask (Fire Pr) A trade name for a canister-type filter mask that can be used in smoke or gas environments in which sufficient oxygen is present to support life. See: breathing apparatus.
NOTE: This type of mask has been outlawed in a number of states.

allyl resin (Chem) A polymer of diallyl phthalate and diallyl isophthalate having good electrical, chemical, and moisture-resistant characteristics. Used in electrical components, furniture, construction, etc.

alnico (Techn) An alloy, principally consisting of iron, cobalt, nickel, aluminum, and copper, used in permanent magnets.

alpha emitter (Nucl Phys) A radioactive substance which gives off alpha particles.

alph particle (Nucl Phys; Symb: a or α^{++}) A small particle with two positive electrical charges thrown off at high velocity by many radioactive materials, such as uranium and radium. An alpha particle consists of two neutrons and two protons and is identical to a helium nucleus. A stream of alpha particles is called an alpha ray or alpha radiation.
NOTE: Of the three common types of radiation (alpha, beta, and gamma) emitted by radioactive materials, alpha particles are the weakest. They can be stopped by a sheet of paper, by two to three inches of air, or the outer layer of the skin, and hence are not particularly hazardous to plants or animals as long as α emitters are not lodged within the body.

alpha ray (Nucl Phys) 1. A stream of alpha particles. 2. Loosely, an alpha particle.

alpha rhythm (Physiol) A low-amplitude brain wave pattern found on the electroencephalograph during periods of relaxed alertness. The rhythm has a characteristic frequency of about ten cycles per second. See: electroencephalogram.

alternate-form reliability (Educ; Psych) Correspondence between the results of alternate (equivalent or parallel) forms of a test. A measure of the extent to which two forms of a test are consistent or reliable, assuming that the examinees themselves do not change in their abilities between the two testings.

alternating current (Electr; Abb: ac; as adjective, a-c) Electrical current which cyclically reverses its direction and whose frequency is expressed in Hertz (Hz).

alternation (Phys) One-half of a complete cycle.

alternator (Electr) An alternating current generator provided on automotive vehicles to maintain battery charge.

altocumulus (Meteor) A cloud, classification Ac. Syn: <u>mackerel sky</u>. See: <u>cloud</u>.

<u>altostratus</u> (Meteor) A cloud, classification As. See: <u>cloud</u>.

<u>aluminum</u> (Metall; Symb: Al) A silvery, light, ductile metal, at. no. 13, sp gr 2.708, mp 660°C, bp 1800°C, used in aircraft structures, building materials, and various consumer goods.
NOTE: Aluminum powder forms flammable and explosive mixtures with air.

<u>alveolar air</u> (Med) Respiratory air within the lung sacs. Normal alveolar pressure is approx 105 mm Hg.

<u>alveolus</u> (Anat) A cavity in the body, such as an air sac in a lung.
NOTE: In the lungs the alveoli are tiny air sacs at the ends of the bronchioles. Blood passing through capillaries in the alveolar walls during respiration takes in oxygen that diffuses through the thin walls and gives up carbon dioxide.

AMA[1] = <u>American Management Association</u>

AMA[2] = <u>American Medical Association</u>

<u>amagat units</u> (Chem; Phys) A measurement system in which the standard atmosphere is the pressure unit and the molar volume (22.4 liters) is the volume unit. Used in the study of the behavior of gases under pressure.

<u>amalgam</u> (Metall) An alloy containing mercury.

<u>amalgamation</u> (Metall) The process of alloying metals with mercury, used in extracting gold and silver from their ores.

<u>ambidextrous</u> (Ind Eng) Equally skilled in the use of both hands.

<u>ambient</u> (Symb: a subscripted) Surrounding, esp pertaining to the environment about a body, as in <u>ambient air</u> and <u>ambient temperature</u>.

<u>ambient air</u> (Phys) The air surrounding an object.

<u>ambient noise</u> (Audiom; Ind Hyg) Total noise present in an environment.

<u>ambient temperature</u> (Phys) The temperature of the surrounding environment.

<u>ambiguous</u> Unclear; having two or more possible meanings or interpretations.

<u>ambulance</u> (Med) An emergency vehicle used to transport sick and disabled persons or accident victims to emergency care facilities.

AMCBO = <u>Association of Major City Building Officials</u>

<u>American Insurance Association</u> (Insur; Abb: AIA) An association of fire insurance underwriters that has absorbed the National Board of Fire Underwriters and conducts broad service programs in fire protection.

<u>American Table of Distances</u> (Safety) A standard of separation distances for the storage of explosives, developed and maintained by the Institute of the Makers of Explosives.

<u>americium</u> (Chem; Nucl Phys; Symb: Am) A silvery-white radioactive, alpha-emitting, metallic element, at. no. 95, mass 241, mp 1079°C, and half-life of 433 years. Used as a radioactive source in certain ionization-type smoke detectors.

AMIA = <u>American Mutual Insurance Alliance</u>

<u>amide</u> (Chem) One of a class of organic compounds of nitrogen containing the characteristic group $-CONH_2$. Acetamide and urea are common examples.

<u>amidships</u> (Navig) At, near, or toward the middle part of a ship, boat, or land vehicle, as distinguished from the ends. Also called: <u>midship</u>.

<u>amine</u> (Chem) A class of basic organic nitrogen compounds derived from ammonia by replacing one or more of the hydrogen atoms by an organic radical.
NOTE: Methylamine is a gas; other amines, such as aniline, are liquids or solids.

aminobenzene = aniline

ammeter (Electr) An instrument for measuring the flow of electric current and provided with a scale, usu graduated in amperes, milliamperes, or microamperes. Current is measured in series with the electrical circuit.

ammonia (Chem) A compound, NH_3, mp -77.7°C, bp -33.35. Ammonia forms an alkaline solution with water, and is widely used as a detergent, fertilizer, and basic chemical.
NOTE: The skin is irritated by liquid ammonia, and mucous tissues by gaseous ammonia. Ammonia fumes are readily absorbed by water spray or fog. See also: sal ammoniac.

ammonia suit (Fire Pr) A rubberized coverall worn with breathing equipment for protection against ammonia fumes. Also called: ammonia wading suit, or simply wading suit.

ammonium ion (Chem) A cation, NH^+_4, produced by the ionization of an ammonium salt.

ammonium nitrate (Chem) A colorless, odorless, crystalline salt, NH_4NO_3, used in explosives and fertilizers.
NOTE: When heated to decomposition (210°C), ammonium nitrate emits toxic fumes. In the molten state ammonium nitrate reacts vigorously with molten aluminum but does not explode.

ammonium phosphate (Chem) A white crystalline compound, $(NH_4)_3PO_4$, used chiefly as a fire retardant and as a fertilizer. See: dry chemical extinguisher.

ammonium sulfate (Chem) A colorless, crystalline salt, $(NH_4)_2SO_4$, used as a fire extinguishing agent, often air-dropped on forest fires.

amnesia (Med) The inability to recall events in one's past, sometimes including one's own identity.

amniotic sac (Med) The fluid-filled sac that surrounds a fetus. Also called: bag of waters.

amorphous (Phys) Having no definite structure, shape, or form; specifically, a lack of crystalline structure.
NOTE: Glass is a common example.

ampere (Electr; Abb: A or amp) 1. A measure of electric current, equal to one coulomb of electricity per second. 2. The quantity of current that deposits 0.001118 gram of silver per second. 3. The quantity of current flowing with a potential of one volt across a one-ohm resistance in one second. 4. A current that produces a force of 2×10^{-7} newtons per meter between two parallel conductors of infinite length and negligible cross section spaced 1 meter apart in a vacuum.

amphetamine (Med; Pharm) One of a class of drugs that stimulate the central nervous system, including Benzedrine, Dexedrine, and Methedrine. Chemical formula $C_9H_{13}N$, used as a spray medication for colds and hay fever.

amphoteric (Chem) Capable of reacting either as an acid or a base.
NOTE: Water, for example, can act as an acid in one system and as a base in another.

amplifier (Electr) An electronic, fluid, or magnetic device in which a low-power, low-current, or low-voltage signal is used to modulate a source of greater energy to produce a related output signal of greater power.

amplitude (Phys) The displacement of a wave or other periodic phenomenon from a reference point.

amu = atomic mass unit

anabatic (Meteor) Upward moving.

anabatic wind (Meteor) A wind blowing uphill, usu caused by surface heating. See: wind.

anaerobe (Bact) A microorganism that grows without oxygen (air).
NOTE: A facultative anaerobe is able to live with or without oxygen; an obligate anaerobe will grow only in the absence of oxygen. See: bacteria.

anaerobic (Bact) Able to live and grow in the absence of free oxygen.

analgesia (Med) Insensitivity to pain.

analgesic (Pharm) A drug that diminishes sensitivity to pain without impairing consciousness; e.g., aspirin, codein, caffein, phenacetin, and morphine. Cf: antipyretic.

analog[1] (Math) A similar function.

analog[2] (Comput) A machine that manipulates continuous quantities, as contrasted to discrete digital quantities.

analogy (Phys) The existence of similarity between certain types of conservation equations and/or transport processes that allows the interchange of dimensionless groups. For example, a correlation for a Nusselt number for heat transfer may be used in mass transfer study provided the Prandtl number is replaced by the Schmidt number.

analysis[1] (Ind Eng) Decomposition into components for study. An examination of a complex system, its elements, and their interrelations.

analysis[2] (Chem) Determination of chemical composition.
  NOTE: Qualitative analysis is used to determine chemical constituents in a compound; quantitative analysis, their quantities.

analysis[3] (Math) The study of functions and limits. See also: operations analysis; systems analysis.

analysis of variance (Statist) Study of the statistical variance of measured results, particularly to determine the contribution of individual factors to the total variation.

analytical chemistry (Chem) The branch of chemistry that deals with the identification of materials. The analysis can be either qualitative, in which the elements, molecular species, or radicals present in a sample are identified, or quantitative, in which their relative quantities are determined. See: chemistry.

analytic system (Fire Pr; Insur) A system of grading the insurance risk of properties, including consideration of fire hazards and fire protection facilities.

anaphylactic shock (Med) A severe allergic reaction that may lead to respiratory failure and death.

anaphylaxis (Med) Heightened susceptibility or sensitivity to a foreign protein resulting from previous exposure to it. Acute anaphylactic shock can lead to respiratory failure. See: allergy; respiratory distress.

anatomy (Abb: Anat) The science dealing with the structure of the body and the relation of its parts, based on dissection.

anchor (Fire Pr) The part of a bed ladder to which the halyard rope is attached. See: ladder.

anchor man (Fire Pr) A firefighter who holds a hose for a nozzle man.

anchor point (Fire Pr) An advantageous point from which construction is started of a fire control line to contain a wildland fire. Usu an existing barrier to fire spread.

anemia (Med) A condition in which the blood is deficient either in quality or quantity (e.g., a deficiency in the hemoglobin and red blood cell, that is, erythrocyte, content of the blood). Any of several pathologic blood disorders that may be due to a large variety of causes and appear in many different forms.

anemometer (Meteor) An instrument for measuring wind velocity. See: Beaufort wind scale.
  NOTE: Similar instruments used in wind tunnels are usu called flow meters or velocity probes.

aneroid barometer (Meteor) An instrument for atmospheric pressure measurements, consisting of a thin-walled, sealed chamber that expands and contracts with variations in atmospheric pressure.
  NOTE: This movement is amplified through levers to move a pointer on a dial. The same instrument is used in height meters called altimeters. In the laboratory a similar, but less sensitive, instrument is called a Bourdon gage. See: barometer.

anesthesia (Med) Loss of sensation;

in particular, the temporary loss of feeling induced by certain chemical agents. The effect may be local, affecting a particular region of the body, or general, if total unconsciousness results. A major part of the body may be rendered insensitive by a spinal block, or the area supplied by a nerve can be anesthetized by local injection of an agent such as procaine in a nerve block. Hypnosis is used to inhibit pain, particularly in childbirth and dentistry.

anesthetic (Med) A substance used to reduce sensitivity to pain. It may be either local, affecting a particular region of the body, or general, if total unconsciousness results. Anesthetics can be administered orally (e.g., codeine), by inhalation (ether, Halothane, cyclopropane), by injection (sodium pentathol), or topically, on the surface (methyl chloride).
NOTE: Most of the common anesthetics are highly flammable and have been a source of hospital explosions and fires.

aneurysm (Med) A localized, sac-like, blood-filled dilatation of a blood vessel.

angina pectoris (Med) A sudden attack of chest pains caused by insufficient oxygenation of the heart muscle, often brought on in arteriosclerotic hearts by physical exertion or emotional stress.

angle beam testing (Testing) A technique in which a beam of ultrasound is passed through a plastic wedge and enters a material at an angle to the test surface.

angle indicator (Fire Pr; Techn) An accessory that measures the angle of a boom to the horizontal. Syn: inclinometer.

angle of approach (Transp) The maximum angle of an incline onto which a vehicle can move from a horizontal plane without interference, for example, from front bumpers.

angle of attack[1] (Fire Pr) The angle of inclination of a body to the direction of an oncoming flow, for example, the angle of impact of a water stream on a surface.

angle of attack[2] (Aerodyn) The angle between the chord of an airplane wing and the horizontal plane.

angle of departure (Fire Pr) The angle between the road surface and a line from the point of contact of the rear tire to the farthest projection of the vehicle behind the rear axle.

angle of incidence (Opt) The angle between a ray of light and the normal to the surface of incidence. See: angle of reflection.

angle of reflection (Opt) The angle between a reflected ray of light and the normal to the surface of incidence. The angle of incidence equals the angle of reflection, and the two angles always lie in the same plane.

angle of refraction (Opt) The angle of deviation of a light ray passing from one medium into another.
NOTE: Snell's law states that the ratio of the sine of the angle of incidence to the sine of the angle of refraction is equal to the ratio of the light velocity in the two adjoining media. This ratio is called the refractive index.

angle of repose (Civ Eng) 1. The maximum angle between the horizontal and an inclined plane at which an object on the plane retains its position without sliding. 2. The maximum slope at which a granular material remains without sliding. Syn: angle of slip.

angle of yaw (Fl Mech) The angle between the direction of a fluid stream and the longitudinal axis of a body immersed in the flow.

Angstrom unit (Symb: A) A unit of length equal to 3.937×10^{-9} in., 10^{-4} micrometer, 10^{-8} cm, or 10^{-10} meter, commonly used to measure wavelengths in the visible and infrared ranges and for atomic and molecular dimensions.

angular distance (Geom) The arc, measured in degrees, between two directions.

angulated roping (Build) A system of exterior building platform suspension in which the upper wire rope sheaves or suspension points are closer to the plane of the building face than the corresponding attachment points on the platform, thus causing the platform to press against the face of the building during vertical travel.

angulation (Med) 1. The misalignment of the bone above and below a fracture. 2. The formation of a short obstructive angle in the intestines.

anhydrous (Chem) Free from water, esp water of crystallization.

aniline (Chem) A highly toxic organic base, chemical formula $C_6H_5NH_2$; sp gr 1.0235; mp 6.2°C; bp 184.4°C; flash point (open cup) 75.5°C. Aniline is a colorless oily liquid that becomes brown when exposed to air and light. It is an extremely hazardous substance; the liquid and vapors are readily absorbed through the skin. In gaseous or vapor form aniline is a chemical asphyxiant. Used mainly in the manufacture of rubber, dyes, pharmaceuticals, and plastics. Also called: phenylamine; aminobenzene.

anion (Chem) A negatively charged ion, usu in an aqueous solution. The ion that moves toward the positively-charged anode in an electrolytic solution.
  NOTE: Typical anions are OH^- (hydroxyl), $SO_4^=$ (sulfate), Cl^- (chloride), $PO_4^{\equiv}$ (phosphate), etc.

anisotropic (Phys) Exhibiting different properties in different directions. An example is double refraction in crystals. Ant: isotropic.

annealing (Metall) The relief of stresses in metals, glass, or other materials, usu accomplished by heating to a temperature at which the molecules have appreciable mobility, followed by gradual cooling. The process softens the material and makes it less brittle.

annual ring (Geom) A layer of wood added to a tree trunk during a growing season.

anode (Electr) 1. The positive pole of an electrolytic cell; the electrode at which oxidation takes place or the one toward which anions migrate. 2. The negative pole of a storage battery. 3. The electron-collecting electrode of an electron tube.

anodize (Metall) To place a protective, usu oxide, coating on a metal surface by electrolytic or chemical action.

anomaloscope (Med) An instrument for distinguishing among various types of color-deficient individuals, esp deuteranopes (who confuse purplish-red and green) and protanopes (who confuse red and blue-green).

anomalous dispersion (Opt; Spectr) A jump or discontinuity in the dispersion curve associated with a line or band in the absorption spectrum. Except for these local increases, dispersion of electromagnetic radiation decreases with increase in frequency.

anomaly Deviation from the norm.

anorexia (Med) Loss of appetite.

anosmia (Physiol) Insensitivity to odors.

anoxemia = hypoxemia

anoxia (Med) Absence of oxygen for physiological use within the body. Sometimes used to mean oxygen deficiency. See: hypoxia.

ANS = American Nuclear Society

ANSI = American National Standards Institute

antagonistic (Physiol) Opposing or acting against another; descriptive, for example, of a muscle that opposes the action of another muscle, or a medication that destroys the effectiveness of another medication.

antemortem (Med) Prior to death. Ant: postmortem.

antenna (Commun) One or more wire conductors, rods, or a reflector for receiving or propagating electromagnetic radiation. Also called: aerial.

anterior (Anat) In front or in the forward part; ahead of or preceding in time or space. Syn: ventral. Ant:

dorsal; posterior.

anteroom (Build) A small waiting room outside a larger room, usu an office.

anthracene (Chem) A crystalline hydrocarbon, $C_{14}H_{10}$, consisting of three benzene rings. The compound is used as a component of smoke screens, and crystals are used in scintillation counters.
NOTE: Anthracene is a volatile solid that has a flash point (121°C) below its melting point (217°C).

anthracite (Mining) A type of hard coal.

anthracosilicosis (Med) Miner's asthma. A complex form of pneumoconiosis; a chronic disease caused by breathing air containing both coal and silica dust. A condition common to hard-coal miners.

anthracosis (Med) A disease of the lungs caused by prolonged inhalation of dust that contains particles of carbon and coal.

anthropology The study of man and his natural history.

anthropometry (Anthro) 1. The study of human body measurements. 2. The collection and application of dynamic and static body measurements as design criteria to improve the ease, efficiency, and safety of the human in a system.
NOTE: Anthropometry provides basic data, for example, for the design of seating arrangements as well as aisle and door widths in theaters for physical comfort, ease of viewing, and freedom of movement during normal or emergency evacuation.

antiagglomerant (Chem) An additive used to prevent clustering or cohesion of particles.

antibacterial agent (Pharm) A compound that inhibits the growth or division of bacteria. Gaseous agents such as ethylene and dye agents such as mercurochrome, acridine, and gentian violet are used. See also: antibiotic; antiseptic. Cf: antimicrobial agent.

antibiotic (Pharm) A substance that destroys microorganisms or inhibits their growth. Examples are penicillin, streptomycin, bacitracin, and tetracyclin. Most of these compounds were originally secretions of microorganisms, but many have been synthesized chemically, and a few are produced synthetically on a commercial scale.

antibody (Med) A substance which develops in the blood serum (the liquid portion of the blood) in the presence of substances called antigens. A serum protein of the blood capable of combatting infection. Antibodies develop to counteract antigens produced by viruses or bacteria, thus immunizing the body to the antigens.

anti-caking additive (Fire Pr) A chemical added to dry chemical extinguishing powders to counteract caking and hardening.

anticipation learning (Psych) A form of rote-learning procedure in which the subject tries to give the next item in the list during each trial. This technique affords a running account of the subject's progress.

anticyclone (Meteor) A wind rotating about a local vertical in a direction opposite to the rotation of the earth. An area of high barometric pressure generally traveling in an easterly direction.

antidetonation agent = antiknock agent

antidote (Pharm) A drug that relieves or counteracts the effects of a poison.

antifoaming agent (Fire Pr) A chemical, such as silicone polymer, that breaks down foam persisting after a fire has been extinguished. Syn: defoamer.

antifreeze agent (Techn) A substance added to liquid to lower its freezing point. Used to protect fire extinguishers and engine cooling systems against freezing. Common antifreeze agents are salts, alcohols, glycols, and hydroxy compounds. The term is used also to describe refrigeration brines, snow melting and deicing agents, etc. See also: additive.

antifreeze extinguisher (Fire Pr) An extinguisher charged with a calcium chloride solution or other compound with a lower freezing point than water.

antigen (Med) A protein, polysaccharide, or lipid that induces the production of antibodies. Bacteria and their toxins, as well as many other substances, can act as antigens in the human body.

antiknock agent (Techn) A chemical added to gasoline to raise its octane number, enabling engines to operate without knock at higher compression ratios. Knock is caused by spontaneous ignition of unburned fuel mixture ahead of the flame front in a combustion chamber. Syn: antidetonation agent. See also: additive.

antilogarithm (Math) The number (argument) corresponding to a logarithm. The inverse function of the logarithm.

antimicrobial agent (Pharm) A chemical compound that destroys (bactericide) or inhibits (bacteriostatic) the growth of microscopic life forms. Antimicrobial agents include antibacterial, antifungal, antiprotozoal, antiparasitic, and antiviral agents. Cf: antiseptic; antibacterial agent; antibiotic.

antimony (Symb: Sb) A hard, brittle, lustrous gray metal often associated with lead and arsenic. The metal is toxic when ingested, and its soluble salts may cause dermatitis. Used in bearing metals and to harden lead for storage batteries.

antioxidant (Chem) A compound used as an additive to retard deterioration by oxidation. Antioxidants for human food and animal feeds, sometimes called freshness preservers, retard rancidity of fats and loss of fat-soluble vitamins (A, D, E, K). Antioxidants also are added to rubber, motor lubricants, and other materials to inhibit deterioration by molecular oxygen (autoxidation).

anti-particle (Nucl Phys) A particle that interacts with its counterpart of the same mass but opposite electric charge and magnetic properties (e.g., proton and antiproton or neutron and antineutron), with complete annihilation of both and production of an equivalent amount of radiant energy. The positron and its antiparticle, the electron, annihilate each other upon interaction and produce a gamma ray.

antipyretic (Med) A drug that lowers the body temperature. Examples are aspirin, quinine, and antipyrine. Cf: analgesic.

antiresonance (Acoust; Electr) The point at which impedance approaches infinity. Ant: resonance.

antiseptic (Pharm) A disinfectant. A substance that destroys or inhibits the growth of microorganisms. Common examples are iodine, mercurochrome, merthiolate, silver nitrate, and alcohol. Cf: aseptic.

antiserum (Med) Serum (the liquid portion of the blood) containing one or more antibodies.
NOTE: Often produced by injecting an animal with bacterial antigens, to which the animal subsequently develops protective antibodies.

antiskid plates (Storage; Transp) Metal plates with sharp projections on each side used between wood members or containers to retard movement.

antisocial behavior (Psych) A pattern of conduct that is in conflict with social order and which brings the individual chronically into difficulties with society.

antisymmetric[1] (Phys) A characteristics of a physical system in which each point has opposite properties to those of a symmetrically opposite point.

antisymmetric[2] (Math) Descriptive of a function that is transformed into its negative when its variables are interchanged in pairs. Ant: symmetric.

antitoxin (Med) A substance found in the blood serum and often in other body fluids that is specifically antagonistic to a particular poison or toxin released by bacteria, as opposed to antibodies, which attach themselves to antigens on the surfaces of the bacteria themselves.

anvil cloud See: cloud

anxiety (Psych) A feeling of apprehension, uneasiness, or fear that a threatening event, either real or imagined, is about to occur.

aorta (Anat) The main trunk which carries the blood from the heart to the rest of the body.

apartment house (Fire Pr) A building or part of one containing three or more dwelling units. See also: highrise.

APCA = Air Pollution Control Association

aperiodic (Phys) Noncyclic; recurring without regularity.

aperture[1] An opening.

aperture[2] (Opt) The diameter of a lens. The relative aperture (or f-number) of a lens is the ratio of the focal length to the diameter.

APHA = American Public Health Association, Inc.

API = American Petroleum Institute

API scale See: Baumé scale

aplastic anemia (Med) A condition in which the bone marrow fails to produce an adequate number of red blood corpuscles.

apnea; apnoea (Med) Cessation of breathing. Asphyxia or suffocation.

apoplexy (Med) Loss of consciousness or sensation due to a break, blood clot, or other obstruction in an artery of the brain. Syn: stroke.

apparatus (Fire Pr) 1. A self-propelled fire vehicle, including tank trucks, ladder trucks, pumpers, crash and rescue vehicles, etc. used to transport personnel and equipment to fires or other emergency incidents. 2. = (Brit) appliance.
NOTE: Vehicles not equipped for firefighting are usu called auxiliary vehicles rather than apparatus.

apparatus floor (Fire Pr) 1. The floor area in a fire station on which the fire apparatus is parked. 2. = (Brit) appliance room.

apparent density (Transp) The ratio of mass to volume of a finely powdered material, under stated conditions, which is always less than bulk density. Also called: loading density.

apparent motion (Phys) Motion relative to a given reference point.

appeal (Law) A procedure by which the decision of a lower court is referred to a higher one for reexamination.

appellate court (Law) A court that has the power to review the decisions of another court.

appliance[1] (Techn) A piece of electrical equipment such as a clothes washer, refrigerator, coffee maker, and the like.
NOTE: A fixed appliance is one that is fastened down during normal use. A stationary appliance is one that is difficult to move from place to place. A portable appliance is one that is normally moved to a place of convenience during use.

appliance[2] (Fire Pr) Any of a variety of tools or devices carried on a fire-fighting apparatus.
NOTE: In British usage "appliance" is synonymous with American "apparatus."

appliance room (Fire Pr; Brit) = apparatus floor.

applicator[1] (Techn) An implement or tool for applying or spreading a material.

applicator[2] (Fire Pr) A special pipe or nozzle used to spray foam or water fog.

applied chemistry (Chem) The application of the science of chemistry and engineering processes to produce and use chemicals in commercial quantities. See: chemistry.

applied research The synthesis of knowledge, material, and techniques directed toward a solution of practical technical or scientific problems.

apposition (Med) Contact of adjacent organs or body parts.

appraisal[1] (Insur) The act of estimating the value of property.

appraisal[2] (Man Sci) Impartial analysis of information, at a responsible management and control level, from which the effectiveness and efficiency of the total process can be measured and appropriate action determined.

approach channel (Techn) One or more passages through which gas must flow to reach a safety relief device.

approved (Fire Pr; Insur) 1. Acceptable to the authority having jurisdiction. 2. Accepted or acknowledged by a recognized professional society or association as meeting appropriate standards or other criteria. 3. Certified by a recognized testing laboratory or designated authority that a method, procedure, practice, material, tool, device, or item of equipment or machinery is appropriate and acceptable for a particular purpose or use.

approved storage facility (Storage) A facility for the storage of explosive materials conforming to requirements and covered by a government license or permit.

approving agency An agency designated or authorized to sanction or certify a procedure or item of equipment as satisfactory for a particular purpose or use.

approximate lethal concentration (Toxicol; Abb: ALC or LCca) The lowest concentration of a substance in the atmosphere that is capable of killing over a given period of time. See: lethal concentration. See also: range finding; toxicity scale.

approximate lethal dose (Toxicol; Abb: ALD or LDca) The least quantity of a substance that is likely to kill when ingested by a route other than the respiratory tract. See: lethal dose.

appurtenance (Techn) 1. An accessory or subordinate part. 2. A device such as a pump, relief device, liquid-level gage, valve, pressure gage, and the like.

apron[1] (Build) The part of a stage in front of the proscenium arch.

apron[2] (Fire Pr) The paved area in front of the apparatus doors of a fire station.

apron[3] (Civ Eng) The paved area next to an airport terminal or hangar.

apron[4] (Civ Eng) The area along the waterfront edge of a pier or wharf.

apron[5] (Civ Eng) A protective covering of concrete, planking, or other material along a river bank, sea wall, or below a dam to prevent erosion.

aptitude (Psych) An ability, capacity, or talent to learn, understand, and perform a task or class of tasks. An aptitude test is often used to determine the probability of a person's success in some activity in which he is not yet trained.

aqueduct (Civ Eng) A conduit or channel built to carry water.

aqueous (Abb: aq) Containing water; made with water, as a water solution of something.

aqueous film-forming foam (Fire Pr; Abb: AFFF) A type of fire-fighting air foam produced by special fluorocarbon surfactant foam concentrates that controls the vaporization of flammable liquids by means of a water film that develops as the foam is applied. The film also serves as a coolant and excludes air from the fuel. Compatible with dry chemical agents and with protein and fluoroprotein foams, but not with foam concentrates. Also called: light water.
NOTE: AFFF is intended primarily for liquid hydrocarbon fuel fires. It is not recommended for liquefied gases such as butane, propane, and butadiene, nor for cryogenic liquefied gases or polar solvents. Because of its water content, AFFF cannot be used on energized electrical equipment nor on materials, such as metallic sodium, that react with water or produce hazardous substances by reaction with water.

aquifer (Geophys) A stratum of rock or gravel that yields enough water to be useful as a source of water supply. See also: artesian well; ground water.

arc (Geom) A portion of a curved line, as of a circle.

arcade (Build) 1. A long passageway covered by an arched roof. 2. A covered pedestrian walk or concourse between rows of shops. See: mall.

arc discharge (Electr) A luminous, high-intensity electrical discharge between electrodes in a gas or vapor, characterized by high current density and low voltage drop. Normally, the current is limited by external constraints. Cf: corona discharge; point discharge; spark discharge; glow discharge.

arc heating (Techn) Raising the temperature of a material by the heat energy of an electric arc discharge. A process used chiefly in smelting and welding.

Archimedes' principle (Phys) The statement that a body immersed in a fluid is acted upon by a vertical force equal to the weight of the fluid displaced. See also: buoyancy.

architectural engineering (Eng) The theory and practice of building design and construction.

arc lamp (Techn) A strong illumination source in which light is generated by an electrical arc between two electrodes.

arc spectrum (Spectr) A spectrum produced by an arc discharge and used chiefly to study atoms or simple, stable molecules. Cf: spark spectrum. See also: spectrum.

arctic front See: front

Arcton (Brit) A trade name for a series of fluorinated hydrocarbons used as refrigerants and fire extinguishing agents. Syn: Halon; Freon.

arc welding (Metall; Techn) A process of joining metal parts by using the heat of an electric arc to either fuse the metal or melt a consumable electrode in the joint between the parts being joined. See also: welding.

are (Metrol) A unit of area equal to 100 square meters. Cf: acre.

area ignition (Fire Pr; For Serv) The setting of several fires simultaneously or in quick succession and sufficiently close together so that they support each other, producing a rapid development and spread of fire. See: simultaneous ignition.
NOTE: A technique used in setting a backfire.

area of major involvement (Fire Pr) The area in a burning building in which the greatest quantity of heat has accumulated. Syn: seat of fire.

area of refuge (Fire Pr) Any portion of a building relatively safe from fire danger; a space not involved in fire and not affected by smoke, toxic gases, or otherwise hazardous to human life and safety.

area sampling (Soc Sci) Inclusion in a survey of all subjects within a given geographical area.

argon (Symb: Ar or A) A noble gas element, bp -185.7°C, mp -189.2°C, and at. wt 39.944. One of the constituents of the atmosphere. Used principally as a filler in electric lamps, as gas shielding for arc welding, and as a protective inert blanket in metallurgical processes. Argon is used also in Geiger counters and ionization chambers to detect ionizing radiation. See also: air.

argument (Math) An independent variable upon which the value of a function depends.

arithmetic mean (Math) The simple average, sometimes called the mean, equal to the sum of a set of values divided by the number of values in the set.

armature (Electr) The current-carrying element in a generator, motor, or magnetic relay. Armatures may be either stationary or moving.

aromatic (Chem) Fragrant; having a marked odor. Referring to compounds containing a benzene ring.

aromatic hydrocarbons (Chem) Hydrocarbons containing one or more benzoid rings. Examples are benzene, anthracene, naphthalene, toluene, styrene, biphenyl, and their derivatives.

around-the-pump proportioner (Fire Pr) A type of liquid foam concentrate proportioning device that feeds the concentrate into the water at the output side of the pump rather than through the pump. See: foam proportioner.

arousal (Psych) A process in which alertness is heightened in response to a sensory stimulus.

array (Math) A collection of objects, numbers, or functions arranged in some order or table. See: set.

arrest¹ To stop, or bring under control. To check, as the spread of fire.

arrest² (Law) To take custody of a person by law.

arrester (Techn) A device that decreases, suppresses, or stops an effect or motion. See: lightning arrester.

Arrhenius equation (Chem) An equation that expresses the rate of a chemical reaction as a function of concentration multiplied by an exponential function of reciprocal temperature.

arrhythmia (Med) Irregularity of the heart beat.

arsenic (Chem; Symb: As) A silvery, brittle, crystalline metal. Toxic when ingested. Its most common compound is arsenic trioxide. Used in metallurgy, glass manufacture, weed killers, and insecticides.

arson (Law) The willful burning of a dwelling, building, structure, or other property, including one's own. Often motivated by profit or other gain or to conceal another crime. Degrees of arson are first degree, burning of dwellings; second, burning of buildings; third, burning of other property; and fourth, attempted arson. Negligent arson is the starting of a fire through carelessness. Cf: incendiarism.

NOTE: The Fire Marshal's Association of North America has developed a model arson law that has been adopted by many states.

arsonist 1. (Law) A person who attempts or commits an act of arson. A firesetter with a deliberate, recognizable motive, such as financial gain, revenge, desperation, frustration, or other personal ends. Also called: firebug; firesetter; incendiarist; torch. Cf: pyromaniac. 2. = (Brit) fire raiser.

arson squad (Fire Pr) A team of police and/or fire department personnel assigned to investigate fires of suspicious or questionable origin.

arteriosclerosis (Med) A condition in which the walls of the arteries are abnormally thickened, hard, and inelastic.

artery¹ (Transp) A main channel in any branching system of communication or transportation.

artery² (Anat) A blood vessel that carries blood away from the heart. One of two major sets of vascular tubes in the body, the other being the veins. Arteries carry blood away from the heart. Pulmonary arteries carry blood to the lungs to be oxygenated and systemic arteries carry the blood to the capillaries, which in turn supply the body cells with needed oxygen and nutrition. Blood is transferred via capillaries to the veins, which return it to the heart.

artesian well (Hydrol) A well fed by an aquifer that lies above the tapping point. The weight of the water creates a pressure, often causing the water to gush out in a geyser-like spout.

articulated (Anat) Jointed.

articulated boom (Fire Pr) A folding power-operated beam used to raise and lower an elevating platform or nozzle. See also: boom.

articulated steering (Transp) A system of steering used in tracked or wheeled vehicles consisting of two or more powered units in which turning is accomplished by yawing the units with respect to each other about a pivot system not located above any of the axles. See also: Ackermann steering.

articulating boom platform (Fire Pr) An aerial platform with two or more hinged boom sections.

articulation (Mech Eng) A movable joint, such as an elbow or swivel, esp one with more than one degree of freedom; a combination of two or more swiveling parts.

artifact[1] (Techn) A man-made object, as opposed to a natural one. A structure mechanically altered from its normal state.

artifact[2] (Psych) Misleading data or incorrect information resulting from improper experimental conditions or procedures.

artifact[3] (Med) An extraneous manifestation introduced by an outside agency; e.g., a spurious wave on ar ECG due to a loose electrode.

artificial abrasive (Techn) A man-made grinding material such as carborundum or emery. See: abrasive.

artificial barricade (Civ Eng) A solid or revetted wall of earth at least 3 feet thick used as protection around munitions stores. See: barricade.

artificial radioactivity (Nucl Phys) Radioactivity induced by bombardment with nuclear particles.

artificial respiration (First Aid; Med) A lifesaving first aid treatment given to victims who have ceased breathing due to drowning, shock, or other accident. Prone pressure and arm lift methods involve periodic external pressure on the rib cage. Mouth-to-mouth resuscitation, or the use of resuscitation equipment, involve filling the lungs periodically by forcing air into the victim's airway. Machines for this purpose are called resuscitators. See also: respiration; breathing apparatus; first aid.

ASA[1] = Acoustical Society of America

ASA[2] = American Standards Association; now ANSI.

asbestos (Mineral) A noncombustible, chemically inert, fibrous gray or greenish silicate mineral used for fireproofing and heat insulation. Asbestos is found in several forms. A variety of serpentine is called chrysolite or fibrous amphibole. Amianthus is a fine silky variety.
NOTE: Asbestos is used also for brake linings; packing and gaskets; protective clothing; and heat, fire, and electrical insulation; as well as in building materials (shingles, siding, tiles). Asbestos dust, and esp the fibers, are hazardous to the respiratory system, causing a lung disease called asbestosis.

asbestos cement board (Build) A noncombustible building finish material made of fibrous material (such as asbestos) and cement, pressed into thin sheets, usu 4 by 8 feet, used as sheathing on frame construction. Syn: Sheetrock.

asbestosis (Med) A disease of the lungs caused by the inhalation of fine airborne fibers of asbestos.

ASCE = American Society of Civil Engineers

ASCII (Comput) The acronym for the United States of America Standard Code for Information Interchange. Referring to an eight-level computer tape code in which seven positions are used for recording information and the eighth position is used for parity (error) checking.

aseptic (Med) Clean and free of infecting microorganisms.

ash (Comb) The solid, noncarbonaceous residue, often powdery, remaining after combustion, usu consisting of metal oxides or silicates.

ASHRAE = American Society of Heating, Refrigerating and Airconditioning Engineers, Inc.

ASIS[1] = American Society for Industrial Security

ASIS[2] = American Society for Information Science

askarel (Electr) A synthetic insulating liquid that evolves nonflammable gases when it is decomposed by an electrical spark.

asphalt 1. A naturally occurring, black or dark-brown bitumen. A viscous mixture of high molecular weight hydrocarbons and other carbon compounds used principally for paving and roofing. 2. = (Brit) bitumen.

asphyxia (Med; Path) Suffocation resulting from lack of oxygen. A state of deficiency of oxygen in the blood and an excess of carbon dioxide caused by mechanical interference with respiration or by inhalation of oxygen-deficient gases.
NOTE: The lowest life-safety limit of oxygen in the air for extended exposure is about 16-17%. Simple asphyxiants act mechanically by excluding oxygen from the lungs when breathed in high concentrations (examples: nitrogen, hydrogen, carbon dioxide). Chemical asphyxiants act through chemical action, preventing oxygen from reaching the tissue or preventing the tissue from using it. Carbon monoxide is a common example. It combines with hemoglobin to reduce the blood's capacity to transport oxygen. Other examples of chemical asphyxiants are hydrogen cyanide and aniline. Hydrogen cyanide destroys enzymes and thus inhibits oxygen intake on the cellular level. Local asphyxia results from interruption of the blood supply to a part of the body and, if prolonged, allows gangrene to develop. Syn: suffocation. See: first aid. Cf: anoxia; hypoxia.

asphyxiant (Med) A gas, vapor, or other atmosphere that deprives the lungs of sufficient oxygen to sustain life.

aspiration[1] (Techn) The drawing of air through a passage.

aspiration[2] (Physiol) The act of inhaling, whether of air, dust, stomache contents, etc.

aspirator[1] (Fire Pr) A device for moving fluids or suspensions by means of a partial vacuum. The vacuum is generated by a pump, such as an eductor or ejector. See also: pump.

aspirator[2] (First Aid) A device consisting of a pump, tube, and bottle, used to clear fluids from body airways and cavities.

ASQC = American Society for Quality Control

assay (Chem; Metall) An analytical procedure used to identify a specific element or compound. Commonly the measurement of the metal content of an ore.

ASSE = American Society of Safety Engineers, Inc.

assembly[1] (Fire Pr; Brit) The deployment of personnel and apparatus on the fireground.

assembly[2] (Techn) A number of parts or subassemblies, or any combination of such items, joined together to perform a specific function.

assembly occupancy (Fire Pr) A building, structure, or any part of one used as a gathering place by people for civic, political, travel, religious, recreational, or entertainment purposes.

assembly space (Techn) An area used for collecting and combining components.

assignable cause (Safety) A causal factor identified as having a relationship to an accident.

assigned risk (Insur) An automobile liability policy written at government request to cover an individual who cannot obtain insurance under normal insurance company rules.

assignment (Fire Pr) 1. The standing designation of individual units that are to respond to alarms in given territories. 2. The entire complement of men and units responding to a given alarm.

assignment card = running card

associationism (Educ; Psych) The doctrine that what is learned consists of associations between ideas, where ideas are determined by experiences in the world.

assumption (Statist) A premise that something is true for the purpose of testing whether or not it is in fact true or for developing a theory. See: *hypothesis*.

assumption of risk (Insur; Law) A doctrine that an employee assumes all the obvious and customarily associated risks when he accepts a job. A common law defense used by employers in litigation over worker injuries.

astatic (Phys) Not stable; showing no tendency to assume a particular orientation.

ASTD = *American Society for Training and Development*

asthenia (Med) Weakness or loss of strength. Syn: *adynamia*.

asthma (Med) A disease, usu of an allergic nature, marked by recurrent attacks of wheezing and breathing difficulty. A condition in which the bronchial tube muscles constrict when irritated.

astigmatism[1] (Opt) A defect in an optical element in which a point source is not brought to a definite focus.

astigmatism[2] (Med) A vision defect due to corneal irregularity, causing a blurred image, similar to that produced by an astigmatic lens.

ASTM = *American Society for Testing and Materials*

asymptote (Math) A limit which a curve approaches. An example is a series in which a number is halved, and succeeding halves are halved. No matter how many times the succeeding halves are divided, half always remains, although its value may become infinitesimally small.

ATA = *American Trucking Association, Inc.*

ataxia (Med) Inability to control voluntary muscles; failure of muscular coordination.

atelectasis (Med) Collapse of air cells in the lung through obstruction of a bronchus by a growth, fluid, foreign body, or other condition.

atherosclerosis (Med) A type of arteriosclerosis characterized by fatty deposits in the arteries.

Athey wagon (Transp) A classification of heavy-duty, unpowered, tracked cargo trailers designed to be towed behind track-laying prime movers for the transport of cargo over soft or rough terrain.

athwartship (Navig) Across a vessel or vehicle; at right angles to the fore and aft center plane.

atmometer (Metrol) An instrument for measuring the evaporation rate of a liquid.

atmosphere[1] A gas mixture used for respiration.

atmosphere[2] (Geophys; Abb: atm) The envelope of air surrounding the earth.
NOTE: Approx half of the atmosphere lies within 3.5 miles and three-quarters within 7 miles of the earth's surface. Almost all weather, clouds, and storms occur in the *troposphere*, which extends to about 10 miles above the earth. The atmosphere consists principally of nitrogen and oxygen, with lesser quantities of other gases, water vapor, and pollutants. See: *air; air mass*. See also: *air pressure; standard atmosphere*.

atmospheric absorption (Acoust) Reduction in the intensity of a sound wave in passing through the air, apart from the normal *inverse square law*, and arising from true absorption.

atmospheric area (Fire Pr) The part of a burning building in which air can circulate freely.

atmospheric ceiling (Fire Pr) The height at which a thermal column in free air ceases to rise and levels off.

atmospheric displacement (Fire Pr) The effect produced when water applied to a fire turns into steam and through expansion drives the heated air out of the burning building.

atmospheric electricity (Meteor) The electrical phenomena occurring in the earth's lower atmosphere, the most spectacular being lightning, which starts a majority of the forest fires. See also: sferics.

atmospheric fallout (Env Prot) The precipitation of small particles through the atmosphere. The precipitate may include nuclear particles, micrometeorites, and cosmic dust. A special case of sedimentation. See also: fallout.

atmospheric interference See: sferics

atmospheric inversion (Meteor) A stratum of warm air above cooler air that tends to trap pollutants and to hinder smoke and other hot combustion products from rising. See: inversion.

atmospheric pollution See: air pollution

atmospheric pressure (Phys) The pressure exerted by the weight of the air, which at sea level is approx 760 mm of mercury, or about 14.7 pounds per square inch under standard conditions. Atmospheric pressure decreases with altitude at the rate of approx 1 inch of mercury for each 1000 feet up to 7000 feet, at which point the rate of decrease becomes steadily less. See: standard atmosphere.
NOTE: Conversions of common pressure units are given in a table under pressure measurement.

atmospherics See: sferics

atmospheric tank (Storage) A storage vessel designed for use at atmospheric pressure to 0.5 psig.

at. no. = atomic number

atom (Phys; Chem) The smallest particle of an element, indivisible by chemical means, that exhibits the characteristic properties of the element. An atom consists of a positively charged nucleus and one or more negatively charged electrons moving around the nucleus in one or more orbits. Atoms of the same species are chemically identical.
NOTE: The smallest part of a compound is a molecule, which is made up of two or more atoms. A drop of water contains about six sextillion (6×10^{21}) atoms.

atomic absorption analysis (Chem; Spectr) A technique of chemical analysis using atomic absorption spectroscopy.

atomic clock (Metrol) A high-precision device that measures time by comparison with the vibrations of molecules or atomic nuclei.

atomic cloud (Env Prot) A radioactive cloud, typically shaped like a mushroom, consisting of hot gases, vapors, dust, and other debris carried aloft after a nuclear explosion.

atomic constants (Phys) A set of constants fundamental to atomic physics; chiefly: e, the electron or proton charge; me, the electron mass; mp, the proton mass; c, the velocity of light; N, Avogadro's number, the number of molecules in a molar mass; and h, Planck's constant, the uncertainty limit of conjugate physical measurements. Many other constants can be derived from this primary set.

$c = 2.997925 \times 10^{10}$ cm sec^{-1}
$e = 4.80298 \times 10^{-10}$ esu
$h = 6.6256 \times 10^{-27}$ erg sec
$me = 9.1091 \times 10^{-28}$ g
$mp = 1.67482 \times 10^{-24}$ g
$N = 6.02252 \times 10^{23}$ (g-mole)$^{-1}$

See: atomic mass; electron mass.

atomic energy = nuclear energy

Atomic Energy Commission (Abb: AEC) 1. A body appointed by the President to direct the independent civilian atomic energy agency responsible for atomic energy matters in the United States. 2. By extension, the agency as a whole.
NOTE: Now a part of the Energy Research and Development Administration.

atomic energy levels (Spectr) The energy states of an atom. See: energy levels.

atomic fission (Nucl Phys) The splitting of the nucleus of an atom into two approximately equal parts, with release of kinetic energy in the process.

atomic mass (Chem; Phys) The mass of a neutral atom, usu expressed in **atomic mass units**.

atomic mass unit (Phys; Abb: amu) A unit of mass defined as one-sixteenth the mass of an oxygen atom of mass number 16. Sixteen grams of O^{16} contain 6.023×10^{23} atoms; hence one atom of O^{16} weighs 2.656×10^{-23} gram. In terms of energy, 1 amu = 931 mev = 1.49×10^{-3} erg. Also called: **absolute mass unit**; **isotopic mass unit**. See: **atomic weight unit**. The international amu is defined as one-twelfth the mass of C^{12}, which is the most abundant isotope of carbon. See: **Avogadro's number**.

atomic number (Symb: Z) The number of elementary positive charges (protons) in a nucleus of an atom. Each element has a characteristic atomic number. From the lightest to the heaviest natural elements, Z ranges from 1 for hydrogen to 103 for lawrencium. Also called: **charge number**.

atomic power (Nucl Phys) Thermal power generated in a nuclear reactor or power plant.

atomic spectrum (Spectr) The characteristic radiation emitted by an excited atom. The distribution of lines of different frequencies can be used in analysis to identify particular atomic constituents of a sample.

atomic waste (Nucl Phys) The radioactive residue paroduced by fission in a nuclear reactor, including products made radioactive in such a device.

atomic weight (Chem; Phys; Abb: at. wt) A number assigned to each element approx equal to the sum of the number of protons and neutrons found in the nucleus of the atom. This sum is also called: **mass number**.
NOTE: The atomic weight of oxygen, for example, is 16, with most oxygen atoms containing 8 neutrons plus 8 protons; aluminum is 27, with 14 neutrons and 13 protons. The mass unit is defined as one-twelfth the mass of C^{12}. See: **atomic mass unit**; **atomic weight unit**.

atomic weight unit (Chem; Phys; Abb: awu) An international mass unit defined as one-twelfth the mass of carbon-12. It was adopted in 1962 for use in both physics and chemistry. See: **atomic mass unit**.
NOTE: Prior to 1962 physicists used a scale based on O^{16} = 16 exactly, while chemists used the isotopic average of natural oxygen as equal to 16.

atomization (Techn) The breaking up of a liquid into small droplets. See also: **aerosol**.

atomized (Techn) The state of a liquid reduced to fine particles.

atraumatic (Med) Noninjurious; not causing injury. Ant: **traumatic**.

atrium¹ (Anat) Either of the two upper chambers of the heart.

atrium² (Build) A courtyard or patio within a house.

atrophy (Med) Arrested development, decrease in size, or wasting away of body cells or tissue.

atropine (Med; Pharm) A drug used to stimulate the heart, to dry up respiratory tract secretions, to paralyze the ability of the eye to accommodate before testing refraction, and to dilate the eye. The compound is a white, crystalline, poisonous alkaloid, $C_{17}H_{23}NO_3$, obtained from belladonna and related plants.

attachment¹ (Fire Pr) The fitting of mating parts together.

attachment² (Phys) The process in which colliding particles stick together to form a single particle.

attack¹ (Fire Pr) The onslaught directed against a fire with decisive committment of personnel and equipment. Vigorous assault on a fire by fire suppression forces.

attack² (For Serv) Activities undertaken to contain a fire by cooling, smothering, applying fire retardants, or removing the fuel from the firepath. See: **direct method**; **indirect method**; **parallel method**.

attack crew (Fire Pr) An operating group of firefighters making a planned assault on a fire.

attack line (Fire Pr) A line of hose used to apply water directly on a fire, as distinguished from feeder lines connecting pumpers to the water supply.

attack time (For Serv) The time direct suppression of a wildland fire commenced.

attendance (Fire Pr; Brit) The total complement of firefighters and equipment at a fire.

attendance time (Fire Pr; Brit) The elapsed time between the receipt of a call and the arrival of the first apparatus (appliance) at the fire. Syn: **response time**.

attention[1] (Safety) Alert focusing of mental and physical abilities on a task to the degree necessary for successful accomplishment. A readiness to respond appropriately to stimuli.

attention[2] (Psych) Selective response to one or more stimuli. Focusing of the consciousness and increase in receptivity toward some object, activity, or situation.

attenuation[1] (Phys; Symb: a) a. A reduction in quantity, intensity, or strength, as of a signal, light, or some other form of energy. b. The reduction in flux or power density with distance from the source, due to adsorption and scattering, as the reduction in electromagnetic wave amplitude. Syn: **extinction**. Cf: **absorption**.
NOTE: Attenuation does not include the **inverse square law** of decrease in intensity with distance from the source.

attenuation[2] (Med) Reduction in virulence or toxicity by repeated inoculation, successive cultures, etc.

attenuation[3] (Acoust) The difference, expressed in decibels, of the sound intensity at a standard reference location as compared with sound intensity at another location which is acoustically farther from the source.

attic (Build) The space between the roof of a building and the ceiling of the top floor. A **blind attic** is one without windows, stairs, or scuttle; a **semi-accessible attic** is one that can be reached by a window, trap door, or scuttle. Cf: **cockloft**; **garret**.

attic ladder (Fire Pr) A small ladder that can be adjusted in width for use in cramped quarters, such as attics, closets, and hallways. Also called: **folding ladder**. See: **ladder**.

attitude (Psych) A semi-permanent cluster of human emotions, feelings, and cognitions about a situation, thing, or idea, which may manifest itself in overt behavior, depending on such factors as intensity, opposing or conflicting forces, the value system applied to a particular situation, motivational forces, and timing. An attitude may produce a predisposition toward behavior in certain ways to a stimulus, depending on its intensity.

attribute A characteristic property.

audibility threshold (Acoust) The frequency above or below which sound cannot be sensed by the ear. The limit of hearing. See: **audio frequency**.

audible range (Acoust) The frequency range of normal hearing, approx 20 Hz through 20,000 Hz. The range above 20,000 Hz is called ultrasonic; below 20 Hz, subsonic.

audio frequency (Acoust) Any of the sound frequencies between 20 to 20,000 Hz that are audible to the normal human ear.
NOTE: The human ear may detect sound vibrations above or below this range, but such frequencies are considered outside the audio range. Frequencies below 20 Hz are felt, rather than heard.

audiogram (Audiom) A record of hearing sensitivity tested at several frequencies between 500 and 6000 Hz, made to determine hearing loss. The audiogram may be numeric or graphic. In the latter case, hearing level is plotted as a function of frequency.

audiologist (Acoust; Ind Hyg) A person trained in problems of hearing and deafness.

audiometer (Audiom) An instrument for testing hearing sensitivity. Essentially a pure-tone audio signal generator that can be tuned to desired frequencies over the audio range and set at given sound intensity levels. Hearing sensitivity is measured in decibels referred to audiometric zero.

audiometric technician (Audiom) A person trained to administer audiometric examinations.

audiometric zero (Audiom) The threshold of hearing, usu taken as 0.0002 microbars (or dyne/cm²) of sound pressure.

auditorium raise (Fire Pr) A method of raising ladders in which the fly is steadied and secured by ropes. Also called: church raise.

auditory (Physiol) Pertaining to the sense of hearing or the organs of hearing.

Aureomicin = chlorotetracyclin

auricle[1] (Anat) Either of the two upper chambers of the heart that receives blood from the veins and forces it into the ventricles.

auricle[2] (Anat) The part of the ear that projects from the head. Syn: pinna. See: ear.

auscultation (Med) The technique of listening by ear or stethoscope to sounds within the body for diagnostic purposes.

authority having jurisdiction (Fire Pr) The duly authorized official or agency having legal enforcement responsibility for a statutory code.

authorized person (Safety) A person approved or assigned by an employer to perform specific duties or to be at a specific location at a jobsite. Syn: designated person.

autoclave (Surg) An apparatus consisting of a pressure vessel into which superheated pressurized steam is introduced to sterilize instruments or other objects.

autoconvection gradient = autoconvective lapse rate

autoconvective lapse rate (Meteor) A lapse rate sufficiently high to cause overturning of a stratified layer of air. Undisturbed air remains static even when the lapse rate exceeds the adiabatic rate of 5.5°F per 1000 feet. The critical lapse rate for overturning is 37.17°C per kilometer, or almost 19°F per 1000 feet. Steep lapse rates cause whirlwinds and dust devils. Syn: autoconvection gradient.
NOTE: Lapse rate in a homogeneous atmosphere is equal to g/R, where g is the acceleration of gravity and R is the gas constant. For dry air this is approx 3.4×10^{-4} °C/cm.

autoexposure (Build) A condition or feature of a building by which it imperils itself to fire spread, as one with large unprotected window areas through which fire can spread upward from floor to floor.

autogenous (Comb) Self-initiating; spontaneous. Now considered obsolete.

autogenous ignition = autoignition

autogenous ignition temperature = autoignition temperature

autograft (Surg) Skin taken from an uninjured portion of the body and grafted on an injury in the same body. Cf: homograft.

autoignition (Comb) Initiation of combustion by external heat but without a spark or flame. Cf: spontaneous ignition.
NOTE: Autoignition occurs when a petroleum product, for example, is heated to its autoignition temperature, at which point the sample ignites and burns.

autoignition temperature (Comb) The lowest temperature at which a combustible material ignites spontaneously in air without an external spark or flame. See: spontaneous ignition.
NOTE: Self-ignition of vapors and gases may be influenced by the presence of catalytic substances.

autokinetic effect (Psych) The apparent movement of a physically stationary spot of light viewed in a dark room due to the motion of the eye.

automatic action (Psych) A well-practiced response that occurs automatically when the appropriate stimulus is presented.

automatic alarm (Fire Pr) An alarm that is actuated by a fire detector or other automatic device.

automatic audiometry (Audiom) A method of administering a hearing test, often used for large numbers of persons. Subjects control the presentation of sound stimuli and record their own responses.

automatic feeding (Techn) Delivery and positioning of a material or part being processed to the point of operation and its removal by a means not requiring intervention by the machine operator.

automatic fire door (Fire Pr) A normally open fire resistant door designed to close by itself in case of fire. See: fire door.

automatic programming (Comput) Any technique whereby a digital computer itself transforms programming from a form that is easy for a human being to produce into a form that is efficient for the computer to carry out.

automatic sprinkler (Fire Pr) A water-spraying nozzle head, held in closed position by a low-melting alloy link, frangible bulb, or chemical pellet, installed in a pipe grid. See sprinkler system.
NOTE: When hot gases or flames weaken or destroy the holding element, the nozzle opens and sprays water or other extinguishing agent on the surrounding area.

automatic vent See: vent

automation (Autom) 1. Commonly, any type of advanced mechanization or technological change involving an increase in mechanization. 2. A continuous-flow production process which integrates various mechanisms to produce a finished item with minimal human intervention.

automaton (Comput) 1. A self-regulating machine that can be programmed to perform precise operations in sequence. 2. An electronic data processing and computing machine.

automobile (Transp) A self-propelled, wheeled vehicle, generally commercially designed and manufactured, for transporting fewer than ten passengers.

automobile fire (Fire Pr) A fire involving an automotive vehicle.

automotive service station See: service station

automotive vehicle (Transp) A general category of self-propelled land vehicles, generally wheeled or track-laying, but including all types of walking and jumping vehicles as well as self-propelled sleds and various air-cushion supported vehicles. An automobile.

autonomic nervous system (Physiol) The self-controlling nervous system that functions largely independently of human volition. The part of the nervous system consisting of sympathetic and parasympathetic branches that regulate body organs controlled by smooth muscles, including cardiac and blood vessel activity.

autopsy (Med) A post-mortem examination of body organs and tissues to determine the cause of death. Syn: necropsy. Cf: biopsy.

autoradiography (Nucl Phys) Images of radioactive sources made by placing radioactive material next to photographic film. The radiations fog the film, leaving an image of the source. See: radiography.

autotransformer (Electr) A transformer in which part of the primary winding is used as a secondary winding, or vice versa.

autoxidation (Comb) Spontaneous oxidation that occurs in air at moderate temperatures without visible combustion. Often the precursor of spontaneous combustion. Also spelled: auto-oxidation
NOTE: Rusting of metal is a common example.

auxiliary (Fire Pr) Referring to supplementary firefighting equipment or manpower, or to forces used in abnormal circumstances to augment first-line units.

auxiliary box (Fire Pr) An alarm box on a branch circuit that actuates a master box, which in turn transmits an alarm signal to the fire department.

auxiliary power unit (Fire Pr; Abb: APU) A small engine used to power a winch, electrical generator, or pump supplying mechanical power, water, or electricity during fire emergencies.

auxiliary pump (Fire Pr) A pump, used on fire apparatus, with a capacity rating below the minimum standard (500 gpm at 150 psi). There are two types: booster pump and high-pressure pump.

available height (Storage) The maximum height at which commodities can be stored above the floor and still maintain adequate distance from structural members and the required clearance below sprinklers.

average (Math; Abb: avg) The sum of a set of values divided by the number of values. Syn: arithmetic mean.

average burning rate (Comb) The arithmetic mean burning rate of pyrotechnic or explosive mixtures at specific pressures and temperatures. The dimension measured is length/time or mass/time.

avirulent (Med) Nonpoisonous; non-deadly. Ant: virulent.

Avogadro's law (Phys) The principle that equal volumes of gases and vapors at the same pressure and temperature contain the same number of molecules. Also called: Avogadro's hypothesis (Brit).
NOTE: Quantities of gases measured in proportion to their molecular weights occupy the same volume: e.g., 2 pounds of hydrogen, 32 of oxygen, and 44 of carbon dioxide each occupy 357 cubic feet at standard temperature and pressure.

Avogadro's number (Phys) The number of molecules contained in one mole or gram-molecular weight of a substance, equal to 6.02252×10^{23} molecules per gram molecular weight. See also: Loschmidt number.
NOTE: The Avogadro number, originally defined as the number of atoms in 16 grams of oxygen, is now defined as the number of atoms in 12 grams of carbon-12.

avoidable accident (Safety) An accident that could have been prevented by proper behavior or by appropriate environmental control or equipment modification.

avulsion (Med) Tearing away of parts or tissues.

awareness (Psych) The state of knowing. Cognizance of the existence or presence of something, esp if not generally known.

awareness indicator (Med) A device used to measure the tension levels in the muscles above the eyes.

awu = atomic weight unit

ax^1; axe a. A cutting or chopping tool. b. A tool with a blade-edged head and pick point fixed to a handle, used for chipping, splitting, or hewing wood. A fireman's ax is a small hand ax carried by firemen. See: flathead ax; forcible entry tool.

ax^2 (Fire Pr) An insignia of rank for some ladder and truck company officers.

ax belt (Fire Pr) A belt used by firefighters to carry an ax and other tools.

axial (Geom) Near, along, or on an axis. Ant: abaxial.

axis^1 (Geom) a. Any one of the lines of a coordinate system connecting all of the points of one variable when the other variables are zero. b. A line of rotational symmetry through a body.

axis^2 (Phys) Any of several imaginary reference lines used to define the planes of a crystal.

axis of rotation (Geom) The axis or line about which a body rotates.

axle¹ (Mech Eng) A shaft or spindle about which or with which a wheel rotates.

axle² (Transp) An automotive undercarriage assembly, including housings, gearing, differential, bearings, and mounting appurtenances. Also called: axle assembly.
NOTE: A bogie is an assembly of two or more automotive-type axles mounted in tandem in a frame so as to divide the load between them and permit vertical oscillation of the wheels.

axle lock (Fire Pr) A lock provided on tractor-trailer axles of aerial ladder trucks to immobilize the chassis and prevent movement prior to raising the aerial ladder. Also called: axle jack. See also: jack.

axle tramp (Transp) The sustained vibration of the axis of a solid axle in a vertical plane.

axman = swamper

azeotrope (Chem) A mixture of two or more liquids having a constant boiling point.

azimuth (Surv) The clockwise angle between a direction in the horizontal plane and true north. Syn: bearing. See: back azimuth.

azimuth circle (Surv) A circle graduated clockwise into 360 degrees.

azotemia (Med) An excess of urea or other nitrogen-containing bodies in the blood brought on by a kidney deficiency.

B

B (Chem) Symbol for boron.

Babbitt (Metall) 1. An alloy of tin, antimony, copper, and lead, used principally as a bearing liner. 2. To line or face with Babbitt.

babbitted fastenings (Mech Eng) A method of wire rope attachment in which the ends of the wire strands are bent back and are held in a tapered socket by means of poured molten babbitt metal.

babilarchy (Info Sci) Structure downward in a hierarchy.

baby Bangor See: Bangor ladder

bacillus (Bact) A rod-shaped, disease-producing bacterium. See: bacteria.

back (Mining) The roof of an underground excavation or the body of ore between one level in a mine and the next level above. The roof of a drift, crosscut, or stope.

back azimuth (Surv) The direction opposite to the azimuth; the azimuth angle plus 180 degrees.

back burn (For Serv; Fire Pr) 1. A prescribed burning fire set to travel against the wind. 2. To set a backfire.

back draft (Fire Pr) A fluctuating flow of air into a burning building in which the fire has depleted the oxygen supply. Smoke flutters back and forth several times during back draft and then may explode into flame. The latter is sometimes called a hot air explosion (erroneous) or a smoke explosion. See: ventilation.

backfill[1] (Civ Eng) a. The excavated material used to build up a road above the original ground level. b. To refill a ditch.

backfill[2] (Mining) To refill mined-out stopes and abandoned excavations in a mine with waste rock, sand, scrap, or other debris.

backfire[1] (For Serv) a. A fire set to burn off fuel in the path of an advancing wildland fire in order to stop its progress. Typically, a backfire is set inside a fire control line to reduce the fuel between the line and the fire edge. Syn: back burn; control fire. See: clean burn. b. = (Brit counter fire. c. To set such a fire. Cf: area ignition; burn out.

backfire[2] (Techn) A loud noise made by an internal combustion engine due to faulty ignition or explosion of fuel in the exhaust manifold.

backfire pot = drip torch

back flushing (Fire Pr) A method of cleaning equipment by flowing clean water in the reverse direction to remove clogged debris, salt, or chemicals.

background (Nucl Phys) The radiation coming from sources other than the radioactive material to be measured. The total effect of natural radiation, cosmic rays, electronic circuitry, and other factors in the environment which may interfere with the radiation measurements of specific interest.

background noise (Electr) The sum of the interference in a system used for the production, detection, measurement, or recording of a signal, independent of the presence of the signal. Energy fluctuations in a detection system produced by random molecular and quantal phenomena, cosmic rays, etc., independent of the signal. Noise is usu associated with a thermal process and is therefore often characterized by an effective temperature.

background radiation (Phys) 1. The sum of natural radiation in the environment, including cosmic rays. 2. Radiation extraneous to an experiment that involves radiation or is affected by it.

backing fire (For Serv) A wildland fire traveling against the wind.

backing wind (Meteor) A wind that changes direction in a counterclockwise sense. Ant: veering wind.

backlash[1] (Mech Eng) Undesired looseness between mechanical parts of a machine or mechanism. Also

BACKLASH • BACTERIOPHAGE

called: play; hysteresis.

backlash² (Fire Pr) The backward thrust of a nozzle in operation.

back pack (Fire Pr) A 5-gallon extinguisher with a built-in pump carried on the back and used for grass and brush fires. An Indian pump. Cf: (Brit) knapsack tank.

back-pack pump (For Serv) A lightweight centrifugal pump and gasoline engine that can be carried by a firefighter into areas inaccessible to motorized equipment. See: pump.

back pressure (Hydraul) The pressure drops occurring in pipes or hose lines due to the hydrostatic head and viscous (friction) losses. Hydrostatic losses are proportional to the difference in height between the pump and the exit nozzle and represent the work required to raise the fluid against the force of gravity.
NOTE: Pressure differences can be positive (pump below exit) or negative (pump above exit). When pumping is not used in the latter case, the system operates as a siphon. In fire practice, head losses are approx 4.5 psi pressure for each 10 feet of lift. Viscous losses are usu negligible compared with head losses, but can become considerable in long lines or rough pipes.

backscattering (Phys) The scattering of particles or radiant energy at angles exceeding 90° from the original direction of motion. Nonelastic x-ray backscattering is used in the analysis of atomic and molecular composition of matter, while elastic backscattering is used to determine the thickness of surface coatings. Syn: backward scattering. Cf: forward scattering.
NOTE: The measurement of backscattering is important for counting beta particles in ionization chambers, in medical radiation treatment, and in industrial radioisotopic thickness gauging.

back step (Fire Pr) A small platform at the rear of a fire apparatus provided for firefighters to stand on while riding to a fire. Syn: tail board.

back stoping (Mining) Removing ore from the back, or roof, of a stope. A method of overhand stoping.

back up 1. To be in reserve. 2. To reverse direction; to move backward.

backup 1. A substitute kept in reserve. 2. An item kept on hand to replace one that fails.

backup line (Fire Pr) 1. A hose laid as a reserve line for use in case the initial attack line proves inadequate. 2. An additional line supporting the attack and protecting men using fog lines for close attack on a flammable liquid fire.

bacteria (Bact; Abb: bact) Microscopic (0.025 to 40 microns) single-celled organisms living in soil, water, organic matter, or plants and animals. Also called germ; microbe.
NOTE: Parasitic bacteria often cause diseases; saprophytic bacteria live on dead plant and animal tissues, causing them to decay and break down into their chemical constituents. Bacteria are classified according to shape: spheres (coccus forms), rods (bacillus forms), and spirals (spirilla forms). Species that can subsist only on organic matter are called heterotrophic; those on inorganic matter are autotrophic, that is, they can build up carbohydrates and proteins out of carbon dioxide and inorganic salts. Bacteria that require free oxygen for existence are obligatory aerobes, those that cannot survive in the presence of free oxygen are obligatory anaerobes, and those that are indifferent to oxygen are facultative anaerobes. Bacteria are destroyed by ultraviolet light and high temperatures. Singular: bacterium.

bactericide (Bact) An agent that destroys bacteria.

bacteriophage (Bact) An ultramicroscopic agent that attacks bacteria. Phages are generally parasitic, filterable viruses, highly specific as to the species of bacteria they attack. Phages, their structure, and mechanism of attack have been visualized by electron microscopy. They generally alter the chromosomes of bacteria in propagating themselves and destroy their hosts in

44

the process.

bacteriostatic (Bact) Interfering with or preventing the growth of bacteria without killing them.

baffle¹ (Techn) A partition. A plate, deflector, or similar device used to regulate flow.

baffle² (Fire Pr) a. A wall or plate installed in a tank to retard sloshing and shifting of liquid loads when the carrier vehicle is in motion. Cf: swash. b. A partition dividing a hose body into sections.

bag house (Techn) A structure containing bag filters for recovery of fumes of arsenic, lead, sulfur, or other substances from flues of smelters.

bag mask resuscitator (First Aid) A breathing apparatus consisting of a face mask, an air bag, and a valve that allows the patient to exhale and the bag to refill with oxygen from an oxygen tank or other source.

bag of waters = amniotic sac

bail¹ A semicircular handle on a pail.

bail² To deliver possession of an item without transfer of ownership.

bail³ To remove undesired liquid by dipping it out.

bail⁴ (Law) a. To accept responsibility for the debt or default of another. b. Security or guarantee of appearance for trial posted for the release of an arrested person.

Bakelite (Techn) A thermosetting plastic produced by polymerizing phenol or cresol with formaldehyde and ammonia. Used for molded plastic wares, electrical insulators, telephones, etc.
NOTE: Bakelite was the earliest commercial plastic.

bakeout (Techn) The cleaning of surfaces by heating during evacuation. The process is often called outgassing.

balance¹ (Techn) a. A state of static and/or dynamic equilibrium of masses. b. A counterbalance.

balance² (Metrol) An instrument for weighing; esp an equal-arm balance on which an unknown mass is compared with weights of known mass.
NOTE: Many special balances exist, including chain balance, torsion balance, density balance (for liquids), single-pan balance, spring balance, etc.

balance³ (Horol) The balance wheel of a clock or watch that regulates the operation of the escapement. Also called: torsion pendulum.

balanced pressure proportioner (Fire Pr) A type of foam concentrate proportioning system in which a pump and appropriate devices feed the concentrate to a metering orifice or controller at the same pressure as that of the water supply. See: foam.

balcony (Build) 1. A platform projecting from a wall of a building, usu enclosed by a low wall or railing. 2. A gallery above an auditorium.

bale (Mat Hand) Articles, materials, or wastes compressed into a shaped form, wrapped, and usu bound with cord or metal ties.

baler (Techn) A machine used to compress and bind materials, including solid wastes.

ballast (Navig) Any solid or liquid weight placed in a vehicle or vessel to change its trim or to improve its stability.

ball blanket (Fire Pr) A layer of plastic balls floating on the surface of a volatile liquid to reduce evaporation and provide support for extinguisher foam in the event of fire. The balls are made of a polyolefin plastic called Allplas.

ball cock 1. A device for regulating the flow of a liquid into a tank, consisting of a lever connected to a guided floating ball that opens and closes a valve at the bottom of the tank. 2. A ball valve.

ballistic body (Phys) A body moving through space or the atmosphere free of all constraints except gravity and air resistance. The path followed by such a body is called a ballistic trajectory.

ballistic movement (Ind Eng) Smooth motion of a limb with follow-through.

ballistic nylon A fabric of high tensile properties designed to provide protection from lacerations.

ballistics (Abb: Ballist) The science of the motion, behavior, and effects of water streams from nozzles, projectiles, esp artillery shells, bullets, missiles, etc.
NOTE: Interior ballistics pertains to the motion of projectiles within the bore of a rifle or gun; exterior ballistics concerns motion of projectiles after they leave the muzzle, as well as the motion of rockets and missiles.

ballistic trajectory See: **ballistic body**

ball lightning (Meteor) A form of lightning described as a small, luminous ball that may move rapidly, float in midair, or have a tendency to enter enclosures. The ball emits hissing sounds, and may explode noisily or disappear silently. Also called: **globe lightning**. See: **lightning**; **St. Elmo's fire**.
NOTE: Statistical studies indicate that ball lightning may be almost as common as ordinary lightning, but is rarely observed because of its small size. It has been suggested that ball lightning is a temporarily stable plasma.

ball mill (Techn) A grinding device in which balls, usu of steel or stone, produce the grinding action in a revolving container.

balloon construction (Build) A type of frame construction in which studs are erected the full height of the building and floor joists rest on ribbon boards nailed to the studs. As a result, interconnecting passages, called **interstitial space**, exist from cellar to roof and, unless firestopped, act as wooden chimneys capable of conducting smoke, heat, and flame throughout the structure in case of fire.

ballooning (Fire Pr; Salv) A technique of trapping air under salvage covers for the purpose of floating them into position.

baluster (Build) A shaped vertical member supporting a railing.

band (Commun) A range of frequencies in the electromagnetic spectrum. Also called: **frequency band**. See: **bandwidth**.

bandage[1] (First Aid; Med) Gauze, elastic tape, adhesive tape, or other material used to keep a dressing in place.

bandage[2] See: **hose bandage**

bandpass filter (Electr) An electronic device that allows electromagnetic waves of certain frequencies to pass through and blocks all others.
NOTE: The lower and upper bounds of the frequency band that is passed are called the lower and upper cutoff frequencies. These limits determine the bandwidth of the filter.

band pressure level (Acoust) The sound pressure level of the acoustic vibrations within a specified band of audio frequencies.

band spectrum (Spectr) A related group of closely spaced lines characteristic of molecular absorption. Under low dispersion, the groupings appear as wide bands.

bandwagon effect (Psych) A social group phenomenon in which more and more individuals ally themselves with the prevailing majority opinion.

bandwidth (Commun) 1. An interval of frequencies, usu indicating the range in which an electronic device operates. 2. The number of Hertz (cycles per second) between the limits of a frequency band. See: **bandpass filter**.

Bangor ladder (Fire Pr) A large extension ladder, developed in Bangor, Maine, equipped with **tormentor poles** for raising.
NOTE: A **baby Bangor** is a small extension ladder, without ropes, used inside buildings. See: **ladder**.

banister (Build) A handrail and its supports along a staircase.

bank (Civ Eng) A mass of soil rising above a digging level.

bank track (Mat Hand) A continuously flexible track, usu an endless bank of rubber reinforced with steel cables.

bar 1. (Phys) A unit of pressure equal to 10^6 dyne/cm², 1000 millibars, or 29.53 in. Hg. 2. A unit of pressure equal to 1 dyne per square centimeter in the cgs system. Prior to 1960, a bar was used as an equivalent to barye (1 dyne/cm²). The preferred unit is Newton/area². One bar is equal to 10^5 N/m² or 10^6 dyne/cm². See: pressure.

Baralyme (Chem) A trade name for a carbon dioxide absorber, consisting of calcium hydroxide and barium hydroxide.

barbiturate (Pharm) A central nervous system depressant or sedative derived from barbituric acid. Barbiturates can be classified into three groups according to the duration of their effects and action: 1) barbital and phenobarbital are long-lasting sleep inducers; 2) amytal, nembutal, seconal, etc. are intermediate-term hypnotics; and 3) thiopentone, pentothal, hexobarbital, etc. are strong, short-term intravenous anesthetics.
NOTE: Barbiturates are habit-forming and are often involved in suicides. Acute barbiturate intoxication results in failure of the involuntary reflexes, depressed respiration, and coma. Death results from respiratory paralysis or vascular collapse. Victims require immediate medical attention.

barge (Navig) An unpowered, flat-bottomed, shallow-draft vessel, including scows and lighters; not including deep-draft barges or towed vessels shaped like ships.

baric (Meteor) Pertaining to pressure; often used as a suffix, as in hyperbaric. See also: barometer; pressure.

bark (For Serv) To remove bark from a log.

bark pocket (For Serv) An opening containing bark between the annual growth rings of a tree.

baroclinic field (Meteor) A state of stratification in a fluid in which isobaric surfaces (surfaces of constant pressure) intersect isosteric surfaces (surfaces of constant density). Air temperature has a gradient on the isobaric surface and the geostrophic wind is in vertical shear.
NOTE: The number of isobaric-isosteric solenoids per unit area intersecting a given surface is a measure of baroclinity, which is often accompanied by intense local winds. Also called: barocliny; baroclinicity. Zero baroclinity is called barotropy. Cf: barotropic field.

barogram (Meteor) A record traced by a recording barometer or barograph.

barograph (Meteor) A recording, aneroid-type barometer in which a stack of flexible sealed boxes is linked to a pen that traces a record of pressure changes on a slowly revolving drum.

barometer (Metrol) An instrument for measuring the absolute pressure of the atmosphere. The common meteorological barometer consists of a glass tube filled with mercury with one end sealed. The other end is open and immersed in a reservoir of mercury. Changes in atmospheric pressure cause corresponding changes in the height of the mercury column.
NOTE: The same device used to measure pressure in a closed system is called a manometer. Liquids other than mercury can be used. The sensitivity is inversely proportional to the density of the liquid. Barometer readings should be corrected for temperature and altitude. See also: aneroid.

barometric damper (Techn) A hinged or pivoted plate that automatically regulates the amount of air entering an incinerator duct, breeching, flue connection, or stack to maintain constant draft.

barometric pressure = **atmospheric pressure**

barometric wave (Meteor) A pressure wave in the atmosphere.

baroscope (Techn) An open-end manometer used for pressure measurement. See also: barometer, aneroid.

baroswitch (Techn) A switch operated by a pressure change. Usually an aneroid device.

barotrauma (Med) An injury to the ear caused by a sudden change in atmospheric pressure.

barotropic field (Meteor) An atmospheric field in which the air density depends solely on pressure; isosteric surfaces do not intersect isobaric surfaces; air temperature is constant on a surface of equal pressure, and the geostrophic wind is not in vertical shear. Cf: baroclinic field.

barrel[1] (Transp) A unit of volume equivalent to 42 US gallons of petroleum or 31 gallons of beverage.

barrel[2] (Techn) a. Commonly a round bulging container made of wooden staves held together with metal hoops. b. Any of a variety of cylindrical or tapered containers used for mixing, grinding, or tumble drying of materials.

barrel[3] (Fire Pr) The body of a fire hydrant.

barretter (Metrol) A device containing a fine wire that has a high temperature coefficient of resistance, used to measure radio-frequency power. Sensitive barretters can measure power as small as 10^{-8} watt. See: bolometer.

barricade (Traffic) An obstruction to deter the passage of persons or vehicles.

barricading (Safety) Screening of an explosives-containing building from a magazine, building, railway, or highway, either by a natural or artificial barrier.

barrier[1] (Fire Pr) A floating boom with a draft of about 12 inches and a 6-inch freeboard used to contain oil spills on water. Also called: boom.

barrier[2] (Fire Pr) a. An obstruction, such as an area bare of fuel, that retards or prevents the spread of a wildland fire. b. Any temporary or permanent structure or earthen dike erected as a flame shield or to prevent fire spread.

barrier guard (Safety) A device that protects operators and other individuals from hazard points on machinery and equipment.

barrier material (Mat Hand) A protective material designed to withstand, to a specified degree, the penetration of water, oils, water vapor, or certain gases. The material may serve to exclude such elements from the package or retain them inside.

Barron tool (For Serv) A fire-fighting, raking tool in which a hoe blade is set at an angle to a row of tines; used to cut light fuels and as a scraper.

barye (Phys; Obsolete) A pressure unit equal to one dyne/cm^2, or 0.001 millibar. The preferred pressure unit in physical measurements is the Newton. See: bar.

basal metabolism (Med) A measure of the amount of energy required by an organism at rest. See: metabolism.

base[1] (Chem) a. An oxide or hydroxide of a metal that yields hydroxyl ions in an aqueous solution. b. A substance that reacts with acid to form salt and water. c. Any species capable of accepting a proton or giving up an unshared pair of electrons to an acid. Syn: alkali. Cf: acid.

base[2] (Math) An integer, the powers of which are used as column values in a numeric system, the base integer being the value of the column to the left of the units column; e.g., the ten's column in the decimal system or the two's column in binary. A system based on five's (biquinary notation) is used in the abacus. Also called: radix. See: number systems.
NOTE: In ordinary decimals, with ten as the base, a number such as 321 represents $3 \times 10^2 + 2 \times 10 + 1$; but this is also equal to $5 \times 8^2 + 0 \times 8 + 1$, so that using base 8, the decimal number 321 is written 501.

base[3] (Aerodyn) The extreme aft end

of a body.

base⁴ (Geom) The line or surface of a figure on which it is assumed to rest, such as the bottom edge of a triangle or the circular base of a cone.

base area (For Serv) A tract of wildland that is representative in terms of fuel and topography of the overall fire protection unit.

base camp (For Serv) The main camp when more than one are established; used as a staging area where personnel, supplies, and equipment are assembled for assignment to specific tasks and areas. See: fire camp.

base fuel model (For Serv) A representation of the vegetative cover and fuel in a base area.
NOTE: Used in the calculation of fire danger ratings.

base injection (Fire Pr) Pumping foam or other light extinguishing agent into the lower part of a burning flammable liquid tank. The extinguishant rises to the surface to smother the flame. Air is sometimes used to raise unheated fuel from the bottom of a tank to cool the surface of a burning liquid below its flammability point. Also called: subsurface injection. See: foam injection.

base line¹ (Geom; Surv) Any reference line used for measurement or angular orientation of other lines.

base line² (Math) a. The abscissa or horizontal (x) axis of a coordinate system. b. The reference set of data against which other measurements are compared.

basement (Build) A level of a building or structure lying half or more below ground level.

basement spray (Fire Pr; Brit) A special nozzle used to fight fires in basements and ships' holds. The spray is designed to throw water at many angles when it is thrust below a floor or deck level to give maximum cooling effect. Syn: cellar spray; cellar nozzle.

base metal (Techn) 1. The parent metal in a brazing, cutting, or welding operation. 2. After welding, the metal which was not melted.

base station (Fire Pr) A fixed central radio station from which the movement of one or more mobile stations is controlled.

basic (Chem) Pertaining to a chemical base. Alkaline. See: alkali.

Basic Building Code (Build) The building code that is maintained and published by the Building Officials and Code Administrators International.

basic research (Science) Research directed toward increase of knowledge, the primary aim of the investigator and sponsor being a fuller understanding of the subject under study.

basic salt (Chem) A salt containing replaceable hydroxyl groups.

basic skills (Educ) Fundamental abilities in reading, writing, and arithmetic needed to progress to higher academic levels of achievement.

basket¹ (Fire Pr) a. The mobile platform of an aerial platform. b. A container, usu of heavy-gage wire or sheet steel, placed on fire apparatus for carrying small hose and tools.

basket² (First Aid) A type of stretcher used for rescue of seriously injured and incapacitated persons or removal of casualties.

battalion (Fire Pr) A fire department district comprising several fire stations.

battalion chief (Fire Pr) A fire department rank, next above captain, and the lowest ranking chief officer. The battalion chief commands one shift or platoon of a battalion district. See: ranks and titles.

battering ram (Fire Pr) A heavy metal bar with handles used to break through walls or other obstructions in order to gain access to fires or to perform rescue work. See: forcible entry tools.

battery¹ (Electr) A device consisting of one or more galvanic cells for converting chemical energy into direct electrical current. Primary batteries are discarded when exhausted. Secondary batteries are rechargeable. Also called: storage battery; power cell; (Brit) accumulator. Cf: fuel cell; solar battery.

battery² (Fire Pr) A heavy stream nozzle.

Baumé scale (Abb: Be) An arbitrary liquid density scale, devised by A. Baumé, a French chemist, based on hydrometer readings for water and for 10% sodium chloride. Two systems are in use. The following relationships are used to find degrees Baumé when the density of a liquid is known: Degrees Be (Bureau of Standards):
for liquids lighter than water
deg Be = (140/density) - 130
for liquids heavier than water
deg Be = 145 - (145/density).
NOTE: The specific gravity of petroleum products is more commonly given in degrees API (American Petroleum Institute). When the specific gravity of a liquid is known, the following relationship can be used to find degrees API:
deg API = (141.5/sp gr) - 131.5.

bay (Storage) A designated area within a section of a warehouse or shop, usu outlined or bounded by posts, pillars, columns, or painted lines.

BCF = bromochlorodifluoromethane. See also: Halon.

beam¹ (Fire Pr) a. The solid or trussed main structural side member of a ladder supporting the rungs or rung blocks. Also called: spar. See: ladder; rails¹. b. = (Brit) string.

beam² (Build) A horizontal structural member resting on supports and carrying loads such as floors, roofs, walls, etc.

beam³ (Phys) a. A stream of electromagnetic energy propagating directionally, such as an electron beam, a radio beam, or a searchlight beam. b. To emit a directional ray of electromagnetic energy.

beam (Metrol) The transverse bar of a balance, supporting the weighing pans.

beam bolt (Fire Pr) A bolt that passes through a truss block and holds the rails of a truss beam ladder together. See: ladder.

beam man (Fire Pr) A man stationed at the beams of a ladder during raising and lowering operations.

beam raise (Fire Pr) A method of erecting a ladder by resting it on one beam and raising it sideways. Used when overhead obstructions interfere with normal raising methods. See: raise.

beam splitter (Opt) A partially reflecting mirror that allows a portion of the incident radiation to pass and reflects the remainder.

bearing¹ (Mech Eng) A mechanical part designed to reduce friction between fixed and moving elements.

bearing² (Navig) A direction, usu expressed in terms of cardinal points or in degrees. Syn: azimuth.

bearing capacity¹ (Soil Mech) The average load per unit area required to produce failure by rupture of a supporting soil mass.

bearing capacity² (Techn) The maximum load that a material, or a structure such as a wall, column, or beam, can support without failing.

bearing pieces (Storage) Material underneath lading, to facilitate loading or unloading, to maintain a clearance below the overhanging portion of a load, and to distribute the weight of the lading over a larger floor area.

bearing wall (Build) A wall that serves as a supporting structural assembly in a building or other structure.

beat (Acoust; Electr) Amplitude or frequency fluctuations caused by the interaction between two or more signals, producing a variation equal to the sum of or difference between the frequencies.

beat elbow (Med) Inflammation of the elbow joint resulting from the use of heavy vibrating tools.

beat frequency (Acoust; Electr) The difference between the frequencies of two or more superimposed sinusoidal signals.
NOTE: If a signal of 1000 Hz is superimposed on a signal of 1100 Hz, the two signals will coincide additively at every 100th wave, thus producing a 100 Hz beat.

beat knee (Med) Inflammation or bursitis of the knee joints due to friction or vibration, a condition common to miners.

Beattie-Bridgeman equation (Phys) An equation of state for real gases. See also: van der Waals equation.

Beaufort wind scale (Meteor; Navig) A scale developed for judging wind force on sails, now used universally for describing wind velocity.

Code No.	Velocity (mph)	Description
0	0-1	calm
1	1-3	light air
2	4-7	light breeze
3	8-12	gentle breeze
4	13-18	moderate breeze
5	19-24	fresh breeze
6	25-31	strong breeze
7	32-38	moderate gale
8	39-46	fresh gale
9	47-54	strong gale
10	55-63	whole gale
11	64-75	storm
12	over 75	hurricane

See also: wind.

becket (Brit) A loop of rope or a flexible ring of other material used as a handle or grip. See also: snotter becket.

Beckmann thermometer (Metrol) A variable-range differential thermometer used to measure small temperature differences.

bed (Mining) A tabular deposit conforming to the local strata.

bed ladder (Fire Pr) 1. The lower section of an extension ladder into which the upper sections retract. 2. The lower section of an aerial ladder that in stowed position rests on the truck frame. Also called: main section. See: ladder.

Beer's law (Phys) Light is absorbed in a solution in proportion to the product of solute concentration and path length. Also called: Bouguer's law.

behavior (Psych) 1. Any aspect of human activity that can be studied. 2. Observable or inferrable responses of an organism to stimuli.

behavioral science (Psych) The science, including psychology, sociology, and social anthropology, concerned with the study of humans and lower animals by observation of activities and through experimentation.

behavior modification (Psych) A therapeutic approach focusing on the alteration of observable and measurable behaviors.

bel (Acoust) A unit of sound level expressed as the logarithm to the base 10 of the ratio of two quantities of power; used to qualify sound levels.
NOTE: Two powers differ by one bel if their actual ratio is 10:1. See: decibel. In Europe the neper is used instead of the bel.

Belding-Hatch index (Safety) An estimate of the body heat stress of an average or standard man for various degrees of activity. Related to sweating capacity.

bell (Fire Pr) 1. A vibrating clapper fire alarm. 2. A warning signal on fire apparatus.

belled excavation (Civ Eng) A bell-shaped shaft or footing excavation. The enlargement of the shaft toward the bottom provides working clearance.

bell jar A bell-shaped glass cover for protecting delicate instruments and for containing gases in vacuum operations. Also called: bell glass.

bellows (Techn) 1. A pumping device for producing a flow of air, consisting of a flexible chamber that draws in air through a valve on expansion and forces it out through another valve on contraction. 2. A flexible connecting element for pipes or shafts.

bell-shaped curve (Statist) A curve of the normal distribution. Syn: **gaussian curve**; **normal distribution**.

belt[1] (Mech Eng) A loop of flexible material used to connect shaft wheels and pulleys for power transmission.

belt[2] (Geophys) A region having distinctive characteristics, such as a **rain belt**; a popular name for **zone**.

belt shifter (Ind Eng) A device for mechanically shifting a belt from one pulley to another.

belt shifter lock (Ind Eng) A device for positively locking the belt shifter in position while the machine is stopped and the belt is idling on loose pulleys.

beltway (Transp) A limited access highway, usu two or more lanes in each direction, circling a city.

Bénard convection cells (Phys) Convection instability produced by heating a fluid from below. More or less regular arrays of hexagonal cells are formed, with fluid rising in the center and falling near the wall of each cell. See: **convection**.

bench (Mining) A step-like formation made in stoping.

bench photometer See: **photometer**

bend 1. To fasten one line to another or tie it to an object. 2. A type of knot. See: **knot**.

bending moment (Build) The sum of all moments on one side of a section perpendicular to the axis of a structural member, such as a floor joist.

benign (Med) Not malignant. Of a mild nature. Often used to characterize a tumor that does not spread throughout the body, but grows locally only.

bent (Build) A transverse frame forming an integral part of a structural unit or supporting another structural unit. Bents are made of steel, concrete, or wood, and carry both lateral and vertical loads in viaducts and industrial buildings. A typical bent consists of a roof truss or girder and its supporting columns.

bent stream tip (Fire Pr) A curved **nozzle** used to direct water streams into hidden spaces or places that are difficult to reach with a direct stream nozzle.

Benzedrine (Pharm) A trade name for a drug used as a central nervous system stimulant. Syn: **amphetamine**.

benzene (Chem) A colorless, aromatic, flammable, liquid **hydrocarbon**, C_6H_6, mp 5.5°C, bp 80.1°C; dissolves oils, fats, rubber, resins; burns with a smoky flame; used as an internal combustion fuel, solvent, and organic intermediate. Derived from coal or petroleum. The simplest member of the aromatic series of hydrocarbons. Benzene is toxic, and prolonged exposure may lead to nausea, fatigue, and anemia. Vapors form explosive mixtures with air. Also called: **benzol**.

berm (For Serv) 1. The downhill side of a ditch or trench. 2. A ledge at the top or bottom of a slope.

Bernoulli's law (Aerodyn) A conservation theorem stating that the sum of the static and dynamic pressures along a streamline in a flow of incompressible fluid is constant. Gravity and frictional forces are disregarded.
  NOTE: Bernoulli's equation that derives from this law states that the sum of the pressure head, elevation head, and the velocity head is constant at every point in a system.

berth (Fire Pr) A stall or parking position of a fire department apparatus on the apparatus floor of a fire station.

beta emitter (Nucl Phys) Any material that radiates beta rays. Beta radiation is easily stopped by a thin sheet of metal, but without protection can cause skin burns. A beta emitter is harmful if it enters the body.

beta particle (Nucl Phys; Symb: β) A high-speed, light-weight, elementary particle ejected by a radioactive nucleus during decay. The particle has a single positive or negative electron charge and a mass equal to 1/1837 of a proton. A positively charged beta particle is called a positron; a negatively charged one is an electron (sometimes called a negatron). Beta radiation is easily stopped by a thin sheet of metal, but without protection can cause skin burns.

Bev (Nucl Phys) A billion (10^9) electron volts. Sometimes written BeV. An electron with 1 Bev of energy travels almost with the speed of light.

bias¹ (Statist) A systematic error in measurements.

bias² (Electr) A voltage applied to the control grid of a vacuum tube to create a voltage relationship with respect to the cathode and thereby establish the operating point of the tube.

bight (Fire Pr) A loose U-shaped loop of rope or line. See: knot.

big line (Fire Pr) A fire hose 2-1/2 inches in diameter or larger, esp one used as a hand line.

big stick = aerial ladder

bilateral Two-sided.

bilge (Navig) The curved section of the hull and underbody between the bottom and side of a ship or boat. The recess into which water drains from cargo and machinery spaces.

bilge pump (Navig) A pump used to remove water and other liquid accumulations from a ship's bilge.

billet¹ (Metall) 1. Any solid metal casting of cylindrical, square, or rectangular shape, made by rolling and cutting an ingot or bloom. 2. A solid, semifinished round or square product that has been hot worked by forging, rolling, or extrusion.

billet² A chunk of wood.

bill of lading (Transp) A document that properly describes a single shipment and, when signed by authorized persons (shipper and carrier), constitutes a contract for movement of material.

bill of rights (Law) The first ten amendments to the Constitution of the United States.

bimetallic strip (Techn) The heat-sensitive element in common thermostats. A strip consisting of two metals that have different coefficients of thermal expansion. Heating causes the strip to bend and thereby make or break an electrical contact. The principle is used also in an instrument called a bimetallic strip gage, which is used for measuring pressure in vacuum technology.

bimodal (Statist) Descriptive of a distribution that has two peaks.

bin area (Storage) A storage area for supply items that are naturally storable in bins.

binary (Math) Based on the integer 2.

binary compound (Chem) A compound in which each molecule consists of two elements.

binary notation (Math) A positional number system expressed by 0 and 1. The positions of the digits are powers of the base 2, such that the number 1111 is equivalent to $2^3+2^2+2^1+1$, which is equal to $8+4+2+1=15$; the number $1010 = 8+0+2+0=10$; etc.

binder¹ (Storage) Any material such as burlap, heavy paperboard, or thin lumber placed between layers of stock to stabilize stacks.

binder² (Transp) A chain, cable, rope, or other approved material used for tying and securing loads.

binding energy (Chem; Phys) a. The force holding molecules, atoms, and atomic particles together. b. The force required to break such bonds. Also called: bond energy.
  NOTE: Neutron or proton binding energies are those required to remove these particles, respectively, from a nucleus. Electron binding energy is that required to remove an electron from an atom or a molecule

biochemistry (Chem) The branch of chemistry dealing with biological processes.

biodegradation (Bact; Env Prot) Decomposition by soil bacteria.

bioengineering (Abb: Bioeng) The science, technology, and practice of designing equipment and systems to match physical characteristics and abilities of people. The discipline draws on engineering principles, together with those of the medical and biological sciences. The field of application of bioengineering is called <u>biotechnology</u>, or more narrowly, <u>ergonomics</u>. Also called: <u>human engineering</u>.

biological agent (Med) Any of four causes of disease or infection: 1) viral and rickettsial, 2) bacterial, 3) fungal, and 4) parasitic, acting on or within the body to produce the disease or infection.

biological death (Med) A state following irreversible brain damage due to a cardiac arrest. See: <u>death</u>.

biological dose (Dosim) The quantity of radiation, measured in rems, absorbed by a biological material.

biological effect (Med) The reaction of the body to a <u>biological agent</u>.

biological half-life (Dosim) The time required for man or animal by natural processes to eliminate half the radioactive material that has entered the body. See: <u>effective half-life</u>.

biological oxygen demand (Abb: BOD) The quantity of oxygen required for the biological and chemical oxidation of water-borne substances.

biological shield (Dosim) An absorbing material installed around a radioactive source to reduce the radiation to a safe level.

biology (Abb: Biol) The science dealing with living organisms, divided into the two broad fields of <u>botany</u> (plants) and <u>zoology</u> (animals).

biomechanics (Bioeng) The theory and practice of designing tools, equipment, and the workplace to meet the physiological and anthropometric characteristics of workers. Design criteria include such factors as error reduction, fatigue reduction, improved safety, increased production quality and quantity, and work efficiency.

biomedical engineering (Bioeng) A branch of science in which engineering methods are applied to biological and medical problems. See also: <u>biotechnology</u>.

bionics (Bioeng) The study of electronic systems that model functions of living organisms.

biophysics (Biophys) A hybrid science involving physics, chemistry, and biology, in which the methods of physics and chemistry are used to study the structures and mechanisms of life processes.

biopsy (Path) The taking and study of a small tissue sample from a living body. A common procedure for determining malignancy of tumors.
NOTE: Examination is usu accomplished by staining the specimen with a dye and viewing it under a microscope.

biosensor (Bioeng) A <u>sensor</u> that provides data on life processes.

biotechnology (Biotech) The application of engineering and technological principles in addition to those of the medical and biological sciences to engineering problems involving life forms. Cf: <u>biomedical engineering</u>.

biotelemetry (Bioeng) Remote transmission of physiological data on life processes and functions.

Biot number (Phys) The ratio of the thermal resistance to the surface film resistance, hL/k, where h is the heat transfer coefficient, L is the distance from the midpoint to the surface, and k is the thermal conductivity. Also called: <u>Nusselt number</u>.

bipropellant rocket (Comb) A rocket in which the fuel and oxidizer are stored separately.

biquinary notation (Math) A numerical system used in counting and computing, in which each decimal digit is represented by a pair of digits, the first being the coefficient of five and the second the coefficient of one. The number 123456789, for example, is written 01 02 03 04 10 11 12 13 14. See: number_systems.

bird_dog = lead_plane

birefringence (Opt) The splitting of an electromagnetic wave front propagating in anisotropic media, such as calcite, quartz, topaz, etc. Also called: double_refraction.

birth_cry (Med) The sound made by an infant when respiration begins immediately after birth.

bismuth (Metall; Symb: Bi) A metallic element, mp 271°C, bp 1630°C, used in the manufacture of low-melting alloys for fusible elements in automatic sprinklers, safety plugs, and automatic shutoff systems.

bistable_element (Electr; Fl Mech) An element having two equilibrium states, each of which is maintained until switched by an external signal.

bit[1] (Techn) A cutter for drilling holes.

bit[2] (Comput) a. A blend of binary and digit, denoting either of the two binary characters (0 and 1) b. A datum that can be recorded in either of two states, for example, by the presence or absence of a magnetized spot, or the direction of magnetization of a magnetic core. c. A unit of storage capacity of a device, expressed as the logarithm to the base 2 of the number of possible states of the device.

bite[1] (Mech Eng) The nip_point between any two inrunning rollers of a machine.

bite[2] (Fire Pr) To attack a fire at a given point in the expectation of locating and cooling the area of major involvement.

bituminous_coal A soft mineral coal that burns with a yellow smoky flame.

bivalent (Chem) 1. Having a valence of two. 2. Having two valence states; mercury, for example, can have a valence of 1 or 2.

black_body (Symb: subscript b) An ideal energy emitter or absorber that radiates or absorbs the maximum energy per unit area at each wavelength for any temperature. Experimental black bodies are hollow cylinders with blackened interiors. Radiation entering a slit in the cylinder is almost completely absorbed by multiple reflections. Radiation exiting from the slit is approximately black body distribution, sometimes called cavity_radiation or total_absorption. See also: gray_body. Also called: total_absorber.

black_box (Techn) 1. Loosely, any unit, usu electronic, that can be installed or removed as a single package. A module. 2. A unit whose output is a specified function of the input but for which the method of input-to-output conversion is not defined.

blackdamp (Mining) Carbon dioxide gas produced in a coal mine by a methane explosion. See: damp.

blacken (Fire Pr) 1. To knock down flames, quench burning embers, and wet down charred fuel. 2. = (Brit) damp_down.

black_light (Opt) 1. Invisible ultraviolet light. 2. Electromagnetic energy just beyond the visible range of the spectrum, in the ultraviolet range of wavelengths between 3000 and 4000 Angstrom units, with a peak at 3650 A, used for flourescent inspection. This wavelength reacts strongly on certain dyes to make them fluoresce visibly to the eye.

black_out (Fire Pr) To suppress flames by water and blacken a previously illuminated area. Syn: darken.

blackout[1] (Commun) A fadeout of radio communications due to ionospheric disturbances.

blackout[2] (Physiol) Temporary loss of vision, accompanied by dulling of other senses, due to decrease of blood pressure in the eye and consequent lack of oxygen. In more serious cases, loss of consciousness results. Cf: grayout; redout.

blackout[3] (Civ Def) The extinguishment of all visible lights in a populated center to deny navigational and aiming points to enemy military operations.

black powder A low explosive consisting of an intimate mixture of potassium or sodium nitrate, charcoal, and sulfur. The powder is easily ignited and is sensitive to friction.

bladder (Fire Pr) A flexible bag, filled with foam concentrate, submerged in an apparatus water tank.
NOTE: The water pressure exerted on the bag forces the concentrate through a metering valve into the water manifold on the suction side of the pump.

blade[1] (Techn) A cutting edge.

blade[2] (Techn) A fan or compressor vane.

blade[3] (Civ Eng) A broad, usu curved plate attached to earth-moving equipment, used to clear land; push, spread, and smooth soil; and for other similar purposes.
NOTE: Blades are often modified for special jobs. A U-blade, for example, has an extension on each side protruding forward at an obtuse angle to enable it to handle a larger volume than a regular blade. A landfill blade is a U-blade with an extension on top that further increases the volume that can be pushed and protects the operator from flying debris.

blade tip circle (Safety) The path described by the outermost tip of a rotating blade.

blank cap (Fire Pr) A cover fitted to a threaded connector to protect the threads, keep out dirt, and close it off when not in use.

blank experiment (Psych) A nonmeaningful experiment introduced at intervals in an experimental series to prevent the subject from becoming automatic in his responses.

blast[1] (Techn) A jet of air injected under pressure into a blast furnace or the like to accelerate combustion, or a jet of steam directed into a smokestack to augment the draft, as in a locomotive.

blast[2] (Mining, Civ Eng) a. The charge of explosives set off at one time in blasting operations. b. To fragment something with an explosive.

blast[3] (Phys) The violent, sudden, and rapid outward movement of a fluid away from a center of high pressure, such as from an explosion.
NOTE: The phenomenon is characterized by an instantaneous rise in fluid pressure, followed by a sudden drop. Properly, the effect of an explosion.

blast area (Civ Eng) The area in which explosives loading and blasting operations are being conducted.

blast cleaning barrel (Techn) A rotating container, or one which has an internal moving tread to tumble pieces within the container in order to expose various surfaces to an automatic blast spray.

blast cleaning room (Techn) An enclosure in which blasting operations are performed and where the operator works inside the room to operate the blasting nozzle and direct the flow of abrasive material.

blast deflector A plate used to divert the exhaust of a rocket.

blast effect (Civ Def) The damage resulting from shock waves generated by an explosion.

blaster (Civ Eng) An individual authorized to use explosives.

blast furnace (Techn) A furnace, commonly used for the reduction of iron ore, in which combustion is augmented by forced air.

blast gate (Techn) A sliding metal damper in a duct, usu used to regulate the flow of forced air.

blast hazards criteria (Med) Standards on which decisions concerning blast injury are based.

blasting agent (Civ Eng) Any material or mixture consisting of a fuel and oxidizer used for blasting, in which none of the ingredients nor the mixture is classified as an explosive, and the finished, unconfined product cannot be detonated with a No. 8 test blasting cap.

blasting cabinet (Techn) An enclosure used for blasting. The operator stands outside and operates the blasting nozzle through one or more openings in the enclosure.

blasting cap (Civ Eng) A small, thin-walled cylindrical case containing a sensitive explosive, such as lead azide. A detonator used to set off another explosive charge. The explosive in the blasting cap is fired either by a burning fuse or by electricity.

blasting frequency (Civ Eng) Cycles per second of shock wave or vibration generated by a blast.

blast wave (Civ Def) A relatively narrow pulse of high pressure at the front of the air mass set in motion by an explosion. Blast waves are followed by rarefaction waves, the two exerting forces in opposite directions. Structural damage initiated by one is helped along by the other.

blaze (Fire Pr) a. A free-burning spectacular fire characterized by a large quantity of flame. b. To burn brightly.

bleachers Uncovered tiered seats made of planks, used for spectators at sports events. Open stands for spectators.

bleb (Med) A blister or skin vesicle filled with fluid.

bleed[1] (Techn) To allow a gas or liquid to escape from a line or tank, usu through a bleed valve or petcock.

bleed[2] (Med) To lose blood, as from a wound or other injury. See: first aid.

bleed-out (Testing) The exudation of a penetrant from flaws onto the surface of a material, due primarily to capillary action.

blend (Chem Eng) An intimate mixture of two or more substances such that the mixture has properties intermediate to those of the individual components.

blepharospasm (Med) Involuntary inability to open the eyes or a spasmodic blinking of the eyelids due to the spasm of the eyelid muscles. See: lacrimator.

bleve = Boiling Liquid Expanding Vapor Explosion Cf: boilover.

blind area (Fire Pr) A forest area that cannot be seen from an observation point. Ant: visible area.

blindness (Physiol) Visual acuity of 20/200 or less as determined by the Snellen test, or vision no matter how good that subtends an angle less than 20°.

blister[1] (First Aid) A thin swelling or sac in the skin containing watery matter or serum, caused by a burn or other injury.

blister[2] (Techn) To raise blisters by excessive heat or direct burns, as on a painted surface exposed to heat.

blizzard (Meteor) A violent wind storm accompanied by severe cold and driving snow. See: wind.

blob[1] (Meteor) A small-scale temperature and humidity inhomogeneity produced by turbulence in the atmosphere.

blob[2] A small glob or drop of liquid.

block 1. A rectangular area bounded by streets and usu occupied by buildings. 2. The distance along one side of a block. 3. A distinctive building or group of buildings.

block a door (Salv) To place a salvage cover over a door to prevent water from passing through.

block and tackle (Mech Eng) A combination of pulleys and ropes used to move heavy objects.

blocked hydrant (Fire Pr) A hydrant so obstructed that apparatus cannot be connected using standard techniques and hoses.

block holing (Civ Eng; Mining) Shattering of a boulder or a large piece of ore by setting off an explosive charge that has been loaded into a drill hole.

block plan (For Serv) A detailed prescription for burning-over a given burning block.

block sampling (Psych) Grouping subjects into categories and selecting the final sample from the categories.

block stacking (Storage) Storing of similar containers or material together in blocks.

blood (Anat) A fluid, consisting of plasma and cells that circulates through the blood vessels of animals and carries nutrients and oxygen to the body cells and removes waste products. Blood contains red and white cells or corpuscles, platelets, and plasma. The plasma contains proteins, aitibodies, dissolved salts, and other substances.
NOTE: Blood makes up about 8 to 9% of body weight.

blood alcohol (Med) The concentration of alcohol in the blood stream, expressed in weight percent.
NOTE: No two people are affected identically, but the general effect and blood alcohol levels for various quantities of 90-proof alcohol consumed are

Quantity (oz)	% Blood alcohol	Effect
1	0.02	Elation, pleasant sociable behavior
2-3	0.05	Inhibitions lessened, impulsive behavior
5-6	0.10	Confusion, staggering gait, slurred speech
16	0.30	Stuperous
24	0.40-0.50	Unconscious

blood count (Med) A count of the number of corpuscles per cubic millimeter of blood. Separate counts may be made for red and white corpuscles. A common procedure in a physical examination.

blood expander = blood extender

blood extender (Med) A solution of whole blood, products derived from whole blood, or dextrans, used to increase the blood volume in living animals. Also called: blood expander.

blood plasma (Med) The fluid portion of the blood.

blood poisoning See: septicemia

blood pressure (Med) The pressure in the arterial system of the body that drives the blood through the circulatory system.
NOTE: Blood pressure is usu measured over the range of systolic (during heart contraction) to diastolic (during heart relaxation) pressures. Systolic values are normally 100 to 140 mm Hg, and diastolic 60 to 90.

blood test (Med) Any of a number of diagnostic procedures used on blood samples.

blood transfusion (Med) The transfer of blood to a patient to renew or supplement the blood supply.

blood type (Med) A classification system for blood that assures compatibility in transfusion. Among the dozen or more subvarieties, the principal groups are named, in order of incidence in the population, O, A, B, AB. Compatibility also depends on an Rh factor, which is either positive or negative, and on subfactors such as M and N. Also called: blood group.

Compatibility for Transfusion

Recipient:	O	A	B	AB
Donor:				
O	Yes	Yes	Yes	Yes
A	Yes	Yes	No	Yes
B	Yes	No	Yes	Yes
AB	No	No	No	Yes

blood vessels (Anat) Tubular channels for the distribution of blood in the body, including *arteries*, *capillaries*, and *veins*. Arteries carry blood from the heart, and veins return it to the heart. Blood is transferred from the arteries to the veins via the capillaries.

blood volume (Med) The quantity of blood in the body; approx 8 to 9% of body weight.

bloom (Techn) The color of an oil in reflected light when it differs from the color by transmitted light.

blow (Techn) The distance an air stream travels from an outlet to a position at which air motion along the axis slows to a velocity of 50 fpm. Also called: *throw*.

blowback (Civ Eng; Mining) Release of explosion products away from the intended direction.

blowdown tunnel (Aerodyn) A transient wind tunnel in which the flow is produced by a stored gas.

blowdown turbine (Techn) A turbine driven by the exhaust gases of a reciprocating engine.

blower (Techn) A centrifugal compressor.
  NOTE: Blowers that generate pressures below 1 lb/in.^{2} are called *fans*; those that generate pressures above 35 lb/in.^{2} are called *centrifugal compressors*. High-speed turbine-driven blowers are called *turboblowers*.

blowoff (Comb) The condition at which a flame becomes unstable and is carried downstream from the flameholder by the flow.

blowoff velocity (Comb) The velocity at which flame *blowoff* occurs.

blowout coil (Electr) An induction coil used to suppress arcs when a

blowtorch (Techn) A hand burner fueled by gasoline, butane, etc.

blowup fire (For Serv) A wildland fire that suddenly increases in intensity or rate of spread beyond control. A blowup fire is usu accompanied by violent convective currents as in a *fire storm*. See also: *conflagration*; *mass fire*.

BLS = *Bureau of Labor Statistics* (US Department of Labor)

blue flame (Comb) The second visible stage of three-stage ignition, the emission being attributed largely to electronically excited formaldehyde.

blue shirt (Fire Pr; Colloq) A paid professional firefighter, so-called because of the blue dress shirt worn with the uniform; distinguished from officers who usu wear white shirts. Cf: *red shirt*.

Board of Certified Safety Professionals (Safety; Abb: BCSP) An autonomous board established to certify the professional and technical competence of individual practitioners in the safety field.

board of review (For Serv) A committee convened to evaluate the performance of a given fire-fighting unit or the action taken on a given fire for the purpose of recommending improvements in performance and fire-fighting techniques.

boat^1 (Chem) A boat-shaped refractory dish used to hold specimens in high-temperature furnaces.

boat^2 = *fireboat*

boatswain's chair (Rescue) A seat supported by slings attached to a suspended rope, designed to accommodate one person in a sitting

position.
NOTE: Used to transfer people between ships, buildings and the like.

BOCA = Building Officials and Code Administrators International

bodily injury (Safety; Insur; Abb: BI) Injury to a human being, as opposed to injury to property.

body bag (Fire Pr) A bag with handles used to remove and transport the remains of fire victims.

body burden (Dosim; Med) The quantity of radioactive material present in the body.

body of fire (Fire Pr) An intense mass of flame accompanied by heavy smoke indicating the center of a fire.

body of revolution (Geom) A symmetrical body described by a plane curve rotated about an axis in the plane.

bogie; bogey (Transp) 1. A four-wheeled, swiveled, railway truck. 2. The four rear wheels of a six-wheeled automotive truck. 3. A weight-carrying and tread-aligning wheel inside the tread of a track-laying vehicle. 4. A type of suspension assembly in which road wheels or axles are interconnected in tandem by a system of arms, links, walking beams, cranks, springs, and the like in such a manner that when one wheel or axle is displaced upward, there is a corresponding change in loading or position of the other wheels or axles of the unit.

Bohr magneton (Phys; Symb: MB) A physical constant, equivalent to the magnetic moment of an electron: 9.27372×10^{-21} erg/gauss. Cf: nuclear magneton.

Bohr radius (Phys; Symb: a⁰) The radius of an electron in the lowest level of a Bohr model hydrogen atom: 5.29167×10^{-9} cm. The radius is expressed as $a^0 = e/(mc^2)$, where m is the electron mass, e is the electron charge, and c is the speed of light.

boil[1] (Chem; Phys) To heat a liquid to its vaporization temperature.

boil[2] (Path) An inflamed sore caused by bacterial infection in which white cells have walled off the invading bacteria.

boiler (Techn) A pressure vessel used to produce steam for heating buildings, driving steam-powered machinery, etc.

boiler codes (Fire Pr) Standards prescribing requirements for the design, construction, testing, and installation of boilers and unfired pressure vessels.

boiling point (Chem; Phys; Abb: bp) The temperature at which a liquid becomes a vapor or gas, or at which the vapor pressure of the liquid is slightly higher than the pressure exerted on it by the surroundings. This temperature is pressure-dependent, and remains constant as long as any liquid remains to boil. Normal boiling point is the boiling temperature of a liquid under atmospheric pressure. For mixtures that do not have a constant boiling point, a distillation may be performed. Also called: vaporization point. See also: vapor pressure; heat of vaporization; ebulliometry.
NOTE: The boiling point of a liquid can be determined by noting the temperature at which bubbles of vapor first appear as the liquid is heated. This is an approximation, since the bubbles may be formed by dissolved gases or other impurities. Every substance that has a liquid state has a more-or-less unique and characteristic boiling point. Hydrogen, for example, has a bp of -252°C; water, 100°C; and titanium, 3260°C.

boiloff (Chem) The vaporization of a liquid as its temperature reaches the boiling point.

boilover (Fire Pr) 1. The spilling over of a burning flammable liquid from a container or tank. Water applied to extinguish the fire may overfill the tank and allow the burning liquid to float and spill over the edge. 2. The more or less violent eruption of crude oil or other liquid from a burning storage tank. The light fractions of the crude oil burn off, producing a heat wave in the residue, which on reaching a water stratum may vaporize it and expel some of the contents of the tank in the form of froth. See: bleve.

bolometer (Metrol) A sensitive thermometric instrument for measuring the intensity of radiant energy by the temperature rise in a heat-sensitive element, often a resistor or thermistor in a Wheatstone bridge.
NOTE: In microwave technology, a barretter and thermistor are sometimes called bolometers. Cf: radio micrometer.

bolt (Mech Eng) A metal rod with a thread on one end and a head on the other, used with a nut to hold objects together.

bolting reel (Techn) A device in which flour is screened through a rotating drum.

Boltzmann constant (Phys; Symb: k) The ratio of the universal gas constant to the Avogadro number: 1.38054×10^{-16} erg/°K. This constant occurs frequently in statistical equations. Also called: gas constant per molecule; Boltzmann universal conversion factor; ideal gas law. The product of k and temperature T in degrees Kelvin is called the thermal energy, and at room temperature (300°K) is equal to 0.0259 electron volt.

Boltzmann equation (Phys) An equation describing the conservation of particles diffusing in a scattering, absorbing, and multiplying medium. The time rate of change of particle density is equal to the rate of production minus the rates of leakage and absorption. Also called: transport equation.

bolus (Med) 1. A single, massive dose of a drug. A large pill. 2. A soft mass of chewed food.

bombardment (Nucl Phys) Firing a beam of alpha rays or other high-energy particles at atomic nuclei, usu in an attempt to split an atomic nucleus or to form a new element.

bomb calorimeter (Comb) A thick-walled vessel used to determine the amount of heat yielded by the combustion of a known quantity of fuel. See: calorimetry.

bomb threat (Law) Expressed intention to cause personal injury or property damage with an explosive device.
NOTE: The threat alone is sufficient to be punishable by law, regardless of whether the threat is real or a hoax.

bond¹ (Phys) The force of adhesion between surfaces.

bond² (Insur) A surety obtained by payment of a fee to a bonding agent to ensure against financial loss for misappropriated funds or goods.

bond³ (Chem) The forces linking two atoms in a molecule. The energy required to break a molecular bond.

bonding¹ (Electr) Interconnection of metal parts of a structure to form a continuous electrical circuit. Used principally to prevent buildup of static charges and arcing.
NOTE: Bonding alone does not ensure that static charge buildup will not occur. A continuous electrical path to ground is required to drain off static electricity. The connection between metal parts is called a bonding jumper.

bonding² (Chem) Chemical forces between atoms in a molecule.

bonding³ (Techn) Cementing or adhering of surfaces together.

bonding jumper (Electr) A conductor that provides electrical conductivity between metal parts that are required to be electrically continuous.

bone conduction (Audiom) Transmission of sound, usu through the bone structure of the head, but also through the scapula and the spine.

bone conduction test (Audiom) A special test conducted by exciting the mastoid process with an audio signal to determine the nerve-carrying capacity of the cochlea and the auditory nerve.

bone marrow (Anat) A soft tissue that constitutes the central filling of many bones and produces blood corpuscles.

bone-seeker (Med) Any element or radioactive species that predominan-

tly lodges in the bone when introduced into the body.
NOTE: Strontium-90, for example, behaves biochemically like calcium.

bonjeans curves (Navig) A plan giving the underwater area for each station or section along a hull as a function of draft.

bonnet (Fire Pr) The dome-shaped protective cover at the top of a fire hydrant.

Boolean algebra (Math) A system of manipulating algebraic operations according to the rules of logic and involving set theory, probability, and the theory of relations. See also: binary notation.

boom[1] (Fire Pr) The elevating beam of an aerial platform. See also: articulated boom; telescoping boom.

boom[2] (Transp) A long spar used for handling cargo or to lift and swing loads over short distances. The boom projects from a mast, is pivoted or hinged at the heel, and its point is supported by chains, ropes, or rods to the upper end of the mast. A rope for raising and lowering loads is reeved through sheaves or a block at the boom point.

boom angle (Techn) The angle between the longitudinal centerline of a boom and the horizon. The boom longitudinal centerline is a straight line between the boom heel pin and the boom point. The angle is usu indicated by an inclinometer or angle indicator.

boom harness (Techn) The block and sheave arrangement on a boom point to which the topping lift cable is reeved for lowering and raising the boom.

boom heel (Techn) The lower, attached end of a boom.

boom hoist (Techn) A hoist drum and rope reeving system used to raise and lower a boom. The rope system may be all live reeving or a combination of live reeving and pendants.

boom point (Techn) The outermost end of a boom.

boom stop (Techn) A device used to limit the angle of a boom at its highest position.

boost (Techn) 1. Additional power or force supplied by an auxiliary device, as by a hydraulic booster, booster pump, or a booster engine supplying extra propulsion to a flying vehicle. 2. To augment with additional power or supercharge. 3. Short for boost pressure. 4. To accelerate, as a ramjet, to self-sustained flight speed by means of a rocket.
NOTE: A blend of boom and hoist.

booster[1] (Techn) Short for any augmenting power device or system, such as a hydraulic booster or booster engine.

booster[2] (Med) A substance, such as a vaccine, usu injected, that prolongs immunity to a given infection. A booster shot.

booster[3] (Electr) A device connected in series in a circuit for the purpose of increasing or decreasing the circuit voltage.

booster[4] (Fire Pr) A booster pump.

booster[5] (Civ Eng) A quantity of explosive that is initiated by a detonator and in turn initiates the main charge.

booster apparatus (Fire Pr) A lightweight vehicle equipped with a water tank and booster reel; used to fight brush fires.

booster engine (Techn) An engine that adds extra thrust to that of the main power plant.

booster line (Fire Pr) A 3/4- or 1-inch water supply line on a pumper that can be quickly charged. Provided for immediate use on fires.

booster pump[1] (Mech Eng) A water, fuel, oil, or other pump that provides extra power and an initial pressure differential to a fluid entering a main pump.

booster pump[2] (Fire Pr) A power take-off pump of less than 500 gpm carried on fire apparatus and used with small-diameter hose.

booster reel (Fire Pr) A reel for carrying small hose (3/4- or 1-inch) on fire apparatus, often equipped with powered rewind.

booster tank (Fire Pr) A water tank mounted integrally on a pumper to supply water to a booster line.

booster unit (For Serv) A portable water tank, hose reel, and pump mounted on a light truck or trailer, used for forest fire control.

boost pressure (Mech Eng) Manifold pressure supercharged above the ambient pressure.

boot-strapping Using a crude measure or procedure to derive a better one.

borane (Chem) A compound of boron and hydrogen, the simplest of which is diborane (B_2H_6), used in the preparation of natural and synthetic rubber and of corrosion-resistant surfaces.

borate solution (For Serv; Obsolete) A compound in slurry form used in the 1950s as a wildland fire extinguishing agent or fire retardant.

borazon See: abrasive

Borda mouthpiece (Hydraul) A reentrant tube in a hydraulic reservoir. As the liquid accelerates to escape from the tube, the jet converges to a cross section smaller than the tube. The smallest cross section of the jet is called vena contracta.

bore (Techn) The circular aperture of a pipe, nozzle, or cylinder, expressed in terms of the inside diameter.

borehole (Mining) A drilled mine shaft. A hole formed by boring.

borehole pump (Fire Pr) A centrifugal pump that operates under water, discharging through a pipe to the surface.

boron (Symb: B) A brown, amorphous element, classified as a metalloid, mp 2087°C, bp 3677°C. Used principally in metallurgy (esp high-temperature technology), fireproofing, fuels, laundry products, etc. See also: abrasive; borane.

borrow (Civ Eng) Road construction material that is taken to another location for use. The source area is called a borrow pit.

borrow pit (Civ Eng) The source of natural road construction material.

boss (For Serv) An officer in wildland fire operations, such as a fire boss, line boss, crew boss, sector boss, etc.

bottom settlings (Abb: BS) A thick sludge consisting of water, tarry materials, and particles that collects on the bottoms of fuel storage tanks.

botulism (Med) Food poisoning caused by a poison (toxin) liberated by an organism known as Clostridium botulinum.

Bouguer's law = Beer's law

bounce (Transp) The upward movement of the sprung mass of a vehicle, away from the unsprung mass, in response to suspension system disturbances. The bounce distance is the maximum upward travel from the free-standing position.

boundary conditions (Math) In the solution of a differential equation and its derivatives, a set of mathematical conditions to be satisfied at the boundaries of the region in which the solution is sought.

boundary layer (Fl Mech; Symb: d) A layer in a moving fluid, adjacent to a stationary surface, in which shear forces are dominant. Fluid particles at the surface are essentially at rest with respect to the surface, and increase in velocity asymptotically with distance from the surface. Boundary layer thickness is arbitrarily defined as the distance from the wall to a point at which particles are moving at 99% of the free-stream velocity. Also called: friction layer. See: skin friction. See also: laminar flow; stagnation point.

boundary-value problem (Math; Phys) A problem in physics described by a differential equation, constrained by certain boundary conditions. See also: eigenvalue.

Bourdon gage (Metrol) A pressure gage consisting of a curved, flattened tube that tends to change its curvature in proportion to pressure differences. This movement is transferred mechanically to a pointer on a dial. Typically, the sensing element is C-shaped, but spiral, helical, and twisted shapes are used in special applications.

bow (Navig) The forward end of a boat or vessel.

box (Fire Pr) A fire alarm signal box. Syn: street box. See also: adjacent box; cottage box; phantom box.

box alarm (Fire Pr) A signal transmitted from a fire alarm box.

box car (Transp) A fully enclosed freight car having doors on both sides or sometimes on the ends. Used for the transportation of general freight.

box circuit (Fire Pr) An electrical circuit connecting fire alarm boxes and fire detection devices to a central alarm station.

box pallet (Storage) A pallet with back and side framework, so constructed that several may be stacked, one upon another, the weight being borne by the pallet and not by the supplies stored inside.

box shop (Storage) An area used for fabricating, manufacturing, assembling, or repairing containers and storage aids.

Boyle's law (Phys) The pressure and volume of a gas at constant temperature are inversely proportional to each other or have a constant product. If the temperature varies, the behavior of the gas is described by the Boyle-Charles' law. Also called: Mariotte's law [in Europe]. See: equations of state. See also: Charles' law.
 NOTE: There are deviations from this law at high pressure.

Boyle temperature (Phys) The temperature for a given gas at which Boyle's law is most closely obeyed.

brace[1] (Build) A rigid tie that holds one structural member in a fixed position with respect to another member.

brace[2] (Fire Pr) A diagonal stiffening member between the rails and blocks of a trussed beam ladder. See: ladder.

bradycardia (Med) A condition of abnormally slow heart beat.

brake[1] (Mech Eng) Any mechanism designed to retard or stop motion, usu by friction.
 NOTE: Disc type: A brake in which the holding effect is obtained by frictional resistance between one or more faces of discs or pads keyed to the rotating member to be held and fixed discs keyed to the stationary or housing member. Pressure is applied axially. Self-energizing band type: An essentially unidirectional brake in which the holding effect is obtained by the snubbing action of a flexible band wrapped about a cylindrical wheel or drum affixed to the rotating member to be held. The connections and linkages are so arranged that the motion of the brake wheel or drum acts to increase the tension or holding force of the band. Shoe type: A brake in which the holding effect is obtained by applying the direct pressure of two or more segmental friction elements attached to a stationary member against a cylindrical wheel or drum affixed to the rotating member to be held. An air brake is a. an automotive vehicle brake operated by air pressure, or b. an airfoil that is rotated or otherwise extended into the airstream to decelerate an aircraft in flight.

brake[2] (Metall) A machine used to bend or fold sheet metal.

brake[3] (Fire Pr) A long lever used to pump hand-powered fire engines.

brake fade (Transp) A temporary failure in a braking system due to excessive temperature. The term is used esp in connection with automotive vehicles.

brake horsepower (Mech Eng; Abb: bhp) The power output of an engine, generator, or motor as measured by the stalling force.

branch (Fire Pr; Brit) 1. A line of hose from a breeching connection or pump. 2. a. A tapered fitting used between the delivery coupling and the nozzle to increase the velocity of the water and provide a solid jet. b. Diffuser branch: A branch yielding a spray or jet of variable size and serving as a shut-off. c. Double revolving branch: A collecting breeching connected to a casting that revolves on ball bearings and carries two diametrically opposed branches. Rotation is provided by the reaction force of water passing through two small jets. d. Foam-making branch: A branch used to generate mechanical foam. e. Hand-controlled branch: A branch that delivers a jet and spray simultaneously, each being controlled separately. f. Radial branch: A special type of branch and holder, requiring only one branch man to operate, for use with jets up to 2 in. diameter. The elevation of the nozzle is controlled by a small winch gear. g. Streamform branch: A short branch with a central tube and guide vanes that reduce turbulence in the flowing water.

branch circuits (Electr) The portion of a wiring system extending beyond the final overcurrent device protecting the circuit.

branch holder (Fire Pr; Brit) A device for counteracting part of the reaction force of a working branch.

branchman (Fire Pr; Brit) = nozzleman

branch pipe[1] (Fire Pr; Brit) A nozzle or playpipe attached to a hose line.

branch pipe[2] (Techn) The part of an exhaust system piping that is connected directly to a hood or enclosure.

brand (Fire Pr; For Serv) A freely burning fragment that becomes airborne either by wind or convection currents and falls at a distance from its origin. A firebrand.
NOTE: Brands are the main cause of spot fires and are one of the principal mechanisms by which conflagrations develop.

brand patrol (Fire Pr) A detail of firefighters or a fire company assigned to patrol areas downwind from a large fire to watch for possible fires ignited by flying brands.

brass[1] (Fire Pr) Brass, nickel, or chrome trim on fire department equipment.

brass[2] (Fire Pr; Colloq) Chief officers of fire departments, from the gold uniform insignia. Company officers and men usu wear silver insignia. See: ranks and titles.

brass and bronze (Metall) Copper-zinc (brass) and copper-tin (bronze) alloys used extensively in plumbing, valves, screws, other hardware, and electrical equipment. Some lead may also be present in these alloys.

brattice (Mining) A temporary partition of cloth or planks constructed in mine passageways to control ventilation. Syn: ventilation curtain.

brazing (Metall) Joining of metal surfaces by flowing a thin capillary layer of nonferrous filler metal between them. Sometimes the filler metal is applied as a thin, solid sheet, as in cladding, and the sandwich is bonded by furnace brazing. The term is used for bonding processes above approx 800°F (1000°F according to the American Welding Society), while soldering is used for processes at lower temperatures. Also called: hard soldering; silver soldering; bronze welding. See: welding.

breach (Fire Pr) 1. An opening made in a wall or other barrier for passing hose lines through or removing people or goods from a burning building. 2. To make such an opening.

breadboard (Electr) 1. An assembly of preliminary circuits or parts to establish the feasibility of a device without regard to the final configuration. 2. To prepare a breadboard model.

break a coupling (Fire Pr) To separate two lengths of hose by unscrewing the coupling.

break a line (Fire Pr) 1. To disconnect a coupling in a hose line for the purpose of attaching a nozzle, a wye, or other fitting. 2. To take a hose line apart into its individual lengths. 3. To disconnect a hose line anywhere along its length for any purpose.

breakaway = flow separation

breakdown potential (Electr) The voltage at which a current flows across an insulating medium between two conductors. Syn: sparking potential.

breaking ground (Mining) The work involved in blasting: locating, drilling, charging, tamping, and firing drill holes.

break-in tool (Fire Pr) One of a variety of instruments, such as axes, shears, chisels, crowbars, mauls, padlock removers, wedges, etc. used to force entry into a building. See: forcible entry tool.

breakover (For Serv) The crossing of a wildland fire over a control line established to confine it. Also called: slopover.

breakover fire (For Serv) A fire that crosses a fireline, esp one that has been declared out but reignites and begins to spread.

breakover point (Fire Pr) The point at which a solid stream begins to disintegrate into droplets to become a spray.

breakthrough A dramatic discovery or major advance in research or theory.

breast (Mining) A vertical face in a mine.

breasting (Mining) Removing vertical slices of ore from a face.

breast stoping (Mining) Removing ore by advancing a vertical face in a horizontal direction.
  NOTE: Breast and bench is a method in which the upper part of a face is broken by breast stoping and the lower part by benching.

breathing apparatus (Fire Pr; Abb: BA) An item of personal equipment with a self-contained oxygen or compressed air supply and a mask worn by firefighters to allow them to work in toxic and noxious atmospheres. See also: mask; SCUBA. Cf: respirator; rebreather.

breathing tube (Fire Pr) A tube through which air or oxygen flows to the facepiece, helmet, or hood of a breathing apparatus.

breech birth (Med) Delivery of a baby in which the buttocks emerge first. See: crown birth.

breeching (Fire Pr; Brit) A type of coupling. a. Collecting breeching: A casting with two inlet hose connections and one outlet connection for supplying one hose line by two others. Also called: Two-in-one breeching. Syn: siamese. b. Control breeching: A breeching fitted with one or more valves to shut off the flow of water. c. Dividing breeching: A fitting used to divide a hose line. Also called: One-into-two breeching. Syn: (US) wye. d. Radial breeching: A collecting breeching fitted with nonreturn valves and used with a Radial branch. See: branch.

breeder reactor (Nucl Phys) A reactor that consumes and produces fissionable fuel at the same time, esp one that creates more fuel than it uses.

breeze (Meteor) A light, moderate wind, blowing at 4 to 31 mph. See: Beaufort wind scale.

Bresnan nozzle (Fire Pr) 1. A trade name for a rotating type of distributor nozzle for use in cellar fires. Cf: cellar nozzle. 2. = (Brit) basement spray.

Brewster angle (Opt) The angle at which a light wave polarized parallel to the angle of incidence to a dielectric surface is totally transmitted.

Brewster's law (Opt) For any dielectric reflector, the tangent of the polarizing angle is equal to the refractive index.

brick (Build) Any of a wide variety of rectangular blocks of fired clay or shale used as building materials.

brick joisted (Build) Descriptive of a brick or masonry building with wooden floors and roof joists. Also called: second class construction.

bridge¹ (Fire Pr) To span an opening in a floor, between structural members, or between buildings with a ladder. See also: hose bridge.

bridge² (Electr) A network of resistors, capacitors, and/or inductors arranged in a ring of four arms; used principally as an oscillator or as a device for measuring voltage, current, or frequency, and, indirectly, resistance, capacitance, and inductance.

bridge plate (Transp) A ramp, usu of metal, used to span the space between freight cars or trucks and loading platforms.

brigade See: fire brigade

brightness (Opt) The visual perception of the quantity of light reflected by a unit area of an illuminated surface. The photometric counterpart of brightness is luminance.

brightness temperature (Phys) The temperature of a nonblack body as determined with an optical pyrometer.

brilliance (Opt) The attribute of any color as compared to a gray scale.

Brinnell hardness number (Metall; Abb: HB or Bhn) A number indicating the relative hardness of a metal; defined as the ratio of the pressure on a standard sphere to the area of indentation produced when the sphere is pressed against the test specimen. See: hardness; Moh's scale.

briquette (Techn) Coal or charcoal dust pressed into oval blocks or pellets.

brisance (Techn) The shattering capacity of explosives, usu measured as the amount of sand crushed in a closed, heavy-walled container. This capacity is a function of the detonation pressure of the explosive.

British candle = international candle. See: candela.

British thermal unit (Abb: Btu) The quantity of heat required to raise the temperature of one pound of water 1°F at the pressure of 1 atmosphere and temperature of 60°F. See: calorie.
 NOTE: A pound of ice at 32°F requires 144 Btu to melt. One Btu = 257.996 International Steam Table calories = 778.26 foot pounds = 1055 joules = about 1/3 watt-hour. One Btu/min = 0.0236 hp.

broaching (Techn) A machining process in which a cutter with a series of teeth is drawn across a surface to remove metal and thereby form an opening of a desired shape.

broadcast burning (For Serv) Intentional burning of a specified wildland area, but not necessarily qualifying as prescribed burning. A method for slash disposal. See: center firing; edge firing; spot burning.

broke (Techn) Paper discarded from a paper-making machine.

broken stream (Fire Pr) A solid stream that has disintegrated into droplets. See also: breakover point.

bromine (Chem; Symb: Br) A corrosive, toxic, diatomic gaseous element, bp 58.78°C, mp -7.2°C. Compounds of bromine are widely used in photography, agriculture, refrigeration, pharmaceuticals, and fire extinguishing agents. See: Halon.

bromochlorodifluoromethane (Abb: BCF) A type of vaporizing liquid fire extinguishing agent. Also called: Halon 1211. See: Halon.
 NOTE: BCF affects the central nervous system, causing dizziness and paresthesia. These effects wear off rapidly in fresh air. BCF appears to be less toxic to animals than to humans. In animal experiments rodents died from exposures of 15 minutes to concentrations >32%. Maximum safe tolerance for humans is said to be 4 to 5% for one minute.

bromochloromethane (Abb: CBM) A vaporizing liquid fire-extinguishing agent. Also called: Halon 1011. See: Halon.

bromomethane = methyl bromide

bromotrifluoromethane (Abb: BTF) A normally colorless, odorless, elec-

67

trically nonconductive gaseous compound, CF₃Br, bp -725°F, mp -270°F; used as a refrigerant (Freon), and as an extinguishing agent for flammable liquids and electrical fires. Also called: Halon 1301. See: Halon.

bronchial (Anat) Pertaining to the airways in the lungs.

bronchial tree (Anat) The complex of branches and subdivisions of the airway or windpipe from the larynx to the alveoli.

bronchiole (Anat) The finest tube of the bronchial tree, which carries air into and out of the air sacs or alveoli.

bronchiolitis (Med) Inflammation of the smallest bronchial tubes.

bronchitis (Med) Inflammation of the bronchi or bronchial tubes.

bronchodilator (Med) An agent that increases the caliber of a bronchial tube.

bronchospasm (Med) A spasmodic narrowing of the openings of a bronchus.

bronchus (Anat) The part of the respiratory system that connects the trachea to the lungs.

bronze See: brass and bronze

Brooke formula (Med) A standard formula for calculating the quantity of fluids to be administered to a burn victim for the first 24 hours after injury: derived by modifying Cope's 75 cc of plasma and 75 cc of saline on the assumption of an average body weight of 75 kg. Originally 1 cc plasma and 1 cc of saline per kilogram per % of burn, later modified to one-half cc plasma and one cc saline solution per kilogram per % of burn. See: blood extender.

brow log (Transp) A log placed parallel to a roadway at a landing or dump to check the wheels of vehicles during loading or unloading.

brownian movement (Phys) Random movement of microscopic particles (<1 micron) caused by random collisions of molecules.

brush¹ (Electr) A wiping electrical contact used in rotating machinery.

brush² (For Serv) Dense growth of shrubs and bushes, as contrasted with forest and grassland.

brush fire (Fire Pr; For Serv) Burning shrubs, brush, and scrub growth, involving heavier fuel than grass. See also: forest fire; grass fire.

brush hook (For Serv) A tool similar to an axe, but with a longer, curved blade, used to cut heavy brush with a pulling motion.

brush rig (For Serv) A pumper apparatus specially equipped to fight wildland fires.

BS = bottom settlings

BTF = bromotrifluoromethane (Halon 1301). See: Halon.

Btu = British thermal unit

bubble chamber (Phys) A chamber containing a liquefied gas such as hydrogen, used to visualize the paths of charged particles passing through the liquid.

bubble gage (Metrol) An instrument used to monitor gas flow by the rate of bubble formation.

bubble pressure (Phys) The pressure required to form a gaseous bubble. The pressure within the bubble exceeds that of the surrounding liquid by twice the surface tension, divided by the bubble radius.

bucket brigade (Fire Pr) One or more lines of people extending from a water supply to a fire, passing buckets of water toward the fire and returning the empties.

Buckingham pi theorem (Math) An important theorem in dimensional analysis: if an equation is dimensionally homogeneous, it can be reduced to a relationship among a complete set of dimensionless products.

buckle (Metall) An indentation on a metal part resulting from expansion of the dip coat.

budget (Man Sci) 1. A plan for the allocation of resources and expenditure of funds for a specified period or time. 2. An estimate of how much money is needed and how it will be spent for a given purpose. 3. A financial statement of an organization covering a given period or time, including estimates of money needed for particular programs, functions, resources, and the like, as well as proposals for financing them.
NOTE: The budget is usu the principal planning and control mechanism of management.

buff (Fire Pr) A dedicated fire department friend, helper, or enthusiast. Syn: **spark**; **sparky**; **ringer**.
NOTE: The term is alleged to have been derived from an early association of young men who called themselves the Buffalo Corps and who were affiliated with various New York Fire Department companies, serving as helpers and errand boys ("go-fers"). According to Webster's Third International Dictionary (1968), the term is derived from the buff colored overcoats worn by volunteer firemen in New York City circa 1820. Another claim is that the term is derived from buffalo skin coats worn by people watching firefighting operations. In some localities buffs are organized into fire reserves and receive training to help in crowd control, assist regular firefighters in advancing lines, and perform other tasks on the fireground.

buffer[1] (Comput) a. A circuit used to prevent the reaction of a driven circuit from affecting the driving circuit. b. A storage device used to compensate for different rates of flow of signals between the various components of a computer system.

buffer[2] (Chem; Pharm) a. An additive that decreases hydrogen ion sensitivity to acidic or alkaline substances. An oxidation-reduction buffer stabilizes the oxidation-reduction potential of a system. b. Any solution that resists change in pH, usu one that contains both a weak acid and a weak base.

buffer solution (Chem) Solution of certain salts whose pH is not appreciably changed by the addition of acid or alkali.

buffeting (Aerodyn) The beating of aerodynamic structures or surfaces by unsteady fluid flow, as encountered in turbulent air or separated flow.

bug (Fire Pr; Colloq) A small Maltese cross in gold, silver, or bronze, worn on the sleeve, awarded for personal risk and heroism in fire rescue work.

buggy (Fire Pr; Colloq) The official vehicle of the chief officer of a fire department.

bugle (Fire Pr) A speaking trumpet (megaphone) shown in relief on insignia designating the rank of fire department officers. See: **trumpets**; **ranks and titles**.
NOTE: Axes are used as rank insignia in some departments.

building (Build) 1. A walled and roofed structure used for dwelling purposes or as a protected space for the conduct of business and other activities. 2. Any structure that stands alone or is separated from others by suitable fire walls.

building code (Build) A set of rules, regulations, and standards governing the construction of buildings intended for particular uses.
NOTE: Building codes are of two types: specification codes, which define the materials that may be used, sizes, tolerances, assembly procedures, and other details; and performance codes, which set forth construction objectives and the acceptability criteria. In the latter, the designer and builder are free to choose any of various alternatives that satisfy the criteria.

building codes (Law) Locally enacted ordinances regulating building construction and safety practices, usu varying from locality to locality. Buildings are generally classified by primary use: 1) public, e.g., theaters, churches; 2) institutional, e.g., hospitals, schools; 3) residential, e.g., homes, apartments; 4) business, e.g., stores, office buildings; and

5) storage, e.g., garages, warehouses.
NOTE: Four model regional codes exist in the US: 1) Uniform Building Code, 2) Standard Building Code (formerly called the Southern Standard Building Code), 3) National Building Code, and 4) Basic Building Code.

building construction (Build) The activity involved in erecting structures. The design, materials, and construction methods used to erect buildings, esp as defined in building codes.
NOTE: There are five common types of building construction classified according to certain firesafe characteristics: fire-resistive, heavy timber, noncombustible, ordinary, and wood frame. **Fire-resistive construction**: A broad range of structural systems capable of withstanding fires of specified intensity and duration without failure. Common fire-resistive components include masonry load-bearing walls, reinforced concrete or protected steel columns, and poured or precast concrete floors and roofs. Although fire-resistive structures do not, in themselves, contribute fuel to fire, combustible trim, ceilings, and other interior finishes and furnishings may produce an intense fire and pose a serious threat to life safety. **Heavy timber construction**: Characterized by masonry walls, heavy timber columns and beams, and heavy plank floors. Although not immune to fire, the great mass of the wooden members slows the rate of combustion. **Noncombustible construction**: Structures in which the structure itself (exclusive of trim, interior finish, and contents) is noncombustible. Exposed steel beams and columns, and masonry, metal, or asbestos panel walls are the most common forms. **Ordinary construction**: Consists of masonry exterior bearing walls, or noncombustible bearing portions of exterior walls. Interior framing, floors, and roofs are made of wood and other combustible material lighter than that required for heavy timber construction. If floor and roof construction and their supports have a 1-hour fire resistance rating, and all openings through floors (including stairways) are enclosed with partitions having a 1-hour fire resistance rating, then it is known as protected ordinary construction. **Wood frame construction**: Characterized by wood exterior walls, partitions, floors, and roofs. Exterior walls may be sheathed with brick veneer, stucco, or metal cladding, asbestos-cement, or asphalt siding.

building materials (Build) Synthetic and natural materials used in the construction of buildings, including brick and masonry, plastics, wood and wood products, metals, and many others, often in the form of specific products.

building of origin (Fire Pr) The building in which an extensive fire originated. Syn: fire building.

buildup[1] (Fire Pr) a. Growth of a fire or increase in heat. b. Increase in the number of men and equipment at the scene of a fire.

buildup[2] (For Serv) The cumulative effects of drying over a period of time, resulting in an increase in fire danger.

bulk density (Transp) The mass per unit volume of a bulk material such as grain, cement, and coal. Used in connection with packaging, storage, or transportation.

bulk handling equipment (Mat Hand) Any of a variety of conveyers, lifts, hoists, trucks, power shovels, and other machines capable of moving bulk material in quantity.

bulkhead[1] (Build) a. A wall, partition, or other structural member at right angles to the longitudinal axis of a structure, serving to strengthen, separate, or give shape to it. b. An enclosure on the roof of a building, over a stairway, providing access to the roof.

bulkhead[2] (Civ Eng) A retaining wall for earth fills and embankments.

bulkhead[3] (Techn) An airtight structure separating a working chamber from free air or from another chamber with a different pressure.

bulk material (Mat Hand) Material stored in aggregate; i.e., not separately and individually packaged.

bulk modulus (Phys) The reciprocal of the coefficient of compressibility.

bulk oxygen system (Techn) An assembly of equipment, such as oxygen storage containers, pressure regulators, safety devices, vaporizers, manifolds, and interconnecting piping, used for the storage and handling of large quantities of gaseous or liquid oxygen.

bulk plant (Techn) A property or part of one where combustible or flammable liquids are delivered by tanker ship, pipe lines, tank car, or tank vehicle and are stored or mixed in bulk for subsequent distribution.

bulk storage (Storage; Transp) 1. Any large quantity of supplies, usu in original containers, waiting to be shipped or removed for processing or other use. 2. Storage of liquids or solids such as coal, grain, lumber, rubber bales, petroleum products, or ores in silos, tanks, or piles.

bull clam (Civ Eng) A tracked vehicle that has a hinged, curved bowl on the top front of the blade.

bulldozer (Civ Eng) The blade assembly mounted on the front of a wheeled or crawler tractor used to push rock, earth, or other material. See: blade.
NOTE: The term is often applied to the entire vehicle.

bulling (Transp) Dragging cargo across a surface without lifting.

bulwark (Navig) The side of a ship above the upper deck.

bund (Safety) A dam or dike used to contain combustible liquids. A wall that creates a moat around a liquid fuel tank to retain spills.

bundle-fold (Fire Pr) A section of hose accordion-folded into a 5-foot length with the couplings tucked inside. The bundle is tied at each end with a short rope.

bungalow station (Fire Pr) A single-story station house.

bunk¹ (Transp) A cross support for a load.

bunk² (Fire Pr) A bed in a fire company dormitory.

bunker (Techn) Any large bin or receptacle for storing fuel or loose material in bulk.

bunker clothing (Fire Pr) Protective clothing, usu fire retardant, water repellent, durable, and insulated, consisting of separate trousers and jacket, esp designed for the needs of firefighters.

bunker fuel (Techn) Viscous and heavy fuels used in large oil burning devices, usu stored in bunkers.

bunk room (Fire Pr) A dormitory in a fire station.

Bunsen burner A common laboratory heater consisting of a straight tube with adjustable air inlet holes and a gas connection at the base. The incoming gas entrains air and mixes with it to produce a premixed flame. If the mixture is fuel-rich, the premixed cone-shaped flame is surrounded by a diffusion flame. The two nested flame cones can be separated by a Smithels separator.

bunsen coefficient (Comb) The volume of gas under standard conditions absorbed by a unit volume of solution.

Bunsen number (Phys) A dimensionless ratio of the heat transferred by radiation to the thermal capacity of a fluid, expressed as $(hr \cdot A)/QC$, where hr is the radiant heat transfer coefficient, A is the characteristic area, Q is the mass flow rate, and C is the specific heat.

bunton (Mining) A strip of wood fastened across a series of planks to strengthen and stiffen the structure.

buoy (Navig) A moored, floating signal used to mark channels, shallow water, and hidden obstructions for navigation. Buoys are coded by numbers, color, and shape; some are equipped with radio beacons, noisemakers, or lights.

buoyancy (Phys) The lifting effect imparted to a body by a fluid in which it is immersed. The force is equal to the weight of the fluid displaced by the body. Also called: Archimedes principle.

burble (Aerodyn) 1. The separation or breakdown of laminar flow past a body. 2. The turbulence resulting from flow separation.

buret; burette (Chem) A graduated tube with a stopcock used to dispense small measured volumes of liquid.

burglary (Law) 1. The act of unauthorized entry into a building with the intent to steal. 2. The act of breaking into a private residence with the intent to commit a felony.

burn[1] (Comb; Fire Pr) a. To be on fire. To be consumed by a rapid, flaming or incandescent oxidation process that produces heat. b. Damage or injury caused by heat.

burn[2] (First Aid; Med) a. Injury by heat, electricity, chemicals, friction, irradiation, or other cause resulting in the destruction of skin and underlying tissues in varying degree. b. A prickling or itching sensation. c. The mark or trace left on the skin by a burn injury. See: scald; skin.
NOTE: Burns are classified as first degree, redness of the skin; second degree, blisters and some broken skin; and third degree, with charring and full thickness skin destruction. Third-degree burns, in contrast to first- and second-degree burns, are not painful, the nerve endings having been killed. Serious burns damage capillary walls, resulting in loss of circulation of blood plasma, and edema (fluid infiltration) in the damaged areas. Skin will blister (second-degree burn) at a temperature of 160°F. A temperature of 250°F can be tolerated with great discomfort for 15 minutes, depending on dryness of the air. At 300°F irreversible damage is done to the skin in one minute. It is reported that dry air up to 300°F can be breathed for a short time, but not moist air, which causes extensive deep lung damage.

burn[3] (Fire Pr; For Serv) a. An area burned over by a wildland fire. b. A working fire. c. A test fire; one set to study the effects of fire or for practice in putting one out.

burnable (Comb) Capable of undergoing combustion, usu under stated or implied conditions. Capable of being consumed by fire.

burnback (Fire Pr) Flames traveling back over an area previously extinguished, due to incomplete foam cover on flammable fuels or degradation of foam by heat or aging.

burned (Fire Pr) The state of having been injured, damaged, or consumed by fire.

burner (Comb) 1. A device for mixing fuel with air for burning. 2. A combustion chamber.

burning (Fire Pr) The state of being on fire.

burning block (For Serv) An area that is sufficiently uniform in stand and fuel conditions that it can be treated under a single burning prescription.

burning brand test (Build) A fire test of roof coverings in which specified burning wooden brands are fastened to a sloping roof deck specimen which is exposed to a specified wind. One of three fire tests normally applied to roof coverings. See also: flame exposure test; flame spread test.

burning conditions (For Serv) The combination of environmental and fuel factors that affect fire in a given situation.

burning glass (Opt) A focusing lens used to concentrate sunlight to ignite combustible material.

burning index (For Serv; Abb: BI) A value determined on a scale of 1 to 10 indicating the relative degree of forest fire danger based on wind, temperature, relative humidity, days since last rainfall, etc.

burning index class (For Serv) A class represented on the burning index scale described as low, medium, high, very high, and extreme fire danger. See fire_danger_rating.

burning_index_meter (For Serv) A calculating device with matching scales used to determine the burning index for different combinations of burning index factors. See fire_danger_rating.

burning_out See: burn_out

burning_period (Fire Pr) The portion of the day, usu from 10 AM to sunset, when wildland fires spread most rapidly.

burning_point (Comb) The temperature, normally a few degrees above the flash_point, at which a combustible material gives off sufficient vapor to ignite and continue to burn.

burning_rate (Comb; Symb: r) The velocity at which a solid or liquid is burned, measured in the direction normal to the surface and usu expressed in inches per second. Also called: regression_rate. Cf: burning_velocity (for gases).
NOTE: Burning rates for flammable liquids will vary if they contain light and heavy fractions. Light fractions tend to burn faster, and after they burn off, the remaining heavier fractions burn at a slower rate. Solid propellants often obey the law $r = a \cdot p$ to the power n, where a and n are constants and p is the pressure of combustion. It has been suggested that burning rate be used to express the combustion of a fuel in terms of mass consumed per unit time (equivalent to mass_burning_rate), and to reserve regression_rate to express the combustion of a fuel in terms of length (depth) consumed per unit time.

burning_velocity = flame_velocity

burnishing (Metall) Surface finishing of a metal by rubbing.

burn_out[1] (Electr) To fail as a result of overheating or burning, as of an electronic component.

burn_out[2] (Fire Pr) To deliberately burn the fuel between a control_line and a wildland fire for the purpose of stopping its advance. Syn: clean_burn; fire_out. Cf: backfire.

burn_out[3] (Metall) To fire a casting mold at high temperature in order to remove pattern material residue.

burnout[1] (Electr; Mech Eng) Failure of a component due to overheating.

burnout[2] (Fire Pr) a. A building gutted by fire. b. An area burned over by a wildland fire. See also: burn.

burnout[3] (Comb) a. The instant the fuel and/or oxidizer in a rocket is completely consumed. b. (Brit) = all_burnt.

burn_toxin See: toxin

burr (Mech Eng) 1. A ragged edge or ridge of metal remaining on the workpiece after a cutting or drilling operation. 2. A rotary milling cutter used in die sinking.

bursitis (Med) Inflammation of the bursa or fluid-containing sac around bone joints.

burst[1] (Electr) A packet of electromagnetic energy.

burst[2] (Techn) To rupture.

burst_disk = rupture_disk

burst_hose_jacket See: hose_jacket

bursting_pressure (Fire Pr) The test pressure at which a fire hose will burst.

bursting_strength (Packaging) The pressure required to rupture a container when it is tested in a specified instrument under specified conditions.

busher = swamper

bushing[1] (Mech Eng) A sleeve or other liner used to reduce the size of a hole, as a replaceable bearing material, a fitting to reduce the size of a pipe, or a smooth ring fitted to the end of an electrical conduit to prevent chafing.

bushing[2] (Fire Pr) A fitting used to decrease the opening on a sprinkler tee to permit replacement of sprinkler heads.

bushwacker (Fire Pr; Colloq) A rural or suburban firefighter who is frequently called on to fight grass and brush fires.
 NOTE: Derived from the practice of beating out grass fires with fire bats and other types of swatters.

butane (Chem) A volatile, flammable hydrocarbon (C_4H_{10}) having two isomers: isobutane ((CH_3)$_2CHCH_3$), bp -11.7°C; and n-butane ($CH_3CH_2CH_2CH_3$), bp -0.5°C. Butane is widely used in liquified petroleum gas (LPG) form as a domestic fuel.

butanol (Chem) One of four isomeric alcohols with the formula C_4H_9OH, used widely in solvents, plastics, and resins. See also: alcohol.

butt[1] (Fire Pr) a. One coupling of a fire hose at the end of a hose section. b. An open-ended hose line. c. The foot of a ladder. Syn: heel.

butt[2] (Transp) A large barrel.

butt a ladder (Fire Pr) To crouch on the lowest rung of a ladder, grasping the third rung, so that the weight of the body anchors the heel and provides leverage to assist in raising. See also: heel.

butt weld (Techn) A butt joint made by welding two pieces butted together.

C

c (Phys) Abbreviation for calorie, centimeter, cubic, curie.

C (Chem; Phys) Symbol for Celsius, capacitance, carbon, speed of light.

cab (Transp) A housing that covers the operator's or driver's station on a vehicle.

CABO = Council of American Building Officials

cadmium (Symb: Cd) A silvery-white, ductile, metallic element, mp 321°C, bp 767°C. Used extensively in low-melting alloys for fusible links in automatic fire sprinklers and safety plugs in water heaters. Also used in batteries, pigments, and atomic reactor moderators. Cadmium fumes, solutions, and compounds are toxic.

cage[1] (Safety) An enclosure fastened to the side rails of a fixed ladder or a structure encircling the climbing space of the ladder for the safety of the person using it.

cage[2] (Build) A system of grounded wires or cables that form a mesh suspended above a structure to serve as a lightning arrester.

cage[3] (Mining) a. A cage-like, man-carrying platform used for hoisting and lowering miners. b. A passenger elevator or hoist in a mine.

caged storage (Storage) A space for the safekeeping of goods or supplies reserved within a building and specially screened or barricaded to prevent theft or to isolate dangerous materials.

caisson (Civ Eng) A wood, steel, concrete, or reinforced concrete air- and water-tight chamber in which men work under air pressure greater than atmospheric to excavate material or perform other tasks below water level.

cake (Techn) To harden or become packed.

caking (Fire Pr; Techn) The formation of adhesion between particles of powder resulting from the absorption of water vapor by the powder.
  NOTE: In severe cases the powder becomes solid and will not discharge from an extinguisher. Anticaking additives are incorporated into the powders to counteract this process.

calcination (Techn) A common industrial process in which solid material is heat-treated to bring about thermal decomposition, usu to a powder but not by melting.

calcium (Metall; Symb: Ca) A silvery, white alkaline metal, at. no. 20, at. wt 40, mp 810°C, bp 1170°C. Used as an alloying agent.
  NOTE: Finely divided calcium will ignite spontaneously in air.

calculation of probabilities (For Serv) Evaluation of all factors pertaining to the probable behavior of a going fire and of the potential of available fire suppression forces to carry out control operations on a given time schedule.

calculator (Comput) A mechanical, electromechanical, or electronic keyboard device used for performing arithmetic operations. Also called: adding machine.

calculus (Math) The branch of mathematics dealing with methods of calculating rates of change (differential calculus) and the summation of infinitesimals (integral calculus).

calender (Techn) A machine equipped with two or more metal rollers revolving in opposite directions and used extensively in the textile, rubber, and paper industries.
  NOTE: In the plastics industry the machine is used for continuously sheeting or plying up rubber and plastics and for frictioning or coating materials with rubber and plastics. Calenders have 2 to 10 rollers, or bowls, some of which may be heated.

caleometer (Phys) An electrical instrument used to measure convective heat loss from a wire, commonly used to determine the concentrations of binary mixtures of gases or as a detector in gas chromatography. Also called: thermal conductivity meter (US); katherometer (Brit).

calibrate (Techn) To check and adjust an instrument or measuring device for uniformity or accuracy.

calibration block (Techn.) A piece of material with specified composition, surface finish, heat treatment, and geometric form, by means of which ultrasonic equipment can be calibrated for the examination of material of the same general composition.

caliper (Techn) A machinist's or carpenter's instrument for measuring dimensions, esp diameter.

calking = **caulking**

call (Fire Pr) A fire or emergency alarm. A signal or message calling for a fire department response.

call box[1] (Fire Pr) A fire alarm signal box. A device capable of transmitting a signal to a central station. One with a sloping, roof-like top is called a **cottage box**.
 NOTE: Box signals are usu coded to indicate the location of the box.

call box[2] (Brit) A public telephone booth.

call firefighter (Fire Pr) A firefighter who regularly follows another occupation but responds to a fire emergency when called by radio, telephone, siren, or house bells. 2. = (Brit) **retained fireman**.
 NOTE: A call firefighter may be paid at an hourly rate for the time spent on the fireground or a flat fee per response.

call man = **call firefighter**

calorie (Abb: cal) A secondary unit of heat, defined approx as the amount of heat required to ce the temperature of 1 gram of water 1°C. Also called **gram-calorie** or **small calorie** (Abb: c). Now defined in terms of joules. One calorie is equal to 4.184 absolute joules. Other definitions are a. **International Steam Table Calorie** = 4.1868 joules, b. **mean calorie** = 4.19002 joules, c. **15°C calorie** = 4.1858 joules, d. **20°C calorie** = 4.189 joules. The **kilogram calorie** (kilocalorie; Abb: C or kcal) is 1000 gram-calories. See: **joule**.

calorific (Comb) Pertaining to the generation of heat. Syn: **caloric**.

calorific potential (Comb) The heat energy that a unit mass of material is capable of releasing during complete combustion. Syn: **potential heat value**.

calorific value (Comb) The heat energy released by a burning material. Syn: **heat value**.

calorimeter (Chem) An apparatus for measuring the quantity of heat liberated or absorbed in various processes and chemical reactions and in the determination of the heat capacities of substances.

calorimetry (Chem) An analytical method used to determine the thermal behavior of materials and reactions, such as average specific heat over a temperature range. Also called: **adiabatic calorimetry**.
 NOTE: In **scanning calorimetry**, the temperature of the sample is varied at a uniform rate, while measurements are taken of heat flowing into or out of the sample. See also: **differential thermal analysis**.

cam (Mech Eng) A contoured, rotating, wheel-like mechanism that imparts a varying linear motion to a follower that rides on its surface, commonly used to open and close valves in engines and to convert rotational to linear motion in machines.

camber (Transp) The setting of a pair of vehicle wheels closer together at the bottom than at the top.

campaign fire (For Serv) A forest fire that requires more than one day's effort.

camp boss (For Serv) An individual responsible for establishing and operating a fire camp, reporting directly to the fire boss or the service chief in a headquarters camp or to a sector boss or a division boss in a line camp.

campfire (For Serv) In reference to forest fire causes, a fire started for cooking, warmth, or light that has spread sufficiently to require firefighting activity.

can (Fire Pr; Colloq) A fire extinguisher.

cancer = carcinoma

candela (Opt; Abb: cd) A unit of luminous intensity equivalent to that emitted from a blackbody having an area of 1/600,000 m² at the temperature at which platinum, under the pressure of 101,325 newtons/m², solidifies (2046°K).
NOTE: The term candela officially replaced candle by act of the US Congress in 1964.

candle = candela

candlepower (Opt) Directional intensity of illumination, expressed in candlas, equal to the illumination in foot-candles falling on a surface normal to the direction of the source, multiplied by the square of the distance in feet to the source.

canister (Fire Pr) 1. A container filled with sorbents and catalysts that remove gases and vapors from the air drawn through the unit. The canister may also contain an aerosol or particulate filter to remove solid or liquid particles. 2. A container filled with a chemical that generates oxygen by chemical reaction. See: chemical cartridge.

cannulization (Med) Insertion of a tube (cannula) into a body cavity or vessel for the purpose of drawing off fluids or introducing medication. See also: catheter.

canopy (For Serv) The topmost branches and leaves of a forest. The overstory of a stand of trees.

cantilever (Build) Generally, a horizontal (sometimes vertical) structural member supported on one end only.

cantilever position (Fire Pr) The position of an aerial ladder supported only at the base by the turntable.

canvas A strong, coarse fabric woven from hemp, flax, jute, or cotton, used for hoses, tents, awnings, webbing, tarpaulins, sails, etc.

cap[1] a. A threaded fitting used to seal off pipes. b. To install such a fitting on a pipe.

cap[2] = detonator. A blasting cap.

cap[3] (Mining) A horizontal timber resting on posts and used as bracing in a mine or as a support for a platform.
NOTE: Caps are set at right angles to a tunnel; a timber used for the same purpose but set parallel to the tunnel is called a girt.

capacitance (Electr) 1. The ability of a nonconducting material to store an electrical charge. 2. The ratio of the electrical charge on one of two electrical conductors, such as the plates of a capacitor, to the potential (voltage) difference.

capacitive reactance (Electr) The opposition offered to the flow of an alternating current by capacitance, expressed in ohms.

capacitor (Electr) An electronic device consisting of two conducting plates separated by a dielectric, used to accumulate and discharge electrical charges for the purpose of limiting current flow, filtering signals, shaping waveforms, tuning circuits, etc. Formerly called: condenser.
NOTE: A common type consists of two or more plates separated by air or other dielectric gaps. Another type is made of strips of foil separated by paper and rolled into tight cylinders (these can be dry or oil-filled). An electrolytic capacitor consists of two plates and a moist electrolytic paste. The dielectric is a thin film of gas formed by electrolytic action.

capacity (Transp) The volume of a container.

capacity operation = volume operation

capillary[1] (Phys) A small-bore tube.

capillary[2] (Anat) The smallest vessels in the cardiovascular and lymphatic systems, forming the blood-transfer link between the arteries and the veins. See also: arteries; cardiovascular system; circulation system.

capillary action (Phys) 1. The tendency of certain liquids to penetrate or migrate through small openings such as cracks or fissures.

2. Rise of liquid in fine-bore tubes due to surface tension.

capital stock (Insur) Shares of stock representing stockholders' equity.

captain (Fire Pr) The officer in charge of a fire department company or station, or one with comparable responsibilities in a fire department. See: ranks and titles.

capture velocity (Techn) Air velocity at any point in front of an exhaust hood required to overcome opposing air currents in order to draw contaminated air into the hood.

Carbenicillin (Med; Pharm) A trade name for a semi-synthetic penicillin used intravenously to combat infection and septicemia in burn patients.

carbide (Techn) A binary compound of carbon and another element, but not including such compounds as CCl_4 and CO. See: abrasive.

carbohydrate (Chem) A large class of organic carbon-hydrogen-oxygen compounds such as sugars, starches, and celluloses.
  NOTE: Most carbohydrates are produced by plants and serve as food for man and animals. About two-thirds of the average daily adult caloric intake is in the form of carbohydrates.

carbolic acid See: phenol

carbon (Chem; Symb: C) A nonmetallic element, mp 3500°C, bp 4830°C, at. wt 12. Carbon is a constituent of all organic compounds, although carbon itself is considered inorganic.
  NOTE: Carbon exists in allotropic forms as carbon black, graphite, and diamond.

carbon black (Techn) The generic name for a commercial pigment composed principally of carbon.
  NOTE: Common soot is identical to carbon black. A variety called channel black contains oxygen, hydrogen, and small quantities of benzene. Furnace and thermal blacks contain smaller quantities of oxygen and hydrogen. Used as a filler in rubber manufacture, inks, and as coloring matter in plastics, carbon paper, and synthetic fibers.

carbon dioxide (Chem) A colorless gas, CO_2, with pungent odor and acid taste, sublimation point -78.5°C at atmospheric pressure, triple point -56.6°C at 1 kg/cm² pressure.
  NOTE: Used as dry ice in refrigeration and as a liquid or gas in carbonated beverages, fire extinguishers, chemical manufacturing, and in hot metal processes (e.g., welding) as an inert gas shield. Pressurized CO_2 cartridges are used to inflate life rafts. The gas is formed during respiration and as a product of combustion of all carbonaceous fuels. In 3 to 5% concentrations, CO_2 is a respiratory stimulant. Concentrations of 8 to 10% by volume are asphyxiating. Higher concentrations are toxic. Maximum allowable concentration is 5000 ppm for 8 hours.

carbon disulfide (Chem; Symb: CS_2) A colorless, highly flammable and toxic liquid, bp 46°C, flash point -30°C, autoignition temperature 100°C, used as a solvent and as an intermediate in the manufacture of viscose rayon, cellophane, and carbon tetrachloride. Syn: carbon bisulfide.
  NOTE: Carbon disulfide is heavier than water and therefore can be extinguished by water floating on its surface.

carbonization (Comb) Thermal decomposition of organic substances in an oxygen-deficient atmosphere, resulting in the formation of carbon. See: pyrolysis.

carbonizing (Techn) The removal of vegetable matter such as burrs, straws, etc. from wool by treatment with acid, followed by heating. The undesired matter is reduced to a carbonlike form that may be removed by dusting or shaking.

carbon monoxide (Chem) A colorless, tasteless, and odorless gas, CO, mol wt 28.01, mp -200 to -205°C, bp -191.5°C. CO is produced during incomplete combustion of a carbonaceous fuel. Mixtures of CO, H_2, N_2 are common fuels called town gas, producer gas, etc. CO + O with traces of water are explosive, and reaction of the gas with chlorine forms deadly phosgene. Syn: sweet

damp; white damp. See: damp.
 NOTE: The gas is extremely toxic, not because of actual damage to body tissues, but because it has a strong affinity for the hemoglobin in the blood, displacing oxygen and causing anoxia in the body tissues, esp the brain. The process is reversible: hemoglobin freed from CO (e.g., by oxygen resuscitation) recovers its oxygen-carrying capacity. Ordinary gas masks afford no protection against CO, and special breathing apparatus should be used. Symptoms of "CO-poisoning" are blurred vision, headache, nausea, dizziness, and finally loss of consciousness. Without protection, 50 ppm (0.005 by volume) is the threshold limit value for prolonged exposure. Concentrations of 1000 ppm (0.1%) can be fatal within 1-hour. See: carboxyhemoglobin. First-aid treatment requires removal of the casualty from the CO atmosphere, administration of oxygen, and absolute rest.

carbon tetrachloride (Chem) A colorless, toxic, dense liquid with a sweetish odor, CCl$_4$, bp 76°C, used in drycleaning, in older types of fire extinguishers, and in Halon-type refrigerants. CCl$_4$ concentrations by volume of 4.8 to 6.3% are lethal, and 50 ppm is the maximum allowable concentration for an 8-hour period. In addition to its own toxicity, it forms phosgene gas upon oxidation in fires. Also called: carbon tet. See: Halon.
 NOTE: Carbon tetrachloride, assigned the value of 100, is used as an evaporation rate standard. cf: ethyl ether, which is also used as an evaporation standard.

carbonyl chloride = phosgene

Carborundum (Techn) A trade name for silicon carbide, widely used as an abrasive.

carboxyhemoglobin (Med; Abb: COHb) Hemoglobin carrying carbon monoxide instead of oxygen. Syn: carbon monoxy hemoglobin. see: carbon monoxide.
 NOTE: A nonsmoker breathing pure air has about 0.4% background COHb in his blood. Cases are known of healthy persons surviving with 20 to 40% COHb. However, experiments indicate that mental processes and judgement tend to deteriorate when COHb levels reach 5 to 10%. Persons in poor health, heavy smokers, and those engaged in strenuous activity while breathing air containing CO are more likely to suffer adverse effects than healthy individuals, nonsmokers, and persons at rest.

carboy (Transp) A large glass bottle, up to about 15 gallons capacity, usu protected by a crate or cushioning wicker work.

carcinogen (Med) A substance that produces cancer.

carcinoma (Med) Cancer. A malignant tumor derived from epithelial skin tissues that tends to invade adjacent tissues and spread to remote parts of the body, the membranes lining the body cavities, and certain glands.

Cardan joint See: universal joint

cardiac (Med) Pertaining to the heart.

cardiac arrest (Med) Sudden, unexpected stoppage of heart action. Cf: heart failure.

cardiac cycle (Med) A complete heart beat, consisting of a systole (contraction) and a diastole (relaxation).

cardiac output (Physiol) The quantity of blood pumped from the heart through the circulatory system, measured in liters per minute.

cardinal points (Cartog; Navig) The principal directions of north, south, east, and west. The intermediate points of northeast, southeast, northwest, and southwest are called collateral points.

cardiograph (Med) An instrument used to record the rate and amplitude of the heart beat.

cardiopulmonary resuscitation (First Aid; Med; Abb: CPR) Artificial ventilation and external heart massage applied simultaneously to counteract cardiac arrest and restore circulation and breathing.

cardiovascular system (Anat) The complex system of the heart, arteries, veins, and capillaries that distributes blood in animals. See also: circulation_system.

cargo handling gear (Transp) Gear that is a permanent part of a vessel's equipment and is used for the handling of cargo other than bulk liquids, ship's stores, gangway, or conveyor-belt systems for the self-loading of bulk cargoes.

cargo tank (Transp) Any container designed to be permanently attached to any highway vehicle to transport a compressed gas in commercial quantities.

Carnot cycle (Thermodyn) A thermodynamic cycle used to describe the working of an idealized heat engine.

carried wet (Fire Pr) Referring to booster hose or pumps carried full of water to avoid filling or priming before use.

carrier[1] (Med) A person in apparent good health who harbors a pathogenic microorganism and therefore can infect others.

carrier[2] (Mat Hand) An industrial truck that straddles a load to be transported and is equipped with mechanisms to pick up the load and support it during transportation.

carrier[3] (Commun) The fundamental frequency for radio transmission that is modulated in amplitude or in frequency.

carry See: hose_carry; ladder_carry; rescue_carry.

carryall (Salv) A salvage cover with rope handles around the edges used to catch or carry debris, such as a smoldering mattress. The carryall can also be used to hold a pool of water for dipping and quenching small burning objects. See also: catchall.

carry cloth (Mat Hand) A large piece of canvas or burlap used to transfer solid waste from a residential solid waste storage area to a collection vehicle. Similar in function to carryall.

Cartesian coordinates See: coordinate system

cartography (Cartog) The science of map-making. See also: map.

carton (Storage) A form of packaging made from a bending grade of paperboard. An acceptable designation for folding paperboard boxes, not shipping containers.

cartridge (Safety) A container with a filter or other purifying medium. See: chemical_cartridge.

cascade[1] (Techn) An arrangement of elements or units in a series such that the output of each feeds into the input of the next.

cascade[2] (Fire Pr) A truck that supplies refill air tanks for breathing apparatus.

case-hardening (Metall) A process by which the carbon content at the surface of a metal is increased to make the surface harder than the interior metal.

case law (Law) Law established by judicial decisions in particular cases, which serve as precedents in subsequent similar cases. Cf: common law.

cask (Transp) A thick-walled container, usu of lead, used for transporting radioactive materials. Also called: coffin.

castellated (Techn) Having the appearance of a castle battlement; descriptive of a nut that has grooves on one side along its diameter to receive a cotter pin.

casting (Metall) a. Making shaped objects by pouring molten material into a mold and permitting it to solidify. b. An object produced by a casting process.

casualty insurance (Insur) Insurance written by companies licensed under the casualty sections of state insurance laws, as distinguished from that written under the fire or marine or life insurance sections. Concerned principally with insurance against loss due to accident or other mishap.

catalysis (Chem) A process by which a chemical reaction is induced or hastened by the presence of a catalyst. If the reaction rate is decreased it is called negative catalysis, and the agent a negative catalyst.

catalyst (Chem) A substance that mediates or accelerates a chemical reaction or process without itself being changed chemically.

catalytic combustion system (Comb) A process in which a substance is introduced into an exhaust gas stream to burn or oxidize vaporized hydrocarbons or odorous contaminants; the substance itself remains intact.

cataract (Med) An opaque cloud that forms in the lens of the eye, obscuring the transparency of the lens.

catastrophe (Insur) A widespread and extraordinarily large instance of injury, death, damage, and destruction, as by a conflagration, earthquake, or hurricane.

cat boss (For Serv) A supervisor of one or more tractor operators and helpers during fireline construction or mop-up.

catch a hydrant (Fire Pr) To drop a man from a moving truck at a hydrant to connect hose during hose-laying operations at a fire. Also called: make a hydrant.

catchall (Salv) A construction of salvage covers to contain water or to serve as a temporary water chute. Also called: shallow bag. See: carryall.

catface (For Serv) A surface defect on a tree resulting from a wound that has not healed completely or has healed imperfectly. See: fire scar; fire wound.

catheter (Med) A slender tube for insertion into a body cavity for drainage of fluids or into narrow canals to probe for obstructions.

cathode (Electr) 1. The negative terminal of an electrolytic cell. 2. The terminal at which chemical reduction takes place. Cf: anode.

cathode ray (Electr) A stream of electrons emitted by a negative electrode, usu a hot filament in a vacuum tube.
NOTE: A television picture, for example, is painted electronically on the phosphorescent screen of the cathode ray tube by a beam of electrons.

cation (Chem) A positively charged ion in an electrolytic solution that moves toward the negatively charged cathode, for example, H+, Na+, Cu++.

catline (For Serv) A fireline constructed by a tractor with a bulldozer blade or scraper.

catwalk (Ind Eng) A narrow, suspended footway, usu for inspection or maintenance purposes.

caulking (Build) 1. To fill seams of boats, cracks in tile, etc. 2. The material used in caulking. Also spelled: calking.

causal factors (Safety) A combination of simultaneous or sequential circumstances contributing to a situation or event.
NOTE: Causes may be direct, early, mediate, proximate, distal, etc.

cause of fire See: fire cause

caustic (Chem) A substance that irritates, burns, corrodes, or destroys living tissue. See: alkali.

cave-in (Civ Eng; Mining) Collapse of excavation walls or the roof of a mine.

caving (Mining) 1. Allowing or causing ore to fall from the roof of a stope in mining. A method of undercutting an orebody and letting it fall and crush under its own weight. 2. Accidental collapsing of an excavation or passageway in a mine.

cavitation (Hydraul) The formation of partially evacuated cavities in a fluid flow due to stresses exceeding the tensile strength of the liquid. Shock pulses induced when such cavities collapse can damage pumping equipment.
NOTE: Cavitation occurs when a pump is attempting to discharge water faster than the supply can deliver it to the input.

CBM = chlorobromomethane

CCTV = closed circuit television

ceiling hook (Fire Pr) A long, wooden or fiberglass pole having a metal point with a spur at right angles. Used for probing ceilings and pulling down sheathing to obtain access to burning material. Syn: pike pole. See also: plaster hook.

ceiling layer (Fire Pr) A bouyant layer of hot gases and smoke produced by a fire in a compartment.

ceiling stick (Fire Pr) A short, single-beam hook ladder, similar to a pompier, used in tight places inside buildings. See: attic ladder; ladder.

cell[1] (Electr) A primary battery or element of a storage battery.

cell[2] (Math) An element of an array.

cell[3] (Comput) An elementary unit of storage in a computer.

cell[4] (Biol) A small biological structural unit, consisting of a semiliquid living substance enclosed in a membrane and generally containing a nucleus.
NOTE: All body tissues are composed of cells, of which there are many different kinds, including nerve cells, muscle cells, blood cells, connective tissue cells, fat cells, and others.

cellar fire (Fire Pr) Fire situated below grade level in a building.

cellar hole (Fire Pr) The excavation remaining after a building has been destroyed.

cellar nozzle (Fire Pr) A special spray nozzle for attacking fires in basements, cellars, ships' holds, and other spaces below floors or decks. Some models can be operated remotely. Syn: basement spray (Brit). Also called: cellar pipe. See also: distributor nozzle; revolving cellar nozzle.

cellar pipe (Fire Pr) Any angled or bent rigid pipe used as an extension to a hose where fire conditions or inaccessibility limit the use of hand lines. See: distributor nozzle.

NOTE: The pipe can be inserted through an opening or through a hole cut in the floor above a fire. The nozzle is directional and can be controlled by the operator.

cellar savers (Fire Pr; Colloq) A derogatory term for badly trained firefighters who leave only a cellar hole after a fire.

cellular plastic See: plastic

celluloid (Chem) A flammable plastic made of camphor and pyroxylin (nitrocellulose). Also called: xylonite (Brit). See also: rayon; polymer; cellulose.

cellulose (Chem) A complex polysaccharide $(C_6H_{10}O_5)n$ occurring as a natural polymer in plants. Cotton is almost pure cellulose. Used extensively in fibers for textiles (rayon), paper, photographic film, and plastics (cellophane).
NOTE: Cellulose in a mixture of sulfuric and nitric acids forms cellulose nitrate, a powerful explosive called guncotton, which is used in the manufacture of cordite.

cellulosic material (Chem) A material or substance containing cellulose or one of its close chemical relatives.

Celsius temperature scale (Symb: °C) A metric temperature scale suggested by Anders Celsius, based on the freezing and boiling points of water (0°C and 100°C, resp), now defined in terms of the Kelvin scale: 100°C = 273.16°K. Formerly called the Centigrade scale. See: temperature scale.

cement[1] (Techn) Any substance that acts as a bonding agent. See also: adhesive.

cement[2] (Build) A finely ground clinker, consisting principally of silicates, lime, alumina, and iron oxides, used in making concrete. Commonly called portland cement or hydraulic cement. See also: concrete.

center firing (For Serv) A method of broadcast burning in which fires are set in the center of an area to create a strong inward draft, then additional fires are set nearer to a control line. The indraft draws the outer fires toward the center. See: simultaneous firing; area ignition.

centerline (Geom) The line extending through the middle of the long dimension of an object.

center of gravity See: center of mass

center of mass (Phys) The point in a body that moves as if the entire mass of the body were concentrated in it. Also called: center of gravity.

center of pressure (Hydraul) A point on a plane surface through which the resultant of pressure forces passes.

centigrade temperature = Celsius temperature

centimeter (Metrol) A unit of length, one-hundredth of a meter, equal to 1/2.54 inch.

centipoise (Phys; Abb: cp) A unit of viscosity. See: poise.

central nervous system (Physiol) The portion of the nervous system consisting essentially of the brain and spinal cord. The system that controls mental activities and both voluntary movements of the body and involuntary body functions.
NOTE: This portion does not regenerate when injured, but heals by scar formation.

central station (Fire Pr) 1. A center for relaying alarms and messages to firefighting and emergency units and services. 2. A headquarters station for chief officers, special units, and the alarm communications system. 3. (Brit) A privately owned security center that upon receipt of an alarm notifies the fire department.

centrifugal force (Phys) A force acting outward on a revolving body equal and opposite to the centripetal force. This force results from the resistance of a body to change its direction of motion. See also: centrifuge.

centrifugal pump (Fire Pr) A pump that has one or more rotating impellers that accelerate water by centrifugal force. Such pumps are used extensively for high-volume operation, because they are adaptable to large capacities, easy control, are economical, and provide a steady, continuous output. See also: positive displacement pump.

centrifuge (Techn) A rotating apparatus that uses centrifugal force to separate components of different densities in liquid or slurry mixtures.
NOTE: Centrifugation, the separation of liquids from solids by centrifugal force, is sometimes called spin drying or wringer drying.

centripetal force (Phys) The inward radial acceleration force required to keep a body moving in a circular path. Cf: centrifugal force.

ceramic (Techn) Narrowly, a product manufactured from a nonmetallic, inorganic, silicate material at high temperature, esp structural clay products, whitewares, glass, porcelain enamel, refractories, cement (portland), and abrasives.

cerebral (Med) Pertaining to the brain.

cermet (Metall; Techn) An intimate mixture of metallic and ceramic components compacted into a desired shape and sintered. Cermets are characterized by high temperature resistance, high strength, special electrical properties, and corrosion and wear resistance.
NOTE: The term is a blend of ceramic and metal.

certificate program (Educ) An educational program in which a student earns a certificate upon successful completion, but no credit toward an academic degree.

certified safety professional (Abb: CSP) An individual who has been certified by the Board of Certified Safety Professionals as having achieved professional competence by training and experience; by remaining abreast of the technical, administrative, and regulatory developments in his chosen field; and who maintains professional integrity and exercises a high standard of ethics with clients, associates, and the public.

cerumen (Med) Earwax.

cesium (Symb: Cs) An alkali metal element, mp 28.5°C, bp 705°C, used in photoelectric cells and various detector devices.
 NOTE: The metal reacts vigorously with oxygen, moist air, and water. The isotope Cs^{137}, a constituent of fallout, has a half-life of 33 years and is an important fission product.

cetane (Chem) Common name for n-hexadecane, $C_{16}H_{34}$, used as the standard reference fuel in determining the ignition quality (cetane number) of diesel fuels.

cetane number (Comb) The percentage by volume of cetane (cetane number 100) in a blend with alpha-methylnaphthalene (cetane number 0) that matches the ignition properties of a given fuel under standard test conditions. Cf: octane number.

cfm = cubic feet per minute

CFO (Brit) = chief fire officer

CFR[1] = Code of Federal Regulations

CFR[2] = crash-fire-rescue

CGA = Compressed Gas Association, Inc.

cgs system (Metrol) A system of units based on the centimeter, gram, and second.

chafing block (Fire Pr) A wooden block placed under or clamped around a charged hose line to protect the hose jacket from wear and damage.

chain[1] (Techn) A flexible line consisting of metal links used for pulling, lifting, binding, restraining, etc.

chain[2] (Chem) Two or more like atoms linked by a bond.

chain[3] (Surv) A measuring line consisting of 100 links of equal length. A surveyor's chain (Gunter's chain) is 66 feet long, with links 7.92 inches; an engineer's chain is 100 feet long, with 1-foot links.

chain branching (Chem; Comb) Autocatalysis in chemical reactions resulting from the formation of more reactive chemical species that are consumed and then replicate themselves as a chain. The reactive species in flames are usu atoms or radicals.

chain breaking (Chem) The process of interrupting the sequence of reactions involved in a chain reaction, either by removing or modifying the chain carriers or agents produced during the reaction and necessary for the completion of the chemical sequence.

chain drive (Mech Eng) A power transmission mechanism consisting of an endless chain and toothed sprocket wheels.

chain hoist (Techn) A power unit used for lifting by reeling in a chain connected to the load, usu with the aid of one or more pulleys.

chain lightning (Meteor) Lightning that appears interrupted or zigzagged. See: lightning.

chain of command (Man Sci) The order of rank and authority in an organization.

chain reaction (Chem; Nucl Phys) A self-sustaining chemical or nuclear process in which one or more of the species or agents necessary for the reaction to continue are regenerated by the reaction itself.
 NOTE: In a nuclear chain reaction, a fissionable nucleus that is split by a neutron releases energy and one or more additional neutrons. These neutrons split other fissionable nuclei, releasing more energy and more neutrons, which makes the reaction self-sustaining as long as enough fissionable nuclei are present. In the case of combustion or a chemical reaction, a chain reaction is a process in which one of the agents necessary for the reaction is produced by the reaction itself (e.g., heat or reactive radicals).

chamber test (Comb) A fire test for floor coverings, developed by Underwriters' Laboratories, in which speed and distance of flame spread are measured.

chamfering (Techn) The operation of

leveling sharp edges and rounding corners of a workpiece; to cut a furrow or a groove.

chance[1] (Statist) a. Random, unpredictable, occurring without observable cause. b. The probability of any outcome when all possible outcomes have an equal probability of occurring.

chance[2] (For Serv) An advantageous condition for a given requirement, such as: gravity chance, water so located that it can be delivered to the desired point by gravity; pump chance, water sufficiently abundant and so located that it can be pumped to the desired point.

change of quarters See: relocate

change of state (Chem; Phys) The transformation of matter from one state (liquid, gas, or solid) into another.
NOTE: Normally, with application of heat, a solid becomes a liquid and then a gas; however, solids can sublime directly into the gaseous state and gases can solidify without passing through the liquid state.

change-over valve (Fire Pr) A transfer valve on a multistage pressure-volume pump used to provide either rated capacity at specified pressure or smaller flows at higher pressure.

char (Comb) 1. The carbonaceous material formed by incomplete combustion of an organic material, such as wood. See: pyrolysis. 2. To change into charcoal or carbon by pyrolysis.

character (Comput) One of a set of elementary marks, such as numerals or alphabetic letters, that may be combined to express information. A character includes all the marks, such as a group of magnetized dots or holes punched in paper, that are necessary to completely identify it.

characteristic A property, capability, or feature of a component or item of equipment.

characteristic curve (Math) A graph that shows the relationship between two variables, for example, pump output at various pressures or speeds.

characteristic impedance (Electr) The ratio of the voltage to the current at every point along a transmission line on which there are no standing waves.

charco (For Serv) A depression, either natural or constructed, used to catch and hold water.

charcoal (Techn) A form of amorphous carbon remaining from incomplete combustion of animal or vegetable matter.

charge[1] (Electr) a. An excess or deficiency of electrons on a body. b. To replenish or restore a secondary battery by applying a suitable direct current in the reverse direction to its normal operation.

charge[2] (Fire Pr) a. The filling in a chemical fire extinguisher. b. To pressurize an extinguisher or hose line.

charged (Fire Pr) 1. The state of a building filled with dense smoke and hot gases and in danger of becoming seriously involved in fire. 2. = (Brit) logged.

charged line (Fire Pr) A filled hose line under pressure.

charged particle (Phys) Any small particle that has a charge, often a positively or negatively charged elementary particle.

charge number (Chem; Nucl Phys) The electric charge of a nucleus, which is proportional to the number of protons. For example, the charge on a deuteron is 1; the charge on a uranium nucleus is 92. Syn: atomic number.

charges (Law) A bill of particulars alleging violation of the law, official regulations, or standards of good conduct.

char length (Comb) The linear measure of burned material of a specimen exposed to a flame. Used in flammability tests.

Charles' law (Phys) A thermodynamic law for gases: at constant pressure, the volume of a given quantity of gas varies directly with temperature (conversely, at constant volume, the pressure varies directly with temperature). The coefficient of expansion derived from this law is approximately 0.003663 (1/273) at 0°C. Also called: ideal gas law; Gay-Lussac's law.

checking station (For Serv) A checkpoint on a main road where traffic is stopped and travelers are instructed in fire-prevention measures. Also called: registration station.

check valve (Hydraul) A valve that permits flow in only one direction. See: clapper valve. Similar to a clack valve (Brit).

check viewer (Mining) A safety inspector who observes activities to detect unsafe working conditions and violations of state and local mining regulations or of contractual agreements.

chemical[1] (Chem) Any element or compound.

chemical[2] (Fire Pr) Referring to a chemical fire extinguisher.

chemical burn (Med) An injury, similar to a thermal burn, caused by a chemical.
 NOTE: After emergency first aid, usu flooding with water for 20 minutes, treatment is often the same as that for thermal burns.

chemical cartridge (Safety) A type of absorption unit used with a respirator to remove low concentrations of solvent vapors and certain gases. Cf: canister; cartridge.

chemical change (Chem) An alteration in the composition of a substance by reaction, e.g., the rusting of iron in air, the combustion of hydrogen in oxygen to produce water.

chemical compound (Chem) A substance composed of definite proportions by weight of two or more chemically united elements and having properties different from those of its constituents. Cf: mixture.

chemical dosimeter (Dosim) A detector that indicates the extent of exposure to radiation by a proportional chemical change, usu through a darkening in color of the radiation-sensitive element. See: film badge; ionizing chamber; dosimeter.

chemical energy (Chem) Energy that is stored in matter and released by a chemical reaction, such as by the combustion of a fuel.

chemical engineering (Chem Eng) The branch of engineering concerned with the production of bulk products from basic raw materials through chemical processes by applying the principles of physics, chemistry, mathematics, and engineering.

chemical equation (Chem) A qualitative and quantitative expression of a chemical change, using chemical notation. Example: $H_2 + Cl_2 \rightarrow 2\,HCl$, meaning two atoms of hydrogen and two atoms of chlorine react to produce two molecules of hydrogen chloride.

chemical extinguisher (Fire Pr) Any extinguisher using a chemical reaction to expel its contents; most commonly a bicarbonate of soda, water, sulfuric acid hand extinguisher. See: fire extinguisher.

chemical foam (Fire Pr) An aqueous froth of CO_2 bubbles used to smother flammable liquids fires; generally produced by reaction between solutions of a carbonate and an acid salt (e.g., bicarbonate of soda and aluminum sulfate), with addition of a stabilizer to promote foaming. See: foam.

chemical inhibition See: inhibition

chemical kinetics (Chem) The branch of chemistry concerned with reaction rates.

chemical line (Fire Pr) A small rubber hose supplied from a fire truck chemical tank.

chemical plant (Chem Eng) A large integrated plant or a portion of a plant, other than a refinery or distillery, where flammable or combustible liquids are produced by or used in chemical reactions.

chemical process industry (Chem Eng) Any manufacturing enterprise that uses chemical reaction at any point in its manufacturing activity.

Contrasted to purely mechanical industries, which merely change the shape and size of materials and assemble them into marketable products. Typical chemical process industries are chemicals and allied products, paper, petroleum, coal products, and glassmaking.

chemical products (Chem Eng) The products of the chemical industry, including basic chemicals and products manufactured chiefly by chemical processes. The three general classes of products are 1) primary chemicals, such as acids, alkalis, salts, and organic chemicals; 2) chemicals to be used in making other products, such as synthetic fibers, plastics, dry colors, and pigments; 3) finished chemical products for ultimate consumption, such as drugs, cosmetics, soaps and detergents, or materials and supplies for other industries, such as paints, fertilizers, pesticides, and explosives.

chemical reaction (Chem) A change in the arrangement of atoms or molecules to yield substances of different composition and properties. Common types of reaction are combination, decomposition, double decomposition, replacement, and double replacement.

chemical reaction rate See: reaction rate

chemical senses (Physiol) The senses of taste and smell, so-called because the stimulus for each is chemical in nature, liquid for taste and gaseous for smell. The two senses interact strongly, to the extent that a taste sensation may be entirely due to smell. The sense of smell is generally much more keen and discriminatory than taste. See: chemoreception; sense organs; smell; taste.

chemical symbols (Chem) A system of notation for chemical elements, their compounds, and reactions.

chemical terminology (Chem) A systematic language for uniquely classifying and naming chemical compounds.

chemi-ionization (Chem) Nonequilibrium ionization produced by reaction, esp in flames.

chemisorption (Chem) The chemical bonding of molecules to surfaces. See also: adsorption.

chemistry (Abb: Chem) The science of matter (its composition, structure, and properties) and the changes that it undergoes, often categorized arbitrarily into five major divisions: 1) inorganic, 2) organic, 3) analytical, 4) physical, and 5) biochemistry. Over 30 branches of chemistry have been recognized, such as pharmaceutical, nuclear, industrial, colloid, and electrochemistry.

chemoreception (Physiol) The response of an organism to a change in the chemical environment, principally with respect to smell and taste. See also: chemical senses.

chemotherapy (Med) Prevention or treatment of disease and infection by chemical agents, such as antibiotics and sulfa drugs.

CHEMTREC = Chemical Transportation Emergency Center, an advisory information center for transportation emergencies involving chemicals, operated by the Manufacturing Chemists' Association. Cf: TEAP.
 NOTE: Assistance in identifying hazardous materials and guidance on how to handle them in case of accident or other emergency may be obtained day or night by calling 800-424-9300.

chemurgy (Chem Eng) The branch of applied chemistry devoted to industrial use of organic raw materials, esp agricultural products, as in the use of pine-tree cellulose for rayon and paper, and soybean oil for paints and varnishes.

cherrypicker = elevating platform

chert (Mineral) A microcrystalline form of silica. An impure form of flint used in abrasives.

Cheyne-Stokes respiration (Med) The peculiar kind of breathing sometimes observed with unconscious patients, who seem to stop breathing altogether for 5 to 40 seconds, start again with gradually increasing intensity, stop breathing once more, and then repeat the performance.

chicken ladder = crawling board

chief (Fire Pr) A fire service officer above company rank, such as department, assistant, deputy, battalion, and district chief. Also called: brass (Colloq). See: ranks and titles.

chief engineer (Fire Pr) A chief officer responsible for the direction of department engineers and care of engines. In some jurisdictions the title for the chief of the fire department. See: ranks and titles.

chief of department (Fire Pr) The highest ranking fire-fighting officer of a department, and often the chief administrative officer as well. Sometimes called: chief engineer. Cf: commissioner. See: ranks and titles.

chill (Physiol) A sudden feeling of cold, accompanied by shivering and chattering of the teeth. See also: wind chill index.

chill space (Storage) A refrigerated warehouse area in which the temperature can be controlled between 32° and 50° F.

chill wind factor = wind chill factor

chimney (Build) A vertical, hollow, open-ended structure used to convey combustion products from a combustion chamber.

chimney effect = stack effect

chimney rods (Fire Pr) Jointed rods used to extend the nozzle of a hand pump when dealing with chimney fires.

chipper (Techn) A machine that cuts material into chips.

chi-square test (Statist) A statistical test used to determine whether a distribution differs significantly from theoretical and is therefore affected by factors other than chance.

CHLOREP = Chlorine Emergency Plan, operated by the Chlorine Institute, provides assistance in emergencies involving chlorine.

chlorine (Chem; Symb: Cl) A greenish, gaseous, chemically active element, Cl_2; at. wt 35.5, bp -34.15°C, mp -100.98°C, widely used for sanitation, laundry, and water purification products.
NOTE: Severely irritating to nose, throat, and lung tissues; gas masks are required. Chlorine can oxidize fuels in a manner similar to oxygen. Iron, for example, will burn in a chlorine atmosphere, and the gas forms explosive mixtures with hydrogen.

chlorobromomethane (Chem; Abb: CBM) A vaporizing liquid chemical compound, CH_2ClBr, used as an extinguishing agent for flammable liquids and electrical fires. See: Halon.

chlorotetracyclin (Pharm) A broad-spectrum antibiotic occurring as yellow crystals, $C_{22}H_{23}ClN_2O_8$. Produced by soil bacteria. Also called: Aureomycin.

chock block (Transp) A concave or mitered blocking piece used to secure objects in position. Also called: bunk block; cheese block.

chocks = wheel blocks

choke[1] (Electr) A coil that impedes the flow of alternating current of a specified frequency range because of its high inductive reactance at that range.

choke[2] (Mech Eng) a. To restrict flow. b. A device that restricts flow, usu of a fluid.

choke[3] (Med) To block or obstruct breathing mechanically by compressing or clogging the windpipe or chemically by contaminating the air with toxic or noxious fumes. Mechanical asphyxiation.

chokedamp (Safety) Unbreathable gas that gathers in unventilated parts of ships, such as tanks and bilges. See: damp.

cholesterol (Med) A lipid present in animal tissue, egg yolks, and oils. A precursor of atherosclerosis of the coronary arteries.

chopper[1] (Phys) A rotating mechanical shutter or electronic device for interrupting an otherwise continuous stream of particles or current.

chopper[2] (Colloq) A helicopter.

chowder fireman (Fire Pr; Colloq) An inactive or honorary fireman.

chromate (Chem) Any of the salts or esters of chromic acid. Used as oxidizers and catalysts. Caustic to mucous membranes and skin, causing burns or ulcers. The salt crystals are often highly colored.

chromatic aberration See: **aberration**

chromaticity (Opt) The color quality of light, based on the attributes of hue and saturation in color vision, taken together.

chromatography (Chem) A procedure used to fractionate a substance for the purpose of analyzing its constituents. Originally, a method in which a solution of a test sample is percolated through a column of porous material which separates and collects it in layers in the column as more solvent is passed through.
NOTE: The method has been extended to colorless materials. Paper, gas, and thin-layer chromatography are other methods based on the same principles.

chromosome (Biol) A rod-shaped constituent of biological cells. Chromosomes contain the genes or heredity-determining units.

chronic (Med) Persistent, prolonged, or recurring.

chronic poisoning (Med; Toxicol) The result of repeated inhalation, absorption, or ingestion of minute quantities of a cumulative poison.

chronic violator (Law) An individual who repeatedly violates established statutes and ordinances, such as traffic laws or posted rules and regulations in a particular business or industrial enterprise. Also called: **persistent violator**.

chronograph (Horol) A laboratory instrument for recording time intervals on a chart mounted on a rotating drum. A time standard that periodically records a reference mark on the chart.

chronometer (Horol) A high-precision clock.

chronoscope (Horol) An electronic instrument that measures short time intervals.

chuck (Fire Pr) A portable hydrant with gated connectors for **flush hydrant** connections. See also: **coffee pot chuck**.

church raise (Fire Pr) Raising a ladder without resting the fly against a building, support being provided by guy lines. Also called: **auditorium raise**. See: **raise**.

churn valve (Fire Pr) A pressure relief valve for bypassing water from the discharge to the supply side of a pump.

chute[1] (Fire Pr) a. A slide-type fire or aircraft crash escape. Syn: **slide escape**. b. = **water chute**.
NOTE: Chutes are generally enclosed in the form of sleeves or tubes; slides are generally open.

chute[2] = **parachute**

cilia (Physiol) Tiny, hair-like organs in the bronchi and other respiratory passages that aid in the removal of dust, debris, soot, or bacteria trapped on the moist surfaces.

ciliary epithelium (Physiol) Cells lining body airways, equipped with hairs or cilia that beat outward, tending to expel foreign matter from air passages.

cinder (Comb) A burned or partially burned fuel fragment.

cinematography (Photog) A technique for taking still pictures of motion in rapid succession and reproducing the motion by projecting the pictures at the same rate on a screen to create the illusion of continuous movement.

circle (Geom) The closed curve formed by points equally distant from a fixed point in a plane. The line bounding the circle is the **circumference**, a line from edge to edge through the center is the diameter, and a line from the center to the edge is the radius. The circumference is 2pi•r, where pi = 3.1416 and r is the radius; the area is pi•r², or 0.7854•D², where D is the diameter.

circuit¹ (Electr) Any electrical network of conducting components through which electrical energy is intended to flow.

circuit² (Fire Pr) An alarm or other signaling channel of a fire alarm system.

circuit breaker (Electr) A protective device for electrical circuits designed to open the circuit when an overload occurs. Breakers are usu designed to permit opening and closing of circuits manually but to disconnect power automatically. See: fuse.

circulating main (Fire Pr) A water main supplied from more than one direction.

circulating valve (Fire Pr) A valve in a fire pump that allows water to be circulated from the pump to the booster tank to keep the pump cool when the hose lines are shut down.

circulatory system (Anat) The complex system of the body consisting of the cardiovascular system and the lymphatic system. The former transports the blood from the heart to the tissues and back to the heart. During the circuit the blood distributes nutrients and oxygen and eliminates wastes. The lymph circulates through one-way vessels and is propelled by muscular and visceral movements carrying invading bacteria to the lymph nodes and restoring excess tissue fluid to the vein system. See also: arteries; heart; veins.

circumference (Geom) The perimeter of a closed curve. See: circle.

circumstance (Safety) Any condition or action accompanying or associated with an accident.

circumstantial evidence (Law) Indirect evidence that tends to support or prove a fact or matter in question.

cirrocumulus (Meteor) A combination of heaped and streaked clouds; classification Ce. See: cloud.

cirrostratus (Meteor) A type of layered or sheet cirrus. See: cloud.

cirrus (Meteor) High, streaked clouds with diverging filaments resembling hair or wool. One of the three basic cloud formations; classification Ci. See: cloud.

cistern (Fire Pr) A water storage container used for emergency water supply and fire protection, esp in rural areas. Cisterns are usu constructed below ground level.

citation¹ (Law) A summons to appear in court or government office. usu for a hearing.

citation² A formal statement in which an individual is recognized for outstanding achievement or meritorious service.

civil defense (Abb: Civ Def) A civilian preparedness program for protecting the population and rendering emergency relief in the event of war or natural disaster.

civil law (Law) The code of law established by a nation or state for the protection of civil and private rights.

civil service (Govt) The public administrative service of a government, excluding armed forces; esp one in which appointments and promotions are regulated by competitive examinations.

clack valve See: check valve

cladding (Metall; Techn) The bonding of a relatively thin facing sheet of metal (stainless steel, for example) to a slab of basis metal (such as copper) for the purpose of obtaining certain mechanical or wear characteristics from the sandwich-like structure.

claim (Insur) The amount that a policyholder believes and declares is due from an insurance company as the result of some insured loss or injury.

claimant (Insur) One who presents a claim.

clamping piece (Techn) A blocking part used on machinery or vehicles to secure movable parts.

clamshell bucket (Mat Hand) A container used to grasp, hoist, and convey materials. A device with two jaws that clamp together when the container is lifted by specially attached cables.

Clapeyron-Clausius equation (Phys) A differential equation relating pressure and temperature in a system in which two phases of a substance are in equilibrium.

clapper valve (Fire Pr) 1. A hinged valve that limits the flow of water to one direction. 2. A check valve in a siamese connection that prevents the escape of water through an unused inlet.

classes of alarm systems See: alarm system

classes of extinguisher See: fire extinguisher

classes of fire See: fire classes

classes of forest fire See: forest fire classes

classes of pumper See: pumper

classes of wall openings See: wall openings

class interval (Statist) The range of scores between two values in a frequency distribution.

clavicle (Anat) The collarbone.

claw tool (Fire Pr) A tool with a hook on one end and a claw on the other, used for forcible entry. A kind of wrecking bar. See: forcible entry tool.

clay (Mineral) Any of a variety of aluminum-silicate rocks that are plastic when wet and hard when dry. Used in pottery, stoneware, tile, bricks, cements, fillers, and abrasives. Some clays may contain appreciable quantities of quartz.

clean air (Env Prot) Air suitable for breathing.

clean burn (For Serv) A backfire that completely consumes the fuel between a fireline and a wildland fire. Syn: burn out.

clean burning See: burn out

clearing up (Fire Pr; Brit) = mop-up

cleat¹ (Fire Pr) A rectangular ladder cross piece placed on edge and used as a foot hold. Also called: rung.

cleat² (Techn) A piece of material, such as wood or metal, attached to a structural body to secure and strengthen it or to furnish a grip.

cleat³ (Transp) Wooden pieces nailed to a floor to reinforce blocking and to secure lading in position.

climate (Meteor) The average weather conditions that characterize a geographic region.

climatology (Meteor) The branch of meteorology that deals with long-term physical states of the atmosphere in specified geographical areas, including averages and variations in atmospheric conditions.

clinch (Build) To bend or turn down the protruding point of a driven nail to make it hold fast.

clinical (Med) Referring to services involving a medical complex such as a clinic or hospital as opposed to those performed by a single physician.

clinical death (Med) The state that occurs when the heart stops.

clinical pathology (Med; Path) The diagnostic evaluation of disease by means of laboratory tests on samples of body fluids, secretions, excretions, exudates, and tissues obtained from living patients.
 NOTE: The most common analyses are blood tests, urinalyses, microbiological studies, and metabolic studies. Small samples of suspicious tumors, taken by biopsy, are examined for cancer or other diseases.

clinical thermometer (Med) An instrument for measuring body temperature.

clinker (Metall) A hard, sintered, or fused piece of residue formed in a fire by the agglomeration of ash, metals, glass, and ceramics. The rock-like product resulting from calcining pulverized stone in a kiln.

clinometer (Surv) An instrument for measuring vertical angles, used principally to determine angles of slope. See: **inclinometer**.

clo (Metrol) A unit measure of the thermal insulation of clothing. The insulation required to maintain a person at rest at a comfortable temperature under normal indoor conditions. One clo is equal to 0.875 ft hr °F/Btu or 0.506 m°C/watts.

closed area (Fire Pr) An area in which burning or other dangerous activities are banned or to which access is restricted because of fire hazard.

closed circuit See: **alarm system**

closed container (Techn) A container sealed by a lid, cap, or other device in such a manner that liquid or vapor cannot escape at ordinary temperatures.

closed-loop system (Autom) A system in which the output, or some result of the output, is fed back for comparison with the input, for the purpose of reducing any difference occurring between input and output.

closing device (Fire Pr) A mechanism, usu automatic, that closes an open fire door and allows it to latch in the event of fire. Also called: **automatic closing device**; **self-closing device**.

closure[1] (Fire Pr) A flap, shutter, door, or any other device or assembly that closes an opening.

closure[2] (For Serv) A legal restriction but not necessarily complete elimination of specified activities in an area, such as camping, smoking, or entry.

cloud (Meteor) 1. A visible concentration of floating particles. 2. Fine water droplets or ice crystals suspended in the air. Particle sizes vary, but must be small enough (typically 0.1 mm) so that winds and air currents can hold them aloft. If critical size is exceeded, precipitation results. Ground level clouds are called **fog**. Clouds are named according to their shapes. The three basic cloud forms are **cirrus** (streaked clouds), **stratus** (layered clouds), and **cumulus** (heaped clouds). The ten main characteristic cloud genera are shown in the table.

International Cloud Classification

Cloud genus	Symbol	Typical height
Cirrus	Ci	High, 7-14 km
Cirrostratus	Cs	(16,500-45,000 ft)
Cirrocumulus	Cc	
Altostratus	As	Middle, 2-7 km
Altocumulus	Ac	(6500-23,000 ft)
Stratus	St	Low, <2 km
Stratocumulus	Sc	(6500 ft)
Nimbostratus	Ns	
Cumulus	Cu	Vertical development
Cumulonimbus	Cb	

cloudburst (Meteor) A sudden torrential rain.

cloud chamber (Phys) An apparatus in which the tracks of charged particles, such as cosmic rays or particle accelerator beams, are displayed. The device consists of a glass-walled chamber filled with a supersaturated vapor, such as moist air. When charged particles pass through the chamber, they produce tracks of tiny condensed liquid droplets, much like the vapor trails of airplanes. Such tracks reveal particle motions and interactions. See: bubble chamber; spark chamber.

cloud cover See: sky condition

cluster (Fire Pr) 1. To group or categorize, esp on the basis of similarity or common characteristics. 2. A set of objects or events that may be grouped for common treatment, such as the analysis of malicious false alarms and the like.

clutch (Techn) A friction, electromagnetic, hydraulic, pneumatic, or positive-action mechanical device used to engage or disengage power in a power train.

CO See: carbon monoxide

CO_2 See: carbon dioxide

coagulase (Med) An enzyme that causes blood plasma to coagulate, often produced by pathogenic staphylococci.

coagulation[1] (Chem) Separation and solidification of dispersed suspensoid particles into clumps.

coagulation[2] (Med) The process of clotting. The settling of a gel or clotting of blood.

coalesce To gather and unite into a whole; to fuse; to grow together.

coal gas (Techn) A combustible gas obtained from destructive distillation of coal, consisting of carbon monoxide, hydrogen, carbon dioxide, olefines, oxygen, and nitrogen. Used as a fuel. See: fuel gas.

coal oil = kerosine

coaming (Navig) A vertical framework surrounding a hatch or cargo well serving to prevent water from flowing into the opening and to strengthen the edge.

coanda effect (Aerodyn) The tendency of a gas jet discharged near a surface to follow the surface.

coarse fuel = heavy fuel

coastal inversion = marine inversion. See: inversion.

cocaine (Pharm) A bitter, addictive crystalline alkaloid, $C_{17}H_{21}NO_4$, obtained from coca leaves; used as a local anesthetic.

coccus (Bact) A spherical bacterium. Plural: cocci.
 NOTE: Streptococci occur in strings, staphylococci in clusters, and diplococci in pairs.

cochlea (Anat) The main auditory part of the inner ear. The organ is spiral shaped, like a snail shell, and contains the basilar membrane on which the end organs of the auditory nerve are distributed. See: ear.

cock (Techn) A faucet or valve for regulating the flow of a liquid.

cockloft (Fire Pr) A low loft or concealed space between the top floor and roof of a building. A garret, esp a small one. See: attic.

code[1] (Info Sci) A system of characters with defined rules for their association in representing information.

code[2] (Build; Law) A set of rules and standards that have been adopted as mandatory regulations having the force and effect of law. See: building code; color code.

code of ethics (Soc Sci) A set of rules stating the ideals, moral principles, or values adopted by an individual or professional group and used as a standard to guide personal conduct.

coefficient (Math) A numerical constant used as a multiplier of a variable quantity in calculating the magnitude of the physical property represented by the variable.

coffee pot chuck (Fire Pr) A portable chuck hydrant with a single outlet, resembling a large brass coffee pot.

cofferdam (Civ Eng) A watertight structure used to expose underwater bottoms for the purpose of laying foundations or making repairs below waterlines.

coffin (Nucl Phys) A shielded shipping container for radioactive materials. Syn: cask

cognition (Psych) Knowing, including perceiving, imagining, reasoning, and judging.

COHb = carboxyhemoglobin

coherence (Phys) Phase coincidence of electromagnetic radiation.

cohesion (Phys) The forces of attraction between two unlike surfaces. Cf: adhesion.

cohesionless soil (Soil Mech) A soil that, when unconfined, has little or no strength when air-dried and insignificant cohesion when submerged.

cohesive soil (Soil Mech) A soil that, when unconfined, has considerable strength when air-dried and considerable cohesion when submerged.

coincidence counting (Nucl Phys) A method for detecting or identifying radioactive materials and for calibrating their disintegration rates by counting events that occur together or in a specific time relationship to each other. See: counter.
NOTE: This method is important in activation analysis, medical scanning, cosmic ray studies, and low-level measurements.

coinsurance (Insur) Secondary insurance carried on a property in an amount equal to a specified portion of the value of the insured property.

coir (Techn) A coarse, natural fiber from coconut husks used for stuffing in furniture, cordage, sailcloth, and matting. Coir rope floats on water. See: grass line.

coke oven chemicals (Chem Eng) Organic compounds derived from bituminous coal during its conversion to metallurgical coke. A major source of raw materials for many chemicals.

cold drawing (Techn) An industrial process for making rods, wires, or fibers from a billet by stretching at or near room temperature.

cold front (Meteor) A boundary between air masses of different temperature at which the cooler air is displacing the warmer air. See: meteorology.

cold smoke (Fire Pr) Smoke generated by a smoldering fire that does not generate enough heat to build up to flaming combustion.
NOTE: Cold smoke can build up pressure inside a closed compartment. When the compartment is opened, the smoke surges out in what is called a cold smoke explosion, and in doing so sometimes blows out the fire.

cold trailing (For Serv) A method of checking a partly dead fire edge by feeling for fire with the hand, digging out every live spot, and trenching any live edge found. See: feeling for fire.

cold work (Metall) Any work that does not involve hot riveting, welding, burning, or other heat-producing operations.

colic (Med) A severe, cramping pain in or near the abdomen.

collapse See: floor collapse

collapsed hose (Fire Pr) Hose in which the lining has separated from the jacket.

collar (Mining) 1. The near or open end of a borehole. 2. The timbering and concrete work around the mouth of a shaft.

collateral circulation (Physiol) A system of accessory blood vessels, often referring to the small arterioles (little arteries) and capillaries supplying an area beyond an obstruction in an artery.

collateral points See: cardinal points

collecting breeching See: breeching

collecting head (Fire Pr) A threaded casting with two or more inlet connections and one outlet connection, used to join one or more lines of hose to the inlet of a pump. Also called: suction collecting head. Cf:

delivery head.

collective protection (Safety) The protection of an area or personnel from toxic agents without the use of individual protection equipment.

collector (Techn) A device for removing particulates from a gas stream.
NOTE: Bag-type: A filter in which the filtering medium is a cylindrical fabric bag. Cyclone: A collector in which an inlet gas stream is made to move vertically; its centrifugal forces tend to drive suspended particles to the wall of the cyclone. Mechanical: A device in which inertial and gravitational forces separate dry dust from gas. Multicyclone: A dust collector consisting of a number of cyclone collectors that operate in parallel; the volume and velocity of combustion gas can be regulated by dampers to maintain efficiency over a given load range.

collector pumping (Fire Pr) A method of increasing the volume of water by supplying water from several pumps to the collecting head of a single pump, which then delivers the water to the fire.

collimated beam (Opt) A parallel beam of light.

collimator (Opt) 1. An optical lens arrangement for producing parallel rays of light, used in various instruments such as spectroscopes, and for aligning optical systems. 2. Any arrangement of apertures used to limit the flow of particles to a directional parallel beam.

collision (Nucl Phys) A close approach of two or more particles, photons, atoms, or nuclei, during which energy, momentum, and charge may be exchanged.

collision diagram (Safety) A location and event diagram of where a strike-type accident occurred. The diagram shows, by the use of designated symbols, the manner of collision and the resting positions of persons and objects after the collision.

colloid[1] (Chem) A dispersion in a fluid of particles so small that surface forces are dominant (usu 10^{-4} to 10^{-7} cm). Though colloids are relatively stable, they are not true solutions, in which the dispersion is on the molecular level. See also: suspension; solution.
NOTE: The term means "glue-like."

colloid[2] (Med) a. A uniform, jelly-like substance occurring in certain diseases. b. Loosely, the protein fraction of blood plasma.

colloid chemistry (Chem) The branch of chemistry dealing with colloids, including sols, aerosols, emulsions, foams, gels, as well as thin films and fibers.

colloid mill (Techn) A machine that grinds materials into extremely fine powder, often simultaneously placing it in suspension in a liquid.

colony (Bact) A visible growth of microorganisms on a solid culture medium.

color code (Ind Eng) A system of coloring piping, wiring, and parts of equipment with various coded colors for identification purposes.

colorimetry (Chem) A quantitative method of measuring the amount of substance in a solution from the intensity of its color.

color point (Comb) The time at which a polymer surface exposed to heat begins to show a change in color or a change in surface characteristics, indicating the beginning of thermal decomposition.

color temperature (Phys) The temperature of a black body radiating energy of the same spectral distribution as that of the surface in question.

column (Build) A vertical load-bearing structural member.

coma (Med) A state of deep unconsciousness from which a patient cannot be aroused. The state can be brought on by a drug, poison, brain injury, or a disease.

combination (Math) A sample of k objects selected from a population of n objects, without regard to order. The number of ways the sample k can be selected is equal to n!/k! (n - k)!, where n! and k! are fac-

torial; e.g., if n = 5, n! = 5 x 4 x 3 x 2 x 1 = 120. See also: <u>distribution</u>; <u>permutation</u>.
  NOTE: As an example, 3 objects (k) can be selected from a population of 5 objects (n) in (5 x 4 x 3 x 2 x 1)/((3 x 2 x 1)(2 x 1)) = 120/12 = 10 ways.

<u>combination apparatus</u> (Fire Pr) A fire apparatus designed to peform two or more functions. Originally a hose wagon and a chemical engine. Later, the addition of an engine-driven pump made a <u>triple combination</u>, ladders a <u>quadruple</u> combination (quad), and addition of an aerial ladder made a "quint."

<u>combination fire department</u> See: <u>fire department</u>

<u>combination nozzle</u> (Fire Pr) An all-purpose nozzle designed to provide a solid stream or a fixed spray pattern.

<u>combined agent unit</u> (Fire Pr) A system supplying two fire-fighting agents, aqueous film-forming foam and potassium bicarbonate dry chemical powder, to two separate nozzles mounted on a common support. Each nozzle is separately controlled. The dry chemical extinguishes the flames and the foam cools the fuel. Also called: <u>twinned agent unit</u>.

<u>combining number</u> (Chem) A small whole number indicating the number of hydrogen atoms that combine with a given atom. Oxygen, for example, has a combining number of 2 in H_2O (water); carbon in CH_4 (methane) has a combining number of 4. Also called: <u>valence number</u>.

<u>combining weight</u> (Chem) The number of grams of an element that combines with or replaces 8 grams of oxygen or its equivalent. Also defined as the <u>molecular weight</u> divided by the <u>valence</u>. Syn: <u>equivalent weight</u>.
  NOTE: The commonly accepted standard is 7.9997 grams of oxygen, which combines with 1.00797 grams (equivalent weight) of hydrogen. No element has a combining weight less than 1.

<u>combustibility</u> (Comb; Transp) The readiness of a substance or material to burn.
  NOTE: Several categories of combustibility are distinguished. <u>Hazardous</u>: Materials that, either by themselves or in combination with their packaging, are highly susceptible to ignition and will contribute to intense and rapid spread of fire. <u>Low combustibility</u>: Materials that in themselves do not normally ignite but which in combination with their packaging will contribute fuel to fire. <u>Moderate combustibility</u>: Materials and their packaging, both of which will contribute fuel to fire. <u>Noncombustibility</u>: Materials and their packaging that will neither ignite nor support combustion under ordinary circumstances.

<u>combustible</u>[1] (Comb) Capable of burning, generally in air under normal conditions of ambient temperature and pressure, unless otherwise specified.

<u>combustible</u>[2] (Fire Pr) Any material or object that can burn more or less readily under ordinary, everyday conditions. Ant: <u>noncombustible</u>.
  NOTE: Combustible is a relative term. Wood, for example, is classified as combustible, but waterlogged wood is markedly less combustible than dry wood. Similarly, a massive wooden timber is less combustible than wood shavings. Most substances will burn given suitable conditions, and, in certain instances, without oxygen. Iron, copper, or zinc, for example, will burn in chlorine. Calcium will "burn" in a nitrogen atmosphere, and zirconium dust in carbon dioxide. Oxides such as sodium nitrate and potassium chlorate generate oxygen for combustion.

<u>combustible gas indicator</u> (Safety) An instrument that samples air and indicates whether there is an explosive mixture present and the percentage of the lower explosive limit of the air-gas mixture that has been reached.

<u>combustible liquid</u> (Fire Pr) A liquid having a flash point at or above 37.8°C (100°F). A Class II or III liquid according to the flash point classification scheme. A Class II liquid has a flash point at or above 37.8°C but below 60°C (100°F and 140°F). Class IIIA includes those having flash points at or above

140°F but below 200°F. When liquids with a flash point at or above 140°F are heated to or above their flash point, they may acquire characteristics of a flammable liquid. Class IIIB includes liquids having flash points at or above 200°F. See: flash point; flammable liquid.
  NOTE: The US Coast Guard defines a combustible liquid as one having a flash point above 80°F and distinguishes two grades: D, any liquid with a flash point between 80 and 150°F; and E, a flash point above 150°F.

combustion (Abb: Comb) A steady-state exothermic, self-catalyzed chemical reaction with the characteristic ability to propagate through a combustible medium, usu a fuel and an oxidizer. These reactants can occur in gas, liquid, or solid phase, alone, or in any combination. The reaction produces heat, generally but not necessarily light, and a variety of products in the form of gases, vapors, particulates (smoke), and residues (ashes). See also: fire.
  NOTE: The most common oxidizer is atmospheric oxygen. The initiation of the process is called ignition; the terminal phase is called extinction. Both of these are transient manifestations of the combustion process. Condensed-phase combustion is usu referred to as glowing combustion, while gas-phase combustion is referred to as a flame. If the process is confined so that an appreciable pressure rise occurs, it is called an explosion. If the combustion wave propagates at supersonic speed, a shock front develops ahead of it, and the process is called detonation. Combustion involves strong interactions between chemical reactions and physical transport processes (diffusion, thermal conduction, thermal diffusion, and in the case of detonation, viscosity as well). This strong interaction is responsible for the unique properties of combustion reactions: self propagation, nonequilibrium atom and radical concentrations in the reaction zone, and the emission of nonthermal light.

combustion air (Comb) Air used for burning a fuel.

combustion chamber (Techn) An enclosed space in which controlled combustion takes place; e.g., a furnace, firebox, internal combustion engine cylinder, etc.

combustion engineering (Comb) The science and technology of burning fuels in various types of combustion chambers, including combustion chemistry, and the physics of heat transfer and fluid flow.

combustion gases (Comb) The mixture of gases and vapors produced by combustion.

combustion products (Comb) Gases, solid and liquid particulates, and residues evolved or remaining from a combustion process. See: smoke.
  NOTE: In general, the products include fire gases, flames, heat, smoke, ashes, and other residues.

combustion suppression (Fire Pr) The process of controlling and halting combustion.

combustion toxicology (Med) The study of the harmful effects on biological systems of the products of combustion. Cf: smoke toxicology.

combustion wave (Comb) The temperature and compositional microstructure associated with a propagating flame. Cf: flame front.

comfort zone (Physiol) The range of temperatures in which most people feel comfortable.

command (Autom) An input variable, such as a pulse, signal, or set of signals, generated to initiate a specific operation.

command and control (Fire Pr) The activity of directing an operation, such as fire suppression, and keeping a running account of the deployment of individual units involved in the operation.

command channel (Fire Pr) A radio channel or frequency reserved for use by chief officers.

command post (Fire Pr) 1. The position of the officer in charge on the fireground. 2. = (Brit) control point.

comminuted fracture See: fracture

commercial credit (Insur) A guarantee of payment to manufacturers and sellers for goods shipped and services rendered. A guarantee of working capital represented by accounts receivable.

commissioner (Fire Pr) A civilian head and highest ranking official of a fire department. Distinguished from the chief of department, who heads the fire-fighting force. See: ranks and titles.

common hazard (Fire Pr) Potential causes of fire, such as smoking, defective electrical circuits, heating equipment, etc.; distinguished from a special hazard, which is unique to a given industry.

common law (Law) A body of laws based on custom, usage, and the decisions of law courts. Contrasted with statute laws, which are enacted by legislative bodies. Cf: case law.

communicable disease (Med) A disease that has a causative agent which is readily transferred from one person to another.

communicated fire = spot fire

communication (Abb: Commun) The science and technology of transmitting information from one point or person to another.

communications center (Fire Pr) The facilities in a building equipped with dispatch and communications devices for use as a command center for firefighting operations.

communications officer (For Serv) The individual responsible for the installation, operation, and maintenance of communications during fire suppression activities. The communications officer usu reports to the service chief and supervises all communications personnel, including messengers.

compaction[1] (Fire Pr) Increase in the bulk density of the powder in a dry chemical extinguisher through vibration or impact (e.g., by dropping) that may adversely affect the quality of discharge from the extinguisher. Syn: packing.

compaction[2] (Soil Mech) The densification of a soil by mechanical means.

compactness of fuel (Fire Pr) Density of fuel arrangement. A stack of logs is tightly arranged; scattered combustibles are loosely arranged. See: arrangement of fuel. See also: connectedness of fuel.

company (Fire Pr) The basic unit of a fire-fighting organization. Normally led by a captain with one or more lieutenants and grades of firefighters who man various types of pumpers, ladders, and special equipment. A company may have several platoons, and several companies may form a battalion.

company journal See: journal

company officer (Fire Pr) A captain, lieutenant, and sometimes a sergeant in command of a fire company or platoon. Distinguished from a chief officer, who commands more than one fire-fighting unit. See: ranks and titles.

compartment (Build) Any enclosed space or portion of a building, such as a room, storage area, or a private or semiprivate accommodation; sometimes referring to passenger quarters in public conveyances.

compartmentation (Build) A construction design in which a building is divided into sections that can be sealed off from other parts of the building to contain fire and smoke in the area of origin and prevent their spread to adjacent compartments. See: fire area.

compass[1] (Geom) An instrument for drawing circles.

compass[2] (Navig) An instrument with a north-seeking pointer for determining directions.

compatibility (Storage) Ability of materials to be stored intimately without reacting chemically. Incompatibility may result in loss of effectiveness, or may be hazardous.

compensable injury (Insur; Safety) An

occupational or work injury for which benefits are payable to the injured worker or his beneficiary under workmen's compensation laws.

compensating eyepiece (Opt) An eyepiece having slight overcorrection to compensate for the chromatic difference of magnification provided by an apochromatic objective.

compensation (Insur) Indemnity paid to an employee for disability sustained in an occupational accident.

competence (Law) The state of being responsible for one's own actions.

competent[1] (Mining) Capable of holding under load. Descriptive of firm ground. Ant: running ground.

competent[2] (Law) Legally qualified to serve as an expert witness.

competent person (Safety) An individual capable of recognizing and evaluating exposure to hazardous substances or to other unsafe conditions and qualified to specify the necessary protection and precautions to be taken.

complement[1] (Fire Pr) a. The men assigned to a working shift of a fire-fighting unit. b. The units, as a whole, responding to a given alarm.

complement[2] (Med) A heat-sensitive complex in blood serum and lymph; ineffective alone, but which with other substances serves as an important defense mechanism against bacterial infection, destroying bacteria, and effective with other substances against foreign blood cells.

complex vibration (Mech) Any vibration that is not a pure sinusoid and whose sinusoidal components are not harmonically related to each other.

compliance[1] (Build; Law) Adherance or conformance to laws and standards.

compliance[2] (Build) The ease with which a material can be flexed. Ant: stiffness.

compo-board (Brit) = wallboard; plasterboard.

component[1] (Techn) An essential part or constituent of an entity or system.

component[2] (Chem) One of the minimum number of chemical entities required to describe a system.

compound (Chem) A chemical combination of two or more elements in definite proportions by weight, such that each molecule of the compound contains the same number of atoms of each constituent element. Separation of the constituents requires chemical operations.

compound fracture (First Aid; Med) A fracture in which the broken end of a bone protrudes from the skin, or one consisting of two or more breaks, or in which there is an open wound communicating with the fracture.

compound gage (Metrol) An instrument capable of indicating pressures both above and below atmospheric. See also: gage; pressure.

compressibility (Phys) The property of a substance to decrease in volume when pressure is applied. See: equation of state.

compressional wave (Acoust) A longitudinal sound wave that squeezes (compresses) an elastic medium.

compression failure (Build) A deformation or buckling of a material or building element resulting from an excessive compression load.

compressor[1] (Fire Pr) a. An air compressor used to power forcible entry equipment. b. A fire truck carrying such equipment.

compressor[2] (Techn) A pump used to pressurize air or other gases.

comptroller = controller

computer (Comput) A machine that can solve mathematical problems and display answers. The two principal types are digital, which operate on variables and constants represented by numbers, and analog, which operate on physical quantities as analogies to variables. See also: data processing.

concealed space (Fire Pr) Space hidden between walls, ceilings, floors, and under roofs, through which fire can spread undetected. Also called: interstitial space.

concentrate See: foam concentrate

concentration (Chem; Phys; Toxicol) The quantity of a substance (by weight, equivalent units, moles, etc.) per unit volume.
  NOTE: In toxicology, toxic gas concentrations are usu expressed in terms of quantity as parts per million (10^6) or by weight per given volume (grams per 100 ml).

conceptual model (Man Sci) A schematic representation of a concept.

concrete (Build) A solid structural material consisting of cement, sand, gravel (or other aggregate) mixed with water and allowed to set and dry. See also: prestressed concrete; reinforced concrete.
  NOTE: Lightweight concrete is commonly made with vermiculite or perlite aggregates. Cinders are also used.

concurrent Acting in conjunction at the same time.

concussion (Med) A jarring injury to the brain, resulting in an impairment of its functions, which is often temporary.

condensate (Phys; Chem) A white cloud of droplets condensing from vaporized water. The steam indicates that the temperature has fallen below the boiling point of water (212°F). Also called: condensing steam.

condensation (Chem; Phys) 1. The change of state of a vapor into a liquid, brought about by cooling. The heat absorbed during evaporation or vaporization (called latent heat) is released during condensation. 2. Moisture droplets accumulating on cold surfaces in contact with humid atmosphere. See also: dew point; boiling.

condensation number (Phys) The ratio of the number of molecules condensing on a surface to the number of molecules striking the surface.

condenser[1] (Mech) A vessel or arrangement of pipe or tubing in which a vapor is liquified by removal of heat.

condenser[2] (Opt) Any lens or system of lenses designed to gather rays of light and cause them to converge at a focus.

condenser[3] = capacitor

condenser[4] (Techn) A vessel or arrangement of pipe or tubing in which a vapor is liquified by cooling.

conditional probability (Statist) The probability that an event occurs on condition that another event has already occurred.

condition of vegetation (For Serv) The stage of growth or degree of flammability of vegetation in a fuel complex. For example, distinctions usu made for the annual growth pattern in a grassy area are: green, curing, and dry (or cured).

condominium (Build) 1. A multi-unit building in which the units are individually owned. Commonly an apartment house in which the residents purchase the apartments in which they live. 2. A unit in such a building.

condual tire (Transp) A tire consisting of two tubes or carcasses of different size, the smaller fitted inside the larger.
  NOTE: The deflection permitted in this type of tire is approx double that of conventional tires and produces desirable soft-soil mobility characteristics.

conductance (Electr; Fl Mech) The property of a system that permits the flow of electricity, fluid, etc. It is the reciprocal of resistance and is defined as the ratio of the flow through a conductor to the difference of potential between its ends.

conduction[1] (Thermodyn) Transfer of heat, usu through a solid or liquid, induced by a temperature gradient. See also: heat transfer.
  NOTE: The direction of heat flow is from higher to lower temperature.

conduction² (Electr) Flow of an electrical charge from a higher potential to a lower one.

conduction deafness (Audiom; Med) An impairment of hearing due to any disorder in the outer or middle ears that prevents sound vibrations from reaching the cochlea. This type of deafness can often be corrected medically or surgically.

conductive (Phys) Capable of transferring heat or electricity.

conduit¹ (Hydraul) A natural or artificial channel that conducts a fluid. A water pipe or the like.

conduit² (Electr) A channel or pipe for enclosing electrical wires.

cone index (Soil Mech) An index of the shearing resistance of a soil obtained with the cone penetrometer, representing the resistance to penetration into the soil of a 30° cone having a 1/2 in.² base area.

confidence interval (Statist) A range of values known to contain the true value of a distribution a certain proportion of the time.
 NOTE: Confidence levels are 95 or 99%. If 99 times out of a hundred trials a value falls between, say, 40 and 43, the confidence interval is 40 to 43, the confidence limits are 40 and 43, and the confidence level is 99%.

confidence limits (Statist) The end points of a confidence interval.

configuration (Ind Eng) Arrangement of parts, form, or figure as determined by the disposition of parts, contour, or outline. An integrated whole with independent properties and functions over and above the sum of the properties and functions of its parts.

configuration management (Man Sci) A discipline applying technical and administrative direction, and surveillance to the control of the total configuration of systems, end items, and equipment.

confined space (Fire Pr) A compartment of small size and limited access such as a double-bottom tank, cofferdam, or other restricted space which, because of its small size and confining nature, can readily support or aggravate a hazardous exposure.

confinement (Fire Pr) 1. Operations undertaken to isolate fire from adjoining areas and structures. 2. Prevention of fire spread beyond the fire zone involved.

conflagration (Fire Pr; For Serv) A fire of large extent, with a moving front, involving a number of buildings on more than one block in an urban area, or a class E forest fire involving structures. In addition to extent, the fire must cross a natural or man-made barrier, such as a street, road, or waterway to qualify as a conflagration. See: fire storm; fire classes. Cf: group fire.
 NOTE: In urban conflagrations the fire spreads from building to building, and in a forest fire, from tree to tree. Spread is accelerated by wind. In extreme cases, high winds can carry burning brands a mile or more ahead of a fire front, often igniting spot fires.

confluence (Meteor) Horizontal air motion toward a line or into a zone. Ant: diffluence.

congenital (Med) Descriptive of a condition resulting from stresses during the embryonic stage of development; existing at or before birth.

congestion (Med) An abnormal accumulation of blood in the vessels of an organ or body part, often associated with edema, an abnormal accumulation of fluid in an organ or in body tissues.

conical strainer (Fire Pr) A removable wire mesh strainer in the end of a suction hose, used to prevent debris from entering the pump.

conjugate solutions (Chem) Two immiscible liquids, each saturated with the other. See also: miscibility.

conjunctivitis (Med) Inflammation of the delicate mucous membrane (conjunctiva) that lines the eyelids and covers the front of the eyeball.

connectedness of fuel (Fire Pr) The degree to which pieces of fuel are in physical contact with each other. See: fuel arrangement; fuel continuity. See also: compactness of fuel.

connective tissue (Anat) Tissues that support and join the parts of a body, including bones, cartilage, tendons and ligaments. See: tissue.

connector (Electr) A terminal or junction that serves to provide electrical continuity for two or more conductors.

consensus standard A standard developed from a common agreement or general opinion among individuals and representatives of various interested or affected organizations.

consequential loss (Brit) Indirect fire loss, other than the material destroyed or damaged, such as loss of revenues, profits, and the like.

conservation of energy (Phys) Energy in a closed system may change in form but cannot be created or destroyed and, therefore, remains constant. Cf: entropy.
  NOTE: In nuclear reactors matter is converted into energy according to the relation $E = (mc^2)/2$.

consistency (Soil Mech) The relative ease with which a soil can be deformed.

consolidation (Fire Pr) 1. Assignment of fire department duties to other agencies or assignment of nondepartment duties, such as police protection, to fire department personnel. 2. Joining of two or more small departments into a single large department.

constant (Math) A parameter that has a single numerical value. An absolute constant is a parameter that always has a single value. An arbitrary constant has a single value under given conditions, and other values in other cases.

constant danger (For Serv) Fire hazard factors normally present, such as topography, prevailing wind, type of fuel, etc., as contrasted with variable danger, such as weather.

constant error (Math) The extent of uniform divergence from a fixed value. Also called: consistent error.

constraint[1] A restriction or a compelling force affecting freedom of action. Forcing into or holding within close bounds.

constraint[2] (Safety) An operational condition that may necessitate work performance in less than an ideal safe environment.

construction See: building construction

contact dermatitis (Med) Skin eruptions or inflammation caused by contact with an irritant or a substance to which the subject is allergic.

contact testing (Testing) A technique in which a transducer is placed in contact with the surface of an object.

container (Storage) Any vessel, tank, cylinder, can, barrel, drum, or sphere used to hold a fluid. A receptacle such as a bag, barrel, drum, box, crate, or package used to hold and to protect contents.

container assembly (Techn) An assembly consisting essentially of the container and fittings for all openings, including shutoff valves, overflow valves, liquid-level gauging devices, safety relief devices, and protective housing.

container train (Transp) Small trailers, hitched in series, that are pulled by a motor vehicle; used to collect and transport materials, often solid wastes.

containment[1] (Fire Pr) Restricting the spreading of fire. Preventing the spread of fire to surrounding structures or areas.

containment[2] (Nucl Phys) Shielding around a reactor for the purpose of confining stray fission products.

contaminant (Env Prot; Ind Eng) Any substance that is likely to cause harm. An injurious, irritating, or nuisance material that is foreign to the atmosphere, material, or object.

contamination¹ (Techn) Any undesirable foreign matter that has an adverse effect on a finished article, material, or product.

contamination² (Nucl Safety) The mixing of radioactive material with part of one's environment, for example, by radioactive fallout contamination of the earth. More precisely, radioactive contamination. Ant: decontamination.

contamination³ (Med) Entry of undesirable organisms into some material or object. Ant: decontamination.
NOTE: Referring to bacteria which have fallen into a wound, but in which bacterial invasion is not yet apparent.

continuance (Law) Postponement of a legal proceeding.

continuous flow system (Safety) A breathing apparatus in which oxygen flows at a steady rate during the breathing cycle.

continuity of fuel See: fuel continuity

continuous pressure system (Safety) Pressure breathing in which a constant pressure is maintained in the mask.

contract (Law) An agreement, enforceable by law, between two or more competent parties for a legal consideration, to do or not to do something not prohibited by law.

contractual liability (Insur) 1. A liability assumed under an agreement by one person for another. 2. Liability set forth between people, as distinguished from liability imposed by law.

contracture (Med) A permanent shortening of a muscle tendon or other tissue, such as from a burn injury.

contributory negligence (Law) An act or omission amounting to lack of ordinary care on the part of the complaining party, which, together with defendant's negligence, is proximate cause of injury.

control¹ (Mech Eng) A system governing the starting, stopping, direction, acceleration, and speed of equipment.

control² (For Serv) The object of a broad program of fire prevention and suppression, as well as containment and extinguishment of working fires. Used largely in reference to Forest Service fire management activities. Equivalent to fire protection.

control a fire (For Serv) To complete a fire control line around a forest fire, including spot fires, burn out fuel between the line and the fire, and take whatever other steps are necessary to hold the control line. See: fireline. Cf: corral a fire.

control center (Fire Pr) A communications center established for the control of emergency operations at disasters, including civil defense purposes.

control force (For Serv) Personnel and equipment used to control a fire.

controlled burning (Fire Pr) 1. Burning undertaken to reduce hazards, such as undesirable undergrowth in woodlands. 2. A backfire set to remove fuel from the path of an uncontrolled forest fire. See: prescribed burning.

controlled experiment (Psych) An experiment involving two or more similar groups; one, the control group, is held as a standard for comparison, while the other, the test group or experimental group, is subjected to some procedure whose effect one wishes to determine. The groups are usu formed by randomized selection.
NOTE: Both groups must be treated in an identical manner except for the single factor under study. The most reliable controlled experiments (double blind experiments) are those in which neither the subjects nor the experimenter know which group is the test group. In a medical experiment, for example, a third party prepares and numbers pills or other medication, surrendering the identity of the numbers only after all results have been tabulated.

controlled humidity area (Storage) A

space that has been esp prepared for and equipped to control humidity.

controller[1] (Comput) An apparatus by which commands are introduced and manipulated. A program controller can compute data, encode storage, perform readouts, process computations, and produce outputs.

controller[2] (Autom) One or more devices, usu contained in a console-type enclosure, that serve to control in some predetermined manner the apparatus to which it is connected.

controller[3] (Man Sci) The chief financial and accounting officer of a business enterprise. Also called: comptroller.

control line (For Serv) A line of natural and man-made barriers and treated fire edges established for the purpose of isolating and halting a forest fire.

controlling nozzle (Fire Pr) A nozzle that can be opened, shut, or adjusted to control the pattern of the water stream.

control point (Fire Pr; Brit) The command post of the officer in charge on a fireground.

controls (Safety) Measures, including devices, to regulate a machine, apparatus, system, or action within prescribed limits or standards of safety and operational effectiveness.

control time (For Serv) Elapsed time from first work on a fire or from corral of a fire until holding of the control line is assured.

control unit (Fire Pr) A vehicle equipped with a mobile control room for use by the commanding officer at large fires. Usu equipped with radio and sometimes with field telephones.

convection (Thermodyn; Meteor) Transfer of heat by the motion of a fluid under the influence of differences in density and gravity. In fluids, convective heat transfer is generally more rapid than conduction. See also: heat transfer; Bénard convection cells.
 NOTE: Convection is commonly a vertical process. A fluid heated at the bottom expands and rises, while cooler, denser fluid above flows downward to replace it. The rate of the convective currents depends on the quantity of heat and difference in temperature. Convection is an important weather factor, associated with wind and cloud formation. Spontaneously occurring convection is called free convection or natural convection.

convection column (Meteor; Fire Pr) Warm air rising above a heat source, such as a fire. Also called: convective column.

convection currents (Meteor) Localized, thermally produced vertical motion of the atmosphere, resulting in the overturning of air layers, associated with density instability and cumulus clouds. See also: inversion; subsidence.

convergence[1] (Meteor) A flowing together of fluids in motion. Ant: divergence.

convergence[2] (Math) The property of having a finite limit.

conveyance (Transp) Any unit used to transport explosives or blasting agents, including but not limited to trucks, trailers, rail cars, barges and vessels.

conveyor (Mat Hand) A fixed or mobile powered or gravity device used to move materials between two points.

cook-off (Fire Pr) Deflagration or detonation resulting from the absorption of heat from the environment.

coolant[1] (Techn) A heat transfer agent used generally in a closed circulating system for removing heat from a high-temperature source and dissipating it from a radiator. The working fluid in refrigeration or emulsions used in metal cutting operations.

coolant[2] (Nucl Phys) A gas or liquid circulated through a nuclear reactor to remove or transfer heat. Common coolants are water, air, carbon dioxide, liquid sodium, and molten sodium-potassium alloy (NaK).

cool flame (Comb) Flames occurring in rich vapor-air mixtures of most hydrocarbons and oxygenated hydrocarbons at temperatures below normal ignition temperatures for the corresponding fuel. The chemistry involves peroxy radicals and is related to three-stage ignition. The visible emission is due to electronically excited formaldehyde.

cooling (Fire Pr) Reducing the temperature. A process by which fire can be controlled by reducing the temperature of the fuel and its immediate environment.

cooling air (Comb) Ambient air that is added to cool hot combustion gases.

cooling sprays (Env Prot; Techn) Water sprays directed into flue gases to cool them and, in most cases, to remove some fly ash.

cooperator See: fire cooperator

Cooper hose jacket See: hose jacket

coordinate system (Math) A set of reference lines used to locate points in a given two- or three-dimensional space by means of numerical quantities. The quantities are the coordinates of the point.
  NOTE: The most common system is the cartesian, which is constructed with two (x, y) or three (x, y, z) mutually perpendicular axes having a common origin. Plane coordinates with two axes are used for the construction of graphs and curves; the three-axis system is used to locate points in three-dimensional space. Other systems, such as the polar, spherical, and cylindrical coordinate systems, are used for special purposes.

copolymer (Chem) A polymer consisting of more than one species of monomer. Block and graft polymers are made up of two or more chains, each containing only one monomer.
  NOTE: A block copolymer has linear connections

A - B - A

while a graft copolymer has branching connections

A - A - A
|
B

cordite (Milit) A smokeless powder compounded of nitroglycerine and guncotton extruded in the form of a filament. see: cellulose.
  NOTE: A military explosive commonly used in World War I.

core¹ (Metall) A shaped, hard-baked cake of sand with suitable compounds that is placed within a mold, forming a cavity in the casting when it solidifies.

core² (Nucl Phys) The central part of a nuclear reactor, made up of fuel elements and a moderator, where the nuclei of the fuel fission (split) and release energy. The core is usu surrounded by a material that reflects stray neutrons back to the fuel.

core³ (Build) The concentration of stairways, elevators, utility and service shafts in a highrise building.

core⁴ (Build) The inner layer or ply of a laminated or sandwich construction, usu consisting of low quality material.

corium (Anat) The layer of skin tissue beneath the outermost epithelium, containing blood and lymph vessels, sweat glands, and nerve endings. Also called: dermis; cutis.

cornea (Physiol) A tough, sensitive, transparent membrane covering the outer portion of the eye.

cornering (Transp) Changing the direction of travel from one straight path to another.

cornering force (Transp) The force, in pounds, measured normal to the longitudinal plane of a wheel or track that is exerted at the ground-contacting area in resisting the centrifugal force developed when a vehicle moves along a nonlinear path.

corner marker (Storage; Safety) A conspicuous marker placed at aisle intersections as a caution to personnel to prevent bumping stacks, racks, bins, or other fixed objects.

<u>corner test</u> (Build; Testing) A full-scale flame spread and smoke generation test in which a simulated room corner is constructed of the material to be tested. Wing walls range from a nominal 2 to 4 feet long to as much as 50 feet in special cases. Ceilings are usu 8 feet high but can be much higher.
 NOTE: A wood crib or other fuel is ignited at floor level in the corner and the rate of flame spread over the walls and ceiling is measured.

<u>coronal discharge</u>[1] (Techn) A phenomenon used to detect voids in otherwise homogeneous dielectric material. An electrical potential ionizes materials within a void causing a burst of current that is easily detected.

<u>coronal discharge</u>[2] (Meteor) A brush-like, luminous discharge of electricity into the atmosphere from charged objects, such as from ship's masts and airplane propellers and wing tips during electrical storms. Also called: <u>St. Elmo's fire</u>.

<u>coronary</u> (Med) 1. Pertaining to the vessels of the heart. 2. Failure of heart action, commonly called <u>heart attack</u>. See also: <u>cardiac arrest</u>.

<u>coronary artery disease</u> (Med) Progressive narrowing and occlusion of the coronary arteries.

<u>coroner</u> (Govt) A local public service officer whose duty it is to look into the circumstances and causes of deaths by violence and those not due to natural causes. Coroners are usu empowered to hold inquests, call witnesses and experts to testify, and to order autopsies.
 NOTE: In some jurisdictions the coroner is not required to have medical training.

<u>corpuscle</u> (Physiol) A red or white blood cell.

<u>corral a fire</u> (For Serv) To surround a fire with a control line that ultimately becomes the line of farthest advance of the fire. Cf: <u>control a fire</u>.

<u>correction</u> (Math) A quantity added to an observed value to compensate for an error and thereby obtain the true value.

<u>corrective lens</u> (Opt) A lens ground to the wearer's individual prescription.

<u>correlation</u> (Statist) Relationship, degree of association, or index of prediction between two scores or sets of data. A measure of the tendency of one function, parameter, score or set of data to vary in relation to another; e.g., the tendency of students with a high IQ to be above average in reading ability.
 NOTE: The existence of a strong relationship (a high correlation) between two variables does not necessarily indicate that one has any causal influence on the other. Usu expressed as a decimal coefficient between -1.00 and +1.00 (<u>Pearson v</u>), where -1.00 indicates a perfect negative relationship, 0 indicates no relationship, and +1.00 indicates a perfect positive relationship. The Pearson v coefficient of correlation can assume any value on a continuum between -1.00 and +1.00.

<u>correlation coefficient</u> (Statist) A number that indicates the degree of association between two variables or between two sets of data.
 NOTE: The correlation coefficient for two variables is obtained by dividing their covariance by the product of their standard deviations.

<u>corrosion</u> (Metall; Techn) Slow deterioration or destruction of metallic materials by chemical or electrochemical processes. Contrasted with <u>erosion</u>, which is a mechanical process.

<u>corrosion preventive</u> (Techn) Any agent such as oil, plastic, paint, or wrap used for surface treatment of metals to prevent, inhibit, or deter rust, pitting, or other corrosion. Usu a compound that can be removed by water or solvent cleaners.

<u>corticosteroid</u> See: <u>steroid</u>

<u>corundum</u> (Techn) An impure form of aluminum oxide, used as an <u>abrasive</u>.

cosmic rays (Geophys) Radiation originating from high-energy particles bombarding the earth's atmosphere. Cosmic radiation is part of the natural background radiation.

cottage box (Fire Pr) A street fire alarm box with a peaked roof-like top.

cotton (Techn) 1. A natural plant fiber, consisting of more than 90% cellulose, that grows on the seeds of a plant belonging to the mallow family. 2. A fabric made from cotton fibers.

Cottrell precipitator (Techn) A device which uses high-voltage static electricity to collect dust and purify air.

couette flow (Fl Mech) The flow pattern of a liquid contained between surfaces moving at different speeds.

coulomb (Electr) The quantity of electricity carried by a current of 1 ampere in 1 second.

council rake (For Serv) A rake with wedge-shaped blades instead of tines; used for trenching, cutting, scraping, and light digging. Also called: fire rake; rich tool.

count[1] (Med) To determine the number, as the number of blood cells.

count[2] (Dosim) The numerical value for the activity of a radioactive specimen determined by a radiation detection instrument.

counter (Dosim) An instrument that detects and counts pulses, used to measure radiation in terms of individual ionizations, displaying them either as the accumulated total or their rate of occurrence. See: Geiger-Mueller counter; scaler.

counter fire (For Serv) 1. A fire set between the main fire and a backfire to accelerate the backfire. Also called: draft fire. 2. To set such a fire. Also called: front fire; strip fire. 3. (Brit) = backfire.

countermeasure A measure or action taken in opposition or retaliation.

counterweight (Mech Eng) A weight used to balance the load of a lifting machine to provide stability.

coupled modes (Mech) Vibrations that influence one another through energy transfer from one mode to the other.

coupling[1] (Mech Eng) Any of a variety of devices used to connect mechanical parts.

coupling[2] (Fire Pr) A threaded fitting for interconnecting pipes and fire hoses.

court (Build) An open space bounded on two or more sides by the exterior walls of a building. See: yard[2]. Cf: atrium.
NOTE: If bounded on four sides it is called an interior court or inner court.

covalent molecule (Chem) A molecule in which the bond between two atoms is a shared electron pair, such as H:H, Cl:Cl, CH_4, etc.

covariance (Statist) A relationship between two variable such that a change in one is associated with a change in the other.

cover[1] (Fire Pr) a. To respond to an assignment in place of another firefighting unit that is unavailable. b. To assume responsibility for a district normally serviced by another fire-fighting unit. c. To protect an assigned area. d. At a working fire, to protect neighboring exposures with hose streams.

cover[2] (Salv) a. A salvage cover. b. To spread salvage covers to protect goods and equipment from water damage.

coverage (Insur) 1. An insured risk or liability. 2. That which is insured, as specified in a policy.

covered space (Build) The area within or beneath any roofed structure.

cover pole (Salv) A pole tipped with a U-shaped hook, used to place salvage covers over high-piled stock.

cover rack (Fire Pr) Shelving used to store salvage covers.

covert Hidden; concealed from view.

cover type (For Serv) Designation of a complex vegetation area by its dominant species, age, and form.

coyoting = gophering

CPR = cardiopulmonary resuscitation

cps = cycles per second or Hertz.

CPSC = Consumer Product Safety Commission

crack a nozzle (Fire Pr) To open a nozzle slightly, releasing a small amount of water for the purpose of clearing air out of a hose.

crack a valve (Fire Pr) To free the stem of a sprinkler valve to ensure proper operation.

cracking (Chem Eng) A process, using heat, pressure, and catalysis, to break down molecules such as hydrocarbons of high molecular weight into smaller molecules.
NOTE: Commonly used for the manufacture of gasoline and industrial chemicals.

cradle (Techn) A movable fixture that carries a part to be cut, ground, or polished.

cramps (Med) Painful muscular contractions.

crane (Mat Hand) A fixed or mobile mechanical device intended for lifting or lowering heavy loads and moving them horizontally. The hoisting mechanism is an integral part of the machine.

cranium (Anat) The skull, esp the brain pan.

crash truck (Fire Pr) A specialized fire-fighting apparatus designed to handle aircraft fires and accidents. Usu equipped with special forcible entry and rescue tools and extinguishing agents to combat large flammable liquids fires. See: apparatus.

crate (Transp) A rigid shipping container constructed of members fastened together to hold and protect the contents.

crawfish (Mat Hand) A bucket-type conveyor system used to unload grain from boxcars, etc.

crawler crane (Mat Hand) A crane with a rotating superstructure and power plant, operating machinery, and boom, mounted on a base, equipped with crawler treads for travel.

crawling board (Build) A plank with cleats spaced and secured at equal intervals for use by a worker on a roof. Also called: chicken ladder.

crawl space (Build) A space, usu a foot or two in height, provided between the bare soil and the floor of a building, primarily to allow for air circulation.

creeping fire (For Serv) A fire that burns with a low flame and spreads slowly. See: smoldering; running fire; spotting.

cremate To reduce to ashes by fire.

creosote (Techn) A distillate oil obtained from coal tar or wood and used as a wood preservative that increases the combustibility of wood.

crew boss (For Serv) The supervisor in charge of a team of forest firefighters. Also called: supervisor.

crib[1] (Comb; Testing) A square or rectangular, self-supporting, box-like construction, generally of loosely-laid wood boards or timbers in which two or more boards are laid parallel and succeeding layers are laid on top, alternately at right angles until the desired height or fuel load is reached. Used to simulate a given quantity of fuel in fire tests.
NOTE: In British practice the crib timbers are fixed together.

crib[2] (Techn) A supply room from which tools, spare parts, and materials are issued.

cribbing (Mining) Lining of a shaft constructed of timbers in parallel pairs at right angles in succeeding layers in the shape of a square.

criminal negligence (Law) Involving or relating to failure to use reasonable care when such failure results in injury or death to another.

criminology (Law) The science of crime, criminals, and penology.

crimp (Techn) To put a bend or crease in metal, often used as a method of sealing or fastening two pieces of metal together.

criterion 1. A standard, rule, or test by which a judgment or evaluation can be made. A basis of comparison in measuring or judging capacity, quantity, content, extent, value, quality, etc. 2. A level or degree of excellence, attainment, etc. regarded as a goal or measure of adequacy. 3. A set of scores, ratings, etc. that a test is designed to correlate or predict.

critical angle (Opt) The maximum angle of incidence at which light rays can pass from a medium with a higher refractive index to one with a lower refractive index. For an angle of incidence greater than the critical angle, the rays are totally reflected back into the medium of higher density.

critical damping (Techn) The minimum damping that will allow a displaced system to return to its initial position without oscillation.

critical density (Phys) The density of a substance at critical pressure and critical temperature.

critical function (Techn) An operation or activity that is essential to the life of a system.

critical humidity (Meteor) The amount of water in the air at which the partial vapor pressure is equal to the saturation vapor pressure. The point at which condensation begins.

critical item (Techn) An item that, having critical surfaces, close tolerances, or fragility, requires special protective measures or handling techniques.

critical mass (Nucl Phys) The quantity of fissionable material necessary to sustain a chain reaction.

critical path method (Man Sci) A planning and operations research technique for scheduling the sequence of steps, operations, or events in a program.

critical point (Thermodyn) The highest temperature at which the liquid and gas phases of a substance can coexist in equilibrium.

critical pressure (Phys; Chem) The pressure of a system at its critical point, when the distinction between gas and liquid phases disappears.

critical size (Nucl Phys) A set of physical dimensions of a nuclear reactor that contains a critical mass.

critical temperature (Phys; Chem) The temperature above which a gas cannot be liquified. The distinction between the liquid and gas phases disappears. If a liquid in a sealed tube is heated, the meniscus disappears when the critical temperature is reached.

critical velocity (Hydraul; Aerodyn) The velocity of a fluid at which steady flow becomes turbulent.

critique A critical discussion or review.

cross aisle See: aisle.

cross-country (Transp) Proceeding on a course directly over countryside, such as across fields, hills, marshes, woods, and not by roads or paths.

crosscut (Mining) 1. A horizontal opening driven across a vein at a large angle to a strike. 2. A passageway connecting a shaft and a vein.

crossed trumpets (Fire Pr) An insignia worn by fire department officers. The number of trumpets indicates the relative rank. See: bugle; trumpet; ranks and titles.

cross patee; cross paty (Fire Pr) A heraldic cross, similar to the cross formee, derived from the cross of the Knights of Malta and used as a badge or an insignia in the fire service.
NOTE: The Maltese cross has triangular arms with deep V-shaped notches pointing toward the center, and thus

it has eight points. The cross patee has flaring arms and squared ends.

cross section¹ (Techn) A section of an object at right angles to its axis.

cross section² (Nucl Phys) A measure of the probability that a nuclear reaction will occur. Usu measured in barns, it is the apparent or effective area presented by a target nucleus or particle to an oncoming particle or other nuclear radiation, such as a photon of gamma radiation.

cross-sectional study (Physiol) A study design in which a representative population is studied at a given point in time, as contrasted to a longitudinal study, in which subjects are observed over long periods of time.

cross shot (For Serv) Intersecting lines of sight from two points to the same object, used to locate fire from lookouts by triangulation.

cross stacking (Storage) The placing of one layer of containers at right angles to those just below to increase the stability of the stack.

crosstalk (Electr) An unwanted signal induced in a channel by a neighboring one.

cross tie (Storage) A crossing layer of items in a stack, used to improve the stability of the stack.

cross wind (Meteor) The flow of air at an angle to a fixed reference line.

crotch lines (Mat Hand) Two short lines attached to a hoisting line by a ring or shackle, the lower ends being attached to loading hooks.

crotch pole (Fire Pr; Archaic) A pole with a U-shaped fixture at one end, used to help erect and brace ladders. Cf: tormentor pole.

crowbar (Fire Pr; Salv) A long, straight, prying instrument, usu with a chisel edge on one end. Loosely, a wrecking bar. See: forcible entry tool.

crowd (Psych) A temporary collection of people who share a common interest.

crown (For Serv; Fire Pr) The top of a tree. Cf: crown fire.

crown birth (Med) Delivery of a baby in which the crown of the head is presented first. See: breech birth.

crown cover (For Serv) The canopy of leaves and branches formed by the crowns of trees in a forest. Syn: leaf canopy.

crown fire (For Serv) A forest fire burning through tree tops. A running crown fire is one that is independent of the surface fire.

crown fuel (For Serv) Tree tops considered as fuel.

crowning (Med) The first appearance of the top of baby's head in the vulva during childbirth.

crowning fire (For Serv) A fire that is advancing through the tops of trees or shrubs.

crown out (For Serv) 1. With reference to a forest fire, to rise from ground level and begin advancing from tree top to tree top. 2. To intermittantly ignite tree crowns as a surface fire advances.

crucible (Chem) A laboratory vessel of heat-resistant material used to hold materials for high-temperature experiments.

crude petroleum (Chem Eng) Hydrocarbon mixtures that have not been processed in a refinery.

cryogenics (Phys) The study of extremely low temperature phenomena.

cryogenic temperature (Phys) Generally, the temperature range below boiling nitrogen: -195°C.

crystal (Phys) A solid in which atoms, ions, or molecules are arranged in a regularly, repeating, symmetrical pattern.

crystallites (Phys) Small-scale, localized crystalline regions in a polymer or other solid.

crystallography (Phys) The study of shapes and structures of crystals.

C-shift (Fire Pr) 1. The third working group of a three-platoon firefighting organization. 2. (Colloq) Off-duty employment (moonlighting) by A- or B-shift (1st and 2nd shift) personnel.

CSP = **certified safety professional**

CTC = **carbon tetrachloride**

CTIF = **Comite Technique International de Prevention et d'Extinction du Feu**

cube[1] (Math) A number raised to the third power; e.g., the cube of five (5 x 5 x 5) is 125.

cube[2] (Transp) A conventional measure of the volume occupied by an item or package in shipping. In principle, it is the product of the three dimensions of the smallest rectangular volume into which the package will fit.

cubicle (Build) 1. A small space partitioned off from a large room. 2. A small sleeping compartment.

cubic feet per minute (Metrol; Abb: cfm, or $ft^3 m^{-1}$) A unit of measure for flow rate. The volume of matter, usu a fluid, passing a given point in one minute.

cul-de-sac A dead-end street.

cullet (Techn) Scrap glass.

culls (Techn) Anything defective or of low grade, that has been sorted and removed.

culture (Bact) 1. To grow and propagate microorganisms by artificial means. 2. A population of artificially grown microorganisms.
NOTE: A pure culture consists of a single kind of microorganism; a mixed culture consists of two or more grown together.

culture medium (Bact) A preparation used for the growth and cultivation of microorganisms. A selective medium is one composed of nutrients designed to allow growth of a particular type of microorganism. A broth medium is liquid; agar is a gel-like solid.

culvert (Civ Eng) A man-made drainage channel or pipe for carrying water under obstructions, such as roads and rail lines.

cumulonimbus (Meteor) A cloud form, classification Cb. See: **cloud**.

cumulus (Meteor) Heaped cloud, classification Cu. See: **cloud**.

cup test (Chem) A test to determine the **flash point** of a flammable liquid. The sample liquid is placed in an open or closed cup, brought to temperature, and then the vapor is ignited by a point source near the surface of the liquid.

cure time (Chem Eng) Time required for polymerization of certain materials to achieve strength or surface hardness.

curettage (Med) The scraping of an organ or other part of the body with a sharp loop or spoon-shaped instrument called a curette.

curie (Nucl Phys; Symb: c) 1. The basic unit used to describe the intensity of radioactivity in a sample of material. The curie is equal to 37 billion disintegrations per second, which is approx the rate of decay of 1 gram of radium. 2. A quantity of any nuclide having 1 curie of radioactivity.
NOTE: Named for Marie and Pierre Curie, who discovered radium in 1898.

curing (For Serv) The drying and browning of grass.

Curling's ulcer See: **ulcer**

current (Electr) The flow of electrons through a conductor, usu measured in **amperes**.

curriculum (Educ) A set of courses or sequence of subjects that must be completed for certification or graduation in an academic program.

curtain[1] (Build) A hanging screen or nonbearing wall.

curtain[2] (Build) The main part of a security door that unrolls to cover a store window, usu consisting of interlocking, corrugated steel slats.

curtain³ (Build) A textile drapery or polymeric film used as a window screen and as an item of interior decoration.

curtain board (Build) A substantial noncombustible curtain extending down from a roof or ceiling at the perimeter of a special hazard area to bank up hot gases and smoke within the curtained area and direct them toward vents. Syn: draft curtain.

curtain wall (Build) An exterior nonbearing wall more than one story in height, usu supported by the structural frame to protect the building interior from weather, noise, or fire. In multistory buildings where such nonbearing walls are one story in height and supported at each floor level, it is called a panel wall.

custodial care Institutional care without therapy.

cut and fill (Mining) A mining method in which waste is placed in a stope as ore is removed from it.

cutaneous (Anat) Pertaining to the skin.

cuticle (Anat) Superficial skin or the outer layer of skin.

cutie pie (Dosim; Colloq) A common radiation survey meter used to determine exposure levels or to locate possible radiation hazards. See: monitor.

cutis (Anat) The layer of skin beneath the epidermis that contains the blood and nerve ends. Also called: dermis or corium.

cutoff¹ (Fire Pr) a. A fire wall, fire door, or other barrier designed to limit fire spread. See: fire stop. b. The point at which a fire is halted.

cutoff² (Electr; Mech Eng) An electrical switch or mechanical device for shutting off power or interrupting an operating cycle.

cutting fluid (Techn) An oil or an oil-water emulsion used to cool and lubricate a cutting tool.

cyanide (Chem; Symb: CN) A salt of hydrocyanic acid, usu extremely poisonous.

cyanogen (Chem) A colorless, poisonous, flammable gas, $(CN)_2$.

cyanosis (Med) A blueness of the skin and lips resulting from insufficiency of oxygen in the blood. Commonly caused by respiratory disorders and gas or drug poisoning that interferes with respiration and the absorption of oxygen by the blood. A common symptom of carbon monoxide exposure. See: anoxia.

cybernetics (Cyber) The science of communication and control in machines, animals, systems, and organizations as it relates to goal-directed activity.

cycle¹ (Mech Eng) A series of operations or states that a mechanical device or process passes through to return to its initial condition.

cycle² (Electr) A single oscillation of frequency. See also: Hertz.

cyclone (Meteor) A circulating system of winds about a low pressure center. Motion is counterclockwise in the northern hemisphere and clockwise in the southern.

cyclone separator (Techn) A mechanical separator used to remove suspended liquid and solid particles from gases, for example, dust from air.

cylinder¹ (Techn) In general, any compressed gas container.

cylinder² (Transp) A container having 1000 pounds of water capacity or less in accordance with Department of Transportation specifications.

cytology (Biol) The branch of biology concerned with cells, their structure, development, growth, pathology, and history. Cf: histology.

cytoplasm (Biol) Cell plasma exclusive of the cell's nucleus.

cytotoxin (Med) A substance developed in the blood serum that has a toxic (poisonous) effect upon specific cells; e.g., nephrotoxin, a poison specifically attacking the kidney.

D

<u>d</u> (Techn) Abbreviation for diameter.

<u>D</u> (Chem) Symbol for deuterium.

<u>Dahill hoist</u> (Fire Pr) A compressed-air power hoist for aerial ladders, named after its inventor.

<u>daily burning cycle</u> (For Serv) A 24-hour period beginning at 10:00 am, coinciding with the <u>fire day</u>, during which wildfires predictably slow down and speed up under the influence of diurnal meteorological variations.

<u>Dalton's law</u> (Phys) The total pressure exerted by a nonreacting mixture of gases is the sum of the pressures of the individual gases.

<u>dam</u> (Fire Pr) A portable water container used for priming pumps.

<u>damage</u> (Fire Pr; Insur) Material injury or loss resulting from fire or other disaster. Loss in value, usefulness, and the like to property or things.

<u>damage risk criterion</u> (Audiom; Med) A measure for determining the risk of hearing loss from sound, including such factors as time of exposure, noise intensity, noise frequency, degree of hearing loss that is considered significant, the population to be protected, and the method of noise measurement.

<u>damp</u> (Mining) Any unknown or unspecified mine gas. Specific kinds of damp have specific names, such as <u>afterdamp</u>; <u>blackdamp</u>; <u>chokedamp</u>; <u>firedamp</u>; <u>stinkdamp</u>; <u>sweetdamp</u>; and <u>white damp</u>.

<u>damp down</u> (Fire Pr; Brit) To wet down and blacken a fire.

<u>damper</u> (Build) A manually or automatically controlled baffle or plate in a duct, or stack, used to regulate a draft or the flow rate of air or other gases. See also: <u>smoke damper</u>.

<u>damper control</u> (Build) A device such as a fusible link or a smoke- or heat-responsive actuator that opens or closes a damper.

<u>damping</u> (Mech) Suppression of oscillations or disturbances in a system.

<u>danger</u> (Safety) The potential for injury, illness, damage, loss, or pain.

<u>danger board</u>[1] See: <u>fire danger board</u>

<u>danger board</u>[2] (Fire Pr; Brit) A sign or placard bearing a symbol and notice of a hazard.

<u>danger class</u> See: <u>fire danger class</u>

<u>danger index</u> See: <u>fire danger index</u>

<u>danger meter</u> See: <u>fire danger meter</u>

<u>danger zone</u> (Safety) A physical area or location within which a danger exists.

<u>dark adaptation</u> (Physiol) The process by which the iris dilates, exposing more of the retina to light, thus enhancing vision in dim light. Cf: <u>light adaptation</u>.

<u>darken</u> (Fire Pr) To blacken an area that had been lit up by flames. Syn: <u>black out</u>.

<u>dark field</u> (Opt; Testing) A system of illumination in spectroscopy in which a sample is illuminated with a hollow cone of light whose minimum aperture is greater than the maximum aperture of the objective, so that only the rays bent by objects in a sample enter the objective.

<u>DASA</u> = <u>Defense Atomic Support Agency</u>

<u>dashpot</u> (Techn) A vibration isolator consisting of a fluid-filled cylinder and a piston.

<u>dasymeter</u> (Comb; Phys) 1. An instrument consisting of a glass ball, used to determine the density of a gas by weighing it in a sample of unknown gas and then in a gas of known density. 2. An instrument for determining the composition of flue gas.

<u>data</u> (plural of <u>datum</u>, qv) Any representations, such as characters, symbols, or analog quantities, to which meaning may be assigned. Data may be expressed in alpha, alphanumeric, digital, graphic, or sym-

bolic form.
NOTE: Processed data are usu referred to as information.

data base (Comput; Info Sci) A collection of data or references to documents that serves as the basis for an information retrieval system, usu supported by a computer.

data communication (Commun) The transmission of information in a specified format or code from one location to another by wire or radio.

data point (Math) An elementary unit of information or quantity derived from measured observations.

data processing (Comput) A systematic sequence of operations on data, performed in accordance with set rules. Operations include sorting, merging, computing, assembling, compiling, translating, extracting, storing, and retrieving information.

data processing system (Comput) Electronic or electromechanical machines used to organize data into a predetermined order or to correlate data for a specific purpose, such as inventory control, reduction of statistics, payroll preparation, and other business and technical applications.

datum A discrete unit of information collected or presented for a particular purpose.

daughter (Nucl Phys) A product nucleus or atom resulting from radioactive decay of the precursor or parent.

day manning (Fire Pr; Brit) A duty schedule in which a fire station is manned only during the daytime.

days of disability (Safety) The number of calendar days on which an injured person was unable to work as a result of injury, excluding the day of the injury and the day of return to work.

dB = **decibel**

dBA (Audiom) Sound level in decibels read on the A-scale of a sound-level meter. The A scale discriminates against very low frequencies (as does the human ear) and is, therefore, preferred for measuring general sound levels.

dBC (Audiom) Sound level in decibels read on the C-scale of a sound-level meter. The C-scale discriminates little against very low frequencies. Because impact noise is mostly in the low frequencies, the C network is used.

DDT = **deflagration to detonation transition**

dead end The end of a street without an exit. A **cul-de-sac**.

dead-end main (Fire Pr) A water main supplied with water from one direction and sealed off in the other.

dead fuel (For Serv) Nonliving natural fuel in which the moisture content is governed primarily by atmospheric humidity and precipitation.

dead load (Build) The weight of a structure itself, in contrast to **live load**, consisting of equipment, furnishings, etc. carried by the structure in service.

deaf (Med) Having deficient hearing or lacking the ability to hear.

deaf mute (Med) Having lost the ability to hear before speech patterns were established, and therefore unable to speak.

deafness (Med) An auditory defect in which hearing is impaired or lost. Hearing may be affected over the entire auditory range or to certain frequencies only.

death (Med) Cessation of vital processes in a living organism. Syn: **fatality**.
NOTE: If the entire organism is involved, the technical term is **somatic death** or **biological death**; if limited to portions of the organism, it is called **necrosis** or **infarction**. See also: **accidental death**; **infarct**.

death rate (Insur; Demog) The number of deaths with respect to a known constant, such as number of deaths per unit of population.
NOTE: Death rates are usu of three

types: **Population**: The number of deaths per 100,000 persons. **Registration** (vehicle): The number of deaths from traffic accidents per 20,000 motor vehicles registered in the area for which the rate is computed. **Mileage** (vehicle): The number of deaths from motor vehicle accidents per 100,000 miles of motor vehicle travel in the area for which the rate is computed.

debility (Med) Decrease or loss of vigor and strength.

debridement (Med) Removal of dead or damaged tissue from the body, usu by cutting.

debris (Fire Pr) Unwanted material; refuse left after a fire.

debris burning fire (For Serv) A fire spreading from any fire set for the purpose of clearing land or disposing of rubbish. A debris burning fire does not include fire hazard reduction along railroad right-of-ways.

decanting (Safety) A method used for decompressing under emergency conditions. The individual requiring decompression is brought to atmospheric pressure, placed in another pressure chamber or lock, and immediately recompressed, after which he is gradually returned to atmospheric pressure.

decay¹ (For Serv) Decomposition or disintegration of wood or other organic matter by fungus. Also called: **dote**; **rot**.
  NOTE: **Incipient decay**: An early stage in which disintegration has not proceeded far enough to soften or otherwise impair the hardness of the wood perceptibly. **Advanced decay**: A later stage in which disintegration is readily recognized because the wood has become punky, soft and spongy, stringy, pitted, or crumbly. See: **dry rot**.

decay² (Nucl Phys) The partial disintegration of the nucleus of an atom by the spontaneous emission of charged particles or photons. What remains is a different element. An atom of polonium, for example, decays to form lead, ejecting an alpha particle in the process.

deceleration (Phys) The time rate of decrease in **velocity**. Ant: **acceleration**. Also called: **negative acceleration**.

deceleration rate (Transp) The rate of reduction of the speed of a vehicle or moving part, expressed in feet per second per second.

decibel (Acoust; Abb: dB) One-tenth of a **bel**. A dimensionless measure of sound expressed as the logarithmic ratio of two values of pressure, power, or intensity, one a measured quantity and the other a reference quantity. The decibel is most often used as the measure of sound intensity level and sound pressure level. Sound intensity is the sound power per unit area.
  NOTE: In common usage, **decibels**, **noise level**, and **intensity level** are used interchangeably. In electronics, the decibel is the unit of relative power level equal to 10 times the logarithm to the base 10 of the ratio of two powers, one of which is usu a reference value. In acoustics, the decibel is often taken as 20 times the logarithm of the sound pressure ratio, with the reference pressure equal to 0.0002 dyne/cm². A whisper is about 10 decibels, conversation about 60, and thunder about 120 dB.

decile (Statist) One of the nine points dividing a frequency distribution into ten equal parts every tenth percentile. The first decile is the 10th percentile, the ninth decile the 90th percentile, etc.

decimal digit (Math) One of the integers 0 through 9 used for counting with the base ten.

decimal system (Math) A number system based on powers of ten. See: **number systems**.

decision making (Man Sci) The process by which a committment is made to a course of action.

deck¹ The floor or platform of a vehicle, equivalent to the floor of a building.

deck² (Build) A plane, composite, horizontal building structure, such as a roof deck.

115

deck boards (Storage) The top or bottom surface of a pallet.

deck gun (Fire Pr) A master stream appliance and nozzle mounted on a pumper or fireboat and connected directly to a pump. Also called: deck pipe.

decompose (For Serv) To separate or break down into constituent parts; to decay.

decomposition (Chem) The breaking down of a complex material into simpler substances. See: pyrolysis.

decontamination[1] (Ind Hyg) Removal of a polluting or harmful substance.

decontamination[2] (Nucl Safety) Removal of radioactive fouling from clothing, equipment, work areas, etc. by cleaning and washing with chemicals. See: radioactive contamination.

deductible insurance (Insur) A reduced rate insurance under which the policyholder pays for losses up to a specified amount, while the insurance company pays the remainder up to the total amount of the insured loss.

deduction (Psych) An inference or conclusion reached from formal premises or propositions.

deep bag (Fire Pr; Salv) An arrangement of tarpaulins to catch dripping water to protect goods, furniture, and equipment from water damage.

deepseated fire (Fire Pr) 1. A fire that has penetrated deep into bulk materials. 2. A persistent fire that is difficult to extinguish.

defect (Techn) 1. Any characteristic or condition that tends to weaken or reduce the strength of the tool, object, or structure. Any imperfection that affects usefulness. 2. Any nonconformance with specified requirements. Cf: flaw.

defendant (Law) An individual sued in a civil case or prosecuted in a criminal action.

defense (Law) Matters of fact or law presented at court by a defendant in support of his case or to refute the prosecution.

defibrillation (Med) Restoration of normal rhythm to a fibrillating heart, accomplished by an electronic defibrillator. See: fibrillation.

defibrillator (First Aid; Med) An electronic device that produces a direct current electrical shock to restore normal rhythm to a fibrillating heart.

deficiency points (Insur; Fire Pr) Points subtracted from community fire protection grading schedules for noncompliance with fire protection standards and good practices. See: grading schedule.

deflagration (Comb) A subsonic gaseous combustion process propagating through unreacted material.

deflection[1] (Build) A measure of the movements by bending or displacement of a structural member by applied forces.

deflection[2] (Metrol) Displacement of a needle or other indicator of a measuring instrument, such as a gage.

deflection mapping (Opt) A quantitative method of analyzing the deflections of parallel light rays by density gradients. See also: shadowgraph; schlieren; interferogram.

degas (Chem; Phys) To remove traces of gas from an instrument or material by heating in high vacuum.

degradation[1] (Chem) Gradual decomposition of a substance.

degradation[2] (Electr) Deterioration in quality, fidelity, or performance.

degreasing (Metall) The removal of oil or grease from metal surfaces prior to further processing or finishing, usu accomplished by dipping a part in solvent or exposing it to hot solvent vapor.

degree program (Educ) A course of study in a college or university leading to the granting of an academic degree.

degrees of freedom[1] (Phys) The set of possible modes of motion of a rigid body, consisting of three linear and three angular directions.

degrees of freedom[2] (Chem) The number of independent variables required to specify the instantaneous position of a body.
NOTE: A diatomic molecule has seven degrees of freedom: the three given for rigid bodies, plus the one in which the atoms move toward and away from each other. A system of three dimensionless particles has nine degrees of freedom. A nonrigid body may have an infinite number.

degrees of freedom[3] (Statist; Abb: df) The number of observations, scores, or trials minus the number of restrictions placed on them.

degressive burning (Comb) Burning of propellant grains in which the surface area of the grain decreases. Also called: regressive burning.

dehiscence (Med) Splitting along a natural line.

dehydrating agent (Chem) A substance that strongly attracts moisture; used for drying materials. Common agents are silica gel, $CaCl_2$, and P_2O_5.

dehydration (Med; Techn) The loss or removal of water and water vapor by any means from a gas, liquid, or solid, including the human body.

dehydrogenation (Chem) The process whereby hydrogen is removed from a compound by chemical means and not replaced.

delay (Civ Eng; Mining) An explosive-train component that introduces a controlled time delay.

delayed response (Fire Pr) An instance of nonimmediate response to an alarm, usu because of commitment to an earlier alarm. See: presignal delay.

deliquescence (Chem; Phys) Attraction and absorption of moisture from the air. Ant: efflorescence.

deliquescent salt (Chem) A salt capable of absorbing sufficient moisture from the air to dissolve; e.g., $CaCl_2$. Cf: hygroscopic.

delivered horsepower (Techn) Power delivered at the output of an engine or motor.

delivery head (Fire Pr) A fitting, usu on fire boats, for feeding a number of hose lines simultaneously. Cf: collecting head.

delivery hose adapter See: adapter

deluge set (Fire Pr) A master stream appliance consisting of a large tripod- or quadrupod-supported nozzle with two or more inlet connections.
NOTE: The device is portable and can be operated unattended from the ground or special vantage point.

deluge system (Fire Pr) A sprinkler system with multiple outlets, some or all of which are open, usu operated by a thermostatically controlled deluge valve.

deluge valve (Fire Pr) A single valve controlling a large array of sprinklers, some or all of which are open, that actuates an alarm when it opens.

demand factor (Electr) The ratio of the power demand on a system to the total load in the system.

demand-type mask (Fire Pr) A breathing mask with a regulator that feeds air or oxygen at a rate determined by the breathing rate of the wearer. See: breathing apparatus.

demising wall (Build) A wall that terminates, borders, or limits a given tenant's space.

demography (Soc Sci; Abb: Demog) The science of social statistics: births, deaths, marriages, and the like. The study of human populations and the social, physiological, psychological, political, and economic factors affecting vital statistics.

demulsification (Chem) Breaking down of an emulsion into its constituents.

denature (Chem) To render a product unfit for human consumption, such as denatured alcohol.

denier (Text) The mass in grams of 9000 meters of yarn.

density[1] (Phys; Symb: p) a. Mass per unit volume. b. Broadly, any quantity related to the volume occupied, such as energy density, electron density, etc. See also: **specific gravity**.

density[2] (Fire Pr) The rate of application of water or foam solution to an area, expressed in gallons per minute per square foot.

density[3] (Transp) The number of vehicles occupying a unit length of the driving lanes of a road or highway at a given instant, usu expressed in vehicles per mile.

density function (Statist) The fraction of the total frequency in any of the infinitesimal intervals into which the range of a variable may be divided.

dep = **deputy**

departure time (Fire Pr; Brit) The time that fire apparatus leaves the fire station in response to a call.

dependent variable (Math) A variable that is a function of one or more independent variables. Changes in a dependent variable are governed by the experimental or independent variable.

depolymerization (Chem) A chemical reaction in which a polymer is broken down into monomers. The reverse of **polymerization**.

depressed car (Transp) A flatcar with the portion of the floor between trucks depressed to provide increased head room for certain classes of lading. Also called: **low boy**.

depth of focus (Opt; Physiol) The region within which objects in the image appear in sharp focus.

depth perception (Physiol) The ability to judge distances of nearby objects by the use of both eyes.

deputy (Fire Pr) A subordinate title, usu meaning a deputy chief. See: **ranks and titles**.

deputy chief (Fire Pr) The second or third ranking chief officer in large fire departments; usu an officer in charge of a division. See: **ranks and titles**.
  NOTE: In some departments a deputy chief ranks above an assistant chief; in others he may rank below.

derivative (Math) A function that expresses the rate of change of a given function.
  NOTE: Velocity may be expressed as the first derivative of a position function (i.e., rate of change of position). Acceleration is the first derivative of the velocity function, or the second derivative of the position function.

derma (Anat) The corium or true skin. Syn: **dermis**.

dermatitis (Med) Inflammation or irritation of the skin. One of two general types of skin reaction: primary irritation dermatitis and sensitization dermatitis. Syn: **eczema**.

dermatophytosis (Med) A fungus infection of the skin, usu of the hands or feet, commonly called athlete's foot.

dermatosis (Med) Any cutaneous (skin) disease or abnormality, including folliculitis, acne, pigmentary changes, and nodules and tumors. A broader term than **dermatitis**.

dermis (Anat) The sensitive layer of skin located beneath the epidermis (the outer layer of skin). See: **corium**.

derrick (Mat Hand) An apparatus consisting of a mast, A-frame, or equivalent member, held at the head by guys or braces, with or without a boom, for use with a hoisting mechanism and operating ropes. A mechanical device intended for lifting and moving cargo or other heavy materials and supplies.

descriptive geometry (Geom) A graphic mathematical method of representing structures in drawings in which each element is shown in exact geometric relation to all other elements. Two projection techniques are used: 1) perspective, in which parallel lines of the object in the drawing converge at a remote point, and 2) orthographic, in which parallel lines of the object are shown parallel on the drawing.

descriptive statistics (Statist) Statistics that describe observed characteristics of a group. Measures that describe a population or sample. Description deals with the collection, tabulation, and summarization of numerical data. It may involve the arithmetic means, standard deviation, coefficient of correlation, and other summary measures.

desiccant (Techn) A drying or dehydrating agent. A material that will absorb moisture by physical or chemical means.

desiccate (Chem) To remove water or moisture from a substance. To dry thoroughly.

designated person (Safety) An individual approved or assigned by an employer to perform specific duties or to be at specific locations at a jobsite. Syn: authorized person.

design load (Build) The weight that can be safely supported by a floor, equipment, or structure as defined by its design characteristics.

design pressure (Techn) Maximum allowable working pressure.

design safety (Safety) The planning of environments, structures, and equipment, and the establishment of procedures for performing tasks, to reduce or eliminate human exposure to injury potential. In product safety, design of the product for safe use.

design standards (Build) Conventions, generally accepted practices and procedures, and formal agreements on materials and construction techniques for products and facilities.

desorption (Chem; Phys) The release of an adsorbed gas. Ant: adsorption.

destructive distillation (Chem) The process of decomposing a substance by heat in the absence of air. Volatile products are condensed and collected. This was once the principal method of producing organic compounds.

destructive test (Ind Eng; Testing) A procedure for quality testing in which the material being tested is destroyed in the course of obtaining desired measurements.

detail (Fire Pr) 1. One or more firefighters assigned to temporary duty to cover special situations, such as a fire watch at a large public gathering, or to alleviate temporary manpower shortages. 2. A team, called a watch detail, assigned to an extinguished fire to guard against rekindling. 3. To assign one or more firefighters to a special task or position, esp on a temporary basis.

detection[1] (Fire Pr) Sensing the existence of a fire, esp by a detector, from one or more signatures of the fire, such as smoke, heat, ionized particles, infrared radiation, and the like.

detection[2] (For Serv) The act or process of discovering and locating a fire Cf: discovery.

detector[1] See: fire detector

detector[2] (Dosim) A material or a device that is sensitive to radiation and can produce a response signal suitable for measurement or analysis. A radiation detection instrument.

detention (Law) Confinement of an individual for any reason other than the serving of a sentence.

detergent (Techn) A solvent or wetting agent used for cleaning. Synthetic detergents are sometimes called syndets or surfactants.

deterioration (Techn) Any impairment in quality, value, or usefulness. Damage caused by erosion, oxidation, corrosion, or contamination.

deterministic function (Math) A function whose value at any time can be predicted from its value at any other time.

detonating agent (Civ Eng; Mining) An explosive used to set off less-sensitive explosives. Includes initial detonating agents and other explosives that may be used as intermediate explosive elements in a detonating train.

detonating cord (Civ Eng; Mining) A flexible fabric tube containing a filler of high explosive initiated by a blasting cap or electric detonator.

detonation (Comb) A supersonic self-propagating combustion reaction or a flame that propagates at supersonic speed. A detonation is a class of explosion that can be initiated in both gas phase and condensed phase. Essentially, the flame is a combustion wave that propagates faster than the local speed of sound so that a shock wave forms with a strong pressure jump. Detonation is often confused with explosion. The pressure rise in an explosion results from confinement (e.g., as gunpowder in a rifle barrel) rather than from a supersonic rate of propagation. The equations describing detonations are the same as those describing flames (See: flame equations), but in the case of detonations the flow is supersonic and viscous losses become important. The rate of advance of the reaction zone is termed detonation rate or detonation velocity. Cf: deflagration.
NOTE: Detonation velocity, once reached, is almost constant for any given gas mixture. Dust will explode, but rarely detonate, because pressure buildup is too slow.

detonation wave (Comb) The shock wave that precedes the advancing reaction zone in a high-order detonation.

detonator[1] (Civ Eng; Mining) A device used to initiate an explosive. In an explosive train, the component which, when detonated by the primer, in turn detonates a less sensitive explosive, usu the booster, or when containing its own primer, initiates the detonation in the train.
NOTE: A detonator can be actuated by either an explosive impulse (a primer) or a nonexplosive impulse. When actuated by a nonexplosive impulse, the detonator contains its own primer. Detonators are generally classified as percussion, stab, electric, or flash, according to the method of initiation. Syn: blasting cap; cap.

detonator[2] (Milit) An explosive charge placed in certain equipment and set to destroy the equipment under certain conditions.

Detroit door opener (Fire Pr) An adjustable tool for forcing locked doors. See: forcible entry tool.
NOTE: Heavy-duty tools used for the same purpose are called door breakers.

deuterium (Symb: ^2H or D) A non-radioactive isotope of hydrogen with a nucleus consisting of one neutron and one proton, making it about twice as heavy as the nucleus of normal hydrogen, which has only one proton. Deuterium, often referred to as heavy hydrogen, is found in water in the proportion of 1 to 6500 atoms of normal hydrogen. See: heavy water; hydrogen.

deuteron (Phys) The nucleus of deuterium, consisting of one proton and one neutron.

development (Mining) Driving of openings toward and into an orebody to prepare it for mining and removing ore.

deviation (Statist) The difference between a score or other item and a reference value such as the mean, the norm, or the score on some other test.

dew (Meteor) Droplets of water formed by direct condensation from the vapor phase. See also: frost; precipitation; rime.

dew point (Meteor) The temperature at which the vapor pressure of the air equals the saturation vapor pressure and moisture begins to condense in droplets on cool surfaces. See also: condensation; vaporization.

diabatic (Thermodyn) Occurring with a gain or loss of heat. Ant: adiabatic. Also called: nonadiabatic.

diagnosis (Med) A procedure used to determine the nature of a disease.

diagnostic routine (Techn) A preventive maintenance check of key system components by use of a special programmed tape or electronic trouble-shooting instruments.

dialysis (Chem; Med) The separation of substance in solution or the removal of impurities by preferential diffusion through a membrane, sometimes facilitated by imposing an electrical potential across the membrane.

diameter (Geom; Abb: dia) The length of a straight line passing from edge to edge of a circle through the center. A measure commonly used to designate the bore or size of pipes, other cylindrical or tubular objects, spheres, and plane circles.

diaphanometer (Metrol) An instrument for measuring the degree of transparency of solids, liquids, or gases.

diaphragm¹ (Techn) A thin partition.

diaphragm² (Anat) The dome-shaped muscle that separates the thoracic and abdominal cavities and drives the breathing cycle.

diastole (Physiol) The period of the cardiac cycle during which the heart is filled with blood and is relaxed. Cf: **systole**.

diastolic pressure (Med) The pressure of the blood during heart relaxation. Cf: **systolic pressure**.

diatomaceous earth (Techn) A soft, gritty amorphous silica composed of minute skeletons of small aquatic plants. Used in filtration and decolorization of liquids, insulation, wax, textiles, plastics, paint, rubber, and as a filler in dynamite. Calcined and flux calcined diatomaceous earths contain appreciable amounts of **cristobalite**.

dibasic acid (Chem) An acid containing two replaceable hydrogen atoms per molecule.

dibromodifluoromethane (Chem) A vaporizable liquid fire extinguishing agent. Also called: **Halon 1202**. See: **Halon**.

dibromotetrafluoromethane (Chem) A vaporizable liquid fire extinguishing agent. Also called: **Halon 2402**. See: **Halon**.

die (Techn) Any of various tools used to shape material. A pair or a combination of pairs of mating members for shaping material in presses.

dielectric (Electr) A material lacking a sufficient number of free electrons to pass an electrical current. An insulator. The material separating the plates of a capacitor.

diesel engine (Techn) An internal combustion engine for large vehicles, electrical generators, ships, etc. The diesel engine is characterized by high-pressure fuel injection as contrasted to carburetion in the Otto engine.

diesel fuel (Techn) An oil commonly used as a fuel in heavy-duty prime movers, such as bus, truck, industrial, and marine engines.

die setting (Techn) The process of placing or removing dies in or from a mechanical power press, and the process of adjusting the dies, other tooling and safeguards for proper operation.

differential equation (Math) An equation that gives the relationship between a function and its derivatives.

differential thermal analysis (Phys; Chem; Abb: DTA) A form of **calorimetry** in which the temperature difference between a sample and a blank are recorded during a heating cycle. Phenomena such as crystallographic rearrangements, phase transitions, melting, boiling, reaction, and glass transitions can be detected.

diffraction (Acoust; Opt) 1. The bending or distortion of a sound wave by an obstacle; a change in direction other than by reflection or refraction. 2. A physical optical edge effect in which radiation bends into the shadow region due to interference.

diffuser¹ (Fire Pr; Brit) A nozzle that can be adjusted to throw a stream or a fine spray.

diffuser² (Aerodyn) A shaped channel that changes the characteristics of the fluid passing through it to a lower velocity and a higher pressure.

diffusion (Chem) A molecular transport process in which a mass transfer of a species is induced by a concentration gradient. In general, the migration of atoms of one material into the bulk of another. The flux of the species is proportional to the concentration gradient, as stated by Fick's law. See also: transport processes.

diffusion burner (Comb) A burner used for the combustion of nonpremixed gases.

diffusion coefficient (Chem) The coefficient used in Fick's law to calculate diffusion flux. See also: diffusion.

diffusion flame (Comb) A nonpremixed laminar flame the propagation of which is governed by the interdiffusion of the fuel and oxidizer. The combustion reaction takes place in the mixing zone and the flame propagation velocity is governed by the diffusion rate. A candle flame is a typical example. See: flame.

diffusion pump (Techn) A type of high-vacuum pump in which a jet of mercury or oil molecules carry off gas molecules diffusing into the pump.

diffusion rate (Phys) The rapidity with which one gas or vapor disperses into another in the absence of mechanical mixing. This rate depends on the density of the vapor or gas compared with that of air, which is given a value of 1. See: Graham's Law.
NOTE: Light gases diffuse more rapidly than heavy ones. Diffusion rate is proportional to the square root of the gas or vapor density.

DIFR = disabling injury frequency rate

digester (Techn) A large steel boiler-like still used to heat materials; for example, to extract and vaporize turpentine from wood pulp.

digit (Math) 1. One of the symbols 0 through 9 when used for numbering in the scale of 10, regardless of position or the type of code in which they appear. 2. A character in any number system.

digital (Comput) 1. Numeric; involving whole or discrete numbers. 2. A discrete state of being, such as the presence or absence of a quantity.

digital computer See: computer

DII = disabling injury index

dilatation[1] (Soil Mech) Increase in volume of soil during shearing deformation.

dilatation[2] (Med) The condition of being abnormally stretched.

dilute (Techn) To thin or weaken by mixing with water or other liquid. To change or weaken in brilliancy, force, etc., by mixing with something else.

dilution[1] (Mining) Undesirable mixing of waste with ore during mining.

dilution[2] (Chem) The process of making thinner or less concentrated by increasing the proportion of solvent or diluent to fluid or particulate matter.

dimensional analysis (Math) A technique used to analyze complex physical systems through a study of dimensional relationships. Particularly used for establishing scaling procedures for model studies. See: dimensionless number; scaling.

dimensionless groups See: dimensionless numbers

dimensionless numbers (Phys) A meaningful ratio of parameters in a system arranged so that the dimensions cancel. This technique is useful in identifying factors in a complex system and in modeling or scaling. A well-known and early example of a dimensionless group is the Mach number, the ratio of the flow velocity to the local velocity of sound, used in describing flow systems. A large number of dimensionless groups are possible. The most complete list of such parameters has been compiled by Catchpole and Fulford, containing 285 dimensionless groups. Several useful and representative dimensionless numbers are given below.

Fanning friction factor: The ratio of the wall shear stress to the number of velocity heads. Used in calculating conduit flow.
$f = (d/4L)(2p/r^0v^2)$,
where
d = characteristic diameter
L = characteristic length
p = pressure drop
r^0 = fluid density
v = mean velocity.

Froude number: The ratio of inertial force to gravitational force.
$Fr = v^2/gL$
where
g = acceleration of gravity.

Grashof number: A number used in calculating free or natural convection heat transfer processes, given by
$Gr = (gr^{0^2}BTL^3)/u^2$,
where
B = thermal coefficient of volume expansion
T = temperature difference
u = viscosity.
Written as
Gr/Re^2,
the group expresses the ratio of buoyancy forces to inertial forces; i.e.,
Gr = (buoyancy forces)(inertial force)/(viscous forces)^2.

Lewis number: The ratio of the energy transported by conduction to that transported by diffusion, given by
$Le = k/r^0DCp$,
where
k = coefficient of thermal conductivity
r^0 = density
D = diffusion coefficient
Cp = specific heat at constant pressure.
The Lewis number is also expressed as the ratio of the Schmidt number to the Prandtl number
$Le = Sc/Pr$.

Mach number: The most important parameter in compressible flow theory, the Mach number is the ratio of the fluid velocity to the velocity of sound
$M = v/a$
where
M < 1 is subsonic flow, and
M > 1 is supersonic flow.

Nusselt number: A number used in calculating forced convective heat transfer, given by
$Nu = hL/k$,
where
h = heat transfer coefficient
k = thermal conductivity.
The Nusselt number (also called: Biot number) may be interpreted as a dimensionless temperature gradient averaged over the heat-transfer surface. It is often expressed as a function of Re, Pr, and geometry
$Nu = Nu(Re, Pr, L/d)$.

Peclet number: The ratio of heat transfer by convection to heat transfer by conduction
$Pe = (Cpr^0vL)/k$.

Prandtl number: The ratio of molecular diffusivity of momentum to diffusivity of heat
$Pr = (Cpu)/k$.
The heat transfer analog of the mass transfer Schmidt number.

Reynolds number: The ratio of inertial forces to viscous forces
$Re = r^0vd/u$.

Rossby number: The ratio of inertial force to Coriolis force
$Ro = v/2WeL \sin \theta$,
where
We = angular velocity, and
θ = angle between the axis of the earth's rotation and the direction of fluid motion.

Schmidt number: The ratio of kinematic viscosity to molecular diffusivity
$Sc = u/r^0D = n/D = u/r^0$
n = kinematic viscosity
The Schmidt number is the mass transfer analog of the heat transfer Prandtl number.

Stanton number: The ratio of the Nusselt number to the product of the Reynolds number times the Prandtl number
$St = Nu/(Re\ Pr)$.

dinitrobenzene (Chem) A yellowish crystalline compound, $C_6H_4(NO_2)_2$, used as a dye intermediate. Hazardous by skin absorption, inhalation, and ingestion.

dip (Mining) The angle between an ore bed or a vein and the horizontal. The angle of incline is called the dip angle.

dipole antenna (Commun) Two metallic elements, each approx one-quarter wavelength long, that radiate the radio frequency energy fed to them by a transmission line.

dip tank (Techn) A tank, vat, or container of flammable or combustible liquid in which articles or materials are immersed for the purpose of coating, finishing, testing, or similar processes.

direct attack (Fire Pr) Playing water streams directly on a fire as compared with indirect application, in which water is applied to generate steam to cool the burning combustibles.

direct cause (Safety) Unsafe behavior or unsafe conditions that contribute to an accident.

direct current (Electr; Abb: dc; as adjective, d-c) Electrical current that flows in one direction.

direct damage (Insur) Damage caused by the direct action of a peril, as distinguished from damage done contingently.

direct drive (Techn) A power transfer arrangement without a clutch. Driving torque is applied or removed by starting and stopping a motor.

directed smoke (Fire Pr) Smoke moving directionally.

direct injury costs (Insur) The sum of compensation payments and medical expenses for an injury.

directional stability (Transp) The property of a vehicle that governs its course-keeping ability.

direct labor (Man Sci) Labor directly applied to production, e.g., to an essential step in a manufacturing process.
  NOTE: **indirect labor** usu includes employees whose jobs cannot be associated directly with a specific product or process. The precise meaning of this term varies from company to company.

direct loss (Fire Pr; Insur) = **fire loss**

direct method (Fire Pr) Application of extinguishing agents directly on a forest fire to smother it or physical separation of burning and unignited fuel, as opposed to indirect methods, such as construction of fire control lines, backfiring, etc.

director (Fire Pr) The principal administrator in certain fire departments, often carrying the full title of director of public safety or director of fire.

disability (Safety) Any injury or illness, temporary or permanent, that prevents a person from carrying on his usual activity. See: **temporary disability**; **permanent disability**.

disabling injury (Safety) 1. An injury that prevents a person from performing a regular job for one full day beyond the day of the accident. 2. Bodily harm that results in death, permanent disability, or any degree of temporary or total disability.

disabling injury frequency rate (Safety; Abb: DIFR) The number of disabling (lost time) injuries per million employee-hours of exposure: DIFR=(Disabling Injuries x 1,000,000)/(Employee-hours of exposure).

disabling injury index (Safety; Abb: DII) An index computed by multiplying the disabling injury frequency rate by the disabling injury severity rate and dividing the product by 1000: DII=(DIFR x DISR)/(1000). This measure reflects both frequency and severity, yielding a combined index of total disability injuries.

disabling injury severity rate (Safety; Abb: DISR) The total number of days charged per million employee-hours of exposure: DISR= (Total days charged x 1,000,000)/(Employee-hours of exposure).

disaster (Fire Pr; Insur) A general catastrophe involving widespread damage and loss of life and property.

disaster control (Civ Def) Advanced planning and established procedures for handling emergency situations.

Major considerations include provisions for protecting personnel, evacuating both injured and uninjured, and care of the incapacitated. Injuries and property damage result from seven basic causes: work accidents, fire, explosion, floods, hurricanes and tornadoes, earthquakes, and civil strife.

disaster unit (Fire Pr; Civ Def) A mobile emergency unit carrying special equipment for coping with disasters, including the trained personnel that man the equipment.

discharge (Fire Pr) The rate of flow of water through a pipe, usu expressed in gallons per minute.

discharge channel (Techn) The passage through which a fluid must pass to reach the atmosphere.

discharge coefficient (Hydraul) The ratio of the actual flow of a liquid emerging from an orifice or pipe to the theoretical flow.

discharge device (Fire Pr) Any of a variety of mechanical outlets used to release a fire extinguishing agent under pressure. A nozzle, sprinkler head, and the like.

discharge gage[1] (Fire Pr) A gage for measuring the exit water pressure from a pump.

discharge gage[2] (Phys) A vacuum gage (10^{-3} to 10^{-13} mm Hg) using a cold cathode discharge. Also called: Phillips gage; Penning gage.

discharge gate (Fire Pr) A gate valve provided to control water flow from an individual pump outlet.

discharge outlet (Fire Pr) An output connection on a fire pump for hose lines.

discipline[1] A branch of knowledge or learning, e.g., engineering, psychology, philosophy, mathematics, physics, chemistry, etc.

discipline[2] (Law) To punish; to enforce obedience, rules of conduct, or patterns of behavior by punishment or penalty.

discontinuity (Phys) An interruption in a regular sequence or continuum.

discovery (For Serv) Finding that a forest fire exists, even though its precise location may be unknown. Cf: detection.

discovery time (Fire Pr; For Serv) The time of day that a fire was found and subsequently reported.
NOTE: The time elapsed between discovery and reporting to fire suppression authorities is called presignal delay.

discrete (Math, Phys) Consisting of distinctly separate elements.

discriminating power (Psych) The ability of a test item to differentiate between persons possessing much of some trait and those possessing little.

discrimination (Psych) The process of distinguishing between two or more things on the basis of details or characteristics.
NOTE: Two-point discrimination is the ability to tell whether one is being touched by one object or two, determined by the spacing of the touch receptors in the skin.

discriminator (Electr) An electronic circuit that selects signal pulses according to pulse height or voltage, used to delete extraneous radiation counts or background radiation, or as the basis for energy spectrum analysis.

disease (Med Path) Any deviation of the body from its normal or healthy state or a particular disorder with a specific cause and characteristic symptoms.

disease vector (Med) A carrier that is capable of transmitting a pathogen from one organism to another; esp an insect that transmits protozoa from one human host to another (as with malaria) and animals that are intermediate hosts (as with trichinosis).

disfigurement (Med) A blemish, defect, or deformity which harms the appearance or attractiveness of the human body or a physical object, including scarring due to burns or other injuries.

disinfectant (Med) An agent used to

destroy or render harmful microorganisms inactive, usu on inanimate objects. Cf: <u>antiseptic</u>.

<u>disinfection</u> (Med) The destruction of pathogenic organisms, esp by means of chemical substances.

<u>disintegrate</u> (Nucl Phys) To emit energy from an atomic nucleus in a spontaneous transformation. To decay through a process or change involving atomic nuclei, accompanied by the release of energy in the form of radiation.

<u>dislocation</u> (Med) Slippage of a bone out of its normal position in a joint. See: <u>first aid</u>.

<u>dismiss</u> (Fire Pr) 1. To relieve a fire-fighting unit from duty. 2. To order a unit back to quarters.

<u>disorderly conduct</u> (Law) Loud, boisterous, disruptive behavior such as fighting in public, using abusive language, and interfering with others. Syn: <u>disturbing the peace</u>.

<u>disparity pay</u> (Fire Pr) A salary differential paid to a member of a fire department for completion of an approved educational program or for participation in community activities in addition to fire service duties.

<u>dispatch</u> (Fire Pr) To order a fire company to respond to an alarm.

<u>dispatcher</u> (Fire Pr; For Serv) An operator in a communications center who receives and confirms alarms and notifies the initial fire-fighting units.
  NOTE: Reinforcements, as needed, are assigned by the dispatcher on orders from the officer in charge at the fire.

<u>dispatcher's meter</u> (For Serv) A device that indicates the strength of attack required for specific fuels to control a fire within a specified time period.

<u>dispersion</u>^{1} (Chem; Phys) The scattering of fine particles of a substance in another medium.

<u>dispersion</u>^{2} (Opt) The scattering of light by an optical process.

<u>dispersion</u>^{3} (Statist) The scatter of scores in a frequency distribution or on a scatter diagram. The distribution of values of a variable about its mean.

<u>displacement</u> (Mech) A vector quantity that specifies the change in position of a body from its mean position of rest.

<u>dispnea</u> = <u>dyspnea</u>

DISR = <u>disabling injury severity rate</u>

<u>dissociation</u> (Chem) Decomposition of a compound into simpler constituents by heat or solvents.

<u>dissociation energy</u> (Phys) The energy required to separate diatomic molecules into individual atoms in the ground state.

<u>distal</u> (Anat) Away from a reference point or from the center of a body.

<u>distal cause</u> (Safety) A behavioral act or condition leading to an accident, but separated from the accident location by time or space. Cf: <u>proximate cause</u>.

<u>distillation</u> (Chem) The process of purifying a liquid by vaporizing and then condensing it.

<u>distillery</u> (Techn) A plant or portion of one in which liquids produced by fermentation are concentrated by distillation.

<u>distortion</u> (Acoust; Electr; Techn) An undesired change in shape, dimensions, or fidelity, such as of a sound or signal.

<u>distraction</u> (Psych) Anything heard, felt, touched, or even thought about that diverts the attention of the mind or eyes away from a task or situation to which close attention should be paid.

<u>distress</u> (Med) A state of misery, pain, or suffering.

<u>distribution</u>^{1} (Chem) The process of dividing and dispersing materials or products. The tendency of a solute to reach an equilibrium concentration between two phases of a mixture of immiscible solvents.

distribution² (Statist) The array of values taken by a variable for a given population.
NOTE: As an example, a tabulation of test scores ordered from high to low, or low to high, showing the number of individuals who obtained each score or who fall into each score frequency interval. Generally called a frequency distribution. The two most common types are normal (Gaussian), in which the mean, median, and mode coincide, represented by a bell-shaped curve; and the Poisson, represented by an asymmetric curve. A sufficient number of measurements of a variable, such as the diameters of a production lot of ball bearings, will fall into a normal distribution. The occurrence of accidents is described by the Poisson distribution.

distribution curve (Statist) A curve on which values are plotted on the base line and their frequencies on the vertical axis.

distribution factor (Med) The modification of the effect of radiation in a biological system attributable to the nonuniform distribution of an internally deposited isotope, such as radium being concentrated in bones. See: absorbed dose; dose equivalent; relative biological effectiveness.

distributor (Fire Pr) 1. A cellar nozzle with a rotating tip that sprays water in a circular pattern. Syn: circulator. 2. A perforated pipe used to spray water omnidirectionally inside concealed spaces.

distributor truck (Fire Pr) A truck equipped with a manifold of inlets and gated outlets, used for distributing water. Syn: manifold truck.

district chief (Fire Pr) A chief fire officer commanding a local district. The rank is approx equivalent to battalion chief. See: ranks and titles.

diuretic (Med) A drug that stimulates the flow of urine.

diurnal 1. Recurring every day or having a daily cycle. 2. Occurring during the daytime.

divided body (Fire Pr) A hose compartment on a fire apparatus that is separated into two sections for storing different types of hose layouts or to permit laying two hose lines simultaneously.

divider (Fire Pr) A vertical partition in a hose compartment that separates stowed hose of different sizes or different loads.

dividing breeching (Fire Pr; Brit) See: breeching.

division (For Serv) A portion of a fire perimeter between designated geographic features and organized into two to four sectors. A division is usu so planned that a division boss can inspect it twice per shift.

division boss (For Serv) A staff supervisor responsible for all fire suppression work in a fire division.

division marshal (Fire Pr) A deputy chief in command of a division of several battalions. See: ranks and titles.

dizziness (First Aid) A whirling sensation accompanied by loss of balance. Faintness, giddiness, lightheadedness, or any similar sensation. Also called: vertigo.

dock (Transp) An open area of water next to a pier or wharf on which a vessel floats while moored.

doctor's case (Safety) A work injury requiring treatment by a physician.

documentation¹ (Info Sci; Comput) The science of collecting, describing, storing, and organizing recorded information or documents for optimum access.

documentation² (Comput) Description of a computer program, including narrative, flow charts, etc.

dog¹ (Techn) a. A mechanical lever-type latch, such as one used to seal a watertight door. b. To seal such a door by engaging and securing the latches.

dog² (Fire Pr) a. To secure a ladder or other equipment. b. A ratchet, pawl, or other device used to lock a ladder in extended position. See:

ladder. See also: Sparky.

dog chain (Fire Pr) A chain with a ring and spike at opposite ends, used to secure ladders to buildings. A fastening device similar to a rope hose tool.

dog house = igloo

DOL = US Department of Labor

domal sampling (Soc Sci; Statist) Home sampling; a form of area sampling in which certain residences are selected and certain household members are surveyed.

donkey engine (Techn) A small stationary steam engine used to drive winches, pumps, and auxiliary machinery.

donut roll (Fire Pr) A length of hose flattened and coiled into a donut shape for ease of handling and storage.

door breaker (Fire Pr) A heavy-duty instrument used to force an entry into a building. Also called: door opener. See: Detroit door opener; forcible entry tools.

door retainer (Fire Pr) A device that holds a fire-check door open under normal conditions and releases it automatically when smoke or flame is detected, usually an electromagnet that is released by a fire or smoke detector.

dormer (Build) A projection, usu with a vertical window in the outer face, built out from a sloping roof. Originally a sleeping chamber.

dormitory (Fire Pr) Fire company sleeping quarters. Also called: bunk room.

dorsal (Anat) Pertaining to the back.

dose¹ (Med) The proper quantity of medicine to be taken at one time.

dose² (Dosim) The quantity of ionizing radiation absorbed, per unit of mass, by the body, by any organ, or other portion of the body.

dose equivalent (Dosim) The quantity of effective radiation when modifying factors have been considered. The product of absorbed dose multiplied by a quality factor and by a distribution factor, expressed numerically in rems.

dose rate (Dosim) The radiation dose delivered per unit time and measured in rems per hour. See: absorbed dose; rem.

dosimeter (Dosim) A detector in which ionizing radiation deflects a needle to record the quantity of radiation received by the wearer. Also called: dose meter.

dosimetry (Abb: Dosim) The science and practice of measuring the intensity of absorbed radiation.

DOT = US Department of Transportation

DOT design pressure (Transp) Maximum allowable working pressure of a vessel or container set forth by the US Department of Transportation.

dote See: decay

double action pump (Fire Pr) A pump that discharges on both the forward and backward strokes. See: pump.

double adapter (Fire Pr) A hose connector with two identical couplings. A double female adapter has two female swivels and is used to connect two male hose nipples. A double male adapter is used to connect two female couplings.

double bank (Fire Pr) Two parallel stacks of ground ladders carried on a ladder truck.

double donut (Fire Pr) Two lengths of hose coiled into a single donut roll or one length rolled into two small coils for ease of handling.

double female adapter See: double adapter

double hydrant (Fire Pr) A hydrant with two outlets.

double jacket hose (Fire Pr) A hose with two fiber jackets covering a rubber lining.

double male adapter See double adapter

do͟u͟b͟l͟e͟-͟w͟i͟d͟e͟ Referring to two mobile homes placed side-by-side to form a single dwelling.

do͟u͟g͟h͟n͟u͟t͟ = d͟o͟n͟u͟t͟

d͟o͟w͟n͟ ͟h͟o͟l͟e͟ (Mining) A hole drilled in the downward direction. Also called: w͟a͟t͟e͟r͟ ͟h͟o͟l͟e͟, because water can be retained in the hole during drilling. Ant: u͟p͟p͟e͟r͟; d͟r͟y͟ ͟h͟o͟l͟e͟.

d͟o͟w͟n͟t͟i͟m͟e͟ (Man Sci) A period of time during which workers are unable to perform their tasks while waiting for materials, repair, setup, or adjustment of machinery. Under incentive wage systems, this term may refer to payment made to employees for such lost time. Also called: d͟e͟a͟d͟ ͟ ͟t͟i͟m͟e͟; d͟e͟l͟a͟y͟ ͟ ͟t͟i͟m͟e͟; w͟a͟i͟t͟i͟n͟g͟ ͟t͟i͟m͟e͟.

d͟r͟a͟f͟t͟¹ (Fire Pr) To draw water from a static source through a pump above the water supply level. The pump removes the air from an internal chamber and atmospheric pressure forces the water into the pump.

d͟r͟a͟f͟t͟² (Comb; Fire Pr) a. The pressure accelerating air toward a fire, supplying oxygen for the combustion process. b. (Brit) = d͟r͟a͟u͟g͟h͟t͟.

d͟r͟a͟f͟t͟³ (Navig) The distance from the static waterline to the keel. The depth of water drawn by a ship.

d͟r͟a͟f͟t͟ ͟c͟u͟r͟t͟a͟i͟n͟ (Build) A barrier suspended from the ceiling around open stairways and elevator shafts to retard or block heat so that it can accumulate more rapidly and actuate sprinkler heads. Syn: c͟u͟r͟t͟a͟i͟n͟ ͟b͟o͟a͟r͟d͟.

d͟r͟a͟f͟t͟i͟n͟g͟ ͟p͟i͟t͟ = p͟u͟m͟p͟e͟r͟ ͟p͟i͟t͟

d͟r͟a͟f͟t͟ ͟s͟t͟o͟p͟p͟i͟n͟g͟ See: f͟i͟r͟e͟ ͟s͟t͟o͟p͟p͟i͟n͟g͟

d͟r͟a͟g͟ (Aerodyn) The retarding force of a fluid medium on a body moving through it. Syn: s͟k͟i͟n͟ ͟f͟r͟i͟c͟t͟i͟o͟n͟.

d͟r͟a͟g͟ ͟c͟o͟n͟v͟e͟y͟o͟r͟ (Techn) A conveyor that uses vertical steel plates fastened between two continuous chains to drag material across a smooth surface.

d͟r͟a͟g͟g͟i͟n͟g͟ (Rescue) Systematic sweeping of a body of water with boats and grappling hooks for the purpose of recovering a drowning victim.

d͟r͟a͟g͟ ͟r͟e͟s͟c͟u͟e͟ See: r͟e͟s͟c͟u͟e͟ ͟c͟a͟r͟r͟y͟

d͟r͟a͟g͟ ͟s͟a͟w͟ (For Serv) A power-driven, reciprocating crosscut saw used for bucking logs.

D͟r͟a͟m͟a͟m͟i͟n͟e͟ (Pharm) Trade name for dimenhydrinate, $C_2_4H_{28}ClN_5O_3$, an antihistamine used to treat motion sickness.

d͟r͟a͟w͟b͟a͟r͟ ͟p͟u͟l͟l͟ (Transp) The amount of tractive effort developed by a vehicle in excess of motion resistance (net tractive effort).

d͟r͟a͟w͟b͟a͟r͟ ͟p͟u͟l͟l͟-͟w͟e͟i͟g͟h͟t͟ ͟r͟a͟t͟i͟o͟ (Transp) An index of the pulling efficiency of a vehicle, similar in concept to the lift-drag ratio for an aircraft. The drawbar pull-weight ratio indicates the effort available for hill climbing, vehicle acceleration, load towing, etc.

d͟r͟e͟n͟c͟h͟e͟r͟ (Fire Pr; Brit) A device that provides a curtain of water across a window or door to block a fire from passing through. A w͟a͟t͟e͟r͟ ͟c͟u͟r͟t͟a͟i͟n͟.

d͟r͟e͟s͟s͟ (Techn) To smooth or finish a surface.

d͟r͟e͟s͟s͟i͟n͟g͟ (First Aid; Med) A sanitary covering, usu of gauze, for a wound to protect it from contamination and infection.

d͟r͟e͟x͟ (Text) The mass in grams per 10,000 meters of yarn.

d͟r͟i͟f͟t͟ (Mining) 1. A more-or-less horizontal passageway in a mine, dug along, on, or parallel to a vein or to the strike of an orebody. 2. A small access crosscut passageway connecting two larger mine tunnels.

d͟r͟i͟f͟t͟ ͟s͟m͟o͟k͟e͟ (For Serv) Smoke that has floated to a distance from its origin and has lost its original form.

d͟r͟i͟l͟l͟¹ (Fire Pr) Practice of firefighting techniques, including laying hose, raising ladders, and operating pumps.

d͟r͟i͟l͟l͟² (Mech Eng) A cutting tool for boring holes.

129

drill manual (Fire Pr) A handbook that explains fire fighting practices and how they are to be performed for a given fire department.

drillmaster (Fire Pr) A department training officer responsible for conducting drills both for basic training and general company proficiency.

drill school (Fire Pr) A school for basic training of probationary firemen and for general company drills.

drill tower = **training tower**

dripping (Textiles) Formation and falling of molten drops; a characteristic of some plastics and textile materials when exposed to heat.

drip room (Fire Pr) A heated space in a fire station for drying items, such as salvage covers.

drip torch (For Serv) A small fuel tank with a handle, nozzle, and igniter used to drip a burning mixture of oil and gasoline on forest fuels to start a backfire or prescribed fire. Also called: **backfire pot**.

drizzle (Meteor) A fine precipitation of water. A light rain. See: **precipitation**.

drop = **aerial drop**

drop ceiling = **false ceiling**

drop lock (Fire Pr) A temporary method of securing doors that have been forced or broken in. Blocks of wood are nailed to the door and the floor and a board is braced between them.

dropout (Educ) A student, esp of high school level, who stopped attending classes.

drop water (Fire Pr) Water draining away during salvage operations through holes opened in the floor.

dross (Metall) The scum formed on the surface of molten metals, largely oxides and impurities.

drought (Meteor) An extended period of dry weather.

drown (Fire Pr; Brit) To extinguish a fire, esp one that is difficult to reach, by applying large quantities of water.

drug abuse (Law) Use of drugs that are not prescribed by a medical doctor and which may be forbidden by law.

drum^1 (Transp) A cylindrical shipping container having straight sides and flat, concave, or convex ends, designed as an unsupported outer package that may be stored or shipped without boxing or crating. Made of metal, plywood, or fiber with wooden, metal, or fiber ends. Drums are also made of rubber and polyethylene.

drum^2 (Mech Eng) The cylindrical members around which ropes are wound for raising and lowering a load or boom.

dry adiabatic lapse rate See: **lapse rate**

dry-bulb temperature (Meteor) The temperature of a gas or mixture of gases indicated by a standard thermometer after correction for radiation.

dry chemical (Fire Pr) A powder mixture used as a fire extinguishing agent, the most common being based on borax, sodium bicarbonate, monoammonium phosphate, potassium bicarbonate (Purple-K), potassium chloride (Super-K), and urea-potassium bicarbonate. Cf: **dry powder**.

dry-chemical extinguisher (Fire Pr) A fire extinguisher containing a dry chemical power that is ejected by compressed gas and extinguishes fire. These extinguishers are used primarily on Class B (liquids) and C (electrical) fires; however, multipurpose dry chemicals are also effective on Class A (ordinary combustibles) fires. Also called: **dry powder extinguisher**.

dry hydrant (Fire Pr) A pipe with connectors permanently installed at a water source such as a river or pond for use by fire department pumpers.

dry ice (Techn) Solid carbon dioxide, CO_2, used as a refrigerant. Nontoxic, noncorrosive, sublimes at -109°F, triple point -69.9°F.

dry lightning storm (Meteor) A lightning storm with little or no precipitation.

dry powder (Fire Pr) A powder mixture used as an extinguishing agent for combustible metal fires. Cf: dry chemical.
  NOTE: A sodium chloride dry powder fire extinguisher is useful for combustible metal fires, such as sodium, titanium, uranium, zirconium, lithium, magnesium, and sodium-potassium alloys.

dry powder extinguisher = dry chemical extinguisher

dry riser (Build; Fire Pr) A normally empty riser connected to a dry sprinkler system. Sometimes erroneously called a dry standpipe, qv. See: riser.

dry rot (Build) A term loosely applied to many types of decay but esp to that in which the wood crumbles easily to a dry powder.
  NOTE: The term is misleading, since the microorganisms responsible for the decay require considerable moisture for growth.

dry sprinkler system (Fire Pr) A sprinkler system without water, pressurized with air, installed in areas subject to freezing conditions. When a sprinkler head is actuated, air pressure drops, releasing a dry valve that allows water to flow into the system.

dry standpipe (Build; Fire Pr) A standpipe normally not connected to a water main. When needed, a fire department pumper supplies water to the standpipe for delivery to outlets on upper floors of a building.

dry tank (Storage) Tanks designed for the storage of materials other than bulk liquids or gases.

dry valve (Fire Pr) The valve that releases water into a dry sprinkler system.

DTA = differential thermal analysis

dual-purpose machine (Fire Pr; Brit) A pumper equipped with a wheeled ladder. Also called: pump escape.

dual-purpose nozzle (Fire Pr; Brit) A hand pump nozzle that can deliver a stream or fine spray by adjusting a slide lever.

dual wheels (Transp) Two wheels mounted on the same end of an axle shaft. Dual wheels are usu counted as one wheel.

duct (Build; Techn) A tube or channel for the passage of a fluid.

ductile (Metall) Capable of being worked without prior melting.

ductility (Metall) The property that allows permanent deformation below the melting point without rupture.
  NOTE: Ductile materials, such as metals, can be drawn into wires or sheet-stamped into various shapes. Ductility is measured by the amount of elongation in a tension test.

duct velocity The velocity of a fluid passing through a duct cross section.

due process (Law) The procedure that is guaranteed by the Constitution, ensuring that no person can be deprived of life, personal liberty, or property except by law.

duff (For Serv) Finely divided organic material littering forest floors, consisting of partially decomposed leaves, twigs, and other debris. Dry duff ignites easily and propagates flame rapidly.

Dulong's formula (Comb) A formula for calculating the approx heating value of a solid fuel based on its ultimate analysis.

dump¹ (Env Prot) A land site where solid waste is discarded.

dump² (Fire Pr) a. To overturn a soda-acid or other chemical extinguisher and thereby actuate the charge. b. To discard any remaining charge in an extinguisher after use, preparatory to cleaning and recharging.

dunnage (Transp) Temporary blocking, flooring or lining, racks, strapping, stakes, planks, pneumatic pillows, or similar bracing or supports used to protect or secure freight.

duplex operation (Commun) Concurrent transmission and reception of messages without switching between speaking and listening, such as in the use of the ordinary telephone.

durability (Techn) The capacity of a component or material to provide lasting service and to withstand abuse. If two or more comparable items are subjected to the same conditions, the one that performs satisfactorily for the longest period of time is the more durable.

dust (Env Prot) Finely divided solids sufficiently light to float in the air but settling under the influence of gravity on surfaces when not affected by air currents or other disturbances. Fine suspended dust can be inhaled, swallowed, and absorbed by the body. See also: aerosol.
NOTE: Dust particles range in size from 0.1 to 25 microns (1 micron = 1/10,000 cm = 1/25,000 in.). Dusts above 5 microns in size usu will not remain airborne long enough to present an inhalation problem. Solid particles are generated by handling, crushing, grinding, rapid impact, detonation, and decrepitation of organic or inorganic materials such as rock, ore, metal, coal, wood, and grain. Dusts do not tend to flocculate except under electrostatic forces. Combustible dusts are classified as: 1) those that can be ignited by a small heat source and propagate flame easily; 2) those that ignite readily but require a substantial heat source; and 3) those that do not form clouds readily and do not burn rapidly.

dust collector (Techn) An air cleaning device that removes particulates from exhaust systems before the air is discharged into the atmosphere.

dust devil (Meteor) A whirlwind created by shearing winds. The whirlwind picks up dust and other light debris from the ground and swirls them upward, sometimes to heights of a hundred feet or more.

dust explosion (Fire Pr) Rapid combustion of a suspension of dust particles such as flour, metal, plastic, etc. See also: explosion.

dust-ignition-proof (Techn) Enclosed in a manner to exclude ignitible quantities of dusts of quantities that might affect performance or rating and which will not permit arcs, sparks, or heat generated or liberated inside the enclosure to ignite exterior accumulations or atmospheric suspensions of dust.

dust loading (Safety) The amount of dust in a gas, usu expressed in grains per cubic foot or pounds per thousand pounds of gas.

dustproof (Electr) Built or protected in such a way that dust will not interfere with normal operation.

dust-tight (Electr) Enclosed and sealed to prevent dust from entering.

dutchman (Fire Pr) A short fold in accordion-folded coupled hoses to allow space for a coupling. Used to minimize snagging when the hose is payed out.

duty (Electr) The time that service is required of an electrical motor.
NOTE: Continuous duty is service required at constant load for an extended period. Intermittent duty is service required at load and no load, or load and rest for alternate intervals of time. Periodic duty refers to regularly recurring loads. Varying duty refers to loads, service intervals, and service demand times that vary substantially.

dwelling (Fire Pr) Any building used exclusively for residential purposes, having not more than two living units, or serving not more than 15 persons as a boarding or rooming house.

dwelling unit (Fire Pr) One or more rooms used by one or more persons living together as a single household with living, cooking, sanitary, and sleeping accommodations.

dynamic balance (Mech) The state of

equilbrium during motion; absence of lateral inertial forces acting on a rotating body.

dynamic load (Mech) Forces imposed by dynamic action, such as wind, acceleration, motion, etc.

dynamic pressure (Mech) The pressure that accelerates a fluid.

dynamics (Mech) The science of the motion of particles and systems under the influence of external and internal forces. A branch of mechanics.

dynamic viscosity (Fl Mech) The ratio of the shear stress to the shear of the motion of a moving fluid.

dynamotor (Electr) An electrical device capable of acting as a motor or as a generator.

dyne (Phys) An absolute unit of force in the cgs system, defined as the force required to accelerate a mass of one gram by 1 cm/sec^2. A dyne is approximately equal to the force exerted by one milligram accelerated by the force of gravity. One gram represents a force of approx 980 dynes.

dysacusis (Med) 1. A hearing impairment in which there is a loss of discrimination between words, syllables, or other sounds. 2. A condition in which sound produces pain in the ears.

dysbarism (Med) Adverse effects on the human body caused by changes in environmental or ambient pressure.

dyspnea (Med) Shortness of breath, difficult or labored breathing. Also called: air hunger.

E

e (Math) The base of natural logarithms, approx equal to 2.71828.

e (Phys) Symbol for the charge of an electron; emissivity.

E (Phys) The symbol for voltage; illuminance; internal energy.

ear (Anat) The organ of hearing and balance in vertebrates.
NOTE: The ear consists of three major parts: 1) The sound collecting outer ear (pinna or auricle and lobe), 2) the sound-transmitting middle ear or tympanic cavity (the ear canal, auditory meatus, and tympanic membrane or eardrum), and 3) the sensory inner ear (auditory nerve, cochlea, semicircular canals, stapes, incus, and maleus).

ear defenders (Safety) Plugs or muffs designed to protect the ears from noise. Syn: ear protectors.

early cause (Safety) An act or condition that leads to a proximate cause or mediate cause or allows one to exist. Also called: distal cause.

eccentric (Mech) Off center; irregularly configured, or not having a common center. Having the axis of revolution outside the geometrical center of gravity.

ECG = electrocardiogram

echo¹ (Acoust) A reflected sound wave.

echo² (Phys) A reflected electromagnetic signal, such as a light beam or radio wave.

ecology (Biol; Abb: Ecol) The branch of biology that deals with the relations between living organisms, esp humans, and their environment.

econometrics (Statist) The application of statistical methods to problems of economics.

economics (Abb: Econ) A social science dealing with the description and analysis of monetary and social problems of production, distribution, and consumption.

economic theory (For Serv) A theory of forest fire control which examines means of minimizing the total cost of prevention, presuppression, suppression, and damage.

economizer (Safety) A reservoir in a continuous-flow breathing apparatus in which exhaled oxygen is collected for recirculation. See: breathing apparatus.

ecosystem (Biol) The complex of organisms and their surroundings.

eczema (Med) A skin disease or disorder, often of allergic origin. Syn: dermatitis.

eddy (Fl Mech) A circulation in fluid flow that retains its identity for a measurable length of time. The fluctuations in turbulent flow are called turbulent eddies.

eddy currents (Electr) Electrical currents induced in a conductor by the variation of an applied magnetic field.

eddy diffusivity (Fl Mech) The augmented diffusion of heat, momentum, and matter due to turbulent transport. See also: diffusion; turbulence.

edema (Med) Swelling, sometimes accompanied by weeping, of tissue due to an accumulation of lymph or other body fluids.

edge firing (For Serv) A method of broadcast burning, in which fires are set along the edges of an area and are allowed to burn toward the center.

EDITH = exit drills in the home

education (Educ) The process of developing mental skills, esp by a program of formal instruction and study in an academic environment.

educational occupancy (Fire Pr) A building or group of buildings used for purposes of instruction, including schools, universities, colleges, academies, nursery schools, and kindergartens.

eductor pump (Techn) A pumping device designed for high flow with low pressure drop for liquids, suspen-

sions, or gases. An eductor has no moving parts, and pumping is produced by entraining the fluid in a high-velocity stream, which can be a jet of water, air, steam, or other fluid. See also: pump.
  NOTE: In fire practice an eductor in a hose line proportions liquid foam or wetting agents into the water stream. An eductor designed for high pressure drop or vacuum operation is called an aspirator or ejector.

effective half-life (Nucl Safety) The time required for a radionuclide contained in a biological system, such as a man or an animal, to lose half of its radioactivity as a result of decay and biological elimination. Cf: biological half-life. See: half-life.

effective horsepower (Abb: EHP) Power actually delivered by a propulsor, written EHP = (DV/C), where D = drag, V = speed, C = constant (the value of C depends on the units of D and V).

effective temperature index (Physiol) An empirically determined index of the degree of warmth or cold felt by the body on exposure to different combinations of temperature, humidity, and air movement.
  NOTE: The numerical value is equal to the temperature of still, saturated air which would induce the identical sensation. Determination of the effective temperature index requires simultaneous measurement of wet- and dry-bulb temperatures. See: wind chill index.

effective value (Electr) The equivalent heating value of an alternating current, as compared to a direct current, equal to 0.707 times the peak value of a sine wave. Also called: root-mean-square value.

effervescence (Chem) The escape of gas from a liquid.

efficiency (Symb: n) The ratio of the useful output of a device or system to the total input, generally with reference to power.

efflorescence (Chem; Phys) The loss of water of crystallization (hydration) on exposure to air. The process occurs to some extent in any hydrated salt, but it is particularly noticeable when the vapor pressure of the salt exceeds the absolute humidity. Ant: deliquescence.

effluent (Env Prot) Anything that flows out, like a stream flowing into a lake. Often in reference to an outflow of a sewer, storage tank, canal, or other channel.

effluent seepage (Env Prot) Diffuse discharge onto the ground of liquids containing dissolved or suspended materials that have percolated through solid waste or another medium.

effusion (Med) 1. The escape of fluid from blood or lymph vessels into the tissues or a body cavity. 2. The fluid that is escaping.

egress (Build) A continuous path of travel from any point in a building to the outside at ground level. Cf: exit; fire escape.

eigenvalue (Math) The characteristic solution of a boundary-value differential equation.

eight-foot tunnel See: FPL tunnel

ejector See: smoke ejector

ejector pump (Techn) A pumping device with a diffuser, mixing tube, and nozzle that mixes a driving fluid, usu water or steam, with the fluid to be pumped and ejects both. The pump is used to lift water from depths beyond the reach of suction pumps, such as from ship's holds, deep basements, etc. See also: eductor pump; pump.

EKG = electrocardiogram; electrocardiograph.

elapsed time (For Serv) The time a given step or phase in forest fire suppression was initiated or the time taken to complete it.
  NOTE: Chronologically, the key time points and periods are: discovery, report time, travel time, first attack time, control time, mop-up time, and patrol time.

elapsed time standards (For Serv) A schedule of maximum times allowed for given phases of wildland fire suppression.

elasticity (Phys) The property of a material or a body that enables it to return to its original shape after removal of a deforming stress.

elasticity coefficient = **modulus of elasticity**

elastic limit (Metall) The maximum stress that a material will withstand without permanent deformation.

elastic medium (Phys) A substance that deforms in direct proportion to applied stress or load.

elastomer (Chem) A synthetic polymer with rubberlike characteristics; a synthetic or natural rubber; or a soft, rubbery plastic with some degree of long-range **elasticity** at room temperature.

elbow-for-nozzle (Fire Pr; Brit) A metal attachment between a branch and a nozzle that delivers a stream at right angles to the branch.

electrical code (Build; Fire Pr) A body of rules governing the installation and operation of electric wiring, devices, and equipment, designed to safeguard life and property from electrical hazards.

electrical insulation (Electr) A material possessing a high degree of resistance to the passage of an electric current, usu of a vitreous or resinous material. A **dielectric**.

electric blasting circuitry (Civ Eng) A system of wires used to detonate blasting charges.

electric delay blasting cap (Civ Eng) A cap designed to detonate at a predetermined time after energy is applied to the ignition system.

electric shock (Safety) The spasm or convulsion that results from the passage of an electric current through the body.

electroacoustics (Phys) The science that deals with the transformation of sound waves into electrical signals and electrical signals into sound.

electrocardiograph (Med; Abb: EKG) A device used to produce electrocardiograms.

electrocardiogram (Med; Abb: ECG) A graphic recording of the electrical pulses generated by the heart, and hence a representation of cardiac functions; used esp in diagnoses of abnormalities in heart action.

electrochemistry (Chem) The science of chemical changes produced by electric current and the production of electricity from the energy of chemical reactions.

electrocution (Safety) Death resulting from electric shock.

electrode (Electr) A terminal at which an electrical current passes from one medium or conductor into another.
  NOTE: In an electrolytic cell, the positive electrode is called the **anode**, the negative electrode is called the **cathode**. In a voltage source, such as a battery, these designations are reversed.

electroencephalogram (Med; Abb: EEG) A graphic recording of the electrical activity of the brain.

electrolysis (Chem) A process in which a chemical compound in solution is plated out by means of an electric current passing through the solution. Positive and negative ions move to the negative and positive electrodes, respectively.

electrolyte[1] (Chem) A substance that dissolves as ions rather than as molecules, and is therefore electrically conductive.

electrolyte[2] (Med) A salt in the body fluids.

electromagnet (Electr) An iron core within a coil of wire that becomes magnetized when an electric current flows in the coil.

electromagnetic compatibility (Electr) The capability of electronic systems, equipments, and devices to operate in their intended environment without excessive electromagnetic interference.

electromagnetic radiation (Phys) Radiation consisting of electric and magnetic waves that travel at the speed of light. Examples are light waves, radio waves, gamma rays, and x-rays. All can be transmitted through a vacuum.

electromotive force (Electr; Abb: emf) The force, measured in volts, that causes an electric current to flow.

electron (Phys; Symb: e) An elementary particle with a unit negative electrical charge and a mass 1/1837 that of a proton. Electrons surround the positively charged nucleus and determine the chemical properties of the atom. Positive electrons, called positrons, also exist. Cf: antimatter.
NOTE: The weight of an electron is so infinitesimal that it would take 500 octillions (500 followed by 27 zeros) of them to make a pound.

electronegative element (Chem; Phys) An element with a tendency to take up electrons.

electron volt (Chem; Phys; Abb: ev) A unit of electrical energy equal to the kinetic energy of an electron accelerated by a potential difference of one volt. The unit is commonly used in describing atomic and molecular energies. An electron volt equals 1.602×10^{-19} joule = 1.602×10^{-12} erg.
NOTE: Energies in ev refer to individual molecules. To compute energies for bulk materials it is necessary to multiply by the number of molecules involved, using the number of moles and Avogadro's number.

electroplating (Techn) The process of applying a thin coating of one metal on another by electrodeposition.

electropositive element (Phys) An element that has a tendency to lose electrons.

electrostatic charge See: static electricity

electrostatic field (Phys) The field between two electrically charged bodies.

electrostatic precipitator (Techn) A device that traps particulates by placing an electrical charge on them and attracting them onto a collecting electrode, which is usu a screen. See: Cottrell precipitator.

element[1] (Chem; Phys) One of the 103 known solid, liquid, or gaseous chemical substances that cannot be reduced to simpler substances by chemical means.
NOTE: The atoms of an element may differ physically but do not differ chemically. All atoms of an element contain a definite number of protons and thus have the same atomic number. Elements are designated by chemical symbols, such as C for carbon, Ni for nickel, etc.

element[2] (Techn) A simple component of a complex entity or system.

elementary particle (Nucl Phys) Any of a number of subatomic particles, most of which are short-lived and do not exist independently under normal conditions except ions, electrons, protons, and neutrinos. Also called: fundamental particle.
NOTE: Originally, this term was applied to any particle which could not be further subdivided; now it is applied to nucleons (protons and neutrons), electrons, mesons, antiparticles, and strange particles but not to alpha particles or deuterons.

elevating nozzle (Fire Pr) A nozzle installed on an articulated elevating boom mounted on a fire department pumper.

elevating platform (Fire Pr) A basket-like platform attached to a mechanical or hydraulic power-operated boom mounted on an apparatus chassis, used to raise firefighters and equipment to upper floors for fire fighting and rescue work. Also called: aerial platform; cherry picker; hydraulic platform; snorkel. See also: basket.

elevation loss (Hydraul) Loss of pressure in a conduit due to the weight of the liquid being raised to a higher point.
NOTE: For water, the loss is roughly 0.5 psi per foot of elevation above the pump.

elevator capture (Build) Taking control of one or more elevators by overriding passenger-operated call buttons.

137

NOTE: Elevator capture is used to eliminate the risk of passengers being trapped in them, and sometimes to make elevators available exclusively for firefighters.

elongation (Metall) The amount of permanent elongation near the fracture in a tensile test, usu expressed as a percent of original gage length.

embolism (Med) Sudden blockage of an artery or vein by a clot (embolus) of blood, air bubble, fat globule, or other substance brought to the site by the blood flow, causing death of the tissue supplied by the blood vessel if occlusion is complete.

embrittlement (Metall) A condition in which a metal loses its strength and tends to crack or fracture easily. Loss of ductility.

embryo (Biol) 1. A developing vertebrate prior to hatching or birth. 2. The human organism from conception to the end of the second or third month of development. See: fetus.

emergency (Law) A situation in which human life or property is in jeopardy and prompt action or assistance is essential.

emergency alarm (Fire Pr; Safety) An alerting or warning device, usu visual or auditory, which indicates the existence of conditions that require action or counteraction.

emergency exit (Build; Safety) Any of several means of escaping from a building, including doors, windows, hatches, and the like, leading to the outside, and not used under normal conditions. Cf: secondary exit.

emergency lighting (Build; Safety) A system of lights installed along escape routes, at exits, in elevators, and at other points in a building to provide illumination in the event the normal lighting system fails.

emergency lock (Safety) An air lock designed to hold and permit the safe passage of an entire shift of employees.

emergency medical technician (First Aid; Med; Abb: EMT) An individual specially trained to render first aid and emergency care to victims of acute illness or injury on the scene and during transport to a medical care facility. Particularly applicable to ambulance and rescue squad personnel who work in communication with a physician.
NOTE: In many jurisdictions an EMT is not trained nor qualified to act as a paramedic, q.v.

emergency procedure A plan for action or steps to be taken in case of an emergency.

emergency stop (Safety; Techn) A switch installed in an elevator or on other moving equipment or machine that cuts off the power to the operating motor.

emergency switch A switch that cuts off or interrupts electrical power to a circuit, network, or to a piece of equipment.

emergency telephone number (Fire Pr) A three-digit telephone number (911) used in many areas to call for emergency assistance, such as the fire department, police, and ambulance.

emergency tender (Fire Pr) An apparatus containing specialized equipment. Examples are: cutting and lighting equipment, breathing and resuscitation equipment, etc.

emergency vehicle (Fire Pr; Police) A vehicle, usu equipped with audible and visible warning devices, that transports personnel and equipment to emergency incidents.

emery (Techn) Aluminum oxide, a natural or synthetic abrasive. See: corundum.

emesis (Med) Vomiting.

emission (Env Prot) Material that is released into the air either by a discrete source (primary emission) or as the result of a photochemical reaction or chain of reactions (secondary emission).

emission control (Env Prot) A rule established to limit the quantity of a given pollutant that is permitted to be discharged into the atmosphere.

emission factor (Env Prot) The statistical average of a specific pollutant emitted from each type of polluting source in relation to a unit quantity of material handled, processed, or burned.

emission inventory (Env Prot) A list of primary air pollutants emitted into a given community's atmosphere in quantities per day by type or source.

emission spectroscopy (Spectr) The spectroscopy of light given off by gas molecules excited by an electrical discharge, flame, or other source. Ant: absorption spectroscopy.

emissivity (Phys; Symb: e) The ratio of the radiation emitted by a substance at a given wavelength and temperature to that of a black body at the same wavelength and temperature, expressed as a number $0 \leq e \leq 1$.

emphysema (Med) A chronic lung disease in which the walls of the air sacs (alveoli) have broken down. Loss of elasticity by the lung tissues.

empirical Dependent on observed or experimental evidence.

employee (Man Sci) An individual engaged by another to perform specified services under given conditions for wages or salary.

employee benefit plan (Man Sci) A system, often insured, that provides disability and death benefits above and beyond those provided by workmen's compensation and is intended to include circumstances not covered by workmen's compensation. Syn: employee welfare plan.

employer's liability (Law) Legal liability imposed on the employer, making him responsible for paying damages to the employee injured because of the employer's negligence. Now replaced by workmen's compensation, which pays benefits to an injured workman regardless of negligence.

EMS = emergency medical services

EMT = emergency medical technician

emulsifier (Chem) 1. A chemical that holds one insoluble liquid in suspension in another. Casein, for example, is a natural emulsifier in milk, keeping butterfat droplets dispersed. 2. A substance, such as soap, that produces an emulsion.

emulsion (Chem) A dispersion of small globules of one immiscible liquid in another. The continuous phase is called the dispersion medium and the droplets the dispersed phase.

enamel (Build) An oily, paint-like substance containing various synthetic or natural resins that dries to produce a glossy finish on surfaces.

enclosed (Safety) 1. Referring to a moving part that is guarded to prevent accidental contact and possible injury. 2. Covered by a protective housing to prevent accidental contact with live electrical parts.

enclosed staircase (Build; Brit) = smokeproof tower

encoding (Info Sci) Transformation of messages into signals or recording data in machine-readable form.

end blocking (Transp) The blocking used to prevent end movement of lading.

endemic (Med) Localized; restricted to one area of the country; said of a disease.

end item (Techn) A final combination of a product, component part, or material that is ready for its intended use.

endocrine (Physiol) Secreting internally; not through a duct or tube. Pertaining to glands that secrete directly into the blood stream or the lymph.

endogenous fire (Comb) A fire originating from within.

endoscopy (Med) Examination of the interior of a hollow body organ or a canal.

endothermic (Chem; Phys) Characterizing a reaction that occurs with the absorption of heat. Ant: **exothermic**. Also called: **endothermal**.

endotoxin (Bact) A toxin retained within the body of a microorganism, as opposed to an exotoxin, which is secreted by the organism into the surrounding medium.

endotracheal tube (Med) A tube inserted into the **trachea** to supply oxygen to a patient to assist ventilation.

endurance (Build) A measure of the time a material, assembly, or structure can withstand failure while subjected to a given increase in temperature. Cf: **fire endurance**.

endurance limit (Metall) The limiting stress below which a metal will withstand an indefinitely large number of cycles of stress without rupture. Also called: **fatigue limit**.

endurance ratio (Metall) The ratio of the endurance limit to ultimate strength. This ratio for most ferrous metals lies between 0.4 and 0.6.

energy (Phys) The capacity to act, produce an effect, or to do work.
NOTE: Various forms include chemical energy, nuclear energy, thermal energy, kinetic energy, potential energy, etc. Potential energy is stored capacity to do work, as represented by a suspended weight. Kinetic energy is the capacity to do work because of motion.

energy deficit (Phys) Insufficiency of energy available for a given purpose.

energy density (Phys) The intensity of electromagnetic radiation per unit area per pulse expressed as joules per cm².

enforcement (Law) The exercise of executive power to require compliance, under penalty, to prescribed standards, codes, regulations, and laws.

enforcing authority (Fire Pr; Law) A fire department officer, government official, or agency legally responsible for enforcing regulations.

engine[1] (Mech Eng) A machine used to convert energy into work to propel vehicles and operate equipment. Also called: **motor**.
NOTE: Some authors distinguish engines as power generators and motors as power consumers. Others restrict the term motor to electrically driven devices.

engine[2] (Fire Pr) A fire department pumping engine.

engine company (Fire Pr) A fire company with a pumping engine.

engineer (Techn) A trained individual who applies the principles of science, technology, and mathematics in designing and developing structures, machines, systems, and manufacturing processes.

engineered performance standards (Abb: EPS) Time values (man-hours) expressed as standard time per unit of work it would take an adequately trained individual or group to perform a defined task as determined by a trained technician using time studies, random sampling, or other appropriate standard time data.

engineering (Abb: Eng) The application of scientific and technological principles to the control of the environment and materials for the purpose of satisfying human needs.

engine house (Fire Pr; Archaic) An obsolete term for **fire station**, originating from the time engines or steamers were housed separately from the men.

enteric (Med) Pertaining to the intestines.

enteric hemorrhage (Med) Bleeding into the intestinal tract.

enterotoxin (Med) A toxin (poison), released by microorganisms in the intestines, that produces gastrointestinal symptoms of food poisoning; a toxin specifically injuring the cells of the intestinal mucosa (lining).

enthalpy (Thermodyn; Symb: H) The sum of the internal energy of a system and the volume of the system multiplied by the pressure exerted on the system by its surroundings. Also Called: heat function; total heat; heat content. See also: entropy.

entropy[1] (Thermodyn; Symb: S) A measure of the unavailable energy in a thermodynamic system.
NOTE: An increase in entropy indicates a loss in the ability to do work. Any change occurring without change in entropy is described as isentropic.

entropy[2] (Info Sci) A measure of the average information content in a communications process.

entry See: forcible entry

entry loss (Techn) Loss in pressure caused by air flowing into a duct or hood.

environment (Env Prot; Soc Sci) 1. The conditions, circumstances, and influences surrounding and affecting the development of one or more organisms. 2. The surrounding conditions, influences, or forces to which an individual is exposed. 3. The aggregate of social and cultural conditions, the work situation, and the community that influences the life of an individual.

enzyme (Biol) A biologically active complex protein that has catalytic properties (i.e., breaks down compounds) and causes biochemical changes.
NOTE: Yeast produces an enzyme that acts as a catalyst in converting sugar to alcohol.

EPA = Environmental Protection Agency

epidemic (Med) An outbreak of disease, with rapid spread.

epidemiology (Med) The study of the incidence, spread, and control of disease in a population. The direct cause of the disease is called the agent, and the organism that transmits the disease pathogen is called a vector.

epidermis (Anat) The outermost layer of the skin. See: corium.

epiglottis (Anat) A flap at the top of the larynx that closes the airway during swallowing to channel liquids and solids to the esophagus and keep them out of the airway.

epilation (Med) The removal of hair by the roots, as opposed to alopecia. Baldness; loss of hair.

epinephrine = adrenalin

epistaxis (First Aid; Med) Nosebleed.

epithelioma (Med) Carcinoma of the epilthelial cells of the skin.

epithelium (Anat) 1. The tissue covering the skin and lining the respiratory, digestive, and genitourinary tracts of the body, providing protective, secretory, absorptive, sensory, and lubricating functions for passages. 2. The superficial or outermost skin layer. Syn: epidermis.

epoxy (Techn) A thermosetting, chemically resistant resin used in laminates, finishes, coatings, and primers. Made from epichlorhydrin and bisphenol.

equation of state (Thermodyn) An equation relating temperature, pressure, and volume of a system in thermodynamic equilibrium; commonly used for gases, but also for condensed phases. In condensed phases the equation of state is often described as a compressibility equation. See also: ideal gas law; van der Waal's equation.

equilibrium (Chem) A state in which forward and reverse reactions are taking place at the same rate and the concentrations of all substances are held constant.

equilibrium constant (Symb: K) The ratio of the product of the active concentrations of the substances produced in a reaction to the product of the active concentrations of the reactants after equilibrium has been reached.

equilibrium flow (Aerodyn) Fluid flow in which energy is constant across streamlines.

equipment officer (For Serv) The person responsible for managing and servicing tools, mechanical ground equipment, and pack equipment required for wildland fire suppression. The equipment officer reports to the service chief and supervises equipment service men, transportation operators, and packers.

equivalent form (Educ) Any of two or more forms of a test that are closely parallel in content and level of difficulty so as to yield similar average scores and measures of variability for a given test group.

equivalent weight = **combining weight**

ERDA = **Energy Research and Development Administration**

erg (Phys) A cgs unit of energy equal to the work done by the force of one dyne acting through a distance of one centimeter.

ergometrics (Physiol) Simultaneous measurement of heart, lung, blood circulation, and muscle functions during the performance of a task.

ergonomics (Biotech; Ind Eng) The science and technology dealing with the optimum use of human labor. Also called: **human dynamics**; **human engineering**. See: **bioengineering**.

erosion[1] (Geophys) The mechanical wearing away of soil and the process by which soil is carried away by the action of wind, rain, and other weather phenomena.

erosion[2] (Techn) The wearing away by the washing action of moving liquids, such as molten slags or metals, or the action of moving gases. See: **ablation**; **corrosion**.

error (Math) The difference between the true value and the observed or calculated value.

error register (Autom) A device for accumulating and signaling the algebraic difference between the quantized signal representing desired machine position and the quantized signal representing the instantaneous position of the machine. Also called: **error counter**.

erythema (Med) Reddening of the skin caused by dilation of the capillaries. A common condition resulting from exposure of the skin to heat, as in sunburn.

erythrocyte (Physiol) A red blood cell.

erythromycin (Med; Pharm) A broad-spectrum antibiotic effective against gram-positive coccus bacteria and used to control respiratory infection. See also: **antibiotic**.

erythron (Med) Red cells in the blood stream and their precursors in the bone marrow. See also: **hemoglobin**.

escalator[1] (Build) A moving stairway.

escalator[2] (Man Sci) Providing for periodic adjustment in proportion to a given index, as wages or salary with respect to cost of living. Referring to such a clause in a labor contract.

escape (Fire Pr; Brit) A wheeled extension ladder carried on a pumper-ladder truck.
NOTE: An **automatic escape** is a device for lowering persons from a height, fitted with an automatic brake to control the speed of descent. See also: **fire escape**.

escape line See: **life line**; **line**.

escape means = **fire escape**

escape route (Build) The egress path which occupants must follow when evacuating a building in the event of fire.

eschar (Med) A scab or crust at the surface of a burn injury. Body tissue destroyed by a caustic chemical or heat.

escharotomy (Med) Surgical removal of an **eschar**.

esophagus (Anat) The muscular tube leading from the throat to the stomach.

espantoon (Law) A policeman's billy club in Baltimore.

essay examination (Educ) An examination in which the student writes extended answers to questions to demonstrate what he has learned and retained from a course of study.

essential oil (Chem) Any of a class

of volatile, odoriferous oils that impart odor to plants. Used in essences, perfumes, flavors, etc.

ester (Chem) An organic compound formed by the reaction of a carboxylic acid and an alcohol with the elimination of water.
NOTE: Simple esters are low-boiling liquids with pleasant odors; high molecular weight esters are often odorless waxes. Esters are used extensively in cosmetics, perfumes, flavoring, plastics, soaps, and as paint and varnish solvents.

estimated design load (Techn) In a heating or cooling system, the sum of useful heat transfer, heat transfer from or to the connected piping and ductwork, heat transfer occurring in any auxiliary apparatus connected to the system.

esu = electrostatic unit

etch (Techn) To cut or eat away material with acid or other corrosive substance.

ethane (Chem) An unsaturated hydrocarbon, formula C_2H_6, derived from petroleum or natural gas; important for organic synthesis and the production of ethylene. An odorless, colorless, alkane; fp -183.3°C, bp -88.6°C. An important constituent of natural gas.

ethics See: code of ethics

ethyl alcohol (Chem) A colorless, volatile liquid, flammable and toxic, formula C_2H_5OH, bp 78.4°C, mp -112.3°C, used as a solvent, antifreeze, alcoholic beverage, and in the synthesis of many organic chemicals. Partial oxidation of ethyl alcohol yields acetaldehyde. Also called: alcohol, ethanol, grain alcohol, industrial alcohol, ethyl hydroxide, methylcarbinol, grain neutral spirits.

ethylene (Chem) A gaseous, olefinic compound, C_2H_4, obtained by cracking petroleum fractions or natural gas. Widely used as the starting material for chemical reactions, notably the manufacture of polyethylene and related plastics.

ethylene glycol (Chem) A colorless liquid, $C_2H_6O_2$. Useful as a humectant because of its affinity for water. Used mainly as an antifreeze. Highly toxic if ingested.

ethyl ether (Chem) A flammable liquid, $C_6H_{10}O$, used chiefly as a solvent and anesthetic.

ethyne = acetylene

etiology (Med) The study of causes, esp of medical conditions or disease.

Eustachian tube (Anat) A canal leading from the back of the throat to the middle ear that equalizes the pressure across the tympanic membrane.

eutectic (Metall) A solution or a composition of two or more substances giving the lowest melting point.
NOTE: Commonly, eutectic alloys of bismuth, tin, lead, and cadmium are used as fusible elements in automatic sprinkler heads, damper releases, and the like.

ev = electron volt

evacuation (Build) The process of emptying a building or clearing an area of people, esp in an emergency.
NOTE: Evacuation can be normal, as in the departure of people at the end of a business day; emergency, as in the case of organized movement of people out of a burning building; or panicky, as in the case of people fleeing in any way they can from some real or imagined danger.

evaporation (Phys) The process by which a liquid is changed into the vapor state and mixed into the air or other surrounding gases. Vaporization of a liquid at temperatures below its boiling point. See also: heat of vaporization. Cf: sublimation.
NOTE: Except for highly volatile liquids, evaporation is a slow process and requires more heat than vaporization at the boiling point of a given liquid. A pound of water, for example, requires 1440 btu to evaporate at 86°F, but only 972 btu to vaporize at 212°F. The heat absorbed in evaporation is given up during condensation, the reverse process.

evaporation rate (Chem; Phys) The ratio of the time required to evaporate a measured quantity of a liquid to the time required to evaporate the same quantity of a reference liquid (ethyl ether). The higher the ratio, the slower the evaporation rate.

evaporator (Techn) The part of a refrigerating system in which the refrigerant is vaporized to absorb heat.

event (Statist) An incident or occurrence that may be recorded as a datum or fact.

evidence (Law) 1. Testimony, documents, records, articles, and expert opinion presented to a court to prove the truth of a matter or disprove an allegation. See: rules of evidence. 2. Something that furnishes proof or substantiates a statement.

evolution[1] (Chem) The release of gas or heat from a substance.

evolution[2] (Fire Pr) A predetermined operational sequence of basic fire-fighting tasks, such as the deployment of equipment and hose lines and the erection of ladders.

exacerbating (Med) Aggravating; increase in the violence of a disease.

examination (Techn) Inspection or investigation, without the use of special laboratory appliances or procedures, of supplies and services to determine conformance to specified requirements.
   NOTE: Examination is generally nondestructive and may include visual, auditory, olfactory, tactile, gustatory, and other observations; simple physical manipulation; gauging; and measurement.

excavation (Civ Eng) Any man-made cavity or depression in the earth's surface, including its walls, or faces, formed by earth removal and producing unsupported faces.
   NOTE: If installed forms or similar structures increase the depth-to-width relationship, an excavation may be called a trench.

excelsior (Transp) Wood shavings, used as a packing material.

excess combustion air (Comb) Air supplied in excess of theoretical air, usu expressed as a percentage of the theoretical air. Also called: excess air.

excess reactivity (Nucl Phys) More reactivity than that needed to achieve criticality. Excess reactivity is built into a reactor by using extra fuel in order to compensate for fuel burnup and the accumulation of fission-product poisons during operation.

excision (Med) Surgical removal.

excitation (Electr) A signal or energy applied to a device or system that causes it to function in a given manner.

excited state (Phys) The state of a molecule, atom, electron, or nucleus in which the particle possesses more than its normal energy. See: ground state.

exclusion area (Nucl Safety) An area immediately surrounding a nuclear reactor where human presence is prohibited for reasons of safety and health.

exclusionary rule (Law) The evidentiary priciple that evidence obtained by means that are forbidden by the Constitution, by statute, or by the decision of another court, is not admissible in court.

excursion (Techn) A sudden, rapid rise in the level of power in an electrical or electronic circuit or in a reactor.
   NOTE: Electrical surges, sometimes caused by lightning, can damage motors, affect electronic equipment, and blow fuses or trip circuit breakers. Excursions in a reactor are usu counteracted by the negative temperature coefficient of the reactor or by automatic control rods.

exempt (Fire Pr) 1. A volunteer fireman who has completed his tour of active duty and retains social ties with his former fire organization. 2. A firefighter who is excused from jury and/or military duty.

exfiltration (Techn) Air flow outward through a porous wall, membrane, or other material.

exfoliation (Med) Sloughing or shedding of dead cells from the body.

exhalation valve (Fire Pr) A device that allows exhaled air to leave a respirator and prevents outside air from entering through the valve.

exhaustion[1] (Techn) The state of being used up, emptied, or worn out.

exhaustion[2] (Med; Physicl) Extreme fatigue brought on by vigorous prolonged physical activity, stress, etc. Loss of energy with inability to respond; depleted. See: fatigue; stress.
NOTE: Heat exhaustion produces a feeling of weakness and is marked by prostration and collapse as a result of dehydration.

exhaust primer (Techn) A device that uses the exhaust from an engine to prime a centrifugal pump.

exhaust system (Ind Hyg) A system consisting of branch pipes connected to hoods of enclosures, one or more header pipes, an exhaust fan, means for separating solid contaminants from the air flowing in the system, and a discharge stack to outside.

exhaust ventilation (Safety) Ejection of indoor air to the outdoors fast enough to keep the concentration of toxic vapor in the air inside a work area within safe limits.
NOTE: A local exhaust system is used to carry off an air contaminant by trapping it near its source, as contrasted with general ventilation, that lets the contaminant spread throughout the workroom, later to be diluted by exhausting quantities of air from the room. A local exhaust is generally preferred to general ventilation because it provides a cleaner and healthier work environment and because a local exhaust system handles a relatively smaller volume of air with less attendant heat loss and with a smaller fan and arrester. A local exhaust system consists of four parts: hoods, into which the airborne contaminant is drawn; ducts, for carrying the contaminated air to a central point; an air cleaning device, such as a dust arrester, for purifying the air before it is discharged; and a fan and motor to create the required air flow through the system. Also called: ventilation.

exhaust ventilation system (Safety) A system for removing contaminated air from a space, comprising two or more of the following elements: enclosure or hood; duct work; dust collecting equipment; exhauster; and discharge stack.

exit (Build; Safety) A way or means for leaving a building and reaching ground level outside. The portion of an evacuation route which is separated structurally from other parts of the building to provide a protected path of travel to the outside.

exit access (Build; Safety) The portion of an evacuation path that leads to an exit; e.g., a door opening into a corridor, stairwell, or the like.

exit capacity (Build) The greatest number of people that can pass safely through an exit per unit time.

exit discharge (Build) An open area between an exit and a public way.

exit distance (Build) The distance between the farthest point in a building and an exit.

exit marking (Fire Pr; Safety) A prominently displayed sign that identifies access to an exit, or one that identifies a door, passage, or stairway that is not an exit but could be mistaken for an exit.

exit width (Build; Fire Pr) A standard measure used for evacuation routes; typically 22 inches, with 12 additional inches considered as one-half a unit of width.

exoskeletal construction (Mech Eng) A construction technique in which the body is a major stressed member.
NOTE: This is the principle of unit construction used by some automotive manufacturers and can result in a sizable reduction in vehicle weight.

exothermic (Chem) Characterized by or formed with the evolution of heat. Pertaining to a chemical reaction in which heat is liberated, as in combustion. Also called exothermal.

exotoxin (Bact) A toxin excreted by a microorganism into the surrounding medium.

expander (Fire Pr) A device for seating expansion rings that secure hose to couplings.

expansion joint (Build) A small space or gap built into a structure to permit sections of masonry, concrete, and the like to expand and contract freely and to prevent distortion or buckling.

expansion ring (Fire Pr) A metal ring that secures a fire hose to the tail piece or shank of a coupling. See: coupling.

expansion wave (Fl Mech) A discontinuity in an isentropic flow of a compressible fluid in which pressure and density decrease in the direction of wave motion. Also called: rarefaction wave.

expectantly (Med) Relieving symptoms and allowing nature to cure the disease. See: palliative treatment.

expectorant (Med) An agent that promotes liquefaction of secretions from the mucous membranes of the air passages and thus makes it easier to raise and expel them.

experience (Psych) 1. An event that has been lived through. 2. Knowledge, skill, or practice derived from first-hand participation or learning.

experience rating (Insur; Safety) An index derived from accident frequency and severity that is used as a basis for determining tax rates or insurance premiums. Also called: merit rating.
NOTE: Workmen's compensation contributions, unemployment insurance, and commercial health insurance premiums are usu based on accident experience to encourage employers to promote safety and thereby benefit from good records.

experiment A procedure for testing a hypothesis under controlled conditions. An experiment is motivated by the prediction of a theory or by qualitative observations of a phenomenon and a desire to find an explanation for it.
NOTE: An experiment is carried out by planning, designing the procedure, carrying it out under controlled conditions, and analyzing and interpreting the results.

expert testimony (Law) The opinion of a witness skilled in a particular art, trade, or profession or possessing special knowledge derived from education or experience.

expert witness (Law) A person who by education, training, experience, or special knowledge is qualified to give evidence concerning a scientific, technical, or professional matter in a court.

expiration (Physiol) Breathing out; expelling air from the chest; exhalation. Ant: inspiration. See: respiration.

expiratory reserve (Med) The volume of air remaining in the lungs after normal expiration.

explode (Comb) To change chemically or physically, usu from a solid or liquid to a gas (as by chemical decomposition or sudden vaporization) so as to suddenly transform considerable energy into kinetic form.

exploratory development (Man Sci) Effort directed toward the solution of specific problems. Specific goals are to define quantitative performance levels and desired future operational capabilities.

explosibility limits = explosion limits; flammability limits

explosimeter (Ind Hyg; Techn) A device that measures the concentration of gas or vapor in an atmosphere in relation to the flammability limits.

explosion (Comb) The effect of an exothermic combustion reaction occurring in an enclosed space, characterized by a destructive buildup of pressure. In popular

terms, an explosion is any rapid pressure rise in a system that ruptures a container or enclosure, causing pressure wave damage. Thus, one may have an explosion due to steam overpressure, as in a boiler explosion. An explosion may be either a homogeneous self-catalyzed reaction or a traveling combustion wave in a confining vessel. A detonation is often called an explosion, though in this case the pressure wave develops without confinement and has the characteristics of an explosion because it propagates at supersonic speed. A detonation is an explosion but not all explosions are detonations. See also: combustion; deflagration; flame.
NOTE: Window panes shatter at blast pressures of approximately 1 psi.

explosion limit (Fire Pr) The highest or lowest concentration of a flammable gas or vapor in air that will explode when ignited.

explosionproof (Electr) Capable of withstanding an internal explosion and preventing ignition of external flammable or explosive atmospheres.

explosion relief venting (Build; Fire Pr) A device or system that provides direct communication to the outside from any duct, vessel, chamber, or building in which an explosive mixture can form. The purpose of the venting is to release overpressures caused by an explosion.

explosion vent (Techn) A diaphragm or flap on a vessel or building designed to yield or open and release pressure in the event of an explosion.

explosion venting (Build) 1. The release of pressure produced by an explosion. 2. The system of vents for releasing explosion pressures.

explosive (Techn) A substance capable of sudden high-velocity reaction with the generation of high pressures. Explosives are classified as deflagrating (low explosive) and detonating (high explosive). Depending on their sensitivity, explosives are set off by fire (heat), by friction, by concussion, by percussion, or by detonation.
NOTE: Examples of common explosives include dynamite, black powder, pellet powders, initiating explosives, blasting caps, electric blasting caps, safety fuse, fuse lighters, fuse igniters, squibs, primacord, instantaneous fuse, igniter cord, igniters, small arms ammunition, small arms ammunition primers, smokeless propellants, cartridges for propellant-actuated power devices, and cartridges for industrial guns. The U.S. Department of Transportation classifies explosives as: Class A: Detonating explosives, such as dynamite, nitroglycerin, picric acid, lead azide, fulminate of mercury, black powder, blasting caps, and detonating primers. Class B: Flammable explosives, such as propellants. Class C: Certain types of manufactured articles which contain small quantities of Class A or Class B explosives as components.

explosive-actuated device (Techn) Any tool or special mechanized device actuated by explosives, such as a nail-driving tool powered by an explosive charge.

explosive atmosphere (Fire Pr) An atmosphere containing a flammable mixture of vapor or gas in any concentration that is within the explosive range.

explosive limit (Comb) The minimum (lower) or maximum (upper) concentration of vapor or gas in air or oxygen below or above which explosion or propagation of flame does not occur in the presence of a source of ignition. The explosive or flammable limits are usu expressed in terms of percentage by volume of vapor or gas in air. See: flammability limits.

explosive mixture (Comb) A mixture of gases or vapors in the proportion in which they combine. Upon ignition, the entire mass combusts rapidly.

explosive powerload (Techn) Any substance in any form capable of producing a propellant force. Also called: load.

explosive range (Comb) The ratios of vapor or gas concentrations in air between the upper and lower limits of flammability. Syn: flammability range.

explosivity See: flammability_index

exponential notation (Math) A mathematical convention for representing very large or very small decimal numbers in abbreviated form. In this notation, a number (or symbol) is superscripted to another number (or symbol) to indicate how many times that number (or symbol) is to be multiplied by itself: e.g., $5^3 = 5 \times 5 \times 5 = 125$. Also called: index.
NOTE: As an example of usage, light travels at the speed of approx 180,000 miles per second, or 1.8×10^5 mps; the wavelength of red light is in the range of 4000 angstroms, or 4.0×10^{-5} cm.

$10^0 = 1$ $10^{-1} = 0.1$
$10^1 = 10$ $10^{-2} = 0.01$
$10^2 = 100$ $10^{-3} = 0.001$
$10^3 = 1,000$ $10^{-4} = 0.000,1$
$10^4 = 10,000$ $10^{-5} = 0.000,01$
$10^5 = 100,000$ $10^{-6} = 0.000,001$
$10^6 = 1,000,000$ $10^{-7} = 0.000,000,1$

exposed (Electr; Safety) Referring to unguarded electrical parts that may be touched accidentally.

exposed to contact (Safety) Referring to an object, material, nip point, or point of operation that a person is liable to come in contact with in his normal course of employment.

exposure^1 (Safety) The length of time that a worker has been in the presence of a given job hazard.

exposure^2 (Dosim) A condition or instance of being acted upon by radiation.

exposure^3 (Insur; Fire Pr) a. The portion or whole of a structure that may be endangered by a fire because of proximity to a hazard and other factors. b. The likelihood of property catching fire from neighboring fire hazards. c. A part or surface that is vulnerable to damage or destruction, as one that faces a nearby hazard.

exposure^4 (For Serv; Meteor) A surface receiving energy from a heat source, such as the side of a hill facing the sun.
NOTE: In the northern hemisphere, southwest exposures receive the most solar radiation during the year.

exposure^5 (Med) a. Lack of shelter or protection from a hostile environment. b. Physical impairment suffered because of inadequate protection from the elements. Debility or other symptoms brought on by the sun or heat, wind and extreme cold, or prolonged immersion in water.

exposure^6 (Photog) a. The time interval required to record a latent image on a photographic film or paper. b. A photographic image, frame, or picture.

exposure fire (Fire Pr) 1. A fire ignited by another fire, usu by radiant heat, but sometimes including spot fires. 2. An external fire that threatens to ignite a nearby building or other combustible object or material.

exposure hazard (Fire Pr) The probability or risk that a building will be exposed to the danger of ignition from a fire in a neighboring or adjoining property. Also called: external_hazard.
NOTE: In general, property within 40 feet of a fire hazard involves an exposure hazard. For large fires the distance is considerably greater, esp if firebrands are carried aloft and are transported by the wind.

exposure hours (Safety) The number of employee-hours worked by all employees, including those in operation, production, maintenance, transportation, clerical, administration, sales, and other activities. Syn: employee_hours.

exposure protection (Fire Pr) Fire protection provided to structures or buildings on property adjacent to special hazards, such as flammable and combustible liquid storage tanks.
NOTE: Protection may be afforded by a public fire department or a private fire brigade.

exposure severity (Fire Pr) The degree of fire intensity that may be expected; classified as light, moderate, and severe, depending on fuel load, flame spread ratings of materials involved, and other factors. See: flame_area.

extensible boom platform (Techn) An

elevating device with a telescopic boom.

extension (Fire Pr) Lengthening a hose line by attaching hose to a nozzle or removing the nozzle and attaching the hose to the coupling. See also: extension of fire.

extension ladder (Fire Pr; Techn) A ladder consisting of two or more sections that can be nested for ease of handling and extended to provide needed height. Also called: Bangor ladder.
NOTE: Long extension ladders are equipped with tormentor poles to assist in raising and steadying them in position.

extension of fire (Fire Pr) Spread of a fire to areas not previously involved.

extension pipe (Fire Pr) A rigid metal nozzle pipe that extends a line of hose and helps to direct water streams into basements, upper floors, etc.

extension trestle ladder (Techn) A self-supporting portable ladder, adjustable in length, consisting of a trestle ladder base and a vertically adjustable single ladder, with suitable means for locking the ladders together. The size is designated by the length of the trestle ladder base.

external hazard = exposure hazard

extinction^1 (Fire Pr) The cessation of burning, or the extinguishment of a fire. The terminal phase of combustion.

extinction^2 (Psych) The process by which a conditioned response is weakened or eliminated by not reinforcing it.

extinguish (Fire Pr) To put out flames. To quench a fire.

extinguishant = extinguishing agent

extinguisher See: fire extinguisher

extinguishing agent (Fire Pr) A substance used to put out a fire by cooling the burning material, blocking the supply of oxygen or chemically inhibiting combustion. The principal extinguishing agents are water; carbon dioxide; dry chemicals; foam; and vaporizing liquids (halogenated compounds). Syn: extinguishant. See also: Halon; foam; surfactant; wetting agent.
NOTE: Additives are often used with water to produce foam, increase its wetting capacity, and reduce runoff.

extirpation (Surg) Complete removal or excision of a body part.

extraction^1 (Chem; Phys) Withdrawal or removal by a chemical or physical process. Separation, such as a metal from an ore, or a soluble fraction of a substance by a suitable solvent.

extraction^2 (Math) A calculation for finding the root of a number.

extractive industry (Techn) An industry that exploits natural resources, such as mining, fishing, lumbering, oil production, and agriculture.

extra hazard (Fire Pr) A condition in which the quantity of combustibles or flammable liquids present is sufficiently large that fires of severe magnitude may be expected. These may include woodworking, auto repair, aircraft servicing, warehouses with high-piled (14 feet or higher) combustibles, and processes such as flammable-liquid handling, painting, dipping, etc.

extra-period fire (For Serv) A fire that continues beyond 10 am of the day following discovery.

extrapolate (Math) To infer or estimate an unknown value outside an interval from known values in the interval; to project a trend from past experience.

extrication (Rescue) The freeing of someone who is trapped. The act of releasing or disentangling someone.

extrusion^1 (Techn) The forcing of a material through a die or a form in either a hot or cold state, in a solid state, or in a partial solution to give it a particular form.

extrusion^2 (Med) A pushing out; the state of being out of position, as a displaced bone or organ.

eye¹ (Techn) A loop formed at the end of a rope by securing the dead end to the live end at the base of the loop.

eye² (Anat) A photoreceptive sense organ capable of image formation.

eye bath (First Aid) A type of water fountain used to flush the eyes after they have been exposed to a hazardous substance.

eye bolt (Techn) A closed-eye fitting with a threaded end, used to secure a wire, rope, or hook.

eyebrow (Build) A horizontal projection above a window on the exterior of a building.

eyepiece (Fire Pr; Safety) A gas-tight, transparent window in a full facepiece, through which the wearer may see.

eye protector (Safety) A device that safeguards the eye in an eye-hazardous environment.

eye-splice (Fire Pr) A loop spliced into the end of a rope, such as a halyard, used to attach the rope to an eye bolt or other anchor. A U-shaped thimble is often inserted in the eye-splice to reduce wear.

eyewitness (Law) A person who has first hand knowledge of an event or situation from having been present.

F

f (Electr) The symbol for <u>farad</u>.

F (Chem) Symbol for flourine.

FAA = <u>US Federal Aviation Administration</u>

fabric (Text) A cloth made of natural or synthetic fibers by weaving, knitting, or felting.

fabrication (Techn) The conversion of materials or parts into manufactured articles.

facade (Build) The front or face of a building, esp an ornamental one.

face (Civ Eng) A vertical or inclined earth surface formed as a result of excavation work.

facepiece (Safety) The part of a breathing apparatus that fits over the face.
NOTE: A half-mask facepiece covers the nose and mouth. A full facepiece covers the eyes as well. The device makes a gas-tight or dust-tight fit with the face and includes the headbands, valve, and connections to an air purifier or respirable gas source.

face velocity (Techn) The air velocity required at the face opening to retain contaminants inside a hood or booth in a spray painting, grinding, welding, or other operation.

facsimile equipment (Commun) A device that transmits and reproduces graphic copy, such as pictures, drawings, and printed matter.

factor[1] A cause or prior condition of an event. Something that contributes to a result.

factor[2] (Math) a. A number or term that divides another number or term a whole number of times with no remainder. The numbers 2, 3, 4, 6 are factors of 12. b. To express a mathematical quantity in terms of its factors, such as 12 = 3 X 4.

factor of safety = <u>safety factor</u>

Factory Mutual (Insur) A group of mutual insurance companies insuring large industrial properties against fire.

Factory Mutual Laboratory (Testing) A large testing and research laboratory that performs full-scale industrial fire-hazard tests. It is associated with the Factory Mutual System.

fading (Commun) The fluctuating strength of a received radio signal, usu caused by variations in the reflectivity of the Heaviside layer in the upper atmosphere; the presence of buildings, structures, or geological formations between the transmitter and receiver; and other factors.

fag station = <u>smoke point</u>

Fahrenheit temperature scale (Metrol) The temperature scale on which 32° is the freezing point of water and 212° is its boiling point. A Fahrenheit degree is 1/180 the difference between the temperature of melting ice in brine and that of water boiling at standard atmospheric pressure. See: <u>temperature scales</u>.
NOTE: The original intent was to set zero at the lowest temperature that could be obtained in practice and 100 at normal body temperature.

failsafe (Techn) Having the property of failing in a safe condition or in a condition no more hazardous than in normal operation.

failure (Ind Eng) Breakdown, fracture, or collapse. A deficiency that prevents a component or subassembly from performing its intended function.

failure analysis (Ind Eng) A logical, systematic examination of an item, component, or assembly, and its place and function in a system, to identify and analyze the probability, causes, and consequences of potential and real failures.

failure rate (Ind Eng) The number of failures of an item per unit measure of life (cycles, time, miles, events, etc.), as applicable for the item.

fainting (Med) A temporary loss of consciousness as a result of a diminished supply of blood to the brain. Syn: <u>syncope</u>.

faintness (Med) A sensation of weakness or dizziness.

fairlead (Techn) An arrangement of rollers that allows free play to wire rope during winching.

fallout (Env Prot) 1. Precipitation of radioactive particles from a nuclear explosion. 2. Airborne radioactive particles that fall to the ground in various patterns following a nuclear explosion.
NOTE: Local fallout reaches the earth's surface within 24 hours after a nuclear detonation. Material injected into the troposphere does not fall out locally but usu is deposited in relatively narrow bands around the earth at about the latitude of injection. Stratospheric injection results in slow, worldwide fallout. Also called: **radioactive fallout**.

fallout shelter (Milit) A structure that affords protection to occupants from the effects of a nuclear explosion.

false alarm[1] (Fire Pr) An alarm for which there is no fire.
NOTE: False alarms may be accidental, when an alarm box operates because of a circuit malfunction, or malicious, when someone deliberately actuates an alarm for individual reasons. A **needless alarm** is one based on evidence that later proves to be unfounded, such as steam mistaken for smoke or a strange odor thought to be coming from burning material. The latter type is usu called a **good intent false alarm** and for statistical purposes is counted separately from malicious false alarms. A **no-fire alarm** includes smoke scares and accidental alarms. A **malicious false alarm** includes bomb scares (no bomb).

false alarm[2] (For Serv) Reported smoke or fire that does not require suppression, such as from controlled brush burning, chimney smoke, **false smoke**, and the like.

false alarm[3] (Police) A complaint or report that proves to be groundless.

false ceiling (Build) A ceiling suspended some distance below the true ceiling for ornamental or other purpose. Also called: **drop ceiling**.
NOTE: The false ceiling usu creates an open, **concealed space** through which fire can propagate freely.

false front (Build) An ornamental facade covering the front of a building and concealing the original front.

false smoke (For Serv) An atmospheric phenomenon mistaken for smoke, such as road dust, fog, etc.

false stull See: **stull**

falsework = **formwork**

family room (Build) A recreation room in a private residence; used also for informal entertaining. Syn: **club room**.

fan (Build) A device consisting of a motor, blades, and housing, used to blow air.
NOTE: A fan used to draw air out of a space is called an **exhauster**. One used to force air under pressure into a space is called a **blower**.

Fanning friction factor See: **dimensionless numbers**

farad (Electr; Symb: f) The unit of capacitance. Equivalent to 900×10^9 electrostatic centimeters. A capacitor has the capacitance of one volt if a charge of one coulomb raises the potential between its plates to one volt.

Faraday equivalent (Chem; Phys) The charge carried by a gram-ion of unit positive valency, equal to 96,487 coulombs per equivalent.

fastener (Techn) Any of a wide variety of hardware items used to hold, pin, attach, or secure things together, such as nails, rivets, snaps, screws, and the like.

fast neutron (Nucl Phys) A neutron with energy greater than approx 100,000 electron volts.

fast water (Fire Pr) Rapid leakage of water through ceilings. Syn: **heavy drip**.

fatigue[1] (Metall) Loss of strength by a metal due to repeated stresses. The tendency of a metal to fracture under cyclic stressing.

fatigue[2] (Med; Psych) A state of

weariness, drowsiness, or irritability following a period of mental or physical exertion or stress, characterized by reduced capacity to work effectively. See: stress; exhaustion.

fatigue crack (Metall) A fracture that starts from a nucleus, which is the focus, or a concentration point of cyclic stresses, then propagates through the metal. Also called: fatigue failure.

fatigue failure (Metall) The maximum stress that a metal will withstand without failure for a specified large number of cycles of stress. Also called: endurance limit.

fatigue limit (Metall) The maximum stress that a metal will withstand without failure for a specified large number of cycles of applied stress. Also called: endurance limit.

fault (Electr) An abnormal condition.

fault warning (Electr) A signal that indicates trouble in an electrical circuit.

FBU = (Brit) Fire Brigades' Union

FCC = Federal Communications Commission

FD = fire department

FDA = US Food and Drug Administration

FDIC = Fire Department Instructor's Conference, an annual meeting of fire department officials and training officers held to exchange information on new developments in fire protection.

fear (Psych) A strong, unpleasant emotion brought on by the awareness of danger, usu accompanied by an overwhelming desire to flee or hide.
NOTE: Physiologically, the body prepares for action: the heart beat, blood pressure, and breathing rate increase, while digestion and metabolic activities are inhibited. In these respects, fear and anger are similar.

feedback[1] (Techn) The return of part of the output of a stage, device, system, or process to the input for the purpose of improving performance or providing self-corrective action.
NOTE: Feedback can be positive or negative.

feedback[2] (Psych) Return to the source of information that is useful for adjusting behavior.

feedback loop (Electr; Techn) The part of a closed-loop system that permits comparison of response with command, or output with input.

feeder (Mining) A gush of methane gas that sometimes spews out from the vein as coal is being mined.

feeder line (Fire Pr) A hose line that delivers water to pumping apparatus from a water source. Cf: attack line.

feeding (Techn) The process of placing materials or parts into or removing them from the point of operation of a machine.

feeling for fire[1] (Fire Pr) Using the sense of touch on walls, partitions, and other surfaces to find hot spots and thereby detect fire in concealed spaces.

feeling for fire[2] (For Serv) Feeling burned materials and ashes with the bare hands to detect live coals.

feldspar (Mineral) Any of a group of minerals composed of silicates of aluminum with either sodium, potassium, calcium, or barium; present in almost all crystalline rocks. Used in ceramics, insulation, and fertilizer.

FEMA = Fire Equipment Manufacturers' Association

female coupling (Fire Pr) A swivelled, internally threaded coupling that connects to an externally threaded male hose nipple.

femoral artery (Anat) The main artery of the leg.

femur (Anat) The thigh bone.

fertilizer A mixture of plant food containing compounds of nitrogen, potassium, phosphorus, sulfur, and traces of other minerals.

fetus (Biol) Any unborn or unhatched vertebrate after it has formed. Esp a developing human organism from the third month after conception to birth. See: embryo.

fever (Med) 1. A condition in which the body temperature is above normal (in humans, above 37°C or 98.6°F). 2. Any of a variety of diseases that cause elevated body temperature.

FF = firefighter

FIA = Factory Insurance Association. An organization of capital stock fire insurance companies that insure large industries that have special fire hazards and usu maintain fire protection services on their premises.

fiber (Techn) Organic or inorganic, natural or man-made threads or filaments used for cloth, rope, and as a filler and binding material.

fiberboard (Build) A building material made of wood or other vegetable fibers pressed into stiff sheets, usu with a binding agent.

fiberboard box (Transp) A rectangular, three-dimensional shipping container, made either of solid or corrugated fiberboard. Distinguished from a carton, which is not a shipping container.

fibrillation (Med) A condition of uncoordinated twitching of individual muscle fibers, esp of the heart, in which individual muscle fibers or bundles take up their own rates of contraction instead of working together in response to nerve impulses from the sinoatrial node, the natural cardiac pacemaker. As a result the heart beat becomes irregular in frequency and force.

fibrin (Physiol) A clotting protein constituent of the blood that upon injury is produced from fibrinogen by thrombin and forms scabs over cuts and other wounds.

fibrosis (Med) Formation of interstitial fibrous tissue; scarring.

field (Electr) The space containing electric or magnetic lines of force.

field intensity (Electr) The strength of an electrical or magnetic field.

FIFI = Fire Information Field Investigation

fifth alarm (Fire Pr) In some jurisdictions, the last and most serious alarm calling for the largest contingent of firefighters and equipment that is printed on the running cards of large fire departments. See: general alarm.

fill hose (Fire Pr) A short length of hose used to fill a booster tank from a hydrant or a tank truck.

fill-in (Fire Pr) To respond to an alarm that can not be answered by the regularly assigned company. Fill-in may be automatic, by prearrangement, or ordered by fire alarm headquarters. See: cover; relocate.

filling density (Transp) The percent ratio of the weight of a gas in a container to the weight of water the container will hold at 60°F.

film (Techn) A thin layer of material, including paint, lubricants, plastic sheets, and foams. Films can form spontaneously on surfaces and often have a profound effect on them, for example, the film of oxide that forms on exposed aluminum.

film badge (Dosim) A light-tight packet of photographic film worn like a badge by workers in nuclear industry or research, used to measure the extent of exposure to ionizing radiation. The absorbed dose can be calculated by the degree of film darkening. See: dosimeter.

film density (Photog) The quantitative degree of blackening on a photographic film, estimated by comparison with a density strip or measured by a densitometer.

filter[1] (Electr) A device for separating frequency components of an audio signal or spectral band of any electromagnetic energy.
NOTE: A pass filter allows one or more frequencies to pass relatively unattenuated and blocks all others. A blocking filter stops one or more frequencies and is indifferent to all others. The essential difference is in the purpose of the filter.

filter² (Ind Hyg; Safety) a. The medium used in a respirator to remove particles from the intake air. b. The element or component containing a filtering medium.

filter³ (Opt) A selectively transparent plane parallel plate that transmits only a specific portion of the spectrum, or reduces the intensity of a light beam passing through it.

filtrate (Chem Eng) A liquid that has passed through a filter.

filtration (Chem Eng) The process of passing a liquid or gas through a porous sheet or mass to separate and remove solid particles or droplets.

filtration rating (Techn) A measure of the effectiveness of a filter in removing particles of a given size and larger. Usu stated in terms of absolute or nominal.
  NOTE: Absolute micron rating implies that all particles larger than a given micron value are removed. Nominal micron rating is generally taken to mean that 98% of all larger particles are removed.

financial management (Man Sci) Methods and procedures used for the administration and control of capital resources.

fine fuel = flash fuel

fingers of fire (For Serv) Long, narrow projections of flame extending from the main body of a wildland fire into unburned fuel.

fire (Fire Pr; Comb) Uncontrolled combustion, usu involving flame. Often accidental and unwanted.
  NOTE: Generally a fire is the rapid oxidation of a gaseous, liquid, or solid fuel in atmospheric air. The fire service classifies fires according to fuel type; the forest service according to size. See: fire classes. Some substances are capable of burning without oxygen. See: combustible.

fire academy (Fire Pr) A comprehensive fire department training facility. Cf: fire college.

fire administration (Fire Pr) 1. The top management of a fire department. 2. The management functions of a fire department.

fire agriculture (For Serv) Farming using fire as the principal means of clearing land, removing crop residues, destroying noxious weeds or diseased plants, and the like.

fire aisle (Fire Pr) An aisle that serves as a firebreak in a storage facility.

fire alarm (Fire Pr) An audible or visible fire emergency signal.

fire alarm box (Fire Pr) A device or communications system that transmits a signal or notification that a fire is in progress and calls for firefighting units to respond. See: alarm system. Usu referring to a device connected to the public fire alarm system, as opposed to a private box. Also called: fire box; cottage box; street box.

fire alarm headquarters (Fire Pr) 1. A fire department communications center that receives and relays alarm signals to appropriate units and keeps records of movements and locations of firefighting units. Syn: central station. 2. = (Brit) fire fall receiving point.

fire alarm system See: alarm system

fire apparatus See: apparatus

fire area¹ (Build) 1. A portion of a building separated from the remainder by fire-resistant construction (floors, ceiling, walls, etc) in which all communicating openings are protected by doors, windows, and other assemblies having appropriate fire resistance ratings. Also called: fire zone. See: smoke zone. Cf: area of refuge. 2. = (Brit) fire cut.

fire area² (Fire Pr) The stories of a building adjacent to fire floors; usu the story immediately below the fire floor and the two stories directly above it.

fire ax See: ax

fireball¹ (Fire Pr) A spherical mass of flame that sometimes occurs when a large quantity of vaporized flammable liquid suddenly combusts.

fireball² (Milit) The luminous, ball-shaped cloud of a nuclear explosion.

fireball³ (Meteor) A brilliant meteorite.

fire bat (Fire Pr) A canvas paddle used to beat out grass fires. A type of flail. Syn: swatter.

fire beater (Fire Pr; Brit) A birch broom or piece of rubber or wire netting attached to a long handle; used to beat out grass and brush fires. A kind of fire bat. Also called: fire broom.

fire bed (Comb) A grate or other support for fuel burning in a combustion chamber.

fire behavior (Fire Pr) The manner in which a fire ignites, develops, and spreads.

fire blanket (Fire Pr) 1. A noncombustible cloth of asbestos or one treated with fire-retardant chemicals, used to wrap around clothing-fire victims to smother the flames. 2. = (Brit) smothering blanket.

fire board (Build) A screen used to close off an unused fireplace.

fireboat (Fire Pr) A motor vessel equipped with pumps, hose, monitor nozzles, and other firefighting gear and appliances and manned as a company in a fire department.

firebomb (Milit) An incendiary device.

fire boss¹ (Fire Pr) The fireground commander responsible for planning, organizing, and directing fire suppression activities at a forest fire.

fire boss² (Mining) A mine employee responsible for inspecting and maintaining fire-safe conditions in a mine.

firebossed (Mining) Inspected regularly for fire safety.

firebox¹ (Techn) A chamber in which fuel is burned.

firebox² (Fire Pr) A fire alarm signal box.

firebrand (Fire Pr; For Serv) 1. Hot, flaming or glowing solids, generally of cellulosic material, raised by strong convective currents and carried by high winds in large-scale fires. Firebrands are the most common sources of spot fires. 2. Any source that is capable of starting a wildland fire.

firebreak¹ (Fire Pr) An open space between buildings, stacks of combustible materials, stored goods, and the like, serving to interrupt the spread of fire. Also called: fire separation.

firebreak² (For Serv) A natural barrier to fire spread or a cleared and sometimes plowed strip of land in a forest serving the same purpose. Also called: fireguard; fire lane; plow line. Sometimes synonymous with fuel break.

firebrick (Techn) Refractory brick, used to line fireboxes, furnaces, fireplaces, chimneys, and the like.

firebridge (Fire Pr; Brit) Any combustible material that is capable of or likely to carry fire across a firebreak or through a fire wall or other barrier.

fire brigade¹ (Fire Pr) a. An organized team of employees in an industrial plant or an institution who are knowledgeable, trained, and skilled in the safe evacuation of people during emergencies and in assisting firefighting operations. Cf: fire guard¹. b. = (Brit) industrial fire brigade; works fire brigade.

fire brigade² (Fire Pr; Brit) A unit of the British public fire service.

fire broom = fire beater

fire bucket = fire pail

firebucket (For Serv) A large water container equipped with a dumping mechanism, carried by a helicopter and used to drop water or other extinguishants on wildland fires.

fire buff = buff

firebug (Fire Pr; Colloq) 1. = <u>arsonist</u>. Also called: <u>fire raiser</u>; <u>incendiary</u>. 2. Loosely, a person who enjoys attending fires but does not set them. Syn: <u>spark</u>.

fire building (Fire Pr) The building or structure in which a fire started. Syn: <u>building of origin</u>.

fire bureau (Fire Pr) An organization, often within a municipal public safety department, that is responsible for fire protection.
NOTE: In addition to fire and police bureaus, a public safety department may include other city agencies.

fire burning index See: <u>burning index</u>

fire call receiving point (Fire Pr; Brit) = <u>fire alarm headquarters</u>

fire camp (For Serv) A camp with accommodations for men and equipment engaged in fighting fire. Also called: <u>base camp</u>; <u>side camp</u>; <u>fly camp</u>, depending on location and function.

fire canopy (Build) A more or less horizontal extension of fire-resistive construction beyond an exterior wall, designed to prevent flames from a window from igniting the contents or structure of floors above. Cf: <u>eyebrow</u>.

fire casualty (Fire Pr) A person who is injured or dies as the result of a fire, excluding those requiring first aid only. Cf: <u>fire injury</u>. See also: <u>fire death</u>.
NOTE: A distinction is made between direct and indirect fire casualties. One is an immediate consequence of a fire, the other is incidental to it.

fire cause[1] (Fire Pr) The agency or circumstance which starts a fire or allows one to start.
NOTE: Although fire causes for statistical purposes are often lumped into broad categories such as 'electrical', 'clothing', and the like, the more exact cause is of the order of 'sparking of frayed wiring', 'clothing too close to a heater', and so on.

fire cause[2] (For Serv) The source of fire ignition. For statistical purposes the US Fire Service uses eight broad categories of fire causes: campfire, debris burning, incendiary, lightning, lumbering, railroad, smoker, and miscellaneous.

fire-check door (Build) A door assembly conforming to certain structural specifications or one that will meet the stability and integrity requirements of the fire-resistance test for a given time, the period for maintenance of integrity being less than that for stability. Its function is to check the spread of fire in a building.

fire chief (Fire Pr) 1. The head or highest ranking officer of a fire department. Syn: <u>chief of department</u>. 2. Loosely, any chief officer of a fire department such as a deputy chief, battalion chief, and the like. 3. = <u>sidewalk fire chief</u> (slang).
NOTE: A number of cities have adopted new titles for the head of the fire department; e.g., Director, Commissioner, Administrator, etc.

fire classes (Fire Pr; For Serv) Categories of fire based on type of fuel or extent of spread that provides a general indication of severity and hazards.
NOTE: For purposes of identification of hazards and to facilitate the control and extinguishment of fires, the fire service refers informally to fires with respect to extent or general category (e.g., friendly or unfriendly, a one-and-one, group fire, conflagration, fire storm, etc.), but formally classifies fires and hazards by type of fuel or combustible.
<u>Class A</u> - ordinary combustibles such as wood, cloth, paper, rubber, and certain plastics.
<u>Class B</u> - flammable or combustible liquids, flammable gases, greases, and similar materials.
<u>Class C</u> - energized electrical equipment [in Britain, Class C involves flammable gases].
<u>Class D</u> - combustible metals, such as magnesium, titanium, zirconium, sodium, potassium, etc.
The US Forest Service classifies fires by size:
<u>Class A</u> - 1/4 acre or less.
<u>Class B</u> - greater than 1/4 but less than 10 acres.
<u>Class C</u> - greater than 10 acres but less than 100 acres.

Class D - greater than 100 acres but less than 300 acres.
Class E - 300 acres or more.

fire clay (Techn) Heat resistant clay used in making fire brick, crucibles, and in other high temperature applications.

fire climate (For Serv) The composite pattern of weather elements over time that affect fire behavior in a given region.

fire coat (Fire Pr) A special waterproof coat worn by firefighters, provided with fasteners, wrist protectors, and often a detachable winter lining. Syn: bunker coat; turnout coat.

fire college = fire school

fire commissioner (Fire Pr) The civilian head or top official of a fire department.
 NOTE: A title adopted in some cities for the chief of the fire department.

fire company[1] (Fire Pr) A fire brigade.

fire company[2] (Insur) A fire insurance company.

fire concentration (For Serv) 1. An instance of several fires burning in the same locality. 2. The rate of fire occurrence per unit area.

fire control (Fire Pr; For Serv) Fire protection and suppression, including efforts to reduce fire losses. Syn: fire protection.
 NOTE: Fire control is used by forest protection agencies to mean fire protection (i.e., prevention, detection, presuppression, suppression, and mopup).

fire control equipment (For Serv) Tools, machinery, special equipment, and vehicles used in fire control.

fire control improvements (For Serv) Structures primarily used for fire detection and control, such as lookout towers, fire guard cabins, telephone lines, and access roads to lookout stations.

fire control planning (For Serv) The systematic technological and administrative process of designing organizations, facilities, and establishing procedures for wildland fire protection.

fire cooperator (For Serv) An individual or agency that has agreed to provide specified fire control services. The cooperator receives advanced training and instruction in providing the service. Also called: cooperator; planned cooperator; fire warden; fire agent; per diem guard.

fire cover (Fire Pr) Personnel and fire apparatus available for firefighting purposes in a given area.

firecracker An explosive charge wrapped in paper, used as a noise maker at celebrations. See: fireworks.

fire curtain[1] See: draft curtain

fire curtain[2] (Build) a. A specially treated curtain that is drawn or dropped across the procenium arch in a theater to form a fire barrier between the stage and auditorium. b. = (Brit) safety curtain.

fire curve = time-temperature curve

fire cutoff = fire stop

fire damage (For Serv) Loss caused by fire, expressed in money or other units, including direct losses such as future value of lost forest products, as well as direct losses of cover, improvements, wildlife, and the like destroyed by fire.

firedamp (Mining) 1. Any unbreathable, flammable, or explosive underground gas such as methane. 2. An explosive gas mixture, consisting chiefly of methane (CH_4), occurring esp in coal mines. See: blackdamp; damp. See also: roof layer.

fire damper (Build) A device that automatically seals off air flow through part of an air-handling system in order to restrict or block the passage of heat and combustion products. See: smoke damper.
 NOTE: The device usu consists of a flap or shutter controlled by a fusible link, though recently a honeycomb partition coated with intumescent paint has been developed.

fire danger (For Serv) The degree of hazard of ignition, spread, and difficulty of controlling a fire, as well as the extent of damage that may result, largely determined by weather factors. See: constant danger; variable danger; burning index.

fire danger board (For Serv) A display of the ratings of variable fire dangers and the values of the factors on which the ratings are based.

fire danger class (For Serv) A segment or interval on a fire danger scale designated by a qualitative term or numerical value.

fire danger index (For Serv) 1. A scale on which relative fire danger can be defined. 2. A number indicating the severity of forest fire danger as determined from variable fire danger factors. See: burning index.

fire danger meter (For Serv) A device, similar in principle to a slide rule, for calculating the ratings of variable factors into numerical classes or ratings of wildland fire danger.

fire danger rating (For Serv) An evaluation of the fire danger, obtained by summing the effects of selected fire danger factors into a qualitative or numerical index that represents existing protection requirements. See: fire danger.
  NOTE: The Forestry Commissioners in Great Britain use a scale of 0 to 100; 0 meaning no danger of fire outbreak and 100 meaning serious fire hazard, such as during a drought in hot weather.

fire day (For Serv) A 24-hour period beginning at 10 am.

fire death (Fire Pr; Insur) A fatality resulting from a fire injury within one year after the fire. See also: fire casualty.
  NOTE: For particular purposes a fire fatality may be specifically defined as one occurring within a certain number of hours or days after the fire.

fire demand zone (Fire Pr) An area in fire hazard analysis in which demands for fire service are similar. The zone, including at most the areas of four phantom boxes, contains the properties to be protected and has a focal point that is approx equidistant in travel time from its boundaries.

fire department (Fire Pr) A public or private fire-fighting organization that provides fire prevention, extinguishment, and sometimes emergency rescue service to a community, municipality, fire district, or other political subdivision, consisting of one or more companies and headed by a chief of department.
  NOTE: A combination fire department is one staffed by both paid personnel and volunteers, a paid fire department is one manned entirely by paid personnel, and a volunteer fire department is one manned entirely by volunteers. Volunteer departments may have a paid house man, q.v.

fire department connection (Build; Fire Pr) 1. A water supply inlet to a building standpipe or sprinkler system that permits the fire department to pump water into the system. 2. Any connection for large fire department hoses, contrasted to small first-aid hose outlets.

Fire Department Instructors' Conference (Fire Pr) An annual meeting of fire department officials and training officers held to exchange information on new developments in fire practice.

fire detection (Fire Pr) The practice and technology of discovering unwanted fire, to generate an alarm and to initiate appropriate fire suppression measures.

fire detector (Build; Fire Pr) A mechanical, electrical, or electronic device that senses the presence of a fire from one of its signatures and generates a signal or trips another device. In simplest form, a temperature-sensor that sounds an alarm or actuates a sprinkler or other protective system at the first sign of a fire. See: fire signature.
  NOTE: Common fire detectors are fusible links and thermostats. More sophisticated detectors measure rate-of-rise in temperature, detect

the presence of fire from an increase in ions or obscuration of a photoelectric beam by smoke particles, or sense infrared and ultraviolet radiation from the fire itself. A laser system has been developed to detect fires. When hot gases, which have a different index of refraction than air, bend the laser beam away from a photodetector, it triggers an alarm.

fire devil (Fire Pr) A comparatively small, rapidly whirling vortex of flame seen in forest and brush fires but sometimes also in freely burning buildings, resulting from hot gases rising and cool air rushing into the low-pressure area. A small fire whirl. Cf: dust devil.

fire district (Fire Pr) A rural or suburban fire protection and suppression organization usu supported by public funds. Syn: fire protection district.

fire division wall = fire wall

fire dog = andiron. See: fire irons.

fire door (Build) A fire-resistive door, including its frame and hardware, which under standard test conditions meets the fire protective requirements for the location in which it is to be used. Usu a self-tripping door that closes automatically when a fire is detected or an alarm is actuated. See: fire resistance classes.
  NOTE: Fire doors are classified according to their fire endurance ratings. Class A: A door having a 3-hour fire endurance rating for use as a closure in Class A openings. Class B: One having a 1-1/2- or 1-hour rating used as a closure in Class B openings. Class C: One having a 3/4-hour rating, used as a closure in Class C openings. Class D: One having a 1-1/2-hour rating and suitable for use in exterior walls as a closure in Class D openings. Class E: A door having a 3/4-hour rating and suitable for use in exterior walls as a closure in Class E openings. Doors are also classified according to construction and material, as tin clad, plywood core with galvanized steel terne plate covering; composite, core with metal or impregnated wood edges, faced with steel, wood, or plastic; hollow, flush or panelled steel; metal clad, metal covered wood; plate steel, steel plates on steel frame; and sheet metal, formed or corrugated steel. Doors are also classed as rolling and counterbalanced types.

fire drill (Fire Pr) 1. A practice exercise by a fire service unit in firefighting procedures and the use of fire service equipment. See: fire training. 2. A practice evacuation of a building in the manner to be followed in case of fire or other emergency.

fire duty (Fire Pr) 1. Physical firefighting, as opposed to staff or maintenance work. 2. The work performed at a fire directly related to its extinguishment. 3. Attendance of trained fire department personnel at a large public occupancy for precautionary and fire safety enforcement purposes. A fire watch. Cf: detail.

fire ecology (For Serv) An area of study dealing with the relationship of fire to the biotic community.

fire edge (For Serv) The boundary of a fire. Cf: live edge.

fire education (Fire Pr) Academic background for work or a career in the fire service. The development, through study or instruction, of intellectual skills for problem analysis and problem solving. The learning of theory and acquisition of factual information for application in practical fire situations.

fire effects (For Serv) Any consequences of a fire, whether good, bad, or indifferent.

fire endurance (Build) 1. A measure of the elapsed time during which an assembly, material, product, element, structural member, or the like maintains its design integrity under specified conditions of test and performance. 2. The capability of a material, structural member, or assembly to retain its design load-bearing capacity or other function and resist failure under given conditions of exposure to fire. 3. = (Brit) fire resistance.

fire endurance rating (Fire Pr) A measure expressed in hours, half-

hours, or quarter-hours, defined by the authority having jurisdiction, or recommended by an independent testing laboratory, during which a building material, structural member, assembly, or structure must retain its design integrity or effectively resist destruction when exposed to fire under given conditions.
  NOTE: Fire endurance rating requirements are usu specified in building and fire protection codes.

fire engine (Fire Pr) A motor vehicle, including pumpers, ladder trucks, aerial platforms, etc. that carries men and equipment to a fire or other emergency. Syn: apparatus.

fire escape[1] (Build) A continuous path of travel to the ground, or other safe area or refuge, for use if fire or other emergency conditions make normal exits unusable.

fire escape[2] (Fire Pr) A means for leaving a building in case of a fire, such as an exterior or interior fire-protected stairway, a chute, etc. An emergency fire exit.

fire escape stairs (Build) Stairs added usu to the exterior of an existing building to supplement its exit capacity.

fire exit See: exit

fire exposure (Fire Pr; Testing) The subjection of a material, assembly, or structure to a heat flux or flame.

fire exposure severity See: exposure severity

fire extinguishant = extinguishing agent

fire extinguisher (Fire Pr) A hand-portable or wheeled device that can be used to extinguish small or incipient fires. Typically a container filled with any of a variety of fire extinguishing agents and means for expelling them.
  NOTE: Extinguishers are usu labeled with instructions for proper use. Some fire extinguishers are effective only on one class of fire; others are suitable on two or three classes. None are suitable for all four classes of fire. Class A: A fire extinguisher for use on fires involving ordinary combustible materials such as wood, cloth, paper, rubber, and many plastics. Class B: For fires involving flammable or combustible liquids, flammable gases, greases, and similar materials. Class C: For fires involving energized electrical equipment where safety to the operator requires the use of electrically nonconductive extinguishing agents. (When electrical equipment is deenergized, Class A or B extinguishers may be effective.) Class D: For fires involving certain combustible metals such as magnesium, titanium, zirconium, sodium, and potassium. CO_2: A fire extinguisher that discharges carbon dioxide gas; suitable for Class B and Class C fires. dry powder: A fire extinguisher that discharges a fine dry powder, usu sodium bicarbonate, potassium bicarbonate, or ammonium phosphate, by the pressure of a compressed gas stored in the same container as the powder; generally suitable for Class B and Class C fires. soda-acid: A fire extinguisher that discharges water by the pressure of carbon dioxide gas that is produced by the mixing of acid and soda when the extinguisher is actuated. Caution: the water may contain unreacted acid or soda.

fire extinguisher symbols (Fire Pr) A set of four letter symbols - A, B, C, and D - used to label the respective classes of fire extinguishers corresponding to the types of fires they are effective against. The letters are white, on colored backgrounds: A on a green triangle; B on a red square; C on a blue disk; and D on a yellow 5-pointed star.
  NOTE: Multipurpose extinguishers are often marked with two or even three symbols. Several of the dry chemical type, for example, carry both B and C designations.

fire extinguishing agent = extinguishing agent

fire extinguishing system (Build; Fire Pr) An installation consisting of one or any combination of automatic sprinklers, foam nozzles, fire hoses, and portable fire extinguishers designed to provide fire extinguishing capability adequate for a room or building, taking into account its size, construction, use, and contents.

fire fan = buff

fire fatality = fire death

firefighter (Fire Pr) An active, participating member of a fire department, including volunteers and part-time firemen. Syn: fireman.
  NOTE: Distinctions are often made to indicate status, as: regular or paid, paid-call, part-paid, call, volunteer, etc.

fire fighting (Fire Pr) The activity and procedures necessary to diminish or halt combustion and flame from fuels. Fires are extinguished by cooling the burning material, removing or interrupting the fuel supply, blocking the air supply, halting the combustion process by chemical quenching materials, or by a combination of such measures.
  NOTE: Cooling is usu accomplished by water. Cutting off or interrupting the fuel supply is sometimes possible, esp when flammable gas or liquid supply lines can be shut, or a firebreak constructed ahead of a wildfire. The air supply can be blocked by blanketing the fire with foam or CO_2. Flaming can be halted by chemical quenching produced when vaporizing liquids and dry chemicals decompose and react in fires.

fire-fighting orders (For Serv) Ten standing fire, safety, and operations rules developed and followed by the US Forest Service:
  1) Keep informed on fire weather conditions and forecasts
  2) Know what your fire is doing at all times
  3) Base all actions on current and expected behavior of the fire
  4) Have escape routes for everyone and make them known
  5) Post lookouts when there is a possible danger
  6) Be alert, keep calm, think clearly, act decisively
  7) Maintain prompt communications with your men, your boss, and adjoining forces
  8) Give clear instructions and be sure they are understood
  9) Maintain control of your men at all times
  10) Fight fire aggressively, but provide for safety first.

fire finder (For Serv) A device used by lookouts to determine the location of a fire. Syn: alidade.

fire floor (Fire Pr) The story of a building on which a fire is burning. See also: affected floor.
  NOTE: The floor immediately below is called the operations floor.

fire flow (Fire Pr) The flow rate of water in the mains of a given area required for fire protection in addition to the normal water consumption for that area.
  NOTE: For purposes of the "Grading Schedule for Municipal Fire Protection," the Insurance Services Office distinguishes between required fire flow and basic fire flow. Required fire flow is the rate of water flow needed to fight and confine a fire to the buildings within a block or within a group or complex of buildings. The required fire flow ranges from a minimum of 500 gpm to a maximum of 12,000 gpm for a single fire. Additional flows of 2000 to 8000 gpm are required for simultaneous fires. Basic fire flow is a practical rate usu somewhat below the required fire flow. It is the fire flow judged to be needed for suppressing fires in important districts.

fire gas (Comb) A gaseous product of combustion, such as carbon monoxide, carbon dioxide, hydrogen sulfide, nitrogen chloride, etc. Syn: flue gas.

fire grading (Build) 1. The procedure of assigning grades of fire resistance to various structural elements of a building as defined in specifications appropriate to the fire hazard arising from the occupancy and according to the height, floor area, or cubic capacity of the building. 2. The investigation and assignment of suitable fire precautions of any kind to attain an adequate standard of safety according to the fire hazard of the building under consideration.

fireground[1] (Fire Pr) a. The area of operations on which a fire is fought. Firegrounds may be marked off by police lines. Also called: fire scene. b. An area defined by law or ordinance requiring that motorists not park within a certain distance of operating fire apparatus.

fireground² (Fire Pr) An area used for fire instruction and training purposes.

fire growth (Fire Pr) The development of a fire from ignition to the point of maximum severity.

fire guard¹ (Fire Pr) Residents or employees designated and trained to fight fires and protect buildings or block areas as a civil defense measure. Syn: fire brigade¹.
NOTE: Fire guard activities are usu intended for small fires when firefighting forces are engaged in large operations.

fire guard² (For Serv) Firefighters, lookouts, fire patrolmen, and others employed to detect, prevent, and suppress fires.

fireguard¹ (Build) A screen placed in front of a fireplace to catch flying sparks. Syn: firescreen.

fireguard² = firebreak

fire hall (Fire Pr; Canadian) A fire station or fire house.

fire hat (Fire Pr) A firefighter's helmet.

fire hazard (Fire Pr; Insur) The relative danger of the start and spread of fire, the danger of smoke or gases being generated, the danger of explosion or other occurrence potentially endangering the lives and safety of the occupants of a building or structure.
NOTE: Fire hazard may be expressed qualitatively as the product of the potential for harm and the degree of exposure.

fire hazard classification (Fire Pr) The designation of a fire hazard as low, ordinary, or high, depending on the nature of the contents and the processes or operations conducted in a building or structure, or on the flame spread rating of the interior finish or other features.

fire hazard test (Testing) A standard experimental procedure used to determine the probable behavior or performance of a material or structure under actual fire conditions.

fire headquarters (For Serv) A fire camp established close to a wildfire, serving as a control center for firefighting and fire-suppression activities.

fire horse (Fire Pr) 1. A specially trained horse once used to pull steamers before the advent of automotive vehicles. 2. (Colloq) A seasoned firefighter who enjoys his profession.

fire hose See: hose

firehouse (Fire Pr) A fire department station housing fire apparatus and on-duty firefighters. Syn: fire station. Also called: fire hall.

firehouse dog (Fire Pr) A fire company pet, usu a dalmatian. See: Sparky.

fire hydrant See: hydrant

fire injury (Fire Pr; Med) Any trauma suffered as a result of a fire or explosion and requiring medical treatment within a certain time period, usu within one year after the incident.

fire insurance (Insur) Protection for losses resulting from fire and lightning, as well as from smoke and water damage.

fire intensity (Build; Fire Pr) The rate of heat liberation per unit volume of the combustion zone. See: fire severity.

fire investigation (Fire Pr) A procedure undertaken to determine the cause of a fire, esp one of suspicious origin. See: arson.

fire irons (Build) Fireplace tools, tongs, pokers, etc.

fire jump (For Serv) Parachuting of crews, supplies, and equipment into a fire area for attack and suppression.

fire lane (Fire Pr) A posted driveway reserved exclusively for fire department vehicles.

fireless cooker A high heat-capacity container that will cook food by retained heat.

fire line[1] (Fire Pr; For Serv) A front or zone in which firefighters operate to extinguish a fire in a building, forest, or grassland.

fire line[2] (Fire Pr) A fire hose, particularly when in use for firefighting.

fire line[3] (Fire Pr) A boundary erected by the fire department or police around a fireground to bar all but emergency personnel and vehicles.

fireline[4] (For Serv) a. A part or segment of a fire control line. b. A trail cleared of combustible material by scraping or digging to the soil in wildland firefighting operations. Also called: fire trail. See: firebreak.

fire load (Build) A measure of the heat that could be liberated during a fire in a given fire area or other building space, derived from the calorific potential of the fuel load. Fire load and fuel load contribute to fire severity and fire duration.
NOTE: The NFPA Fire Protection Handbook uses "fire load" as the preferred synonym of "fuel load." The definition given above follows the International Standards Organization convention. Fuel load is thus released to express the quantity of fuel in a fire area.

fire lookout (For Serv) 1. A forest fire lookout tower. 2. A member of a fire suppression unit manning a lookout tower.

fire loss (Fire Pr; Insur) 1. The cost, in monetary units, of restoring or replacing a building and its contents that have been destroyed or damaged by fire. Syn: direct loss. Cf: total loss. 2. = (Brit) material damage. Syn: direct loss.

fireman[1] (Fire Pr; For Serv) Any member of a fire department, including regular, call, and volunteer firefighters. Syn: firefighter.

fireman[2] (Transp) One who tends the fire on a steam locomotive. A stoker.

firemanic (Fire Pr) Referring to the fire service.

fireman's carry (Fire Pr) 1. A technique of lifting and carrying a disabled person to safety. The casualty is carried over the shoulders and held by one arm and hand, leaving the firefighter one arm and hand free for climbing down a ladder. See: rescue carry. 2. = (Brit) fireman's lift.

firemanship (Fire Pr) Referring to the skills, performance, and professionalism in the fire service.

fireman's lift (Brit) = fireman's carry

fire mark (Fire Pr; Insur) A metal plaque sometimes found on old buildings identifying the fire insurance company or fire protection organization extending protection to the building.
NOTE: Introduced in Philadelphia by Benjamin Franklin's fire insurance company. Now a collector's item.

fire marshal (Fire Pr) The principal fire prevention officer of a state, county, or city.
NOTE: In some cities the fire marshal is the head of the fire prevention bureau or serves as an assistant or deputy chief. See: ranks and titles.

Fire Marshals' Association of North America (Fire Pr) A professional association of state, provincial, county, and city fire prevention officers; affiliated with the National Fire Protection Association.

firemen's lift (Fire Pr; Brit) An elevator which may be commandeered from predetermined floor levels for the exclusive use of firefighters in an emergency. See: elevator capture.

Firemen's Memorial Sunday (Fire Pr) A day, usu the second Sunday in June, on which commemorative services are held for deceased firefighters. See: Fire Service Recognition Day.

fire nomenclature (For Serv) The names assigned to the parts of a wildland fire.

NOTE: Four sides are normally distinguished in a typical free-burning fire: 1) the <u>head</u>, which is the most rapidly advancing edge of a fire; 2) the two <u>flanks</u>, which are the sides adjoining the head at more-or-less right angles; and 3) the <u>rear</u>, which is the slowest-moving edge of the fire.

<u>fire pack</u> (For Serv) A set of tools, supplies, and miscellaneous equipment prepared in advance and carried as a back pack by forest firefighters.

<u>fire pail</u> (Fire Pr) A bucket, containing sand, water, or other extinguishing agent, painted red and sometimes having a rounded bottom and set into a special frame to discourage use except for firefighting. Syn: <u>fire bucket</u>.

<u>fire partition</u> (Build) A partition that serves to restrict the spread of fire but does not qualify as a fire wall.

<u>fire patrol</u>[1] (Insur) A fire salvage organization maintained by fire insurance underwriters in New York City and Philadelphia.

<u>fire patrol</u>[2] (Fire Pr; For Serv) A special fire watch or continuous inspection. Cf: <u>brand patrol</u>.

<u>fire penetration test</u> (Build) 1. A means of assessing the resistance of a foamed plastic to penetration by a blowtorch flame. 2. (Brit) A means for measuring the capacity of a representative section of a roof to resist penetration by fire when the external surface is exposed to radiation and flame.

<u>fire performance</u> (Build) The behavior of materials or structural elements under real or test fire conditions.

<u>fire picket</u> (Fire Pr; Brit) = <u>fire watch</u>

<u>fireplace</u> (Build) A structure of stone or brick for holding a fire, and a chimney for carrying off smoke. A hearth.

<u>fireplace insert</u> (Build) An open-flame, radiant heater mounted in a decorative metal panel installed in a fireplace. Cf: <u>gas log</u>.

<u>fireplace stove</u> (Build) A stove with a fire chamber opened partly to the room.

<u>fire plow</u> (For Serv) A heavy-duty share or disk type plow designed for rough work in the woods; drawn by horses or a tractor to construct firebreaks and fire lines.

<u>fire plug</u> = <u>hydrant</u>

<u>fire point</u>[1] (Comb) The lowest temperature at which a liquid gives off sufficient flammable vapor to produce sustained combustion after removal of the igniting source.

<u>fire point</u>[2] (Fire Pr) The point at which fire extinguishing equipment is located.

<u>fire prevention</u> (Fire Pr) Precautionary fire protection activities carried out before fires occur, aimed at preventing outbreak of fire, early detection, and minimizing life and property loss if fire should occur. Activities undertaken include public education, inspection, law enforcement, and reduction of hazards.

<u>fire prevention bureau</u> (Fire Pr) An agency of a county, city, or other political entity engaged in fire prevention or investigation and not in actual firefighting. Also see: <u>inspection bureau</u>.

<u>fire prevention code</u> (Fire Pr) The set of fire safety ordinances and regulations adopted by a city, county, or other jurisdiction.

<u>fire prevention week</u> (Fire Pr) A week proclaimed by the President of the United States and sometimes by State Governors, devoted to publicizing and encouraging fire prevention activities. The week usu includes the anniversary of the Great Chicago Fire of 9 October 1871.

<u>fire progress map</u> (For Serv) A situation map maintained on large fires to show the extent of the fire, deployment of firefighting forces, and progress being made in extinguishment.

<u>fireproof</u> (Build) 1. The property of a material, such as concrete or

iron, not to burn or decompose when exposed to ordinary fire. Syn: fire resistant. Cf: refractory. 2. To treat a combustible material with a fire retardant to reduce the fire hazard.
  NOTE: The term fire resistive is recommended by the NFPA in preference to fireproof because no material can withstand heat of sufficient intensity and duration.

fireproofing (Build; Obsolete) 1. A substance, material, or treatment applied to provide increased fire endurance to structural elements or assemblies, usu serving no structural function itself. Syn: fire retardant. 2. Treating a material or surface with a fire retardant.
  NOTE: Fire retardant material is often applied by spraying it directly on structural elements or on specially provided bases, such as metal lath.

fire propagation¹ (Comb) The spread of fire through or over a combustible medium. A phenomenon that depends on the ignitability, flame spread, and heat release characteristics of a material.

fire propagation² (Fire Pr) The penetration and travel of fire through openings and concealed spaces.
  NOTE: Fire propagation carries the sense of fire traveling through a gas or vapor in three-dimensional space; fire spread is more appropriately descriptive of fire travel across surfaces.

fire propagation test (Build; Brit) A procedure for assessing the rate of heat release from a combustible material when it is subjected to specific heating conditions.

fire protection¹ (Fire Pr) The theory and practice of reducing life and property loss by fire, fire extinguishment, fire control, and fire prevention. Narrowly, the field concerned with the detection and extinguishment of fires. See: fire prevention.
  NOTE: Fire protection is provided by proper building design, by installing appropriate detection and extinguishing systems in buildings, by establishing adequate firefighting services, and by training building occupants in fire safety and evacuation procedures. Two types of fire protection are distinguished: active and passive. Active fire protection is represented by sprinklers and other automatic fire suppression systems. Passive is represented by fire walls, fire and flame retardant treatments, and the like.

fire protection² (Fire Pr) Materials, devices, fire apparatus, and other equipment used to prevent fire or to minimize loss of life or property resulting from fire.

fire protection³ (Insur) Insurance against fire loss.

fire protection district (Fire Pr) A rural or suburban district that contracts for fire protection services from a nearby fire department or maintains its own fire protection service and equipment. See: fire district.

fire protection engineer (Fire Pr) A trained graduate engineer specializing in fire protection problems by applying scientific and technical principles to the reduction of loss of life and property by fire, explosion, and related hazards. See: Society of Fire Protection Engineers.

fire protection engineering (Fire Pr; Ind Eng) The field of engineering concerned with the safeguarding of life and property against loss from fire, explosion, and related hazards.

fire protection map (Fire Pr) A map that shows travel distances to various points in a protected area. Cf: travel time map.

fire protection rating = fire endurance rating

fire pump (Fire Pr) The main pump of a fire department pumper or a stationary power-driven water pump used by private fire protection services.

fire raising (Fire Pr; Brit) Firesetting; committing arson.

fire report (Fire Pr) A comprehensive description of a fire prepared by fire department personnel or fire protection officials, specifying all details pertinent to a fire incident.

<u>fire research</u> (Build; Fire Pr) Systematic investigation into anything connected with unwanted fires and explosions.

<u>fire resistance</u> (Build; Fire Pr) The attribute or capacity of a material, structural component, or assembly to withstand fire without losing its fire-separating or load-bearing strength. Fire resistance is usu expressed in terms of a time period, as a 1-hour fire resistant door. Cf: <u>fireproof</u>; <u>fire retardant</u>.
NOTE: As applied to elements of buildings, it characterizes the ability to confine a fire or to continue to perform a given structural function, or both. A wall assembly, for example, has a certain degree of fire resistance if for a specified time and under standard conditions of heat intensity it does not fail structurally and does not permit the side away from the fire to become hotter than a specified temperature.

<u>fire resistance classes</u> (Fire Pr) A system of letters, A through F, used to classify fire doors, fire windows, roof coverings, openings, fire extinguishers, interior finishes, places of assembly, and the like to indicate gradations of fire safety requirements.
NOTE: Not all A-F classifications are used for all things being classified.

<u>fire resistance grading</u> (Build; Testing; Brit) The time period which is nearly equal to, but does not exceed, the test period for which the element fulfills all the relevant test requirements.

<u>fire resistance rating</u> (Build; Testing) A fire performance classification, usu in integral hours or half-hours, that has been assigned to a structural element on the basis of a fire test having been applied to a prototype.

<u>fire resistance test</u> (Build; Testing) A test to enable elements of construction to be assessed according to their ability to retain their stability, to resist the passage of flame and hot gases, and to provide the necessary resistance to heat transmission.

<u>fire resistant</u> (Fire Pr) 1. Having the quality or capability of withstanding fire for a specified period, usu in terms of hours with respect to temperature, without loss of fire separating or loadbearing function. 2. Possessing the quality of resisting damage or destruction by fire.

<u>fire resistant building</u> (Build) A building that will withstand a major fire or even complete burnout without collapse.

<u>fire resistive</u> (Build); Fire Pr) Pertaining to the relative capability of withstanding fire for a specified period, usu in terms of hours with respect to temperature.

<u>fire-resistive construction</u> (Build) A type of construction in which the structural members, including walls, partitions, columns, floors, and roof structures, are made of materials, primarily structural steel and reinforced concrete, having fire endurance ratings not less than those specified by the authority having jurisdiction.

<u>fire retardant</u> (Build; Fire Pr) 1. Possessing or imparting low flammability or flame spread properties. 2. A wall, partition, door, or other structure that holds back the spread of fire. See: <u>cutoff</u>. 3. A substance or treatment, such as monoammonium sulfate, that reduces the flammability of a material.
NOTE: The term <u>fire retardant</u> indicates a greater resistance to fire than <u>flame retardant</u>.

<u>fire retardant chemical</u> (Build) A chemical that is impregnated into or compounded with a material to reduce its flammability.
NOTE: Some common fire retardants used for wood are ammonium phosphate, ammonium sulfate, borax, boric acid, zinc chloride, and sodium dichromate.

<u>fire retardant coating</u> (Build) A material applied to building components, usu interior surfaces, to interfere with or decrease the rate of flame spread. See: <u>intumescent</u>

paint. Cf: flame retardant.
  NOTE: Coatings are applied as paints or mastics by brush, roller, sprayer, or trowel.

fire risk See: risk

fire safety (Fire Pr) Security against fire and its effects.

fire safety officer (For Serv) A staff officer responsible for the health and safety of fire suppression forces and for advising the fire boss on how to minimize hazards and accidents.

fire sale Sale of merchandise that has been damaged by smoke, water, or flames in a fire.

fire scar (For Serv) 1. A fire wound on a woody plant that is partly or completely healed. See: catface. 2. A burned-out area on the landscape.

fire scene = fireground

fire school (Fire Pr) 1. A school maintained by a fire protection organization or a fire department for the purpose of training officers, prospective officers, and firefighters in methods of fire extinguishment and control. 2. An abbreviated intensive course in fire science subjects given by a university for members of the fire service and safety professionals. Syn: fire college.

fire science (Fire Pr) The systematic body of knowledge and principles drawn from physics, chemistry, mathematics, engineering, administration, management, and related branches of arts and sciences required in the practice of fire protection and prevention.

firescreen = fireguard[3]

fire season (For Serv) A period of time during the year when weather and other conditions are particularly favorable for the outbreak and spread of fire. See: fire weather.
  NOTE: A normal fire season is one with an average number of fire outbreaks.

fire separation[1] (Build; Fire Pr) A floor or wall, either without openings or with adequately protected openings, that meets specified fire endurance requirements for a barrier against the extension of a fire from one area of a building to another.

fire separation[2] (Fire Pr; Storage) A space provided between objects, such as goods in storage, buildings, or structures, that serves as a firebreak and as a fire lane or area for firefighting operations.

fire service (Fire Pr) 1. An organization that provides fire prevention and fire protection to a community. 2. The members of such an organization. 3. The firefighting profession as a whole.

Fire Service Recognition Day (Fire Pr) The second Saturday in May, honoring the fire service. A day promoted by the International Association of Fire Chiefs for improving fire service and community relations. See: Firemen's Memorial Sunday.

fire service training See: fire training

fire setting[1] (Fire Pr) Committing an act of arson. The starting of a fire maliciously and illegally. Cf: firing.

fire setting[2] (Mining) The technique of breaking large rocks by heating them and then quenching them with cold water.

fire severity (Build) The potential of a fire to inflict damage. A measure of the intensity of a fire over time, expressed in degree-hours. Depicted graphically as the area beneath a time-temperature curve and above a base line that represents the maximum temperature to which given materials may be exposed without damage.
  NOTE: Fire severity is determined by the nature and quantity of the fuel, its burning rate (which is affected by its arrangement), temperature, and other factors. Consideration should be given to possible temperature sensitivity of certain materials and structures. For example, long-term exposure of a given material to moderate temperature may have little or no effect, whereas a brief exposure to a higher temperature may cause severe damage

to it.

fire shutter (Build) A fire-resistive assembly, including frame and hardware, that under standard test conditions meets the fire endurance requirements for the window or other opening in which it is to be used.

fire signature (Comb; Fire Pr) Any product or emission of a fire that can be used to detect it. A sensible sign of a fire.
  NOTE: Typical signs are smoke; invisible aerosols; infrared, visible and invisible ultraviolet radiation; and various gases. The flickering of flames is characteristic and detectable, and thus serves as a signature.

fire spread (Comb) The development and travel of fire across surfaces. Cf: fire propagation; flame spread.

fire station (Fire Pr) A building housing firefighters and fire equipment, serving as a headquarters for a fire department company. Syn: fire house; fire hall.

fire statistics (Fire Pr) Data obtained from fire incident reports.

fire stone[1] (Build) A stone that endures high heat. A fire-resistant building stone.

fire stone[2] (Mineral) Flint or pyrite. A stone that produces sparks when struck against metal.

firestop (Fire Pr) Fire-resistant material, barrier, or construction installed in concealed spaces or between structural elements of a building to prevent the extension of fire through walls, ceilings, and the like. Syn: fire cutoff.

firestopping (Build) 1. The act of installing firestops during construction. 2. The complex of firestops in a building.

fire storm (Fire Pr) A large, rapidly developing fire ignited roughly simultaneously over a large area. Characteristically, accompanied by fire-generated winds and turbulent convective currents.
  NOTE: There is debate over the conditions required for such systems or whether they have occurred. Such events took place in the bombing of Hamburg, Germany, during World War II and in very large forest fires.

firestream (Fire Pr) 1. A stream of water delivered from a nozzle and used to control and extinguish a fire. 2. The portion of a water stream from the nozzle tip to the point of impact.

fire suppression (For Serv) The activity involved in controlling and reducing a fire by constructing fire breaks, backfiring, applying fire retardants, etc., as opposed to direct fire extinguishment. See: suppression.

fire suppression organization (For Serv) A group of supervisory and service personnel assigned to fire suppression duty.

fire test (Build; Testing) A standard experimental procedure for assessing the reaction of a material, product, or system to a specified fire exposure. The behavior observed may or may not relate to other fire exposure conditions, such as full-scale fires.

fire-test-exposure severity (Testing) 1. A measure of the degree of fire exposure in a fire test. 2. The ratio of the area under the average furnace temperature curve to the area under the standard time-temperature curve, each from the start of the test to the end or time of failure, and above the base temperature of 20°C (68°F).

fire-to-hydrant lay (Fire Pr) Stretching of hose from a fire to a water supply; the opposite of the usu foreward lay. Syn: reverse lay. See: hose lay.

fire tool cache (For Serv) Sets of fire tools and equipment assembled in planned quantities or standard units and stored at a strategic point for use in fire suppression.

fire tower[1] (Fire Pr) a. A fire department training tower. b. A fire department communications and control center. c. A fire department water tower truck or aerial ladder used as a water tower.

fire tower[2] (For Serv) A forest fire lookout tower.

fire tower[3] (Build) a. A structurally independent stairway attached to a building and used as a fire escape and as a means of access for fighting fires. b. A vertical enclosure in a building, the walls and doors of which have sufficient fire endurance to qualify the enclosure as a fire escape. Syn: smokeproof tower.

fire trail (For Serv) A narrow strip of exposed mineral soil, often constructed in a firebreak.

fire training (Fire Pr) Development of basic manipulative skills through exercizes or drills, aimed at improving proficiency in manual dexterity, physical strength and agility, and mechanical aptitudes, including conditioning of automatic responses of individuals and teams to given situations.

firetrap[1] (Fire Pr; Colloq) A building lacking adequate exits or fire protection equipment, or one which because of interior layout presents a major hazard to life in case of fire. Syn: taxpayer (Colloq).

firetrap[2] (For Serv) An accumulation of flammable fuels in wildlands or conditions which make firefighting extremely hazardous.

fire trench (For Serv) A firebreak in the form of a ditch dug into the soil to interrupt the ground fuel.

fire triangle (Fire Pr) The three factors necessary for combustion: fuel, oxygen, and heat.
 NOTE: A fire tetrahedron has been proposed to account for the chemical chain reaction in combustion processes.

fire truck (Fire Pr) A firefighting apparatus such as a ladder truck, hose wagon, pumper, and the like. See: apparatus.

fire tube test Cf: FPL tunnel test.

fire wall (Build) A fire-resistant solid wall of masonry or equivalent material provided to delay or restrict the horizontal spread of fire for a specified period of time; generally a self-supporting assembly designed to maintain its integrity even if the structures on both sides collapse. If a wood roof is involved, the wall extends through the roof and above it.
 NOTE: Fire walls subdivide buildings into fire areas.

fire ward (Fire Pr; Archaic) An individual appointed to organize and command a bucket brigade at a fire.

fire warden[1] (Civ Def; Fire Pr) A fire prevention officer or a civil defense worker assigned to local fire prevention and protection duties.

fire warden[2] (For Serv) A forest fire protection officer. See also: fire guard[2].

fire watch (Fire Pr) 1. One or more firefighters detailed to be present at a public gathering as a fire safety precaution. Syn: fire duty. 2. = (Brit) fire picket.

firewatch (Fire Pr; Brit) = watch detail

fire weather (For Serv) Atmospheric conditions up to 10 miles above the earth, as they affect the state of wildland fuel and favor fire ignition and spread.
 NOTE: Fire weather includes considerations of temperature, humidity, wind, and other factors. See also: fire season.

fire weather forecast (For Serv) A weather forecast prepared esp for use in forest fire control.

fire weather station (For Serv) A forest meteorological station specially equipped to record weather factors that have an important bearing on forest fire control.

fire whirl (Fire Pr) A revolving mass of flame in the air caused by strong convective currents and drafts in an intense fire. Cf: fire devil.
 NOTE: A well-defined fire cyclone was observed and photographed on 5 June 1974. It formed above a tenement roof in Brooklyn, N.Y., during test burns of condemned buildings.

fire wind (Fire Pr) Movement of air caused by a large fire depleting the oxygen from the atmosphere and thereby creating a partial vacuum.

fire window (Build) A fire-resistive window assembly, including frame, wired glass, and hardware, that meets the fire protective requirements for the location in which it is to be used.

firewood Wood intended for use as fuel, for heating or cooking.

fireworks Explosive and combustible devices used to produce colored lights, flashes, smoke, streamers, and noise for celebration and amusement.

fire wound (For Serv) An injury caused by fire to the cambium of a woody plant. See: fire scar.

fire zone = fire area

firing (For Serv) The act of igniting a fuel, for prescribed burning or other purpose.

firing out = burning out

first aid[1] (First Aid; Med) Emergency assistance given to a victim of sudden illness or injury until medical attention is obtained. See: artificial respiration.

first aid[2] (Fire Pr) a. Referring to firefighting equipment, such as hose reel, hand pump, chemical extinguishers, buckets of water or sand, etc., used in the first stages of a fire, or to a small tank apparatus used for small fires. b. Measures taken to suppress a fire before the arrival of fire department forces.

first aid box (First Aid) The medical first aid kit carried on fire apparatus.

first aid injury (First Aid) An injury requiring first aid treatment only.

first alarm (Fire Pr) The initial alarm of fire, or the first signal calling for the first alarm response.

first arrival (Fire Pr) The first unit or officer to report at the scene of a fire. Syn: first in. Cf: first due.

first attack (For Serv) The first work done to suppress a fire. Syn: initial attack.

first attendance (Fire Pr; Brit) 1. The first apparatus to go to a fire. 2. = (US) first due.

first-degree burn See: burn[2]

first due (Fire Pr) 1. Referring to the pumper, ladder truck, and officer listed on the first alarm assignment. 2. = (Brit) first attendance.

first in (Fire Pr) Referring to the personnel and equipment first to arrive at a fire. Syn: first arrival.

first line (Fire Pr) Referring to the apparatus and firefighting units normally responding to fire alarms, as contrasted to reserve units.

first water (Fire Pr) The first hose stream to be applied to a fire.
NOTE: At one time an achievement of great honor and a matter of fierce competition among fire companies to apply first water.

first work period (For Serv) The time between discovery of a fire and 10 am of the following day. Succeeding work periods are each 24-hours long, beginning at 10 am. Also called: initial shift.

fish plate (Build) A metal or wooden plate used to cover a butt joint.

fission (Nucl Phys) The splitting of a heavy nucleus into two approximately equal parts, which are nuclei of lighter elements, accompanied by the release of a relatively large amount of energy and generally one or more neutrons. Fission can occur spontaneously but usu is caused by nuclear absorption of gamma rays, neutrons, or other particles.

fissionable (Nucl Phys) Pertaining to radioactive material capable of undergoing fission.

fission fragment (Nucl Phys) The two medium-weight, radioactive nuclei formed by fission from a heavy nucleus. Also called: primary fission products.

fission material (Nucl Phys) Any material that can be fissioned by slow neutrons. The three basic ones are uranium-235, plutonium-239, and uranium-233.

fission products (Nucl Phys) The nuclei or fission fragments formed by the fission of heavy elements plus the nuclides formed by the radioactive decay of the fission fragments.

fitting (Fire Pr) Any of a variety of standard hardware accessories used to space or connect mechanical parts, as a bushing, sleeve, adapter, etc.

fixed carbon (Comb) The ash-free carbonaceous material that remains after volatile matter is driven off.

fixed ladder (Build; Safety) A ladder permanently attached to a structure, building, or equipment.

flail (Fire Pr; For Serv) Any of a variety of swatters, such as a fire bat, used to beat out grass fires and the like.

flake (Fire Pr) 1. A loose fold of fire hose. 2. To pay out folds of hose from a flaked hose load.

flaked hose (Fire Pr) 1. Loosely folded hose. 2. (Brit) = (US) accordion hose.

flame[1] Popularly, the hot, light-emitting portion of a combustion process; the heat-producing, usu visible part of a fire. Hot, incandescent, chemically reacting gases.
NOTE: In combustion engineering some fuels are burned with nonluminous flames.

flame[2] (Comb) A rapid, gas-phase exothermic combustion process characterized by self-propagation, which results from a strong coupling between chemical reaction and the physical transport processes of matter and energy (i.e., diffusion, thermal conduction, and thermal diffusion). Flames usu involve reactions between a fuel and an oxidizer (generally air).
NOTE: Several types of flames occur. If fuel and oxidizer are premixed, a premixed laminar flame results, typified by the laboratory bunsen flame. Propagation of such a flame is controlled by the interaction between chemical reaction and the transport processes. A rigorous set of so-called flame equations can be written to describe such systems. It is characteristic of premixed flames that they propagate at a velocity determined solely by initial conditions (pressure, composition, and temperature). This velocity, called burning velocity, can be identified with the eigensolution to the flame equations. If the fuel and oxidizer are not premixed, the propagation rate of the flame is determined by the mixing rate. If the flow is laminar, propagation is limited by diffusion and the result is a diffusion flame. If the flow is turbulent, mixing is dominated by eddy diffusivity, and the result is a turbulent flame. Turbulent flames may be either premixed or nonpremixed. If the flame propagates at supersonic speed, it is referred to as a detonation. Flame reactions are generally rapid and are principally those of atoms or radicals with molecules. Radical and atom concentrations generally exceed thermal equilibrium because of rapid radical-producing reactions. The subsequent recombination of these radicals produces nonthermal light emission, which is the most conspicuous flame phenomenon, and charged species. Despite these striking nonequilibrium processes, flames are overall adiabatic processes, so that the final maximum flame temperature can be closely calculated from thermodynamic considerations.

flame area (Build; Fire Pr) An area in a vertical plane, determined from the height and width of expected flames. Used in estimating fire exposure severity.
NOTE: In the case of a building, the flame width is taken as the sum of the window widths between vertical fire walls. Flame height is estimated as equal to the height of the roof above ground. For a tank with a free-surface burning liquid, the flame area is estimated as an equal-sided trapezoid, the height and base of which are equal to the diameter of the tank and the top as equal to half the diameter.

flame arrester (Comb) A screen of wire gauze, compressed wire mesh, metallic foam, ribbons, or slats that halts the propagation of flame in a gaseous atmosphere. Syn: flame trap; flash arrester.
  NOTE: Flame arresters are installed in the vents of flammable liquid or gas tanks, storage containers, cans, gas lines, or flammable liquid pipelines to prevent flashback (movement of flame through the line or into the container) when a flammable or explosive mixture is ignited. Wire screen of 40 mesh is used on small openings. On large openings parallel metal plates or tubes are more effective.

flame detector (Comb) A device that senses the presence or the absence of a flame and produces an appropriate signal.
  NOTE: A flame detector may be used, for example, as a flame failure device to shut down the gas supply to the main burner of a furnace when the pilot flame goes out.

flame equations (Comb) The set of mathematical and chemical equations that describe strongly coupled reacting flow systems and apply both to flames and detonations.
  NOTE: The equations consist of an aerodynamic description, the constraints of conservation of energy, conservation of each elemental species individually, and the differential equations for reaction, diffusion, thermal conduction, and thermal diffusion of low molecular weight species. In the one-dimensional case, which applies quantitatively to bunsen-type flames, the system is treated as a boundary value problem with an eigensolution that can be identified with propagation velocity or burning velocity.

flame front (Comb) The temperature and compositional microstructure associated with flames considered in a stationary reference frame. Cf: combustion wave.

flame-front thickness (Comb) A parameter giving the thickness of a flame front or combustion wave.

flame ionization (Comb) A phenomenon whereby electrically charged particles and electrons impart electrical conductivity to hot combustion gases. It can be used as a sensitive means of detection in gas chromatography or as a means of altering the flow pattern in combustion devices.

flameproof[1] (Fire Pr; Brit) Descriptive of an explosion-proof enclosure.

flameproof[2] See: flame resistant

flameproofing (Techn) Surface treatment or impregnation of wood products, textiles, and other materials with fire-retardant chemicals.
  NOTE: The terms flame-retardant and flame-resistant treatment are recommended by the NFPA.

flame propagation (Comb) 1. The motion or travel of flame through a combustible gas mixture. 2. Spread of flame from region to region in a combustible material, esp in a combustible vapor-air or gas-air mixture. Cf: flame spread.
  NOTE: A vapor-air concentration below the lower flammability limit may burn at the point of ignition without propagating.

flame resistance (Testing) The capacity of a material or object to resist ignition and rapid flame involvement, usu in reference to textiles, sheet materials, or other light section products.

flame resistant (Fire Pr) Characterizing a material that does not conduct flame or continue to burn when an ignition source is removed. Cf: fire resistant.

flame retardant[1] (Build; Comb) Having the property of delaying flame spread.

flame retardant[2] (Comb) A flame-inhibiting chemical compound, such as inorganic salts, Lewis acids, and free radical initiators, used on surfaces as well as in bulk to reduce the flammability of a product or structure.
  NOTE: Flame retardants are used on textiles, in plastics, paints, etc. Surface treatments include glazes, such as sodium silicate or borax; coatings that evolve inert gas when heated; endothermic salts that dissipate heat or carry it away;

chemicals that produce carbon dioxide and water; and intumescent paints that form an insulating carbonaceous blanket over a surface exposed to high heat. Among chemicals that control the decomposition of cellulosic material catalytically at flame temperature are borax-boric acid, borax-boric acid-diammonium phosphate, ammonium bromide, stannic acid, antimony oxide, and combinations containing formaldehyde.

flame-retardant finishing (Text) Application of fire retardants to textiles.

flame safety lamp (Mining) A device used to test atmospheres for sufficiency of oxygen. The flame goes out when the oxygen concentration in the air drops to about 16%.

flame speed = flame velocity?

flame spread (Comb) The progress of flame over a surface. Cf: flame propagation.

flame spread classification (Build; Abb: FSC) A measure of the surface flame spread characteristics of a material. For building materials, the FSC is usu determined by the 25-ft tunnel test (ASTM E-84). Common classification ranges are 0-25, 25-75, 75-200, and over 200, which is the most hazardous.

flame spread index (Comb) A number or classification indicating a comparative measure derived from observations made of the advance of a flame boundary under defined test conditions. See also: flame spread rating.

flame spread rate (Build; Testing) The distance traveled per unit time of the leading edge of a flame across a surface.
  NOTE: The rate of flame spread, for various reasons, may be different at different points along the leading edge of the flame.

flame spread rating (Comb) A number indicating the relative rate at which flame spreads over the surface of a given material as compared with a scale on which flame spread over asbestos board is zero and over red oak is 100 (flame travels 3.56 fpm on red oak). The rating number is obtained by test in a Steiner tunnel, and is a relative index and not the actual rate of flame spread or of the fire resistance of a material. See also: flame spread index.
  NOTE: The NFPA flame spread rating classes are: A, 0 - 25; B, 26 - 75; C, 76 - 200; D, 201 - 500; and E, >500.

flame spread test (Build; Testing) A procedure used to measure surface flame spread or burning characteristics of a material or assembly.
  NOTE: A flame spread test for a roof covering, for example, is one in which a flame of specified characteristics is played on a test roof deck continuously while exposing it to a wind of specified velocity.

flame temperature (Comb) The measured or calculated intensity of the heat of a flame. Flame temperature can be calculated on the basis of thermodynamic considerations, assuming that the flame is an overall adiabatic process.
  NOTE: A straightforward calculation is difficult because the high temperatures produce appreciable radical and atom concentrations, and the usual stoichiometric equations become poor approximations. Flame temperature calculations are tedious when done by hand, but computer programs are available that make these computations easily and economically. There is generally good agreement between such calculated temperatures and measured flame temperatures. Also called: adiabatic flame temperature.

flame trap (Techn) A device consisting of screens or grids spaced closer than the quenching distance so that flame propagation stops but gas is allowed to flow through. Syn: flame arrester.
  NOTE: The principle was discovered by Davy and was used in his miner's safety lamp.

flame velocity (Comb) The speed with which a combustion wave propagates into unburned gas. Syn: burning velocity. Also called burning rate (incorrect); combustion velocity; propagation velocity. See: combustion; flame. Cf: burning rate (for solids and liquids).
  NOTE: In steady state, flame such

as a bunsen flame propagates at a rate that is balanced by the flow velocity opposing the propagation. For premixed flames, the velocity depends only on the initial conditions in the cold gas (temperature, pressure, and composition). For simple systems the flame velocity can be determined from the solution of the flame equations. This property is often used to characterize combustion systems. See: burning velocity

flammability (Comb) 1. The capacity to burn with a flame. 2. Readiness to ignite and become rapidly involved in flame.
  NOTE: The flammability of a given material is usu determined by a test that measures ignition, flame spread, or burning rate. Different tests usu produce conflicting values of flammability for the same material. It is important, therefore, to know which aspect of the flammability of a material is meant.

flammability index (Comb; Symb: E) The ratio of the fuel vapor concentration by volume in a fuel-vapor and air mixture to that at its lower flammability limit.
  NOTE: The percent flammability index is set at 100E. If E < 1 (or %E < 100%), the vapor-air mixture is nonflammable. If E ≥ 1 (%E ≥ 100%) the mixture is flammable unless its concentration is at its upper limit. The percent flammability index is also called: percent explosiveness; explosivity; percent of lower limit.

flammability limit (Comb) The maximum or minimum concentration of combustible gas and air mixture that will ignite. The lower (lean) limit is the point of fuel deficiency, the upper (rich) limit is the point of oxidizer deficiency to sustain combustion. Cf: explosion limit.
  NOTE: A vapor-air mixture below the lower limit of flammability may react at an ignition source without propagating.

flammability range (Comb) The scale of mixture ratios between the upper and lower flammability limits. Cf: explosive range.
  NOTE: As an example, the flammability range of acetylene is 2.5 to 80% by volume with air; for gasoline the range is 1.6 to 6.0%.

flammability temperature limit (Comb) One of two temperatures at which the vapor pressure of a liquid is such that the equilibrium vapor composition above the liquid is equivalent to that at the lower concentration limit (lower flammability temperature limit, or flash point), or that at the upper concentration limit (upper flammability limit, or flash point).

flammable[1] (Comb) Capable of burning with a flame. Easily ignited or highly combustible. Syn: inflammable (Obsolete). Ant: nonflammable.

flammable[2] (Fire Pr) Subject to easy ignition and rapid flaming combustion.

flammable aerosol (Transp) An aerosol which is required to be labeled flammable under the Federal Hazardous Substances Labeling Act.

flammable gas (Fire Pr) Any gas that can burn.

flammable liquid (Comb) A liquid that has a flash point below 140°F and a vapor pressure not exceeding 40 psia at 100°F. In contrast, a combustible liquid is one that has a flash point above 140°F. Sometimes in fire practice a flammable liquid is loosely defined as one that ignites and burns at temperatures below 100°F. See: flash point.
  NOTE: The US Coast Guard defines a flammable liquid as one that gives off flammable vapor at or below 80°F. Three grades are distinguished: A, a flammable liquid having a Reid vapor pressure of 14 pounds or more; B, 8.5 to 14 pounds; and C, <8.5 pounds. According to NFPA Standard No. 30, there are three classes of flammable liquids, depending on whether their flash points are above or below 22.8°C (73°F) and boiling points above or below 37.8°C (100°F).

flammable range (Comb) The range of mixtures of air and gas, vapor, or dust through which flame will propagate.

flammable storage space (Storage) A warehouse area designed for the storage of highly flammable material.

flammable vapor (Comb) A concentration of vapor in air from a flammable liquid between the lower and upper flammability limits.

flank fire (For Serv) A fire set along a fire control line parallel to the wind and allowed to spread at right angles to the wind.

flanking attack 1. (Fire Pr) Fighting a fire by working along its sides rather than attacking it head-on. 2. (For Serv) Attacking a fire by working along the flanks either simultaneously or progressively from anchor points and attempting to connect the two lines at the head of the fire.

flanks (For Serv) The sides of a fire that are more or less parallel to the main direction of fire spread. See: fire nomenclature.

flap = swatter

flareback (Techn) A burst of flame from a furnace in a direction opposed to the normal gas flow; usu occurring when accumulated combustible gases ignite.

flare stack (Ind Eng) An exhaust chimney with an igniter at the top, used to vent and burn off volatiles from petroleum and petrochemical processes.

flare-up (Fire Pr) A sudden eruption or outburst of flame.

flash (Comb) A flame, spark, or intense light of short duration.

flash arrester = flame arrester

flashback (Comb) Propagation of a flame from an ignition source to a supply of flammable liquid.
  NOTE: The vapor of the liquid acts as a fuse or wick to carry the flame. See: reflash; rekindle.

flash burn (Med) A skin burn caused by a brief exposure to intense thermal radiation. Such burns are distinguished from flame burns by the fact that they occur on parts of the body exposed in a direct line with the thermal radiation. See: ionizing radiation; thermal burn.
  NOTE: Flash burns may occur under dark clothing while areas under light-colored clothing may be spared.

flash fire (Fire Pr) A fire that spreads with extreme rapidity, such as one that races over flammable liquids or through gases.

flash fuel (For Serv) Grass, leaves, draped pine needles, fern, tree moss, and slash which ignite readily and burn rapidly when dry. Syn: fine fuel. See: heavy fuel.

flashing (Build) Sheet metal used to deflect water; used for waterproofing roofs, joints, and seams exposed to the weather.

flashover[1] (Fire Pr) A stage in the development of a contained fire in which all exposed surfaces reach ignition temperature more or less simultaneously and fire spreads throughout the space and flames appear on all surfaces. Cf: blowup fire.
  NOTE: Prior to flashover, exposed surfaces are absorbing heat; after flashover they radiate it.

flashover[2] (Electr) Arcing between electrical conductors or terminals.

flash point (Comb) The minimum temperature at which a liquid vaporizes sufficiently to form an ignitable mixture with air. Flash points are determined in the laboratory by cup tests. See: combustible liquid; flammable liquid.
  NOTE: In fire practice flammable liquids are sometimes classified as high and low flash point liquids, depending on whether they flash above or below 100°F (37.8°C). To facilitate the management and extinguishment of flammable and combustible liquids, they have been divided into flash point classes; Classes I and II called flammable and Class III combustible liquids:
  Class I - Liquids having flash points below 100°F (37.8°C).
  Class IA - those having flash points below 73°F (22.8°C) and bp below 100°F.
  Class IB - those having flash points below 73°F and bp at or above 100°F.
  Class IC - those having flash points at or above 73°F and bp below 100°F.

Class II - Liquids having flash points at or above 100°F and below 140°F (60°C).
Class III - Liquids having flash points at or above 140°F.

flat (Build; Brit) An apartment.

flatcar (Transp) A freight car that has no sides or body extending upward.

flathead ax (Fire Pr) A chopping and splitting tool consisting of a head mounted on a handle; the head has a cutting edge on one side and a flat surface on the other that can be used as a hammer or maul. See: ax; forcible entry tool.

flat hose load (Fire Pr) Hose folded and laid flat instead of on edge as in the accordion horseshoe load.

flat raise (Fire Pr) A common method of raising a ladder in which one or more men butt the ladder heel and one or more other men lift the ladder at the other end and walk in toward the heel, elevating the ladder as they approach the butted end. See: raise.

flaw (Ind Eng; Testing) An imperfection in the normal physical structure or configuration of a part, such as a crack, lap, seam, inclusion, or porosity.
NOTE: A flaw does not affect the usefulness of a part. An imperfection that does is classified as a defect.

flexor (Anat) A muscle that bends a limb.

flick (Fire Pr) A small, easily extinguished fire.

float (Techn) A hollow sphere that rests on the surface of a liquid in a container to control the quantity of liquid or to indicate its level.

floc (Chem) A loose, fluffy mass formed by the aggregation of fine particles.

flocculator (Techn) A device for aggregating fine particles.

flood (Fire Pr; Brit) To extinguish a fire, esp one below ground, by applying large quantities of water.

floodlight apparatus = searchlight unit

floor¹ a. A level base, surface, or bottom plane. b. The inside bottom surface of a cavity, enclosure, or other hollow structure. c. The lower, usu horizontal walking surface of the interior of a building compartment.

floor² (Build) A story of a building.
NOTE: In American usage the ground floor and first floor are synonymous. In British usage the first floor or story is the one next above the ground floor or story.

floor area (Build) The size, in square feet or square meters, of the bottom plane of a room or building.
NOTE: Gross floor area is the total area within the perimeter of the outside walls of the building, including hallways, stairs, closets, thickness of walls, columns, or other features, used mainly for the purpose of determining the classification of occupancy. Net floor area is the actual occupied area, not including accessory unoccupied areas or thickness of walls, used mainly for the purpose of determining the number of persons for whom exits are to be provided.

floor collapse (Fire Pr) Fall of a building floor when weakened by fire or when its supports give way.
NOTE: A pancake collapse occurs when a floor drops more-or-less flat onto the floor below; a V-type collapse is one in which the center gives way and the opposite edges lean against the walls; a tent collapse is an inverted V-type collapse; a lean-to collapse is one in which one edge of the floor falls and the opposite edge leans against the wall.

floor furnace (Build) A self-contained furnace or heating unit set into a well in the floor.

floor load (Build) The weight that can safely be supported by a floor, expressed in pounds per square foot of floor area.
NOTE: For design purposes floor loads are usu assumed to be evenly distributed.

floor loading (Build) The weight and distribution of live loads on a floor.

floor opening (Build; Safety) An opening, such as a hatchway, stairwell, ladder opening, pit, or manhole in any floor, platform, pavement, or yard, through which a person may fall.
  NOTE: Floor openings occupied by elevators, dumb waiters, conveyors, machinery, or containers are excluded.

floor saw (Fire Pr) A small hand saw with a curved blade used to cut a surface by working the saw parallel to the surface being cut.

flotation[1] (Phys) Pertaining to the state or condition of floating.

flotation[2] (Mining) A method of ore concentration in which pulverized particles are floated by chemical frothing agents while the impurities sink.

flotation[3] (Transp) a. The characteristic of a tire not to sink into snow or soft ground. b. The ability of a vehicle to traverse unimproved terrain without bogging down.

flotation reagent (Chem Eng; Mining) A chemical used in flotation separation of pulverized ore.
  NOTE: Added to a mixture of solids, water, and oil, it causes certain solid particles to remain dry and to float, while the wet particles sink.

flotsam See: jettison

flow capacity (Hydraul) The quantity of fluid that can be passed per unit time.

flow measurement (Ind Eng) Determination of the quantity of fluid (gas, vapor, or liquid) passing a given point in a pipe or other channel, expressed as a volume (e.g., ft^3/sec) or a quantity (e.g., gallons/min).
  NOTE: A wide variety of mechanical and electrical devices are used to measure flow.

flue (Build) A channel in a chimney through which smoke and combustion gases are exhausted into the atmosphere.

flue effect = stack effect

flue gas (Comb) Gaseous products of combustion, including nitrogen, carbon dioxide, water vapor, excess air, and carbon monoxide passed up a chimney from a combustion chamber or fire box. Syn: fire gas.

fluid (Phys) A state of matter, including gases, vapors, and liquids, capable of continuous motion under external shearing forces. Gases and vapors completely fill a vessel in which they are contained, and, unconfined, are capable of indefinite expansion. Liquids, in contrast to gases and vapors, are relatively incompressible, assume the shape of their containers, and present an upper surface.
  NOTE: Three types and two classes of fluids are distinguished in fluid mechanics. Types: 1) ideal or perfect fluid - one that is incompressible, has zero viscosity, and can transmit normal pressure forces only. 2) inviscid or nonviscous fluid - one that has zero viscosity and may be compressible or incompressible. 3) real or viscous fluid - one that has finite viscosity. Classes: 1) Newtonian fluid - one having a coefficient of viscosity that is independent of the rate of shear at a given temperature and pressure. 2) Non-Newtonian fluid - one for which viscosity at a given temperature and pressure is a function of the velocity gradient.

fluid dynamics (Phys) The science of fluids in motion, including gasdynamics (gases) and hydrodynamics (liquids).

fluidity (Phys) The property of being able to flow. The reciprocal of viscosity.

fluid insulation (Fire Pr) A layer of foam or bubbles applied to surfaces exposed to fire to protect them from heat and direct flame.

fluidized bed (Techn) A layer of powder fluffed by air from below; used for dip-coating parts.

fluid mechanics (Fl Mech) The science of fluids, including fluid dynamics and fluid statics, concerned with pressure, velocity, acceleration, compression, and expansion effects on fluids.

fluid meter See: flow measurement

fluid statics (Fl Mech) The study of pressures of fluids at rest. See: hydrostatics.

flume (Mat Hand) A long trough with running water used to transport materials.

fluoralkane (Chem) A saturated fluorinated hydrocarbon used as a fire extinguishing agent and as a propellant in pressurized aerosol dispensers. Syn: Freon. See: Halon.

fluorescence (Phys) The emission of light (other than reflected light) by a substance under illumination.

fluorescent screen (Phys) A screen coated with a fluorescent substance that emits light when irradiated with X-rays.

fluoride (Chem) A gaseous or solid compound containing fluorine.
NOTE: Inorganic fluorides are highly irritating and toxic.

fluorine (Chem; Symb: F) A highly reactive element, bp 84.93°C, mp 53.54°C.
NOTE: Fluorocarbon compounds have high chemical and thermal stability and are used as refrigerants, aerosol propellants, and polymers for plastics. Combustible gases mixed with fluorine are explosive.

fluorocarbon surfactant See: fluorochemical

fluorochemical (Fire Pr) A synthetic chemical compound containing fluorine, carbon, oxygen, and other elements.
NOTE: One type, when added to water, gives it surface-active properties. Another type is used as a foam concentrate. Syn: aqueous film-forming foam; fluorosurfactant; Light Water. See: Halon.

fluoroprotein (Fire Pr) A type of firefighting foam concentrate containing fluorocarbon surfactant compounds mixed with protein foaming agents.
NOTE: Foam generated from this concentrate resists mixing with flammable liquids. Commonly used for subsurface injection in fighting fuel-tank fires.

fluoroscope (Med) An instrument with a fluorescent screen suitably mounted with respect to an x-ray tube, used for viewing internal organs of the body and examination of the internal structures of metals by means of x-rays. See: x-ray.

flush hydrant (Fire Pr) 1. A hydrant recessed in a well with a cover even with the ground level. Used where post-type hydrants present undesirable obstacles, such as at airports. See: hydrant. 2. = (Brit) ground hydrant.

flutter (Med) Irregular, rapid beating of the heart, in which the auricles beat more rapidly then the ventricles.

flux^1 (Chem) a. A substance that enhances the fluidity and "wetness" of another material, lowers its melting point, or removes impurities. Widely used in the smelting of metals and in the manufacture of ceramics. b. An agent used to clean surfaces and promote fusion in soldering.

flux^2 (Phys) A measure of the quantity of energy (heat, light, magnetism) or mass passing through a given area per unit time.

flux^3 (Nucl Phys) A measure of the intensity of neutron radiation: the number of neutrons passing through 1 square centimeter of a given target in 1 second, and expressed as nv, where n = the number of neutrons per cubic centimeter and v = their velocity in centimeters per second.

fly (Fire Pr) 1. One or more extendable sections of an extension ladder. 2. The top extremity of a ladder. Syn: (Brit) head.

fly camp = fire camp

flying squad (Fire Pr) A team of firefighters that responds in a van or personnel carrier to a fire to provide added manpower to attack forces on the fireground. Cf: tactical squad.

fly rope (Fire Pr) A rope used to extend the fly sections of an extension ladder. See: line. Also called: halyard.

FM¹ = Factory Mutual

FM² = frequency modulation

FMA = Fire Marshals' Association of North America

FM label (Testing) A seal or other identification attached to a building material or product, with the authorization of Factory Mutual Engineering Corp., indicating that nominally identical products have been subjected to appropriate fire or other tests and approved for normal use. See: approved.

foam¹ (Fire Pr) A dispersion of gas in a liquid, consisting of a mass of air or inert gas bubbles formed chemically by various compounds or mechanically by air introduced into a stream of water by a special valve, orifice, or proportioner and foam maker to develop a blanket capable of smothering a fire, esp one involving a flammable liquid.
NOTE: Foam has a density of one-third to one one-thousandth that of water, depending on the type of foaming agent used. The various firefighting foams commonly used include chemical, protein, synthetic, wetting agent, and aqueous film-forming types. The smaller the bubbles, the more persistent the foam. Foam can be destroyed by antifoam agents, such as silicone polymers and alcohol.

foam² (Fire Pr) A compound used to produce foam; more properly called a foam concentrate. Also called: foam agent; foam solution; foam stabilizer.

foam³ (Fire Pr) A portable or wheeled foam extinguisher.

foam application rate (Fire Pr) A measure of the quantity of liquid applied per unit area per unit time. Usu stated in gallons of water-in-foam applied per square foot of surface per minute. See: foam expansion.

foam applicator See: applicator

foam blanket (Fire Pr) A relatively thick layer of foam applied to a burning or vulnerable surface to smother flames or prevent ignition.

foam branch pipe See: foam nozzle

foam breakdown (Fire Pr) The process of foam collapse and disintegration due to heat or mechanical stress. Breakdown also occurs through aging, as water drains from the foam bubbles.

foam breaker (Fire Pr) Any substance that destroys foam.

foam carpet (Fire Pr) The layer of foam on an aircraft landing strip; the result of foaming the runway.

foam chamber (Fire Pr) A boxlike enclosure attached to the top edge of a fuel storage tank and housing foam-making equipment or foam application devices.

foam chute (Fire Pr) A sloping channel for the application of foam, consisting of a series of baffles in a hollow enclosure that gently deposit a layer of foam on the fuel surface, independent of the fuel level in the tank. See: foam trough.

foam classification (Fire Pr) Fire fighting foams are divided into four main categories: 1) proteins, 2) synthetics, 3) surface film-forming agents, and 4) chemical foams. Each category may be further subdivided into type, such as regular, alcohol-resistant, and fluorocarbon. Another classification categorizes foams into high-, medium-, and low-expansion types.

foam compatibility (Fire Pr) The ability of a foam to coexist in contact with other substances without collapsing or breaking down due to surface forces. Usu used in reference to dry chemical action on foam.

foam compound (Fire Pr; Archaic) A liquid concentrated solution of foam forming materials used for the production of air foam. Syn: foam liquid; foam concentrate; foaming agent.

foam concentrate (Fire Pr) A concen-

trated liquid solution of foam forming materials used for the production of air foam. Depending on the type, the concentrate is usu mixed with water in 3% or 6% concentrations by volume. Often referred to as <u>air-foam concentrate</u>. Syn: <u>air-foam liquid</u>; <u>stabilizer</u>; <u>foam compound</u>; <u>air-foam forming concentrate</u>; etc.
    NOTE: There are seven principal types of foam concentrates: 1) regular protein, 2) alcohol resistant, 3) fluoroprotein, 4) regular synthetic, 5) wetting agent, 6) high-expansion, and 7) aqueous film-forming foam concentrate.

<u>foam drainage</u> (Fire Pr) Drainage of water from individual bubbles in a mass of foam to the lower surface on which the foam rests.
    NOTE: Foam drainage is a qualitative measure or comparative characteristic of foams for firefighting purposes. Foam drainage bears a direct relationship to foam viscosity and spreading characteristics.

<u>foamed concrete</u> (Build) Lightweight concrete containing bubbles of air or other gas.

<u>foamed plastic</u> See: <u>plastic</u>

<u>foam eductor</u> (Fire Pr) A venturi type device for drawing foam concentrate from a container in correct proportions and introducing it into a water stream. Also called: <u>foam inductor</u>; <u>induction device</u>.

<u>foam expansion</u> (Fire Pr) The volume occupied by foam as compared to the volume of its water content. Low-expansion foams have an expansion factor of up to 10, medium-expansion 10 up to 200, and high-expansion 200 up to 1000.
    NOTE: Expansion 10 foam (10X) denotes 1) one volume of water expanded to 10 volumes of foam, 2) an expansion ratio of 1 to 9, 3) foam which contains 10% water, or 4) a foam with a density of 0.10 referred to water. Expansion is an important characteristic of foam.

<u>foam generator</u> (Fire Pr) A device that produces firefighting foam from water solutions of foaming agents.
    NOTE: A continuous chemical foam generator mixes dry foam chemicals, either single powder or dual powder, to produce foam at its outlet. In the latter case the two powders are kept separated until used. A <u>chemical foam generator</u> consists of an inverted, cone-shaped hopper containing dry powdered (A and B) chemicals. At the apex of the cone a water jet mixes the chemicals with water, and a reaction occurs that generates carbon dioxide, which produces the foam. A pressure-type generator is a closed device that contains foam chemicals and has provisions for admitting water as needed. Air foam is produced by mixing air mechanically with the foam-forming solution. The three principal methods of making air foam are 1) air aspiration using venturi eductors, 2) positive displacement pumps, and 3) air blowers or fans.

<u>foam hopper</u> (Fire Pr) A device for adding chemical foam powders to a stream of water for purposes of generating chemical foam. The hopper is constructed in the form of a hollow inverted cone with a jet-venturi water stream at the lower (apex) end of the cone, where admixture and reaction take place.

<u>foam hose stream</u> (Fire Pr) A foam stream applied by a hand-held hose and nozzle.

<u>foam inductor</u> (Fire Pr) A piece of equipment that introduces the proper quantity of foaming agent into a water stream. A type of <u>proportioner</u>.

<u>foaminess</u> (Fire Pr) A measure of the foaming of a liquid, defined as vt/V, where v is the volume of foam, t is time in seconds and V is the volume of air passing through the liquid.

<u>foaming agent</u> = <u>foam concentrate</u>

<u>foam injection</u> (Fire Pr) A method of applying foam in which the foam is introduced into a burning liquid at a point below the fuel surface. Also called: <u>base injection</u>; <u>subsurface injection</u>.

<u>foaming the runway</u> (Fire Pr) Applying foam to an aircraft landing strip to prevent or lessen the hazard of sparking when an aircraft is forced to crash-land.

foam inlet adapter (Fire Pr) An adapter fitted to a fixed foam inlet to enable a foam-making branch or nozzle to feed into a fixed installation.

foam maker (Fire Pr) A device that aspirates air into a pressurized and premixed water and foam concentrate solution, all in correct proportions, to produce firefighting foam.

foam nozzle (Fire Pr) A special playpipe, usu combined with a foam maker that aspirates air and mixes it into a water stream to produce foam. The nozzle may be portable or fixed. Also called: air-foam nozzle; nozzle foam maker; foam branch pipe; aspirator foam nozzle.

foam powder (Fire Pr) A granulated blend of acidic and basic constituents (A and B powders) used to produce chemical foam by mixing with water. Called single powder when premixed and packaged together.

foam premix (Fire Pr) A solution of water and foam concentrate, usu stored in a tank ready for foam-making operations.

foam proportioner (Fire Pr) A device using any of several means for mixing correct proportions of foam concentrate with water to generate air foam. See: around-the-pump proportioner; foam eductor; water-motor proportioner; balanced orifice proportioner.

foam pump (Fire Pr) 1. A rotary, positive displacement pump with sliding plastic vanes that sweep out a constant volume per revolution when an off-center slotted bronze rotor turns. In operation, a foam-forming solution is metered into the inlet, which is open to the atmosphere. As each vane sweeps across the inlet, a specific quantity of air and solution is inducted. As the mixture is progressively squeezed out of the pump through a tapered cavity, a homogeneous foam is produced at the discharge port. 2. (Erroneous) A pump for pumping foam concentrate.

foam pump proportioner (Fire Pr) A system for metering foam concentrate into water in the recommended ratio. The pump supplies proportioning orifices, etc., with foam concentrate under positive pressure. See: balanced pressure proportioner.

foam quality (Fire Pr) The degree to which foam possesses desirable characteristics.
NOTE: Among the more important are tenacity, the ability to cling to inclined or verical surfaces; stability, longevity, and resistance to rupture from wind and flames; continuity, toughness, resistance to breakdown, flowability, spreadability, high density, abilility to heal after rupture; etc.

foam screen (Fire Pr) A mesh inserted in piping or equipment at (or downstream from) the point where air is mixed with foam-forming solution in order to produce or refine air foam.

foam solution (Fire Pr) 1. A water solution of foam concentrate in correct proportions (usu 3 to 6% by volume) for making air foam mechanically. 2. The solution draining from air foam during aging.

foam stabilizer (Fire Pr; Archaic) 1. A material, compound, or concentrated solution capable of forming relatively stable firefighting foam. 2. A substance such as saponin or licorice extract added in powdered form to chemical foam formulations for the purpose of stabilizing the foam produced in a "seltzer" reaction. Syn: foam liquid; foam-forming concentrate; foam concentrate; foam compound; foaming agent.

foam system (Fire Pr) An apparatus consisting of all necessary operating components for applying foam to a fire hazard. The system is usu connected to an external water supply.

foam tender (Fire Pr) An appliance largely or entirely used to carry foam-making equipment.

foam tests (Fire Pr) A series of physical property tests conducted on foams for the purpose of judging their quality and comparing their characteristics. See: foam quality.

foam tower (Fire Pr) A portable, extendable (usu telescoping) tube which can be erected at the site of a fuel tank fire for the purpose of acting as a conduit for foam to be applied to the burning fuel surface.

foam trough (Fire Pr) A channel for foam application to a fuel surface independent of the fuel level. The trough spirals down the side of the tank at a gentle angle to the bottom from the top foam inlet. See: foam chute.

foam truck See: foam vehicle; aircraft crash-rescue vehicle.

foam vehicle (Fire Pr) A specially designed mobile unit with integral foam system equipment for control of fuel fires of all types. See: aircraft crash-rescue vehicle.

foam volume (Fire Pr) Equivalent to foam expansion. A foam volume of 10 = expansion 10.

foam-water spray nozzle (Fire Pr) A type of spray nozzle esp designed to be used with foam solution or water as an overhead applicator installed in a pipe grid for area protection.

foam-water sprinkler (Build; Fire Pr) A type of sprinkler head designed to generate and distribute foam when supplied with air-foam solution under pressure and to discharge it in a pattern similar to that of water alone.

FOBFO = Federation of British Fire Organizations

focal length (Opt; Photog) The distance from the optical center of a lens to its focus.

focus[1] (Opt) The common point to which light rays converge in an optical system.

focus[2] (Med) The seat of an injury or disease.

foehn wind See: wind

fog[1] (Meteor) A cloud of water droplets sufficiently thick at ground level to reduce visibility below 1 km (5/8 mile). See: precipitation.

fog[2] (Fire Pr) A jet or cloud of fine water spray discharged by a fog or spray nozzle.

fog cone (Fire Pr) The angle at which a spray leaves a fog nozzle; usu 30°, 60°, or 90°.

fog foam (Fire Pr) Foam in the form of a spray.

fog foam nozzle (Fire Pr) A playpipe and tip for applying foam as a spray. See: nozzle.

fogging (Techn) The use of compounds in the form of vapor or mist.

fog gun (Fire Pr) A nozzle with a pistol grip used to throw a small, high-pressure spray of water on a fire. See: nozzle.

fold (Salv) Any of several methods of folding salvage covers for ease of handling and storage.

folding ladder (Fire Pr) 1. A hinged ladder that can be folded to a width of a foot or less for use in cramped quarters. See: attic ladder; ladder. 2. = (Brit) pole ladder.

foliage (For Serv) Collectively, the leaves, flowers, and branches of plants.

follicle (Anat) 1. A small excretory or secretory sac or gland. 2. A small anatomical cavity or deep depression. 3. The depression from which a hair grows.

follow up (For Serv) The commitment of additional men and equipment to assist or back-up those first-in at a fire. Also called: reinforcement.

food chain (Dosim) The pathways by which any material, such as radioactive material from fallout, passes from the first absorbing organism, through plants and animals, and finally to man.

food poisoning (Med) A stomach or intestinal disturbance due to eating food containing pathogenic bacteria or bacterial toxins. Also called: ptomaine poisoning. See: enterotoxin. See also: gastritis.

foot[1] (Metrol) A unit of length of 12 inches, one third of a yard, equal to 30.48 centimeters.
NOTE: A square foot is equal to $0.0929 m^2$; a cubic foot is equal to

183

0.283m³.

foot² (Fire Pr) The base of a ladder. Syn: heel.

foot candle (Opt) The illumination on an area of one square foot exposed to a flux of one lumen. Also defined as the illumination on a surface of one square foot, all points of which are one foot from a point source of one candela.

foot head (Hydraul) The pressure at the base of a column of water, equal to 0.434 psi for each foot of column height.

footing (Build) A thickened base of a wall, provided to spread the load over a larger area.

foot-lambert (Opt) A unit of photometric brightness (luminance) equal to 1 lumen per square foot.

foot pound (Phys) A unit of energy or work performed by a force of one pound through a distance of one foot.

force (Phys) An effect on a body that causes it to accelerate if it is free to move. Force is defined as the product of mass and acceleration and is expressed in dynes or poundals.

forced air system (Build) A central warm-air heating system in which air is circulated by a fan or blower.

forced burning See: simultaneous ignition

forced convection (Techn) The motion of a fluid caused primarily by an external agent, as in the circulation of air by a fan or movement of a fluid through a pipe by an imposed pressure gradient. See: convection.

forced vibration (Mech) A vibration that takes place due to excitation by external forces. It occurs at the frequency of the exciting force and is independent of the natural frequencies of the system.

force pump = positive displacement pump. See: pump.

forcible entry (Fire Pr) Methods and techniques used by firefighters to enter closed buildings quickly and with minimum damage to the structure.

forcible entry tool (Fire Pr) Any of a large assortment of axes, door breakers, pry bars, mauls, chisels, saws, and the like used to gain entry into closed buildings or through walls and other obstructions as required to carry out fire-fighting operations.

fording (Transp) Referring to the ability of a vehicle to negotiate a water obstacle of a specific depth.

foreman (Fire Pr; Archaic) A fire department rank equivalent to the rank of captain.

forensic (Law) Legal; pertaining to courts of law.

forensic chemistry (Chem; Law) The branch of chemistry that specializes in chemical information, experiments, and analyses related to legal matters.
NOTE: Originally devoted to crime detection and toxicology, legal chemistry in addition now concerns the legal aspects of pure foods, water supplies and sanitation, chemical patents, marketing of chemicals, transportation of hazardous chemicals, approval of new drugs, use of pesticides, waste disposal, etc. Also called: legal chemistry.

forensic medicine (Med) The science that deals with the application of medical facts to legal problems and court procedures. Legal medicine.

forepoling = spiling

forest (For Serv) A large tract of densely-wooded land, characterized by the presence of mature trees.

forest fire (For Serv) Any wildland fire not prescribed for the area by an authorized plan. Cf: wildfire.

forest fire classes (For Serv) A classification scheme for indicating the size or extent of a forest fire. The classes are: A, 1/4 acre or less; B, 1/4 acre to 10 acres; C, 10 to 100 acres; D, 100 to 300 acres; and E, over 300 acres. See: fire classes.

forest fire control (For Serv) Measures, procedures, and activities established and undertaken to prevent the outbreak of fire in forest and wildlands.

forest floor (For Serv) The layer of organic debris on the soil surface of a forest.

forest fuel factors (For Serv) The three factors governing the nature of a possible forest fire: the size of combustible material, its arrangement, and its volume.

forest land (For Serv) A tract of land at least an acre in size and at least 10% of which is canopied by trees.

Forest Products Laboratory tunnel test See: FPL tunnel test

forest protection (For Serv) Prevention and control of any factor that can cause damage to a forest.

forest ranger (For Serv) An officer responsible for the care, management, and protection of a public forest.

forest utilization fire (For Serv) A wildland fire resulting directly from the harvesting of forest products or from forest and range management activities.

fork-lift truck (Mat Hand; Transp) A powered truck with vertical, elevating back plates and horizontal forks for raising loads.
NOTE: Such trucks are used for short-distance hauls within warehouses, for van and trailer loading, and for stacking palletized items.

form drag (Fl Mech) The resultant force set up by an adverse pressure gradient and acting in a direction opposite to the motion of a body.

formula (Chem) A graphic expression consisting of one or more symbols representing chemical elements or compounds. If more than one atom of a given element is present in a molecule, the number is shown by a subscript. The formula shows the elements present and the number of atoms of each; e.g., H_2O (water), CCl_4 (carbon tetrachloride).

formwork (Build) 1. The total system of support for freshly placed concrete, including the mold or sheathing which contacts the concrete, as well as all supporting members, hardware, and necessary bracing. Syn: falsework. 2. = (Brit) shuttering.

forward lay (Fire Pr) 1. The stretching of hose from a hydrant toward a fire. 2. Laying of hose in the manner envisioned when the hose was loaded on the hose bed. Syn: straight lay. See: hose lay.

fossil fuel (Techn) A combustible solid, gas, or liquid hydrocarbon formed under the surface of the earth in past geological ages.

four-company box (Fire Pr) A fire alarm box or location calling for four engine or pumper companies, two ladder companies, rescue squad, and other units on the first alarm.

four-eleven (Fire Pr) A fire alarm signal code indicating a fourth alarm fire.

four-hour run-in test (Fire Pr) A factory test conducted by the manufacturer on new pumping engines.
NOTE: As part of the test, the pumper must deliver rated capacity for one hour at 150 psi, 70 per cent for 1/2 hour at 200 psi, and 50 per cent for 1/2 hour at 250 psi net pump pressure.

Fourier number 1. (Thermodyn) A number used in mathematical treatment of time-dependent heat conduction studies:
Fourier number = kt/dC_pL^2
where
k is the thermal conductivity
t is time
d is the density
C_p is the specific heat at constant pressure, and
L is a characteristic dimension.
2. (Mass Transfer) A number used in the mathematical treatment of time-dependent mass transfer studies:
Fourier number = $Dt/L^2 = k_mt/L$
where
D is the diffusion coefficient
k_m is a mass transfer coefficient.

four-to-one ratio (Safety) An arbitrary ratio frequently used to

compare the indirect costs of an accident to the direct costs.
NOTE: Generally considered invalid since no fixed ratio is known to exist among various types of accidents.

four-way valve (Fire Pr) A hydrant valve that permits connections to pumpers without shutting down previously connected hose lines. Syn: four-way gate.

fovea (Anat) A pit in the middle of the macula lutea; the area of sharpest vision in the center of the retina of the eye.

fox lock (Build) A lock in the center of a door securing bars that extend into the door frame.

foyer (Build) An anteroom at an entrance. A form of lobby.

FPA = Fire Protection Association (UK)

FPL = Forest Products Laboratory

FPL tunnel test (For Serv; Testing) A fire-test, developed at the Forest Products Laboratory, used to determine the rate of flame spread and the quantity of smoke developed, temperature, and heat produced from the surface of an 8-ft (2.44-m) specimen burning in a convective draft. Syn: eight-foot tunnel test.
NOTE: This test, as adopted by ASTM, is specified for research and development tests of materials rather than for the establishment of ratings or approvals.

FPRI = Fire Protection Research International

Fr = firefighter

FR = flame retardant

fraction (Math) A proportional part of a unit or a whole, usu expressed as a ratio.

fractional distillation (Chem) A process of separating two or more liquids having different boiling points. As each fraction vaporizes, it is collected in a separate container.

fractionation (Chem) Separation of a mixture into different portions or fractions, usu by distillation.

fracture (Med) A broken bone.
NOTE: A break in which the bone ends are slightly displaced is a simple fracture; one in which the bone ends penetrate the skin is a compound fracture. A greenstick fracture is one in which the bone is split, or partly broken and bent. A comminuted fracture is one in which the bone is broken in several pieces.

fractus (Meteor) A ragged cloud type. See: cloud.

frame (Build) 1. A class of building structure with wooden frame exterior walls, typically sheathed with wooden shingles, stucco, composition siding, etc. 2. The wood or metal supporting structure of a building.

frame of reference (Math) A base used to locate points in multi-dimensional space, consisting of a system of coordinates against which the points, representing physical values, are plotted.
NOTE: Time is a common variable used in plotting.

frame spacing (Build) The longitudinal center-to-center distance between transverse frames in a building.

frangible disc (Safety) A disc, usu of metal, which seals a safety relief channel under normal conditions. The disc is designed to burst and relieve any unsafe overpressure that may occur.

frangible disc-fusible plug (Fire Pr) A pressure-relief device consisting of a disc and a plug that prevents the disc from rupturing at its predetermined fracture pressure unless the temperature is high enough to melt or soften the plug.

freeboard (Navig) The distance from the waterline to the weather deck.

free burning[1] (Fire Pr) a. Unrestricted combustion of flammable materials. b. Vigorously burning.

free burning[2] (For Serv) Descriptive of a wildland fire that is burning unchecked by natural barriers or fire suppression forces.

free convection See: convection

free electron (Phys) An electron which is loosely held and consequently tends to move at random among the atoms of a material.
NOTE: A common phenomenon in solid-state physics.

free energy[1] (Thermodyn) A measure of the energy in a system free to perform work at constant pressure and temperature.

free energy[2] (Chem) A measure of the extent to which substances can react.

freeing port (Navig) An opening in a bulwark or coaming that permits water to drain overboard.

free path (Phys) The average distance a molecule or atomic particle travels between collisions. Also called: mean free path.

free radical (Chem) An atom or molecule with one or more unpaired electrons. Considered a molecular fragment with a short lifetime and high chemical reactivity.

free vibration (Mech) A vibration that occurs due to the action of forces inherent within an elastic system and in the absence of external impressed forces.
NOTE: Such a system vibrates at one of its natural frequencies.

freeze space (Storage) Refrigerated warehouse area where temperatures can be controlled below 32º F.

freezing point (Chem) The temperature at which a liquid changes into a solid. More exactly, the temperature at which the solid and liquid phases are in equilibrium at atmospheric pressure. Syn: ice point.
NOTE: If the point is reached by cooling, it is called the freezing point; if by heating, it is called the melting point. In practice, freezing point refers to temperatures below the freezing point of water.

freight (Transp) Any material, product, commodity, express, or mail, shipped by rail, water, highway, or air.

Freon (Techn) A trade name for derivatives of methane and ethane containing fluorine, and often chlorine or bromine, used as a refrigerant and as a fire-extinguishing agent. Freon is widely used as an aerosol propellant. Other trade names include Genetron; Isotron; and Ucon. See: Halon.

frequency[1] (Acoust, Electr; Phys) The time rate of repetition of a periodic phenomenon, specifically with reference to sound or electromagnetic energy, expressed in cycles per second or Hertz.
NOTE: The frequency is the reciprocal of the period. In acoustics, frequency is sometimes called pitch, or the highness or lowness of sound. A number of adjacent frequencies is called a band. An array of wave bands is called a spectrum, as in radio spectrum, or electromagnetic spectrum.

frequency[2] (Psych; Statist) The number of times an event, score, or phenomenon occurs. The array of such frequencies is called a frequency distribution.

frequency distribution (Statist) A set or sample set of quantitative data. See: frequency[2].

frequency polygon (Statist) A curve drawn by connecting the midpoints of the class intervals of a frequency distribution.

friction (Phys) The resistance to relative motion between two bodies or surfaces in contact. The science of friction is called tribology.

friction drag 1. Resistance to fluid motion. 2. The tangential component of force acting on a body, resulting from shear stress on the body.

friction factor (Phys) A mathematical expression for the resistance to flow. Generally a simple function of Reynolds number and system shape.

friction flow (Fl Mech) Flow accompanying the motion of an inviscid (i.e., zero-viscosity) or perfect (i.e., inviscid and incompressible) fluid past a surface. The assumption of frictionless flow occurs frequently in the classical theory of hydrodynamics.

friction loss (Hydraul) Pressure loss in hose lines or pipes due to friction between the flowing liquid and the walls.
  NOTE: Friction losses are directly proportional to the length of the pipe, vary directly as the square of the flow velocity, vary inversely as the fifth power of the hose diameter, and are independent of pressure for a given flow velocity. They are directly affected by the interior wall roughness of the hose.

friendly fire (Fire Pr) A fireplace, furnace, or other fire used for heating or cooking with a properly confined flame. Ant: unfriendly fire.

fringe benefit (Man Sci) A pension, paid holiday, health insurance, and the like provided by an employer in addition to salary or wages.

front[1] (Build) The facade of a building.

front[2] (Fire Pr) The first position, generally facing the street, taken by firefighting units upon arrival at a fire.

front[3] (Meteor) A sloping boundary between two air masses of different temperature. A polar front (arctic front), for example, is the forward side of an advancing mass of polar air.

front mount pump (Fire Pr) A pump mounted on the front end of a pumper.

frost (Meteor) Ice deposited on surfaces directly from the vapor phase, often called hoar frost when deposited in the form of needles, fans, feathers, etc. See also: ice; rime; precipitation.

frost valve (Fire Pr) A valve incorporated into a hydrant that opens when the hydrant valve is closed, thereby draining the water from the neck of the hydrant.
  NOTE: In cold weather, water remaining in the hydrant could freeze and prevent opening of the hydrant valve.

frothing (Fire Pr) A violent foaming of a liquid mixture that may occur when water or foam is applied to a burning, viscous flammable liquid which has been heated above the boiling point of water.

frothover (Fire Pr) Overflow of a hot, viscous oil from a tank when water at the bottom of the tank boils.

Froude number (Phys) The ratio of inertial forces to gravitational forces; a diminsionless number used in the study of wave and surface behavior:
$Fr = V^2/gL$
where
  V is the fluid velocity
  g is the acceleration of gravity, and
  L is a characteristic length.

FSC = flame-spread classification

fuel[1] (Comb) A material that yields heat through combustion.
  NOTE: Fuels can be gaseous, liquid, or solid material capable of igniting and burning.

fuel[2] (Nucl Phys) Fissionable material used to produce energy in a reactor. Also applied to a mixture, such as natural uranium, in which only part of the atoms are readily fissionable if the mixture can be made to sustain a chain reaction. See: fissionable material.

fuel arrangement (For Serv) Pertaining to the manner in which fuel is distributed.

fuel bed (Comb) The mass of fuel burning in a furnace grate.

fuel break (For Serv) A strip of land in which natural vegetation has been largely removed or permanently modified so that fires burning into it can be extinguished more easily. See: firebreak.

fuel cell (Electr) An electrochemical cell in which energy from a reaction between a fuel and oxygen or air is converted directly into electrical energy.

fuel class (For Serv) Fuels grouped according to common characteristics. Dead fuels are grouped according to time lag (1-, 10-, or 100-hour) and live fuels are grouped into herbaceous and woody.

fuel compactness (For Serv) Referring to the number of fuel particles per unit volume or the degree of compaction.
NOTE: Loosely compacted fuels generally burn more vigorously than densely compacted fuels.

fuel continuity (For Serv) The arrangement of fuel or the degree of connectedness of combustible materials.
NOTE: Uniform continuity means that combustible materials are physically in contact with each other; patchy continuity means that fuel is connected only here and there.

fuel element (Nucl Phys) The unit of fuel introduced into a reactor, commonly in the form of rods, plates, or pellets.

fuel gas (Techn) Any of a wide variety of gases produced from natural (fossil) fuels such as petroleum and coal, including natural gas, acetylene, hydrogen, LP-gas, and other liquefied and nonliquefied flammable gases used for heating, cooking, and industrial purposes. Solid fuels are often reduced to gaseous or liquid form to take advantage of economical distribution by pipeline.
NOTE: Hydrocarbon fuels with 5 or more carbon atoms per molecule are liquid or solid.

fuel injector (Transp) A device that first atomizes a fuel and then introduces the spray under pressure into a combustion chamber, as in a diesel engine.

fuel load (Build; Fire Pr) The total quantity of combustible contents of a building space or fire area, including interior finish and trim, expressed in heat units or the equivalent weight in wood. See: fire load.
NOTE: The fuel load per unit area is called the fuel load density.

fuel model (For Serv) A simulated fuel complex in which all of the fuel is described appropriately for a mathematical fire spread model.

fuel moisture analog (For Serv) A representation of a specific class of dead forest fuels that simulates moisture content when exposed to the same environment.
NOTE: A basswood slat behaves like a 1-hour time lag fuel and a half-inch ponderosa pine dowel like a 10-hour time lag fuel.

fuel moisture content (For Serv) The quantity of water contained in a fuel, expressed as percent by weight and determined by weighing a quantity of the fuel before and after thorough drying at 100°C (212°F).

fuel moisture indicator stick (For Serv) A fuel moisture analog. A wooden stick of known dry weight continously exposed to the weather and weighed periodically to determine the probable moisture content of forest fuels. Also called: wood cylinder.

fuel oil (Techn) A liquid petroleum product supplied in six standard grades and used for heating and power production.

fuel reduction See: hazard reduction

fuel type (For Serv) An identifiable association of fuels of distinct species, form, size, arrangement, or other characteristic that will burn with a predictable rate of fire spread under given weather conditions.

fuel type classification (For Serv) Identification of fuel types by characteristics for the purpose of dividing a forest area into units on the basis of the predictable rate of fire spread and the difficulty of establishing and holding a control line and containing the fire.

fugacity (Thermodyn) 1. The effective pressure in imperfect gas theory; equal to the absolute pressure in the limit of low pressure. 2. Volatility; the tendency of a molecular species to escape. Fugacity can be identified with the vapor pressure and a condensed phase corrected to ideal gas conditions.

full assignment (Fire Pr) The total complement of companies, apparatus, special units, and officers listed on the assignment card for a given location.

fully involved (Fire Pr) A state of development of a fire in which an entire building or structure is in flame or engulfed in smoke so that immediate access or entry to the interior is impossible.

fulminating (Med) Running a rapid course, as a rapidly worsening disease or condition.

fume (Toxicol) An airborne irritating, noxious, or toxic smoke, vapor, sol, or any combination of these produced by a volatile substance or a chemical reaction.

fume fever (Toxicol) An acute condition caused by brief exposure to the fresh fumes of metals, such as zinc, magnesium, or their oxides.

fumigant (Techn) A fume used for disinfection or destruction of insects and rodents.

function¹ The purpose for which something exists.

function² (Statist) A quantity that varies with another quantity.

functional method See: progressive method

fundamental frequency (Acoust; Electr) The lowest frequency harmonic component of a complex wave. Usu the one with the highest amplitude.

fundamental particle = elementary particle

fungus (Biol) A simple parasitic plant form that lacks chlorophyll and feeds on living or dead organic matter.
 NOTE: Fungi include molds, rusts, mildews, smuts, mushrooms, and certain bacteria.

furnace (Techn) A heating device consisting of an enclosed combustion chamber supplied with fuel and air and equipped with a means for distributing heat to desired locations and a chimney for discharging the waste products of combustion.

fuse¹ (Electr) An overcurrent device designed to interrupt the current flow when it exceeds the capacity of the conductor.
 NOTE: Among the many types of fuses are link, expulsion, plug, and cartridge. Link Fuse: A strip of fusible metal between two terminals of a fuse block. Expulsion Fuse: One so designed that when it blows, the gases generated aid in quenching the arc. Plug Fuse: A fusible metal strip in a screw-type base. Used on circuits that do not exceed 30 amperes at not more than 150 volts to ground. Cartridge Fuse: A fusible metal strip enclosed in a tube.

fuse² (Civ Eng) An igniting or explosive initiating cord, consisting of a flexible fabric tube and a core of explosive. Used in blasting and demolition work, and in certain munitions.
 NOTE: The types of fuses are distinguished by modifying terms that form part of the name.

fused head (Fire Pr) An automatic sprinkler head that has melted.

fusee (For Serv) A pyrotechnic candle, used as a flare or a firing device to start a backfire.

fusible alloy (Metall) A low-melting (60 to 180°C) alloy of bismuth, tin, lead, and other metals, used in electrical fuses to break the circuit when an excessive current flow occurs. Used similarly in automatic sprinklers to turn them on in the event of fire. See: fusible link.

fusible link (Fire Pr) A link of low-melting alloy that holds an automatic sprinkler head in closed position and melts at a predetermined temperature, thereby allowing the valve to open and release water or other extinguishing agent. A similar link may be used to hold a fire door or fire damper in open position, to shut off the fuel supply to an oil-fired boiler in case of fire, etc.

fusible plug (Techn) An operating part in the form of a plug of suitable low-melting material, usu a metal alloy, which holds a safety relief device shut under normal conditions and yields on melting at a predetermined temperature to release the safety device.

fusion¹ (Metall) The melting of a substance.

fusion² (Nucl Phys) The formation of a heavier nucleus from two lighter ones (such as hydrogen isotopes), with the attendant release of energy (as in a hydrogen bomb). Cf: fission. See: nuclear reaction; thermonuclear reaction.

fuze = fuse

G

g¹ (Phys) An acceleration equal to that imparted by gravity, approx 32.2 ft/sec² at sea level (980.665 cm/sec²).

g² (Metrol) An abbreviation for gage; giga-; gram.

gable (Build) A triangular wall beneath the ridge of a roof.

gage¹ (Techn) An instrument that indicates pressure.
 NOTE: Vacuum gages indicate pressure below atmospheric (negative pressure), straight pressure gages indicate pressures above atmospheric (positive pressure), and compound gages indicate both positive and negative pressures. Zero on pressure gages usu represents atmospheric pressure. Oil-pressure, fuel, and temperature gages are often electrical. Mechanical pressure gages generally operate on the Bourdon principle. See: Bourdon gage.

gage² (Techn) An instrument for determining the size or shape of an object and whether it falls within an allowable tolerance.
 NOTE: The spelling "gage" is often used for indicating instruments while "gauge" is used for nonadjustable instruments, including feelers, depth, go-no-go, wire, and thread gages that measure thickness, length, or tolerance. This distinction seems to be disappearing in favor of the single form "gage."

gage³ (Build) a. The proportion of plaster of paris added to mortar to hasten setting. b. To mix plaster in definite proportions.

gage pressure (Techn) The direct reading of a pressure gage that registers pressure in excess of atmospheric, expressed as psig (pounds per square inch gage). Thus, psig = psia (pounds per square inch absolute) - 14.7 psi.

gaging (Metrol) The measurement of the thickness, density, or quantity of material by a fixed standard instrument or by the quantity of radiation it absorbs.
 NOTE: The latter application is the most common use of radioactive isotopes in industry. Also spelled: gauging.

gale (Meteor) A wind of 32 to 63 mph. See: Beaufort wind scale.

gallery¹ (Build) a. A roofed promenade. b. An outdoor balcony or veranda.

gallery² (Navig; Techn) A railed catwalk around an engine compartment.

gallery³ (Mining) A working drift or level in a mine.

gallon (Metrol) A unit of liquid measure containing 4 quarts (8 pints) and having a volume of 231 cubic inches.
 NOTE: A gallon of pure water weighs 8.336 pounds. A British imperial gallon (equivalent to the space occupied by 10 pounds of distilled water) has a volume of 277.42 cubic inches or 0.16 ft³, and is equal to 1.201 US gallons or 4.546 liters. One US gallon = 0.804 imperial gallon = 3.7853 liters.

gallons per minute (Hydraul; Abb: gpm) The standard measure of water flow in firefighting, used to measure the output of pumpers, hose streams, nozzles, hydrants, water mains, etc.
 NOTE: One gpm is equal to 3.785 liters/min.

gallows (Fire Pr; Brit) A support at the front of a pump escape appliance or turntable ladder, provided to bear the weight of the ladder. Syn: headrest.

galvanizing (Techn) A method of providing a protective coating for metals by dipping them into a bath of molten zinc.

gambrel roof (Build) A double-sloping roof with the lower part steeper than the upper part.

gamma rays (Nucl Phys) High-energy, short-wavelength electromagnetic radiation. Gamma radiation frequently accompanies alpha and beta emission and always accompanies fission.
 NOTE: Gamma rays are highly penetrating and are best stopped by dense materials, such as lead. Gamma rays are essentially similar to

x-rays, but are usu more energetic and are nuclear in origin. Cf: x-ray. See: radioactive decay.

gangrene (Med) Local tissue death usu occurring in an arm or leg due to insufficient or blocked blood supply.
  NOTE: Infection of damaged tissue by specific organisms known as clostridia causes gas gangrene, which spreads rapidly.

gangue (Mining) Useless chipped rock.

gangway (Navig) Any ramp-like or stair-like means of access, including accommodation ladders and gangplanks, provided to enable personnel to board or leave a vessel.

gang welding (Techn) A welding technique in which several welding operations are conducted simultaneously by a machine called a gang welder.

gantry (Fire Pr; Brit) A fitting at the rear of an appliance (apparatus) to carry a ladder.

garage (Build) A building or part of one used for the storage, repair, or maintenance of automotive vehicles.

garden apartment (Build) 1. An apartment building surrounded by open space, such as lawns, gardens, and walks. 2. A living unit in such a building.

garret (Build) An unfinished room under the roof of a house. Cf: attic; cockloft; loft.

gas^1 (Phys) a. A fluid. One of the four states of matter (gas, liquid, solid, plasma), characterized by low density, high fluidity, high compressibility, and indefinite expansion. See: fluid. b. A substance that produces a poisonous, irritating, or asphyxiating atmosphere.
  NOTE: Combustible gases are extinguishable only by oxygen starvation or by chemical interaction using dry chemicals. Dense water fogs, changing to steam that depletes the oxygen supply by displacement, may also be used.

gas^2 (Toxicol) An aeriform fluid that is poisonous, irritating, or asphyxiating.

gas^3 (Colloq) = gasoline

gas barrier (Techn) Any device or material used to exclude or divert the flow of gases.

gas constant (Symb: R) The constant in the ideal gas law, expressed as $R = PV/T$, where R is the energy per degree-mole. The equation of state approached by all gases at low pressure. The units for R are energy per degree-mole: $R = 82.06$ cc-atm/$°$-mole $= 1.987$ cal/$°$K-mole. See also: Boyle's law; Charles' law.

gas container (Transp) A tank or cylinder used to store or transport compressed or liquefied gas.

gasdynamics (Abb: Gasdyn) The science that combines fluid mechanics and thermodynamics and deals with the motion of gases undergoing thermal processes.

gaseous hydrogen system (Techn) A system in which hydrogen is delivered, stored, and discharged in gaseous form to consumer's piping.
  NOTE: The system includes stationary or movable containers, pressure regulators, safety relief devices, distribution manifolds, interconnecting piping and controls. The system terminates at the point where hydrogen at service pressure first enters the consumer's distribution piping.

gas-forming organism (Bact; Med) 1. A bacterium that causes gas gangrene. 2. Any bacterium that produces gas (CH_4, NH_3, H_2S) in decomposing organic matter.

gasification (Chem Eng) The process of converting a solid or liquid fuel into a gaseous fuel.

gasket^1 (Mech Eng) A flexible, compressible lining used to provide a seal between mechanical components. See also: gland; packing.

gasket^2 (Fire Pr) A rubber, cork, or synthetic sealer ring in a hose coupling used to make the connections watertight.

gas kit (Fire Pr) A small, portable set of miscellaneous hand tools.

gas laws (Thermodyn) A set of thermodynamic laws applied to perfect gases, specifically Boyle's law, Charles' law, Dalton's law, Amagat's law, van der Waal's law, and the equations of state.

gas log (Build) An unvented, open-flame, gas-fed heater consisting of a simulated log supported on a metal base and installed in a fireplace. Cf: fireplace insert.

gas mask (Safety) A face mask used to protect the eyes and respiratory tract from noxious fumes, gases, and vapors. See: breathing apparatus.
 NOTE: Three principal types exist: 1) protective masks that filter out contaminants in the air (called respirator); 2) rebreathing masks that operate as closed systems, absorbing carbon dioxide and generating oxygen; and 3) air masks that are supplied with oxygen from a cylinder or with pure air from a remote compressor.

gasoline (Techn) 1. Common automotive fuel. Also called: gas; 2. = (Brit) petrol; benzine; motor spirit.
 NOTE: Five gallons of gasoline can explode with a force equal to 414 pounds of dynamite.

gastight (Techn) The property of a structure that prevents gas or vapor from entering or leaving except through vents or through piping and valves provided for removal or addition of material.

gastritis (Med) Inflammation of the stomach as the result of alcohol, chemicals, poisons, or excessive acid secretions.

gastrointestinal tract (Anat) A digestive system within the body consisting of the stomach and the intestines.

gas vent (Build) A channel for conveying flue gases from a gas appliance to the atmosphere.

gas welding (Techn) A process by which metals are united by heating them with the flame from the combustion of fuel or gases. Commonly oxyacetylene welding.

gate[1] (Techn) A type of valve closure.

gate[2] (Metall) A groove in a mold that serves as a passage for molten metal.

gate a hydrant (Fire Pr) To attach a gate valve to a hydrant before turning the hydrant valve on. The gate permits use of the outlet without having to close the main hydrant valve.

gated wye (Fire Pr) A Y-shaped hose connector with one female and two male couplings. Each male coupling has a gate valve for controlling the flow through it independently of the other. The opposite of a siamese.
 NOTE: A wye is usu used to divide one large line into two smaller ones.

gate valve (Fire Pr) A shut-off valve for a hose, pump outlet, or a large-caliber nozzle. See also: valve.

gauge See: gage

Gaussian distribution = normal distribution

Gay-Lussac's law See: Charles' law

gear (Mech Eng) A toothed machine part used to transmit motion between rotating shafts.

gear pump See: pump

Geiger-Mueller counter (Nucl Phys) A radiation detection and measuring instrument consisting of a gas-filled tube containing electrically charged electrodes. When ionizing radiation passes through the tube, ions are produced, and a short, intense pulse of current jumps from the negative to the positive electrode. The number of pulses per second indicates the intensity of radiation. Also called: Geiger counter.
 NOTE: Named after its inventors, Hans Geiger and W. Mueller.

gel (Chem; Phys) 1. Gelatin. 2. A semisolid colloidal dispersion of a liquid and a solid. Cf: colloid; slurry.

gene (Biol) The ultimate biological unit of heredity.

general alarm (Fire Pr) A fire alarm turning out all available first-line fire apparatus and personnel, as well as off-duty firefighters. An alarm for a mass disaster. Cf: fifth alarm.

generalization 1. The extension of a fact or conclusion to a large class. 2. The process of applying a well-known idea to a new situation or event. 3. A broad principle or law in science.

general order (Fire Pr) A standing order that remains in effect until specifically amended or cancelled; contrasted with a special order that is issued for a limited purpose or temporary situation.

general practitioner (Med) A physician in general practice who attends to a variety of medical cases.

genetic effects of radiation (Biol) Mutations or other changes caused by radiation that can be transferred from parent to offspring. Any radiation-caused changes in the genetic material of sex cells. Cf: radiomutation.

gentamicin sulfate (Pharm) A broad-spectrum, topical, antibiotic medication used for treatment of burns.

geometric mean See: mean

geometric progression (Math) A series of the form $1, x, x^2, x^3, \ldots$, such that each term has the same ratio to its immediate predecessor.

geometry¹ (Techn) The spatial configuration, pattern, or relationship of components in an experiment or apparatus.

geometry² (Math) The branch of mathematics dealing with the properties, relations, and measurements of lines, angles, surfaces, and solids.

geostrophic (Meteor) Pertaining to the deflective force due to the rotation of the earth, as in geostrophic wind.

geostrophic wind (Meteor) A hypothetical wind corresponding to exact balance between coriolis acceleration force and horizontal pressure. The concept is used to approximate wind fields from pressure data for large areas in which few wind observations are available. See also: gradient wind; wind.

germ (Bact) A microorganism; a microbe usu thought of as pathogenic. See: bacteria.

germicide (Ind Hyg) An agent capable of killing germs.

getaway time (Fire Pr; For Serv) The elapsed time between receipt of a report of a fire by a fire suppression unit and its departure to the fire.
NOTE: Getaway time plus travel time is equal to response time.

gingivitis (Med) Inflammation of the gums.

gin pole (Mat Hand) A standing derrick constructed of a pole or ladder held in vertical position by guylines fastened to the top end and firmly staked to the ground at a distance from the base. Used with a block and tackle to lift heavy objects. See: hoist.

girder (Build) A large main structural beam used in building and bridge construction.

girt (Mining) A piece of wood mortised between a cap and post to secure the joint, or a timber connecting the caps and sills of two four-piece stull sets.

gland¹ (Mech Eng) Soft packing and mechanical retaining components provided around rotating or reciprocating shafts to prevent leakage of fluid under pressure or in motion. Cf: gasket.

gland² (Physiol) Any body organ that produces a secretion, whether excreting through a duct (exocrine, such as the salivary glands) or emptying directly into the blood stream (endocrine, such as the thyroid gland). The testes and ovaries are called cytogenic glands because they form and secrete living cells.

glare Any brightness within the field of vision that causes discomfort, annoyance, interference with vision, or eye fatigue.

glass (Techn) An amorphous (structureless) super-cooled liquid composed of silicates of sodium, calcium, etc., used for window panes (flatware), decorative building material, bottles and jars (container ware), and as fibers for reinforcement of plastics or in fabrics. See also: plate glass.

glaze¹ (Meteor) Freezing rain. Cf: sleet. See: precipitation.

glaze² (Techn) A thin, impervious glassy layer fused on a ceramic surface.

glazing (Build) Glass in any form used to fill an opening and to pass light.

globe lightning = ball lightning

globe thermometer (Ind Hyg) A temperature measuring instrument consisting of a thin-walled, blackened copper sphere, six inches in diameter, with a temperature sensor at the center.

globulin (Med) A group of complex organic compounds belonging to the proteins and related to the albumins. Found widely in plant and animal tissues.
  NOTE: Gamma globulins play a role in immunity.

glory hole¹ (Metall; Techn) A viewing port in a glass-making or steel furnace.

glory hole² (Mining) A shaft in the center of a series of concentric circular benches of ore. As the ore is mined in all directions, it is broken and dropped down the hole, loaded on cars, and hauled out of the mine.

glove box (Techn) A sealed box in which workers, using gloves attached to and passing through openings in the box, can handle hazardous materials safely from the outside.

glowing combustion (Comb) Oxidation of solid material without a visible gas phase flame. A stage in the ignition of a solid occurring before sufficient volatile fuel has been generated to sustain a gas-phase flame.
  NOTE: Glowing combustion appears to be a heterogeneous surface reaction related to pyrolysis. Radiative transfer is important in such systems. For example, glowing combustion can be initiated by a thermal pulse but will die out when the radiation is removed. Steady-state glowing combustion of coal or charcoal requires a coating of ash, which reduces surface radiation loss while allowing oxygen to diffuse in. A pile of such material will also burn in the absence of ash when radiation losses are compensated by radiation absorbed from other parts of the pile. Glowing combustion rarely initiates a fire, so a fire retardant in many cases favors glowing combustion rather than flaming combustion. Cf: smolder.

glucose (Med) A syrupy liquid obtained by incomplete hydrolysis of starch, composed mainly of dextrose and various dextrins, used to replenish body fluids and for intravenous nourishment.

glue See: adhesive

glycerol (Chem) A clear, sweet, colorless, syrup-like liquid (solid at lower temperatures) derived as a by-product of soap manufacture by fermentation or synthesis. Used in medicines, soaps, antifreeze, and as a solvent or reagent, as a humectant in foods and tobacco, and in the manufacture of explosives.

go (Fire Pr) A call or alarm to which a firefighting unit is assigned to respond.

gob (Mining) Waste mineral material, such as from coal mines, but containing sufficient coal that a fire may arise from spontaneous ignition.

go-fer (Fire Pr; Colloq) A fire buff or volunteer helper who assists firefighters by running errands. See: buff.

goggles (Safety) Spectacles or glasses to protect the eyes against dust, impacting objects, strong light, or other harmful environmental influences.

going fire (Fire Pr) An active fire from the time an alarm is received until the fire is declared out. A working fire.

gondola (Transp) A freight car with sides and ends but without a top covering. May be equipped with high or low sides, drop or fixed ends, solid or drop bottoms, and is used for shipment of any commodity not requiring protection from the weather.

gong (Fire Pr) A large bell used to turn out a fire company in response to a fire alarm. Also called: house gong.

goniometer (Metrol) An instrument for measuring angles. A direction finder.

goodness of fit (Statist) A measure of how well a statistical result agrees with or conforms to a standard or theoretical value.

Good Samaritan Law (Law; Med) One of a set of laws in some states and countries that protect medical personnel who administer emergency care.

goose neck (Fire Pr) 1. An early type of hand engine carrying a discharge pipe in the shape of a goose neck on top of the engine. 2. An early type of portable playpipe. 3. (Brit) A U-shaped fitting used in place of a branch for filling dams.

gophering (Mining) Mining in irregular drifts, following the ore. Irregular, unsystematic workings. Also called: coyoting.

governor (Techn) A device used to control the speed of an engine according to the demands made upon it or to limit its top speed.
NOTE: In fire practice, a governor regulates engine speed when nozzles are shut down to keep pressure in the other hose lines and pump within safe limits.

gpm = gallons per minute

grab bars (Safety) Individual handholds placed adjacent to or as an extension above ladders for the purpose of providing access beyond the top of the ladder.

gradation (Soil Mech) Proportion of material of each grain size present in a given soil.

grade (Civ Eng) The slope of a surface, such as a roadway. Also, the average elevation of a real or planned surface, adjacent to a structure, measured at the center of each exterior wall.

gradeability (Transp) The slope-climbing ability of an automotive vehicle.

grade equivalent (Educ) A level of performance equal to that of a given primary or secondary school grade, determined from the results of one or more tests.
NOTE: As an example, an individual may be shown by test to be capable of performing eighth-grade-level work, even though he may not have completed the seventh grade. In some school districts equivalency certificates are awarded to school dropouts who successfully complete a battery of tests.

grade resistance (Transp) The motion-resisting force acting on a vehicle traveling up a grade. For a vehicle going down a grade, the grade resistance force is negative.

gradient (Phys) 1. The spatial change of a physical quantity, such as temperature or pressure. 2. A gradual and continuous change in a variable.

gradient wind (Meteor) A hypothetical wind blowing on a tangent to a contour line of constant pressure, such that pressure, coriolis force, and centrifugal force are in balance. The gradient wind is a good approximation to the true wind, and is often better than the geostrophic wind. See: wind.

grading schedule (Fire Pr; Insur) The scale by which insurance underwriters evaluate community fire protection.
NOTE: The National Bureau of Fire Underwriters' schedule has 5000 deficiency points and 10 classes. The water supply is graded on a 1700 point scale; fire departments, 1500 points; alarm system, 550; etc. Each 500 deficiency points assessed drops the community one class level. Class 1 is equivalent to fully protected and class 10 to unprotected. The

American Insurance Association also has a grading schedule for larger cities for the purpose of establishing insurance rates.

graft (Med; Surg) 1. Skin or other tissue emplaced on or in a body part to repair an injury or correct a defect. See also: autograft; homograft. 2. To perform a grafting procedure.

Graham's law (Phys) The rates of diffusion of two gases are inversely proportional to the square roots of their densities.

grain (Metall) A crystal of metal.

gram (Abb: g) An elementary unit of mass in the cgs system, equal to the weight of one cubic centimeter of pure water at 4°C. One gram is 1/1000 of a kilogram and itself contains 1000 milligrams. One g = 0.03528 ounce (avoirdupois); 1 ounce = 28.35 g. A US dime weighs about 2 grams. The British and French spelling is gramme.
NOTE: To convert grams to ounces, multiply by 20 and divide by 567, or for a fair approximation, simply divide by 30. Reverse this procedure to convert ounces to grams.

gram-atom (Chem) A mass in grams of an element, numerically equal to its atomic weight.

gram-ion (Chem) A mass in grams of an ion, numerically equal to its molecular weight.

gram-molecule (Chem; Abb: g-mol) A mass in grams of a compound, numerically equal to its molecular weight. See: Avogadro's law.
NOTE: Equal gram-molecular weights of all elements and compounds contain the same number of molecules.

grandstand (Build) A large, usu roofed, structure with chairs or benches used for seating spectators at outdoor events. Cf: stands.

granny knot (Fire Pr) An improperly tied square knot.

granular snow See: snow

granulation[1] (Chem) The formation of crystals during constant agitation of a supersaturated solution.

granulation[2] (Med) Fleshy protrusions formed on an injury during the healing process.

granuloma (Med) A mass or nodule or chronically inflamed tissue with granulations; usu associated with an infection.

graph (Math) A diagram that represents the changes in one variable as compared with changes in one or more other variables, usu shown as points, curves, or proportional areas.

graphic analysis (Math) The use or comparison of graphs to discover significant relationships between or among variables.

graphite (Techn) A soft, pure form of carbon that is heat-resistant and conducts electricity. Used as brush material in electrical motors, in lead pencils, for crucibles, as a lubricant, and as a moderator in nuclear reactors. See: carbon.

grapnel = grappling hook

grappling hook (Salv) A device similar to an anchor but with 4 or 5 flukes, used in dragging operations to snag and recover sunken objects.

Grashof number See: dimensionless numbers

grass fire (Fire Pr) A fire involving grass, weeds, brush, and other similar ground cover. Cf: brush fire.

grassland (For Serv) Land on which the dominant cover is grass and herbs.

grass line (Fire Pr; Brit) A rope made of coir, capable of floating on water.

grate (Comb) A platform made of bars and used to elevate a bed of fuel to allow air to reach the fuel more freely and thus improve the rate of combustion or the burning process.

graticule = reticle

gravel (Build; Civ Eng) Rock fragments, pebbles, and small stones, used as fill and as aggregate in concrete.

gravity (Phys; Symb: g) 1. The force of mutual attraction between masses. 2. The weight of a body, with g representing the force of gravitational acceleration. See: **Newton's laws**. 3. The force of gravity causing a body to accelerate while falling. Acceleration due to gravity is generally taken as 32.2 ft per sec^2.

gravity chance See: **chance**

gravity sock (For Serv) A canvas cone used as a funnel to channel flowing water into a hose attached to its apex.

gravity system (Build) A central warm-air heating system in which air is circulated by heat and gravity.
NOTE: A hot-water heating system may also be of the gravity type, although not common today.

gravity tank (Hydraul) An elevated water storage tank provided for fire protection or used as a community water supply.
NOTE: Water raised to a level of 100 feet produces a static pressure head of 43.4 psi.

gray body (Phys) A heat radiator whose spectral emissivity is less than that of a black body. The spectral emissivity of a gray body remains constant across the spectrum.

gray iron (Techn) Cast iron; in general any iron with a high carbon content.

grayout (Physiol) Partial loss of consciousness.

greater alarm (Fire Pr) Any alarm subsequent to the first, calling for assignment of additional fire companies and units. Syn: **additional alarm**.

greater circulation (Physiol) Blood circulation through the entire body, including head, extremities, trunk, and internal organs. Cf: **lesser circulation**.

green fuel (For Serv) Live vegetation with high moisture content that will not burn until it is dried.

greenhouse effect (Meteor) The trapping of energy within a system in the manner of a botanical greenhouse. The glass is transparent to incoming radiation but opaque to the longer wavelength energy, in the form of heat, reradiating outward, thus effectively trapping the energy or heat inside the enclosure.
NOTE: The earth's atmosphere behaves similarly in trapping solar radiation.

greenstick fracture See: **fracture**

grid^1 (Civ Eng) A water main or piping system with lateral feeders provided to improve water distribution.

grid^2 (Cartog) The coordinate system of a map.

gridiron (For Serv) To conduct an aerial search for a small fire by systematically passing over an area along parallel grid lines.

grief (Psych) A deep, emotional distress, caused by bereavement or other serious loss.

grievance (Man Sci) A complaint related to conditions of work, as unfair or unequal treatment, a situation affecting safety or health, and the like.

grillage (Build) A system of crossed beams used underneath load-bearing columns to spread the load over a wide area.

grizzly (Mining) An arrangement of parallel beams across a chute to prevent excessively large pieces from falling through into ore cars below. Grizzlies can also be made of crossed rails, which form a grid of square holes.

gross tractive effort (Transp) The propelling force that can be developed by the ground-contacting elements of a vehicle on a given type of support.

gross vehicle weight (Transp; Abb: gvw) The total weight of a vehicle, including fuel, water, crew, and all other operating equipment and loads.
NOTE: The rated gross vehicle weight is the maximum loaded weight

of an apparatus as specified by the manufacturer.

gross weight (Transp) The weight of the container plus its contents.

ground (Electr) 1. A conductor that provides an electrical path for the flow of current into the earth or to a conductor that serves as the earth. 2. An unintentional electrical path to earth.

ground clearance (Transp) The height above the ground of the lowest part of the vehicle.

ground cover (For Serv) Grass, brush, and other low-growing vegetation.

ground fire[1] (For Serv) Fire beneath the surface litter of a forest floor, such as burning peat and other natural organic matter.

ground fire[2] (For Serv) A fire at ground level as contrasted to a **crown fire** or a fire burning in the upper levels of a structure.

ground floor (Build) The floor of a building most closely level with the ground.

ground fuel (For Serv) Combustible materials, such as grass, shrubs, dry leaves, twigs, branches, duff, logs, and the like, scattered about on the ground. Cf: **aerial fuel**.

ground hydrant (Fire Pr; Brit) = **flush hydrant**

ground jack See: **jack**

ground level (Build) The base of a building at the level of the ground or street.

ground monitor (Fire Pr) A device, consisting of a multiple siamese connector and a tripod, used to fix a fire hose and nozzle in a stationary position so that it throws a large hose stream unattended. Syn: (Brit) **portable director**. See: **monitor**.

ground plate (Fire Pr) A load-distributing plate placed under outrigger jacks when the ground is uneven or soft; used to stabilize an aerial ladder or aerial platform apparatus. See: **jack**.

ground spill (Fire Pr) Discharge, esp if accidental, of a flammable liquid on the ground or other surface where it presents a fire hazard.

ground state (Phys) The normal state of a nucleus, atom, or molecule at its lowest energy level. Cf: **excited state**.

ground water (Hydrol) Water saturating rocks and soil below ground level, forming a **water table** and serving as a water supply for wells and streams. See also: **aquifer**.

ground zero (Civ Def) The point on the surface of land or water vertically below or above the center of a nuclear explosion.
 NOTE: For a burst over or under water, the term **surface zero** is used.

group[1] (Fire Pr) Members of a fire company or department working the same tour of duty.

group[2] (Math) A number of entities or things considered together or as being related in some way.

group acceptance (Psych) The willingness of a group to take in a new member.

group behavior (Psych) 1. The actions of a group as a whole. 2. The actions of an individual as influenced by his membership in a group.
 NOTE: An individual's behavior when isolated from the group may be quite different from his behavior as a member of the group.

group consciousness (Psych) The feelings that an individual has toward other members of a group or for the group as a whole.

group dynamics (Psych) The cause and effect of relationships within a group and between its members. The sociological study of the interactions within a group.
 NOTE: The study of group dynamics includes the emergence of leadership, development of rules of conduct and their enforcement, decision-making, formation of cliques and subgroups, as well as other processes.

group fire (Fire Pr) An extensive fire involving several buildings, usu in the same block, and threatening to spread to neighboring blocks. Cf: conflagration; mass fire.

group identification (Psych) The sense that one is a member of a group and that he shares a common bond, ideals, standards, and goals with the other members of the group.

grouser (Transp) A projection (often chevron-shaped) on the shoe of a track-laying vehicle, normal to the tread surfaces, provided to improve traction in off-the-road operations. Also called: spud.

grout (Build) A thin mortar used for filling cracks, joints, and small voids in masonry, concrete, and the like.

gr wt = gross weight

GSA = US General Services Adminstration

guard (Safety) 1. A fixed enclosure that blocks access to the point of operation of a machine. 2. A movable device that functions as a fixed guard. 3. A body restraint that serves as a guard. 4. A sensory system (usu electronic) that prevents operation until certain predetermined conditions are fulfilled.

guarded (Safety) Covered, shielded, fenced, enclosed, or otherwise protected by means of suitable covers, barriers, safety bars, or screens, to eliminate the possibility of injury.

guardrail (Safety) A railing erected to prevent individuals from falling or contacting hazards. A handrail.

guard room The main office of a security force at a plant or other installation.

guidance (Educ) A procedure designed to assist an individual in achieving success and satisfaction from an academic or vocational career, usu aimed at developing a systematic program.
NOTE: Guidance generally includes consideration of alternatives and counseling.

guide line (Fire Pr) A rope life line tied to a firefighter wearing breathing equipment to enable him to find his way out of a smoke-charged building.
NOTE: The line is used for communication: one tug, ok; two tugs, allow slack; three tugs, take up slack; four tugs, help! See: line.

guides (Fire Pr) Strips of wood or metal that guide a fly section or erecting section of a ladder while it is being extended. See: ladder.

guillotine damper (Build; Techn) An adjustable plate, installed vertically in a duct, used to regulate the flow of gases.

gun (Fire Pr) 1. A fixed monitor nozzle mounted on a fire apparatus or fireboat and supplied directly by a pump or with multiple inlet connections for large lines to produce a heavy stream. 2. A small gun-type nozzle with a pistol grip. 3. A life gun.

Gunite (Mining) A sand concrete applied by a spray gun, which mixes dry material with water to form a paste and sprays the mixture as a thin coating. Cf: shotcrete.

gusset (Build; Mech Eng) A flat plate connecting two or more structural members at the point of joining, used to strengthen and stiffen the joint.

gust (Meteor) A brief rapid fluctuation of wind velocity. See: wind.

gutter (Civ Eng; Build) 1. A channel for water runoff. 2. A trough along the edge of a roof to carry off rainwater.

gutter trench (For Serv) A ditch dug into a slope below a fire to catch burning material rolling or sliding down.
NOTE: Sometimes part of an undercut line.

guy (Mat Hand) A rope or line used to steady or secure a mast, spar tree, boom, or other member in the desired position. Also called: guyline.

guyline (Fire Pr) 1. A line attached

to a turntable ladder for stability. 2. A line attached to another line to keep it free of obstructions when loads are being lowered from high places. See: line.

gvw = gross vehicle weight

gypsum (Build) A mineral, calcium sulfate, $CaSO_4 \cdot 2H_2O$, used in plaster and plasterboard.

gypsum board (Build) A building interior wall and ceiling finish material made principally of hydrated lime and gypsum, faced with paper, and manufactured in standard 4- by 8-foot sheets 3/8-, 1/2-, and 5/8-in. thick. Also called: plaster board.

H

<u>h</u> (Phys; Techn) Abbreviation for hardness, height, henry, humidity, Planck's constant.

<u>H</u> (Chem; Phys) Symbol for enthalpy, hydrogen, magnetic field strength.

<u>habeas corpus</u> (Law) A writ ordering that a person in custody be brought before a court to determine whether his detention is lawful.
  NOTE: The literal meaning in Latin is: 'you should have the body.'

<u>habit</u> (Physiol) 1. An acquired response manifested in a characteristic pattern of behavior. 2. An automatic action or activity formed through previous experience or training.

<u>habitual offender</u> (Law; Safety) An individual whose record, during a given time period, shows repeated violations of laws or regulations.

<u>hachure</u> (Cartog) Shading in the form of short lines drawn parallel to slopes to depict topography on a map.

<u>HAD</u> = <u>heat-actuated device</u>

<u>hail</u> (Meteor) Solid pellets of ice falling from clouds in a shower. See: <u>precipitation</u>.

<u>half-life</u>[1] (Chem) The measure of the halfway point of a process in which a substance undergoes a change. The time required for one-half of a given chemical species to react.

<u>half-life</u>[2] (Nucl Phys) 1. The time during which half the atoms of a given radioactive substance disintegrate into another form. See: <u>radioactive decay</u>. 2. A means of classifying the rate of decay of radioisotopes according to the time it takes them to lose half their strength.
  NOTE: Half-lives range from fractions of a second to millions of years. Thorium^{232}, for example, has a half-life of over a trillion years; polonium^{212}, about 0.3 ten-millionths of a second. A sample of radioactive material has lost half its strength when its age is equal to its half-life. The term is often applied to other things that gradually lose vitality or usefulness with age.

<u>half-space landing</u> (Build; Brit) A landing, usu between floors, where a staircase reverses direction.

<u>half-thickness</u> (Dosim; Nucl Safety) The thickness of any given absorber that reduces the intensity of radiation to one-half its initial value.

<u>half-track vehicle</u> (Transp) A vehicle in which some wheels (usu the front steered wheels) run without tracks while the others run on tracks.

<u>half-value layer</u> (Nucl Safety) The thickness of a given material necessary to reduce the dose rate of an x-ray beam to one-half its original value.

<u>Halligan tool</u> (Fire Pr) A forcible entry tool, consisting of a claw at one end and two spikes at the other, projecting at right angles.
  NOTE: The tool was designed by Chief Halligan of the New York City Fire Department. See: <u>forcible entry tool</u>.

<u>hallucination</u> (Med; Psych) Hearing, seeing, or feeling things that do not exist in reality. A false perception.

<u>halo effect</u> (Psych) The tendency to rate an individual too high or low because of one outstanding trait.

<u>halogen</u> (Chem) An element in the fluorine, chlorine, bromine, iodine, and astatine group.
  NOTE: The term literally means "salt former."

<u>Halon</u> (Fire Pr) A coined and generic name for several <u>hal</u>ogenated hydrocarb<u>on</u> compounds, three of which (carbon tetrachloride, monobromochloromethane, and bromotrifluoromethane) have found significant use as fire-extinguishing agents. Halon 1211, 1202, and 2401 also have fire extinguishing properties, but are not used as extensively as Halon 1301 (see table below). The fire-extinguishing mechanism is that heat breaks down the halogenated compound into reactive hydride radicals which interrupt the chain reaction of the

combustion process. Carbon tetrachloride is a nonflammable, volatile, colorless, chemically inert liquid used as a solvent and extinguishant, but it is being phased out because it is toxic and evolves toxic gases, such as carbonyl chloride, $COCl^2$ (phosgene), when used in the presence of hot metals. Bromochloromethane is less toxic but still is hazardous to life in enclosed areas. The most effective extinguishing agent is Halon 1301, a noncorrosive, colorless, odorless, electrically nonconducting gas at room temperature, commonly stored as a liquid under pressure. It vaporizes instantly when discharged from its container. Halon 1301 is effective against gas and liquid fires and fires involving live electrical equipment. It is not effective against burning fuels that produce their own oxygen, such as nitrocellulose, or metal hydrides, or active metals such as sodium, potassium, magnesium, titanium, zirconium, uranium, and plutonium. Halon 1301 is 3 to 10 times (or more) effective than CO_2 as an extinguishant. Prolonged exposure to 10% concentration has an anesthetic effect and greater concentrations can asphyxiate. Thermal decomposition produces hydrogen fluoride (HF), free bromine (Br^2), and carbonyl halides, which are not as toxic.

NOTE: A five-digit number system is used to indicate the chemical composition of the compounds. From left to right, the digits stand for the number of atoms of carbon, fluorine, chlorine, bromine, and iodine in the molecule. Terminal (rightmost) zeros are dropped. The number of hydrogen atoms in the molecule, if any, is equal to the first digit times 2, plus 2, minus the sum of the remaining digits. For example, bromochloromethane, CH^2BrCl, is Halon 1011, having 1 atom of carbon, no fluorine, 1 chlorine, 1 bromine, no iodine, and $((2 \times 1) + 2) - 2 = 2$ hydrogen atoms.

Number	Chemical Name	Formula
11	Methyl fluoride	CH_3F
101	Methyl chloride	CH_3Cl
104	Carbon tetrachloride	CCl_4
122	Dichlorodifluoromethane	CCl_2F_2
1001	Methyl bromide	CH_3Br
1011	Bromochloromethane	CH_2BrCl
1202	Dibromodifluoromethane	CBr_2F_2
1211	Bromochlorodifluoromethane	$CBrClF_2$
1301	Bromotrifluoromethane	$CBrF_3$
2402	Dibromotetrafluoroethane	$CBrCBrF_4$
10001	Methyl iodide	CH_3I

halyard (Fire Pr) A rope or cable used to elevate the fly sections of an extension ladder. Also called: fly rope. See: ladder; line.

hammer See: water hammer

hammermill (Techn) A broad category of high-speed equipment consisting essentially of hammers or cutters rotating on an axle inside a cylinder to crush, grind, chip, or shred.

hand-and-foot counter (Dosim) A radioactivity contaminator monitoring device arranged to give a rapid radiation survey of the hands and feet of persons working with radioactive materials.

hand counter (Fire Pr) A device for checking the speed of the pump and engine of a pumping apparatus.

hand-feeding tool (Techn) Any hand held tool designed for placing or removing material or parts to be processed within or from the point of operation.

handhold (Transp) A handle or grip, attached to a strap, that can be grasped by the passenger to provide a means of maintaining a balance. Also called: handgrip.

handicap (Med) A physical or functional defect.

Handie-Talkie (Commun; Tradename) A portable, two-way radio set used for

handline (Fire Pr) 1. A small hose line handled manually by a hoseman. 2. A light rope used to hoist fire tools and to secure ladders and equipment. See: line.

handling (Transp) The maneuvering and course-keeping characteristics of an automotive vehicle, including the ease and precision with which it can be steered or maintained on a desired course, its overall response to control, and its stability.

hand pump (Fire Pr) A small pump, delivering from 1 to 3 gpm, worked by hand and supplied with water from buckets.

handrail (Build; Safety) A rail designed to assist personnel in using stairways or as a guard around floor and wall openings, or at the edges of platforms to prevent falls.

hand tub (Fire Pr) An early hand-operated, pumping engine used by volunteer fire companies.

hanger (Build; Fire Pr) A metal support for suspending sprinkler piping, generally consisting of a rod or strap with a loop at one end to hold the pipe and a bolt, screw, clamp, or other fastener at the other for attachment to a ceiling structural member.

hanging wall (Mining) The upper wall of an inclined raise. The ore limit or wall of rock on the upper side of a dipping orebody. In bedded deposits, the roof.

hangover fire = **holdover fire**

hardenability (Metall) The property that determines the depth and distribution of hardness of a ferrous metal induced by quenching.

hardening (Metall) A process that increases the hardness of metal, usu involving controlled heating and cooling.

hardhat (Techn) Protective headgear used to guard against falling objects.

hardness[1] (Nucl Phys) The penetrating quality of radiation.

hardness[2] (Metall) The resistance of a material to cutting, abrasion, indentation, or plastic deformation, often used to denote rigidity and strength.
NOTE: Several tests of hardness exist for different purposes. The file test and Mohs test are scratch tests; the Brinell, Rockwell, Vickers, Tukon, Eberback, and others are penetration tests in which a ball or diamond pyramid indenter is pressed against a material and the size of the indentation is the measure of hardness. In the scleroscope hardness test the kinetic energy of a falling metal tup is absorbed by indentation of the specimen, and the hardness is determined from the height of the tup rebound.

hardness number (Techn) An arbitrary measure of the hardness of a solid, expressed as a number on one of various scales, such as the Mohs, Brinell, Knoop, Meyer, Rockwell, Shore scleroscope, and Vickers.
NOTE: None of the scales are readily convertible one to another. The oldest, the Mohs scale, which is based on 15 minerals arranged by hardness in such a way that each can scratch the ones that come before it, includes: talc (1), gypsum (2), calcite (3), fluorite (4), apatite (5), feldspar (6), vitreous silica (7), quartz (8), garnet (9), topaz (10), zirconia (11), fused alumina (12), silicon carbide (13), boron carbide (14), and diamond (15).

hardpan (Soil Mech) A hardened, compacted, or cemented soil layer.

hard soil (Soil Mech) Compact earth materials not classified as running or unstable.

hard suction (Fire Pr) A length of rigid suction hose, ranging from 2-1/2 to 6 inches inside diameter, used to draft water from static sources below the level of the pump.

hard vacuum (Phys) Extremely high vacuum or pressures of less than 10^{-7} torr. Syn: very high vacuum. See: vacuum.

hard water (Hydrol) Water that does not readily form a lather with soap due to dissolved calcium and magnesium salts.
NOTE: A degree of water hardness is equivalent to one grain of calcium carbonate, $CaCO_3$, in one gallon of water.

harmonic (Acoust) A frequency that is an integral multiple of the fundamental oscillation.

harmonic mean (Math) The reciprocal of the arithmetic mean of the reciprocals of a series of values.

harmonic motion (Phys) A motion in which the displacement is a sinusoidal function of time.

hate (Psych) A strong, aversive attitude toward someone or something with a desire to inflict injury.

hat shield (Fire Pr) The plaque on a helmet identifying the fire department and unit of the wearer. Syn: **hat front**.

haulageway (Mining) A passage through which mined materials are transported for removal from a mine.

hawser (Mat Hand) A rope of six inches or more in circumference, usu consisting of three strands with a right-hand **lay**. The strands themselves are usu laid left-handed. See: **line**.

hazard[1] (For Serv) A fuel complex defined by type, arrangement, volume, physical state, location, and ambient conditions that presents a special ignition threat or suppression difficulty.

hazard[2] (Fire Pr) Ignitability, explosibility, instability, radiation, toxicity, and the like of materials.
NOTE: Common hazards are those encountered in ordinary properties; special hazards are peculiar to given industries, processes, or operations, such as chemical or explosives manufacturing. Fire hazard refers to the probability of fire outbreak; **life hazard** refers to the probability of loss of life from fire.

hazard[3] (Safety) A potentially or inherently dangerous condition which may interrupt or interfere with the orderly progress of an activity.

hazard classification (Fire Pr; Insur) A designation of relative accident potential based on probability of an accident occurring. See: **hazard level**.
NOTE: Hazards are often classified as low, ordinary, or high.

hazard identification (Safety) Marking, labeling, or placarding applied to containers, lading, objects, areas, or premises to inform or warn of the presence of a hazardous substance, material, item, process, or condition.

hazard index (Safety) A number that represents an expected loss, obtained by multiplying the hazard probability by the hazard loss.

hazard level (Safety) A qualitative measure of hazards stated in relative terms.
NOTE: Hazard levels are grouped into four categories: **Category I** - Negligible. Will not result in personnel injury or system damage. **Category II** - Marginal. Can be counteracted or controlled without injury to personnel or major system damage. **Category III** - Critical. Will cause personnel injury or major system damage, or will require immediate corrective action for personnel or system survival. **Category IV** - Catastrophic. Will cause death or severe injury to personnel, or system loss.

hazard of contents (Fire Pr; Insur) The relative danger of the contents of a building or structure igniting and spreading fire, smoke, or gases; of exploding; or otherwise threatening the lives and safety of the occupants.

hazardous act (Safety) Any act which deviates from a standard or norm and which has a potential to cause damage, injury, illness, or death.

hazardous area (Fire Pr) An area in a structure, building, or part of one used for processes that involve highly combustible, flammable, or explosive products or materials or which may produce explosive dusts, or poisonous fumes or gases, including highly toxic, noxious, or irritant chemicals.

hazardous cargo (Mat Hand; Storage; Transp) A commodity in transit classified as hazardous by Federal regulations, including explosives, combustible liquids, and other dangerous articles.

hazardous commodity (Mat Hand; Storage; Transp) A material which, because of its inherent composition, may endanger life or property.

hazardous condition (Safety) The physical condition or circumstance that is causally related to an accident. The hazardous condition is related directly to both the accident type and the agency of the accident.

hazardous material (Fire Pr) A material that presents a potential danger to life, health, or property through fire, explosion, toxicity, etc.
NOTE: A material is considered hazardous if it: 1) Has a flash point below 140° F, closed cup, or is subject to spontaneous heating. 2) Has a threshold limit value below 500 ppm in the case of a gas or vapor, below 500 mg/m³ for fumes, and below 25 mppcf in case of a dust. 3) Has a single dose oral LD^{50} below 500 mg/kg. 4) Is subject to polymerization with the release of large quantities of energy. 5) Is a strong oxidizing or reducing agent. 6) Causes first-degree burns to skin on short exposure or is systemically toxic by skin contact. 7) Produces dusts, gases, fumes, vapors, mists, or smokes.

hazardous substance (Ind Eng; Safety) Any substance that by reason of being explosive, flammable, poisonous, corrosive, oxidizing, irritating, or otherwise harmful, is likely to cause death or injury.

hazard pay (Man Sci) Extra payments to workers in dangerous occupations or in work where the chances of injury are greater than normal.

hazard recognition (Safety) The perception of a hazardous condition.

hazard reduction (For Serv; Fire Pr) Any treatment of a fire hazard that reduces the threat of ignition and spread of fire, such as prescribed burning.

hazards code (Fire Pr) A numeric coding system for categorizing the type and degree of danger to be expected from various materials.
NOTE: The codes, ranging from 0 (no special hazard) to 4 (extreme hazard), are displayed on a label in the shape of a rhombus divided into four equal squares: the left square is a blue field, for health; the top, red, for flammability; the right, yellow, for reactivity; and the bottom, white, for unusual reactivity with water, indicated by the letter W with a horizontal bar through it.

Hazchem (Fire Pr; Brit) A hazardous cargo placarding system used in the United Kingdom for rail and highway transportation.
NOTE: The code used in this system identifies the material and the nature of the hazard, including the firefighting agent to be used, personal protection, explosive risk, spillage action, and evacuation warning.

haze (Meteor) A suspension of fine particles, close to the ground, that appears as a thin mist and reduces visibility. See also: precipitation; smoke. Cf: smog.

hazemeter (For Serv) An instrument for measuring the distance at which a standard smoke column can be seen by eye under existing conditions of haze. Also called: visibility meter.

HD = heavy duty

head¹ (For Serv) The advancing front of a forest fire.

head² (Hydraul) The energy of a fluid derived from its height, velocity, or pressure. The static head for water equals about 0.434 psi for each foot of elevation.

head³ = sprinkler head

header (For Serv) A conduit or channel that supplies water or other fluid to branch lines.

header pipe (Build) A pipe that connects one or more branch pipes to the remainder of an exhaust system. Also called: main pipe.

head fire (For Serv) A wildland fire advancing with the wind or one set to spread with the wind.

head harness (Safety) A device for holding the facepiece of a protective mask or eyeshield securely in place on the wearer's head.

heading (Mining) A horizontal drift or tunnel, or the end of a drift or gallery.

head of a fire (For Serv) The most rapidly advancing part of a fire edge, usu upslope or with the wind. See: fire nomenclature.

headquarters (Fire Pr) 1. The main administrative offices of a fire department. 2. The office of the chief of department. 3. A fire alarm or communications center.

headrest (Fire Pr; Brit) A support located at the front of a pump escape, turntable ladder, or pumper to take the weight of the head of an escape or other ladder. Syn: gallows.

health insurance (Insur) Insurance coverage that pays benefits, including reimbursement for loss of income in case of illness, injury, or accidental death.

health officer (Safety) An individual who investigates suspected cases of exposure to communicable diseases to prevent further spread.

health physicist (Nucl Safety) A professional who is trained in radiation physics, esp in problems of radiation damage and protection.

health physics (Nucl Safety) The discipline concerned with recognition, evaluation, and control of health hazards from ionizing radiation.

hearing (Physiol) The sensory capacity to detect sound waves.

hearing conservation (Ind Hyg) Prevention of noise-induced deafness through the use of hearing protection devices and the control of noise.

heart (Anat) The pumping organ of the body that moves the blood through the circulatory system. See also: artery; vein; cardiovascular system.

heart attack (Med) A sudden blockage of a coronary artery that restricts the flow of blood to the heart muscles.

heart block (Med) A condition in which the auricles and ventricles beat independently, resulting in decreased cardiac output.

heart failure (Med) Inability of the heart to maintain blood circulation at the level required by the body, resulting in damming back of fluid into the lungs, abdomen, and dependent extremities (e.g., pulmonary edema, ascites, and ankle edema). Cf: cardiac arrest.

hearth (Techn) The surface in a combustion chamber that holds fuel during combustion.

heat (Thermodyn) A form of energy characterized by chaotic vibration of molecules and capable of initiating and supporting chemical changes and changes of state.
NOTE: Technically, heat is the form of energy that travels from a higher temperature source to a lower temperature sink. Available heat is the quantity of useful heat produced per unit of fuel if it is completely burned minus the heat values of the dry flue-gases and water vapor.

heat-actuated device (Fire Pr; Abb: HAD) A detector that senses heat or rate of rise in temperature and sets off an alarm and/or a fire extinguishing system.

heat-affected zone (Metall; Techn) The portion of the base metal which was not melted during brazing, cutting or welding, but which is altered in microstructure and mechanical properties by the heat.

heat capacity (Thermodyn) The quantity of heat required to raise a unit mass of substance one degree in temperature without phase or chemical changes.

heat conduction (Thermodyn) Flow of heat energy through a continuous medium or from one body to another in direct contact.

heat convection See: convection

heat cramp (Med) Painful spasms of voluntary muscles due to the loss of body salt, brought on by physical exertion in a hot environment and by profuse sweating, associated with dilated pupils and weak pulse. Less common than heat exhaustion.

heat detector (Fire Pr) A device that senses the presence of heat or a change in temperature. See: rate-of-rise detector.
NOTE: A common heat detector is one that has a fusible link that melts at a predetermined temperature and sets off an alarm or fire extinguishing system. Another type operates on the bimetallic strip principle, such as the ordinary thermostat.

heat exchanger (Thermodyn) Any device that transfers heat from one fluid (liquid or gas) to another or to the environment.

heat exhaustion (Med) Physical distress characterized by fatigue, nausea, heavy perspiration, pale clammy skin, subnormal temperature, vacant eyes, and general weakness brought on by overexposure to high temperature and deficiency of salt. Syn: heat prostration. See: heat stroke.
NOTE: First aid includes rest and, if the victim is able to swallow, three or four doses of 1/2 teaspoon of salt in 1/2 glass of water every 15 minutes.

heat exposure (Physiol) Open to the effects of a hot environment.

heat flow (Thermodyn) The movement of heat energy from a region of higher temperature to one of lower temperature.

heat flux (Thermodyn) The intensity of heat transfer across a surface, expressed in watts/cm^2, joules/m^2, or Btu/in^2/sec.

heat fusion joint (Techn) A joint made in themoplastic piping by pressing the parts together and fusing them with heat.

heating (Build) 1. Raising the temperature of a space or object. 2. Referring to the type of equipment or method used to generate heat.

heat of combustion (Comb) The quantity of heat evolved by the complete combustion of one mole (gram molecule) of a substance. Cf: heat of reaction.

heat of decomposition (Chem) The quantity of energy liberated or absorbed by a substance when it is broken up into one or more products of lower formula weights.

heat of formation (Chem) The number of calories of heat absorbed or liberated during the formation of one mole (gram molecule) of a compound from its elements.

heat of fusion (Chem) 1. The quantity of heat required to melt a unit quantity of a substance. The heat of fusion for a pound of ice at 32°F is 144 Btu. 2. The quantity of heat released in the conversion of a unit quantity of a gas to a liquid or of a liquid to a solid.

heat of neutralization (Chem) 1. The quantity of heat, in calories, liberated in the formation of a mole of water from hydrogen and hydroxyl ions. 2. The heat liberated during the neutralization of an acid by a base to form one gram equivalent of salt.

heat of oxidation (Comb) The quantity of heat evolved in an oxidation process.
NOTE: For many common substances 537 Btu of heat is liberated for each cubic foot of oxygen consumed, regardless of the heat of combustion value for the particular substance.

heat of reaction (Chem) The quantity of heat evolved or absorbed in a reaction between combining weights of two reactants. Cf: heat of combustion; heat of formation.

heat of solution (Chem) The quantity of heat evolved when a substance is dissolved in a large surplus of water.
NOTE: If water is added to sulfuric acid, sufficient heat may be genera-

ted to produce violent boiling.

heat of vaporization (Chem) The quantity of heat required to change a unit quantity of a liquid into a vapor. One pound of water at 212°F requires 972 Btu to vaporize. See also: <u>evaporation</u>.
NOTE: The ratio of the heat of vaporization to the boiling point of different liquids is approximately a constant.

heat prostration (Med) A form of heatstroke characterized by low body temperature, collapse, and, in severe cases, coma and death. Also called: <u>heat exhaustion</u>.

heat pump (Techn) A reverse refrigerating system used to transfer heat into a space or substance. The condenser provides the heat while the evaporator is arranged to pick up heat from air, water, etc.
NOTE: By shifting the flow of air or other fluid, a heat pump system may also be used to cool the space.

heat radiation (Thermodyn) One of three basic mechanisms of heat propagation (the others being conduction and convection), in which heat is transmitted by electromagnetic waves.

heat release rate (Comb) The quantity of heat liberated during complete combustion; usu expressed in Btu's per hour per cubic foot of the internal volume of the space in which the combustion takes place.

heat sink (Phys) 1. A body in which heat from a system can be dissipated or stored. 2. A region in a system toward which heat migrates. 3. A material or device that absorbes unwanted heat, esp in electronic equipment and industrial processes.
NOTE: Parts of the environment may serve as heat sinks; for example, the atmosphere, or a body of water.

heat strain (Physiol) The physiological responses to heat stress. The severity of heat strain includes, in addition to the degree of exposure, considerations of age, physical fitness, degree of acclimatization, degree of dehydration, etc.

heat stress (Ind Hyg) The aggregate of factors that impose a heat load on the body, including air temperature, water vapor pressure, radiant heat, air circulation, level of physical activity, type of clothing, etc.

heat-stress index (Ind Eng) A measure relating the ambient temperature, humidity, and level of activity to the stress of a person working in a hot environment.

heat stroke (Med) A condition of dry skin, high body temperature, and collapse due to long exposure to high temperature. A more serious condition than <u>heat exhaustion</u>.
NOTE: If untreated, heat stroke may lead to delerium, convulsions, coma, and even death.

heat transfer (Thermodyn) The movement of thermal energy by <u>conduction</u>, <u>convection</u>, or <u>radiation</u>. As a rule, heat travels most rapidly by radiation.

heat transmission (Thermodyn) The passage of heat energy from one solid to another in contact with it.

heat treatment (Metall) A combination of timed heating and cooling operations applied to solid metals to provide desired properties.
NOTE: Treatments involve any of several processes of metal modification, such as quenching, homogenizing, annealing, or cyaniding. Particularly the thermal treatment of steel by normalizing, hardening, tempering, etc.

heat value = <u>calorific value</u>

heave (Transp) Motion of a vehicle in the vertical direction. Ant: <u>jounce</u>.

heavy drip (Fire Pr) Water dripping rapidly and persistently through floors and ceilings. Syn: <u>fast water</u>.

heavy-duty apparatus (Fire Pr) A master stream apparatus with large-bore nozzles designed to deliver a high volume stream.

heavy fuel (For Serv) Slow-burning fuel of large dimensions, such as logs, tree stumps, and large tree limbs. Also called: <u>coarse fuel</u>.

heavy hydrogen = deuterium

heavy stream = master stream

heavy timber construction (Build) A type of construction in which bearing walls, bearing elements, and nonbearing exterior walls are of noncombustible construction, the columns, beams, and girders are of heavy solid or laminated timber, and the floor and roof are of wood without concealed spaces. See: building construction.

heavy water¹ (Fire Pr) Deep water collected on the floor.

heavy water² (Nucl Phys) Water containing significantly more than the natural proportion of heavy hydrogen atoms to ordinary hydrogen atoms, or water enriched with deuterium.
 NOTE: Heavy water is used as a moderator in some reactors because it slows down neutrons effectively and also has a low cross section for absorption of neutrons.

heel¹ (Fire Pr) a. The bottom end of a ladder. b. To take a position at the foot of a ladder for raising or lowering.

heel² (For Serv) The rear edge of an advancing forest fire; the slowest moving edge, or the edge opposite the head. See: fire nomenclature.

heel³ (Techn) The quantity of liquid in a storage tank at the time of refilling. See: ullage.

heel man (Fire Pr) A firefighter who carries the heel of a ladder and holds it in position during raising operations. See: ladder.

heel plate (Fire Pr) A plate attached to the base of a ladder to protect the heel and provide support. Also called: ladder shoe. Cf: stirrup.

Heimlich maneuver (First Aid) An emergency method used to clear a bolus of meat or other blockage from the throat. By compressing the residual air in the lungs with an arm squeeze from behind the victim and forcing the diaphragm upward, the obstruction may be expelled.

held line (For Serv) The portion of a control line that still contains a fire after mop-up has been completed, excluding lost line, natural barriers that were not backfired, and unused secondary firelines.

helical spring (Mech Eng) Round, square, or rectangular wire, wound in the form of a helix, offering a resistance to a force applied along the axis of the coils.
 NOTE: When wound with space between coils, the spring may be loaded in compression. When the force is applied to stretch the coils, it is termed a helical tension spring. When force is applied in a twisting direction in the plane of the coils, it is called a helical torsion spring.

helijump (For Serv) Exit from a hovering helicopter without a parachute, at heights of less than 10 feet.

helijumper (For Serv) A firefighter specially equipped and trained to jump from helicopters in areas where helicopters cannot land.

heliport (For Serv) A helicopter base.

helispot (For Serv) A temporary landing area for helicopters.

helitack (For Serv) Use of helicopters to transport men and equipment to attack wildland fires.

helitanker (For Serv) A water-carrying helicopter used for wildland fire suppression.

helix (Anat) The curved outer part of the ear; the margin of the pinna.

helmet (Fire Pr) Protective headgear for a firefighter, usu carrying the insignia of his rank and unit. fire hat.
 NOTE: Traditionally fire helmets were made of leather, but now often of aluminum or synthetic materials.

hematocrit (Med) 1. The ratio of the volume of red blood cells to the volume of the whole blood. 2. Rarely, a centrifuge for separating blood cells and other particles from the plasma.

hematology (Med) Study of the blood and the blood-forming organs.

hematoma (Med) A localized mass of clotted blood that has exuded from a blood vessel and has collected in an organ or tissue. Syn: blood tumor.

hematuria (Med) The presence of blood in the urine.

hemiplegia (Med) Paralysis of a lateral half of the body.

hemisphere (Anat) Either of the two halves of the brain.

hemoglobin (Med; Symb: Hb) The oxygen-carrying pigment of the red blood cells; a protein-containing compound capable of combining with oxygen to form oxyhemoglobin and able to release oxygen in areas where oxygen tension is low. Also capable of combining preferentially with carbon monoxide.

hemolysis (Med) The alteration or destruction of red blood cells, releasing hemoglobin into the surrounding medium.

hemolytic (Med) Destructive to blood cells; commonly descriptive of streptococci, which are capable of destroying cells that attempt to wall off infection.

hemoptysis (Med) Spitting or coughing up of blood from the respiratory tract.

hemorrhage (Med) Profuse bleeding; leakage of blood from a blood vessel, caused by disease or physical damage.

hemorrhagic shock (Med) A condition brought on by loss of blood or plasma, characterized by cold, clammy skin, nausea, and fainting.

hemostasis (Med) Arrest of bleeding.

hemothorax (Med) Free blood or body fluids in the chest cavity.

henry (Electr; Symb: h) The basic unit of inductance.

hepatitis (Med) Inflammation or infection of the liver.

heptane (Chem) A hydrocarbon, C_7H_{16}, belonging to the alkane family; n-heptane is a straight-chain arrangement of heptane.

NOTE: Assigned an octane value of zero, heptane is used as a standard for rating the anti-knock properties of automotive gasoline.

herb (For Serv) Any nonwoody plant that is relatively soft and succulent.

herbaceous fuel (For Serv) Undecomposed living or dead plant material.

hermaphroditic coupling (Fire Pr) An interlocking coupling both parts of which are identical.

hertz (Electr; Phys; Abb: Hz) A unit of frequency equal to one cycle per second; i.e., 1 Hz = 1 cps.

heterogeneous Consisting of assorted ingredients, or of two or more distinct phases, such as gas-solid, or gas-liquid.

heterogeneous combustion (Comb) A heat-producing reaction in which the reactants are not mixed uniformly and exist in different phases. Usu applied to the burning of solid particles or liquid droplets in air.
NOTE: In rocket propulsion a dispersion of a solid oxidizer in a plastic binder is also considered a heterogeneous combustion system.

heterograft (Surg) A transplant of tissue from one species to another, as the transplant of pigskin to man as a temporary dressing.

heuristic (Math) Leading to discovery; assisting in the solution of a problem, for example, a partial result helping to formulate the next step in problem solving.

HEW = US Department of Health, Education, and Welfare

hexane (Chem) A colorless flammable liquid, C_6H_{14}. Any of the six hydrocarbon alphatics, saturated hydrocarbon. Normal hexane consists of six carbons joined linearly. It is a common constituent of gasoline.

Hg (Chem; Phys) Symbol for mercury, used in pressure measurements to indicate a mercury barometric column.

hierarchy (Info Sci) An arrangement

according rank or order, each element being subordinate to the one above it, as a series of terms, each successively broader in meaning. Ant: babilarchy.

Higbee cut (Fire Pr) A technique for tapering coupling threads to prevent cross-threading during connection.

Higbee indicators (Fire Pr) Marks provided on couplings to show where the threads start. Used to line up couplings when connecting hoses.

high air (Techn) Air pressure used to supply power to pneumatic tools and devices.

high-energy fuel (Abb: HEF) Liquid or solid chemical propellant providing more energy than conventional fuels. Exotic chemical fuels used to propel rockets and missiles; e.g., hydrocarbons, hydrogen, hydrazine, aniline, etc.

higher education (Env Prot) Education provided at the level of colleges, universities, graduate schools, and advanced technical schools.

higher-level skills (Psych) Abilities that apply to a variety of tasks.

higher mental processes (Psych) Mental activity that involves thinking, imagination, judgement, and intelligence.

high-expansion foam See: foam expansion

high explosive (Civ Eng; Techn) An explosive that detonates, rather than deflagrates or burns; that is, the rate of advance of the reaction zone into the unreacted material exceeds the velocity of sound in the unreacted material. See: low explosive.
NOTE: High explosives are divided into two classes: primary high explosives, and secondary high explosives, according to their sensitivity to heat and shock. Whether an explosive reacts as a high explosive or as a low explosive depends on how it is initiated and confined.

high-hazard contents (Build; Insur) Building contents that are liable to burn with extreme rapidity or that may explode or generate poisonous fumes or gases in the event of fire.

highjacking (Law) Seizure of a vehicle in transit or its contents by threat or violence.

high-pressure fog (Fire Pr) A small capacity spray jet produced at very high pressure, discharged through a small hose with a gun-type nozzle. See: fog.

high-pressure main (Fire Pr) A water main with sufficient pressure to provide high-volume fire streams without pumping.

high-pressure pumper (Fire Pr) 1. A pumper with special high-pressure pumps and equipment to supply small, high-pressure fire streams. 2. A pumper capable of developing higher than standard pump pressures for use in areas occupied by tall buildings.

high-rack storage (Storage) A type of warehousing in which goods are placed on open metalwork racks up to 70 feet high.

high-radiation area (Nucl Safety) Any area, accessible to personnel, in which radiation exists at such levels that a major portion of the body could receive a dose in excess of 100 millirem in one hour.

highrise (Build) A multistory building, the upper floors of which are beyond the reach of fire department aerial equipment for firefighting and rescue, and in which fires must be fought from inside.
NOTE: Rapid emergency evacuation of such a building is not practical, and significant interior-exterior pressure differences or stack effects can be expected.

high-school equivalency certificate (Educ) A document certifying that the individual to whom it is issued, while not a high-school graduate, has met the requirements for high-school graduation.

high service (Fire Pr) A system of high-pressure water mains supplying hydrants in the higher elevations of a city and certain hydrants in high-value districts. See: high-pressure main.

213

high-temperature head (Fire Pr) An automatic sprinkler head designed to operate at 212, 286, or 360°F.
NOTE: Respectively, these heads are designated and color-coded: intermediate (white), hard (blue), and extra-hard (red). Higher-temperature extra-hard heads are available for 400°F (green) and 500°F (orange).

high-tension bands (Transp) Steel strapping of various widths and thickness, each with a standard load strength, used to secure lading.

high-value company (Fire Pr) A fire company serving a high value district.

high-value district (Insur) A section of a city in which values of property and building contents are high. Usu the central business district with multistory buildings.

high-velocity tool (Techn) A tool or machine that propels or discharges a stud, pin, or fastener.

highway (Transp) Any public street, public alley, or public road.

hip (Build) The line of intersection between two sloping roofs.

histogram (Statist) A graph of a frequency distribution in which proportional vertical bars represent the number of cases in each class interval.

histological (Med) Pertaining to the microscopic structure of biological specimens.

histology (Anat) The branch of anatomy that deals with cells and microstructure of tissues and organs. Cf: **cytology**.

hit-and-run (Law) Failure to stop after a motor vehicle accident which involves bodily injury or property damage to others.

hitch[1] (Fire Pr) a. One of several methods of securing rope to objects. See: **knot**. b. To wrap a loop of hose around a fire hydrant so that the hose will not pull away when the apparatus rolls forward toward the fire. See also: **quick hitch**.

hitch[2] (Fire Pr) A period of duty.

hitch[3] (Mining) An indentation or recess in a footwall, hollowed out to hold the toe of a stull.

hoar frost See: **frost**; **precipitation**

hog (Techn) A machine for cutting or grinding slabs and other coarse residue from a mill.

hogging (Build) Bending upward in the middle.

hoist (Techn) 1. A device for lifting equipment, materials, and other loads, as the raising mechanism of an aerial ladder. See also: **gin pole**. 2. To lift or raise, esp a heavy object mechanically.

hoistway (Build) A vertical passageway or shaft designed to enclose and provide support for an elevator, platform, or other lifting device.

hoistway door (Build) A door across the opening between an elevator shaft or hoistway and the floor landing, normally closed except when the elevator is stopped at that floor.

holdout device (Safety) A mechanism on a machine that prevents the operator's hands from entering the point of operation. Also called: **restraint device**.

holdover fire (For Serv) A fire that remains dormant for a long period of time. Also called: **hangover fire**; **sleeper fire**.

hollow-core construction (Build) A common method of panel construction in which an open lattice of wood (sometimes paperboard rings) is faced with plywood, hardboard, or similar facing material.

homeostasis (Med) Maintenance of internal dynamic equilibrium by a living organism independent of the environment. As an example, body temperature tends to remain constant while ambient temperature varies widely.

homeowner's insurance (Insur) A package type of insurance for homeowners, covering losses due to fire, theft, and personal liability.

homo- A prefix, meaning self or same.

homogeneity Uniformity in structure or composition.

homogeneous atmosphere (Meteor) A hypothetical atmosphere in which density is constant with height. See: atmosphere.

homogenizer (Techn) A machine that forces liquids under high pressure through a perforated plate to blend or emulsify them.

homograft (Surg) Tissue provided by a donor of the same species and used to repair injuries where skin has been lost; i.e., a transplant of skin from one person to another. See: graft. Cf: autograft.

homografting (Surg) Transplanting skin or other tissue from one person to another.

homologous series (Chem) A series of compounds in which the members have similar functional groups and may be represented by a general formula; e.g., the methane series: C_nH_{2n+2}.

homologue (Chem) A compound considered as a member of a homologous series.

honeycomb (Build) A cellular structure that resembles a honeycomb.

honeycombing (Techn) The storing or withdrawing of supplies in a manner which results in vacant space that is not usable for storage of other items.

honorary member (Fire Pr) A title conferred on former firefighters or special friends of a fire department, frequently with social or other privileges, such as passage through fire lines. Some fire departments confer honorary officer titles.

hood¹ (Ind Hyg) A protective device, used to carry away dangerous fumes or dusts by forced ventilation through an exhaust system.

hood² (Ind Hyg) The partial or complete enclosure around a wheel or disc through which air enters an exhaust system during operation.

hook¹ (Fire Pr) Any of a variety of fire-department implements, such as a plaster hook, ceiling hook, pull-down hook, etc.

hook² (Fire Pr) The operating lever of a fire alarm box.

hook and ladder (Fire Pr; Archaic) A ladder truck, deriving from the time large pull-down hooks were commonly used in fighting fires.

hook and loop fastener (Techn) A quick-fastening closure for a garment, consisting of a tape with plastic loops and another with plastic hooks which engage the loops when the two tapes are pressed together. The tapes are opened by peeling them apart.

hook belt (Fire Pr; Brit) A firefighter's belt equipped with a hook for use in climbing, to secure a purchase in order to free the hands for work aloft. Cf: hook ladder belt.

hooking up (Fire Pr) The activity involved in connecting a pumper to a hydrant and connecting hose lines.

hook ladder (Fire Pr) A scaling ladder with one or two hooks at the top end for hooking over window sills, etc.

hook ladder belt (Fire Pr) A belt with a close-fitting snap hook for use with a hook ladder.

hooks (Fire Pr) Swivel hooks mounted on the beams at the tops of roof ladders for hooking the ladders over the peaks of gabled roofs or over window sills.

hopper car (Transp) A freight car with sides and ends but without a top covering; the floor slopes from the ends and center to permit the entire load to be discharged by gravity through the hopper doors.

horizon (Geophys; Mining) 1. The geological deposit of a particular time, usu identified by the presence of distinctive fossils. 2. One of the layers of a soil profile distinguished principally by its texture, color, structure, and chemical content.

horizontal channel (Storage) Any uninterrupted space between horizontal layers of stored commodities. Such channels may be formed by pallets, shelving, racks, or other storage arrangments.

hormone (Physiol) A chemical substance secreted by the endocrine glands which, when carried by the blood stream to another organ, excites it to functional activity.

horn[1] (Fire Pr) A compressed air device used to sound coded signals to volunteer firefighters and call men at a distance from the station.

horn[2] (Commun) An intercommunications device. See: bullhorn.

horsepower (Abb: hp) A unit of power equal to 33,000 ft-lb/min, 550 ft-lb/sec, or 746 watts.
NOTE: The power output of an engine measured in brake horsepower at the output shaft under standard conditions.

horse scaffold (Build) A scaffold for light or medium duty, composed of horses supporting a work platform.

horseshoe hose load See: accordion horseshoe load

hose (Fire Pr) A flexible conduit used to convey water under pressure. Fire hose is available in various diameters from 3/4 to 4 inches or more and in standard lengths of 50 feet. Usu constructed of a rubber tube covered with one or two woven fiber jackets that protect the inner liner. Double-jacketed hose is now the most commonly used in the fire service.
NOTE: The first use of hose is said to have been in the Netherlands in 1672.

hose bandage (Fire Pr) 1. A piece of canvas, shaped like a bandage, used to stop minor leaks temporarily in charged hose. See: hose jacket. 2. = (Brit) hose gaiter.

hose becket (Fire Pr) A leather strap with a handle that can be used to carry a hose line or to make a hose line fast to a ladder, fire escape, or other object. See: hose tool; snotter becket.

hose bed (Fire Pr) A compartment in a fire apparatus for carrying connected sections of fire hose.

hose belt (Fire Pr) A strap used to carry hose ropes, lines, spanners, etc. or in conjuction with hose ropes to hold charged hose lines.

hose-body (Fire Pr) A fire apparatus body with one or more hose compartments.

hose bowl (Fire Pr) The threaded cavity of a female coupling into which a male threaded coupling is inserted.

hose bridge (Fire Pr) 1. A double-wedge ramp with one or more channels or archways used to protect hose lines from damage when vehicles cross the lines. Syn: hose jumper; (Brit) hose ramp. 2. A structure for elevating hose above traffic.

hose cabinet (Build; Fire Pr) A case recessed into a wall and used to enclose a wall hydrant and a connected length of hose.

hose cap (Fire Pr) A female threaded fitting used to close off a hose line or a pump outlet. See: hose plug.

hose carry (Fire Pr) Any of several methods of folding hose for convenience in carrying.

hose clamp (Fire Pr) A manually-operated mechanical or hydraulic clamping device used to pinch charged hose lines to stop the flow of water temporarily in order to make repairs or connections.

hose compartment (Fire Pr) 1. Any of the sections of a divided hose body used for stowing hose on a fire apparatus. 2. (Brit) = A case or locker used for storing first-aid fire-fighting equipment.

hose control (Fire Pr) An adjustable tripod appliance used to hold hose nozzles in fixed position on the ground unattended.

hose coupling See: coupling

hose dryer (Fire Pr) A metal cabinet with racks and electric heater used to dry wet hose.

hose gaiter (Brit) = hose bandage

hose hoist (Fire Pr) A special hook used to raise and lower hoses, ladders, and appliances.

hose jacket[1] (Fire Pr) The woven fabric outer covering of a lined fire hose.

hose jacket[2] (Fire Pr) A hinged clamping device, more properly called a burst hose jacket, used for temporary repair of leaking couplings or burst hose.
 NOTE: A typical hose jacket for burst pipe consists of a brass casing and leather liner fastened and tightened around the damaged hose with a large wing nut. The Cooper hose jacket is used to seal leaking couplings or to temporarily join mismatched couplings.

hose jumper = hose bridge

hose lay (Fire Pr) The evolution or method used to stretch hose lines at a fire. In forward lay (also called: straight lay or hydrant-to-fire lay) the hose is stretched from the water source toward the fire. In reverse lay (also called: fire-to-hydrant lay) the hose is stretched from fire to water source. In the first case the hose-carrying apparatus stops nearer to the fire, in the other nearer to the hydrant.
 NOTE: The terms forward and reverse lay are sometimes used to indicate the deployment of hose with respect to the manner in which the hose load is stowed in the hose compartment. If the coupling that comes off first is the one required, it is called a forward lay. If a double female or double male must be used to "reverse" the hose coupling, it is a reverse lay.

hose layout (Fire Pr) The evolution or positioning of hose lines stretched at a fire.

hose line[1] (Fire Pr) Two or more lengths of hose connected and ready for use or in use.

hose line[2] (Fire Pr) A short rope used to secure hose and other objects. See: line.

hose load (Fire Pr) 1. The total quantity of hose carried on a fire truck. 2. One of several methods of folding and stowing hose in a fire truck, such as a flat hose load, accordion hose load, etc.

hoseman (Fire Pr) 1. A firefighter assigned mainly to hose-handling duties. 2. Any firefighter handling hose at a fire.

hose mask (Ind Hyg) A mask supplied with unmodified air through a hose or piping. See: respirator.

hose plug (Fire Pr) A male threaded fitting used to close off female couplings on hose lines or the auxiliary suction intakes on a pumper. Similar in function to a hose cap.

hose pulsation (Fire Pr) A throbbing effect in hose lines caused by variation in water pressure.

hose rack (Fire Pr) A rack for drying or storing fire hose.

hose ramp (Fire Pr; Brit) = hose bridge

hose reel (Fire Pr) 1. A large spool of small fire hose permanently mounted on a fire apparatus. The hose on the reel is supplied by a water tank on the apparatus, and only the required length of hose needs to be drawn off the reel. Syn: booster reel. 2. A hand-drawn spool of hose on wheels used by industrial fire brigades.

hose roll See: donut roll; potatoe roll.

hose roller (Fire Pr) A metal device with rollers used to protect hose from abrasion when it is pulled over cornices and window sills. Chafing blocks are sometimes used for the same purpose.

hose rope (Fire Pr) 1. A rope with an eye-splice or metal ring at one end and a hook at the other, used to secure hose to ladders, fire escapes, etc. Also called: rope hose tool. See: hose tool. 2. A loop of rope threaded through the eye of a hook, used for the same purpose.

hose sling (Fire Pr) An appliance made of rope or webbing, used to lift a hose or secure it to something. Cf: hose becket.

hose spanner (Fire Pr) An instrument with a C-shaped jaw used for tightening or loosening hose couplings. A spanner can be used also to open valves, as a forcible entry tool, etc.

hose station (Fire Pr) A standpipe outlet, usu equipped with a hose and nozzle. Often installed in a recessed wall cabinet with a glass door.

hose strap (Fire Pr) A strap, similar to a short belt, with a hook on one end and a hook and handle on the other, used to carry, drag, or hoist hose lines or to secure hose to ladders, fire escapes, and other objects. See: hose rope.

hose stream = fire stream

hose-stream test (Build; Testing) A fire test conducted on walls, partitions, and doors in which, after a period of fire exposure, the specimen is subjected to the impact, erosion, and cooling effects of a stream of water from a fire hose directed first at the middle and then at all parts of the previously fire-exposed face.

hose tender (Fire Pr) A fire apparatus used principally for carrying hose, and also equipped with a pump, hose connections, and turret nozzles for throwing water streams.

hose thread See: thread

hose tool (Fire Pr) An appliance such as a hose strap, hose sling, or hose rope. A rope hose tool.

hose tower (Fire Pr) A tall structure attached to a fire station and used for drying hose. Wet hose lengths are suspended vertically in the tower and are allowed to drip and air dry.

hose tray (For Serv) A 4- by 8-foot carrier platform used to transport hose loads by helicopter in forest-fire operations. A hose line can be laid directly from the tray while in flight.

hose wagon (Fire Pr) 1. A fire apparatus used primarily to carry fire hose. 2. A pumping engine used as a hose wagon.

hose washer (Fire Pr) A sleeve-shaped device that can be connected to a hydrant and used to wash hose by passing the hose through the sleeve.
NOTE: Cabinet-type washers are available that scrub hoses with detergent, rinse, and drain them automatically.

hospital (Fire Pr; Med) A building, or part of one, used continuously for medical services, obstetrics, or surgery. An institution for the care of the sick and injured.

hospitalization benefits (Insur) A plan that provides workers, and in many cases their dependents, with hospital room and board or cash allowances toward the cost of such care for a specified number of days, plus the full cost of specified services.
NOTE: Usu part of a more inclusive health and insurance program.

host (Biol) A plant or animal harboring another as a parasite or as an infectious agent.

hostile fire = unfriendly fire

hostility (Psych) The feeling of anger toward others and the desire to inflict harm or injury.

hot (Nucl Phys) Highly radioactive.

hot-air explosion See: smoke explosion.

hot cell (Techn) A shielded enclosure in which radioactive materials can be handled remotely by manipulators.

hotel (Fire Pr; Insur) One or more buildings under the same management, including inns, clubs, motels, and apartment hotels, in which sleeping and incidental living accommodations are offered for public hire.
NOTE: Hotels are ordinarily used by transients for stays of a few days or weeks. Apartment hotels provide accommodations for longer stays.

hotel raise (Fire Pr) A method of raising a fire department ladder vertically near a row of hotel windows so that persons on several floors can escape simultaneously.

See: ladder raise.

hot laboratory (Nucl Phys) A laboratory designed for the safe handling of radioactive materials.

hotline (Commun) A special telephone line reserved for use in emergencies.

hotshot crew (For Serv) A specially trained firefighting team used primarily to follow up the first attack forces.

hot spot[1] (Nucl Phys) a. A surface area of higher-than-average radioactivity. b. A part of a fuel element surface that has become overheated.

hot spot[2] (For Serv; Fire Pr) a. A particularly active part of a fire. b. A location in a forest fire that requires special attention or is particularly difficult to cope with.

hotspotting (For Serv) Halting the spread of fire at the points of most rapid spread or exceptional threat.
 NOTE: Usu the initial step in prompt control with emphasis on priorities.

hot station (Fire Pr) A particularly active fire station; one that responds to many working fires.

hotwork (Techn) Riveting, welding, burning, or other fire or spark-producing operations.

hour control (For Serv) A time standard in fire-control planning specifying the maximum time, usu from the time of fire origin to the first attack, that should limit the burned acreage to an acceptable maximum for a given unit or type of cover.

house gong See: gong

household (Demog; Soc Sci) The individuals, collectively, living together in a residence. Usu blood-related members of a family.

housekeeping (Fire Pr) Providing for the cleanliness, neatness, and orderliness of an area, maintaining everything in its proper place.

house lights (Fire Pr) Illumination lights throughout a fire station that are switched on at the patrol room or watch desk when the company or unit is turned out at night.
 NOTE: House lights are sometimes switched on by a fire-alarm headquarters operator.

house line (Fire Pr) Privately owned fire hose attached permanently to a standpipe in a building, contrasted with fire department hose attached at the time of a fire.

house man (Fire Pr) An individual who remains in the fire station to maintain the quarters and receive calls in a fire department manned largely by volunteer or call firefighters. The house man also stays in the station to receive calls when the company is out of quarters responding to an alarm.

house watch (Fire Pr) 1. A tour of duty at the watch desk. 2. The firefighter on duty at the watch desk.

Hugoniot curve = Rankine-Hugoniot curve

human engineering (Ind Hyg) The study of natural human capabilities and limitations related to machines and systems, specifically directed toward the design of machines, tools, and environments compatible with human capacities and preferences. Syn: ergonomics; biomechanics; psychotechnology.

human-factors engineering (Ind Eng) The application of the principles, laws, and quantitative relationships of anthropometry, physiology, or psychology to the analysis and design of machines and other engineering structures, so that they are suitable to ordinary human capabilities.

human relations (Man Sci; Psych) A broad area of managerial effort and research dealing with the social and psychological relations among people at work.

humectant (Chem) A hygroscopic substance, such as glycerin, that has a strong affinity for water, used as an additive to maintain moisture in foods, plastics, cosmetics, and other articles to keep them from drying out.

humidifier (Ind Hyg; Techn) A device that diffuses water vapor into the atmosphere of an enclosure or compartment.

humidify (Ind Hyg; Techn) To add water vapor to the atmosphere; to add water vapor or moisture to any material.

humidity (Meteor) The quantity of water present in a given volume of air. See also: dew point.
  NOTE: The weight of water vapor per 1000 cubic feet of air is absolute humidity. The percent ratio of the quantity of moisture in a volume of air to the total quantity the volume can hold at a given temperature and pressure is relative humidity, which can also be expressed as the ratio of the actual vapor pressure to the saturation vapor pressure.

humilis (Meteor) A small, flat cloud type. See: cloud.

humping (Transp) The switching of railroad cars in classification yards where the cars are pushed over a mound (hump). The slope of the hump provides momentum for the cars to roll as they are switched onto desired tracks.

humus (For Serv) Decomposed organic material.

hunting (Phys) Oscillation about a given value. Fluctuating about a midpoint due to instability.

hurricane (Meteor) A tropical storm with wind velocities of 75 mph or more. A hurricane is part of a cyclonic wind system that has very low pressure at the center and rapid rise in pressure toward the periphery. Winds are especially strong in the quadrant in which the circular wind and direction of motion of the hurricane are additive. See: wind.

hux bar (Fire Pr) A forcible-entry tool used for prying.

hydrant (Fire Pr) A cast metal fitting attached to a water main below street or grade level. The hydrant normally has a control valve and one or more gated or ungated threaded outlet connections for supplying water to fire department hoses and pumpers. Usu refers to the common post-type hydrant. Also called: fire plug; plug. Cf: flush hydrant.

hydrant adapter (Fire Pr) A coupling for connecting suction hose to hydrant outlets.

hydrant-cover key (Fire Pr) A tool used to lift covers off certain types of hydrant.

hydrant gate (Fire Pr) A valved connection that permits control of a hydrant outlet independently of the main control valve. A gate valve is usu attached to a hydrant outlet before the main control valve is turned on.

hydrant house (Fire Pr) A small protective structure built around a yard hydrant for storage of hose, nozzles, axes, spanners, lanterns, and other firefighting tools.

hydrant key (Fire Pr) A nonadjustable tool used to open a hydrant and turn on the valve. See: hydrant wrench.

hydrant nipple (Fire Pr) A threaded brass outlet connection on a hydrant barrel to which fire-hose couplings may be connected.

hydrant pit (Fire Pr) A recess for a hydrant below grade or street level.

hydrant stem (Fire Pr) The valve-operating rod extending from the top of a hydrant barrel to the valve near the base. The valve is usu positioned well below the frost line.

hydrant-to-fire lay = forward lay. See: hose lay.

hydrant wrench (Fire Pr) An adjustable tool used to remove caps from hydrant outlets and to turn the hydrant valve on and off. See: hydrant key.

hydrate (Chem) A compound containing water of hydration.

hydration (Chem) A reaction of a substance with water, in which the H-OH bond is not split but water modecules are incorporated into the overall molecular structure; in contrast to hydrolysis, in which the water molecule is split.

hydraulic cement See: concrete

hydraulic hoist (Fire Pr) A mechanism that supplies power by pressurized fluid to operate aerial ladders and platforms. A full hydraulic hoist is designed to raise the ladder, rotate the turntable, and extend the fly sections. Cf: hydromechanical hoist.

hydraulic operation (Fire Pr) Use of hydraulic power to operate aerial ladders, large-capacity deluge or turret nozzles, etc.

hydraulic platform (Brit) = aerial platform

hydraulic radius (Hydraul) A factor governing friction losses in a pipe, defined as the cross-sectional area of the flow divided by the wetted perimeter of the conduit cross section. The hydraulic radius of a circular pipe is 1/4 the diameter.

hydraulic ram (Fire Pr) A jacking device used to level an aerial ladder apparatus on a steep grade.

hydraulics (Phys) The branch of fluid mechanics, including hydrostatics and hydrokinematics, concerned with the study of liquids in motion. Syn: hydrodynamics.

hydrocarbon (Chem) A gaseous, liquid, or solid organic compound that contains only hydrogen and carbon.
 NOTE: The major sources of hydrocarbons are petroleum, natural gas, and coal.

hydrocortisone sodium succinate (Med; Pharm) An intravenous medicine sometimes effective against bronchospasm related to pulmonary burn injury by reducing inflammation.

hydrodynamics (Fl Mech) The branch of fluid mechanics that deals with the laws of motion of incompressible fluids and their interactions with boundaries.

hydrogen (Chem; Symb: H) A colorless, odorless, tasteless, monatonic gas that burns in air with a light blue flame at about 1085°F. The lightest element in the atomic series. It has two natural isotopes of atomic weights 1 and 2. The first is ordinary hydrogen, or light hydrogen; the second is deuterium, or heavy hydrogen. A third isotope, tritium, atomic weight 3, is a radioactive form produced in reactors by bombarding lithium-6 with neutrons.
 NOTE: The flammability limits of hydrogen-air mixtures are 4 to 74% by volume. Liquid hydrogen is classed as a flammable compressed gas by the US Interstate Commerce Commission.

hydrogenation (Chem Eng) A reaction of molecular hydrogen with organic compounds. An example is the hydrogenation of olefins to paraffins or of the aromatics to the naphthenes or the reduction of aldehydes and ketones to alcohols.

hydrogen bromide (Chem) An irritating gas, HBr, that fumes in moist air.

hydrogen chloride (Chem) A poisonous gas, HCl, that forms hydrochloric acid when dissolved in water
 NOTE: A combustion product of polyvinylchloride.

hydrogen cyanide (Chem) A poisonous gas, HCN, with a bitter almond odor.
 NOTE: In an aqueous solution, called hydrocyanic acid.

hydrogen-ion concentration (Chem; Symb: pH) The degree of acidity or alkalinity of a solution, measured on a scale ranging from 1 to 15.
 NOTE: pH is a function of the reciprocal of the hydrogen ion concentration. Strong acids have a pH near 1 and strong bases near 15. Distilled water is neutral and has a pH of 7.

hydrogen sulfide (Chem) A flammable poisonous gas, H_2S, that smells like rotting eggs.
 NOTE: The gas is released by burning substances that contain sulfur, such as wool and rubber.

hydrokinematics (Fl Mech) The motion of liquids, considered apart from the propelling force.

hydrokinetics (Fl Mech) The study of forces produced by liquids in motion, esp in pipes and through openings.

hydrology (Hydrol) The study of water, its properties, distribution, and circulation in the earth and on its surface.

hydrolysis (Chem) 1. The chemical decomposition of a compound by water. 2. A reaction between a salt and water in which an acid and base are formed, one or both of which are only slightly dissociated. 3. The reaction of water with another substance to form new substances; e.g., the conversion of starch to glucose by water in the presence of a catalyst.

hydromechanical hoist (Fire Pr) An aerial ladder mechanism that uses hydraulic power to elevate the ladder but mechanical means to rotate the turntable and extend the fly sections. Cf: hydraulic hoist.

hydrometallurgy (Metall) The science of metal recovery by a process involving treatment of ores in an aqueous medium, such as acid or cyanide solution.

hydrometeor (Meteor) Any form of precipitation, including rain, hail, snow, sleet, etc. See: precipitation.

hydrometer (Techn) A graduated tubular instrument used to measure the specific gravity of liquids by floating the tube in the liquid and reading the density value directly on a scale inscribed on the tube. The denser the liquid, the higher the instrument floats. More complex instruments incorporate electrical circuits that record readings automatically.

hydroquinone (Chem) A strong, reducing phenol in the form of colorless hexagonal crystals, $C_6H_6O_2$, used as a photographic developer and as an antioxidant.
NOTE: Contact with the skin may cause sensitization and irritation. Excessive exposure to dust may cause corneal injury.

hydrostatic pressure (Hydraul) The pressure exerted by the weight of a layer of liquid.

hydrostatics (Fl Mech) The branch of fluid mechanics dealing with equilibrium states of fluids at rest and the pressures exerted by them.

hydrothorax (Med) The presence of fluid in the pleural cavity.

hydroxyl radical (Chem) A reactive odd electron radical with the formula OH. In flames it is the major reactant in attack of carbon-hydrogen fuels.

hygiene (Med) The science of health and its maintenance. Systems and principles for the preservation of health and prevention of disease.

hygrometer (Meteor) An instrument used to measure the relative humidity of the air. The moisture-sensing elements in simple instruments are usu organic materials (hair, paper, wood, etc.) that expand and contract in response to variations in humidity. Syn: psychrometer.

hygroscopic (Chem) Capable of absorbing moisture from the air. Cf: deliquescent salt.

hypalon = chlorosulfonated polyethylene

hyperbaric (Phys; Med) Pertaining to environmental pressures above atmospheric. Ant: hypobaric.

hypercalcemia (Med) Elevated levels of calcium in the blood.

hypercapnia (Med) A condition of excessive carbon dioxide in the blood, resulting in extra-stimulation of the respiratory center and overbreathing. Also called: hypercarbia.

hypercarbia (Med) Excessive carbon dioxide in the blood. Also called: hypercapnia.

hyperemia (Med) An excess of blood in an organ or tissue. Syn: congestion.

hypergolic (Chem) Referring to substances that ignite spontaneously when mixed with each other.

hyperkalemia (Med) Excessive potassium in the blood.

hypernea (Med) Abnormally rapid or deep breathing.

hyperon (Nucl Phys) One of a class of short-lived, elementary particles with a mass greater than that of a proton and less than that of a deuteron. All hyperons are unstable and yield a nucleon as a decay product. See: baryon; elementary particle.

hyperopia (Physiol) Farsightedness.

hyperoxia (Med) A rare condition of excess oxygen in the body. Ant: hypoxia.

hypertension (Med) Abnormally high blood pressure.

hyperthermia (Med) Exceptionally high fever or body temperature.

hyperventilation (Med) Breathing at a rate higher than normal. Syn: over-breathing. Ant: hypoventilation.

hypnotic (Med; Pharm) Any agent that induces sleep or that produces a hypnotic effect. A soporific.

hypobaric (Phys) Pertaining to pressures below atmospheric. Ant: hyperbaric.

hypocalcemia (Med) Low calcium in the blood.

hypocapnia (Med) A condition of dizziness and confusion brought on by a deficiency of carbon dioxide in the blood and body tissues.

hypocarbia (Med) Low carbon dioxide in the blood.

hypokalemia (Med) Low potassium in the blood.

hypopharynx (Anat) The lowest part of the pharynx.

hypothesis (Statist) A tentative proposition concerning an unknown advanced as an explanation of its occurrence or existence. A hypothesis may be a provisional assertion to help guide an investigation or one that is widely accepted as probable in the light of known facts.

hypoventilation (Med) Breathing at a rate less than normal. Syn: under-breathing. Ant: hyperventilation.

hypovolemia (Med) Low quantity of blood in the body.

hypoxemia (Med) Deficiency of oxygen in the arterial blood. Also incorrectly called: anoxemia, which means absence of oxygen in the blood.

hypoxia (Med) Deficiency of oxygen in the body tissues. Also incorrectly called: anoxia. Ant: hyperoxia. See: asphyxia.

hysteresis (Electr; Phys) A retardation of an effect when the forces acting upon a body are changed, as if from viscosity or internal friction. Specifically, the magnetization of a sample of iron or steel lags behind the magnetic field that induced it when the field varies.

Hz = Hertz

I

I (Chem; Phys) Symbol for electric current, iodine.

IAAI = International Association of Arson Investigators

IAFC = International Association of Fire Chiefs

IAFF = International Association of Fire Fighters

iatrogenic (Med) Inadvertently caused by a doctor.

IBPFFA = International Black Professional Fire Fighters Association

ICBO = International Conference of Building Officials

ice (Meteor; Techn) The compact, solid phase of water below 0°C or 32°F. See: precipitation.
NOTE: If frozen in contact with a bulk liquid, compact solid ice is formed. Frozen rain droplets, called hail, are ice particles. In contrast, water vapor freezing directly from the gas phase forms open crystals of snow, or if in contact with a solid, lacy patterns of frost. Several other crystalline phases of ice can exist at pressures >10^3 atmospheres.

ice needles See: precipitation

ice point (Phys) The temperature at which the liquid and solid phases of a substance are in equilibrium at normal atmospheric pressure. Syn: freezing point. See also: triple point.

ID = inside diameter

ideal gas law (Phys) The equation of state, $PV = nRT$, approached by all gases at low pressure (P is pressure, V is volume, n is the number of moles, R is the gas constant, and T is the temperature). The law combines two others, the Charles law, concerning the effect of temperature on gas pressure at constant volume, and Boyle's law, concerning the effect of pressure on gas volume at constant temperature. Also called: perfect gas law. See: equation of state.

idealize To attribute perfect characteristics. To simplify or assume regularity and the like in order to gain a general understanding of a complex process or system.

idiopathic (Med) Of a disease originating spontaneously or from an unknown cause.

idiosyncrasy (Med) A special susceptibility to a particular substance introduced into the body.

idler (Transp) 1. On track-laying vehicles, the wheel at the end opposite the driving sprocket over which the track returns. It maintains track tension and reduces track skipping. 2. A freight car used to protect overhanging loads or used between carrying cars loaded with logs on bearing pieces or pivoted bolsters.

IFCA = International Fire Chiefs Association

IFPA = International Fire Photographers Association

igloo¹ (Storage) A temporary, dome-shaped storage building for explosives. Also called: dog house.

igloo² (Techn) A kiln shaped like a beehive, used for calcining ceramics.

ignitability (Comb) The ease of ignition, esp by a small flame or spark.
NOTE: Ignitability is a function of the material, its physical form and state, and sometimes the point of ignition. For example, a piece of paper is ignited more easily at an edge or corner than at the center. The same holds for shavings cut from a timber as compared to the timber itself.

ignite (Comb) To initiate combustion.

ignition (Comb) The initiation of combustion.
NOTE: Ignition usu takes place when the temperature of a fuel is raised sufficiently by some mechanism, such as a flame or electric spark. See also: autoignition; spontaneous ignition; three-stage ignition.

ignition component (For Serv) The probability that a fire will ignite if a firebrand falls in fine fuel.

ignition delay (Comb) The time interval between the instant an oxidizer contacts a combustible and ignition takes place. Also called: **ignition lag**.

ignition energy (Comb) The quantity of heat or electrical energy that must be absorbed by a substance to ignite and burn.

ignition temperature (Comb) The lowest temperature at which sustained combustion of a substance can be initiated.

ignition time (Comb) The time, in seconds, between the application of an ignition source to a material and the instant self-sustained combustion begins.

igpm = **imperial gallons per minute**

illuminance (Opt; Symb: E) The luminous flux, in footcandles, falling on a surface, expressed as E = F/A, where F is the luminous flux in lumens and A is the area.
  NOTE: The typical unit is the foot candle, the illumination falling on a surface 1 foot from a 1-candlepower source, which is equal to 1 lumen per square foot. Another expression is $E = I/d^2$, where I is the source intensity in candlepower and d is the distance to the incident surface.

illumination (Opt) The density of radiant energy falling on a surface and the effectiveness of the radiation in producing a visual effect. Illumination units are the **lumen**, **foot candle**, **phot**, and **lux**. One foot-candle = 1 lumen/ft²; one phot = 1 lumen/cm²; and one lux = 1 lumen/m². Syn: **illuminance**.

imaginary box = **phantom box**

immersion heater (Techn) An electrical device containing a resistance element that is immersed in a liquid to heat it.

immersion testing (Testing) The ultrasonic examination of a body while it is immersed in a liquid, usu water, by causing the ultrasonic energy to propagate through the liquid and into the body and vice versa.
  NOTE: This technique, which is relatively easy to automate, is useful for testing irregularly shaped parts.

imminent danger (Safety) An impending or threatening dangerous situation which could be expected to cause death or serious injury to persons in the immediate future unless corrective measures are taken.

immune (Med) Protected from or esp resistant to disease. Capable of producing antibodies to given antigens.

immunity (Med) Resistance of the body to an infectious agent.
  NOTE: Immunity can be inherited, congenital, or acquired naturally or artificially.

impact of water (Fire Pr) The force of a water stream exerted on an object or surface.

impact resistance (Ind Eng; Testing) The toughness of a material or structure to shock, often determined by a drop test. The height of fall, the weight of the item being tested, and the number of shocks before failure indicate the relative impact resistance.
  NOTE: A pendulum is used to determine the fracture resistance of materials in which a sample bar, sometimes notched, is broken by one blow.

impact strength (Testing) The resistance of a material or article to shocks such as from dropping and hard blows, usu expressed as foot-pounds of energy required to break a standard specimen.

impedance[1] (Electr) Total opposition, consisting of ohmic resistance and reactance, to the flow of an alternating electrical current.

impedance[2] (Phys) The ratio of a force-like quantity to a velocity-like quantity, such as pressure to flow rate, temperature to heat flow, or sound pressure to particle displacement.
  NOTE: The term is used similarly in

electronics, mechanics, and acoustics.

impedance coil See: choke

impeller (Techn) A vaned, rotating member of a centrifugal pump that imparts motion to the fluid. See: pump.

impeller eye (Techn) The intake opening in the center of a centrifugal pump impeller.

imperial gallon (Metrol) A British unit of liquid measure equal to 1.201 US gallons. See: gallon.

impervious (Techn) Resistant to penetration by fluids or gases.

implantation (Med) Grafting of bits of tissue, such as skin on burn wounds.

implied consent (Law; Med) An assumption that an injured, unconscious person would desire to be transported for emergency medical attention and would consent to receive treatment. See: Good Samaritan Law.

implosion (Techn) A sudden inrush of air when an evacuated container is breached or shattered. Cf: explosion.

impregnation (Techn) Penetration of a solution into the surface layer or bulk of an absorbent material.

improper fraction (Math) Any fraction in which the numerator is greater than the denominator; e.g., 5/3 or 12/5.

improvised deluge set (Fire Pr) A means of rigging a heavy stream device by using standard pumper equipment, such as suction hose, siamese connection, reducer, and playpipe.

improvised water tower (Fire Pr) A raised ladder supported by guylines or ropes, used to elevate a charged line and nozzle to reach upper stories with a hose stream.

impulse (Phys) The product of a force and the time during which the force is applied.

impulse shock (Testing) A particular type of shock for which a simple waveform can be assumed.

IMSA = International Municipal Signal Association

inby (Mining) Inward; toward the workings side of a mine. Ant: outby.

incandescence (Comb) The emission of light by a substance due to its high temperature.

incendiarism (Fire Pr; Law) An act of arson, q.v.

incendiarist (For Serv; Psych) One who sets fires in protest or rebellion against the existing community social structure and political institutions. A firesetter motivated by political and social ends, who uses fire as a means of exciting factions and disrupting the prevailing order. Cf: arsonist; pyromaniac.

incendiary^1 (Fire Pr) a. Referring to a fire believed to have been set deliberately. b. = arsonist; a person who deliberately sets fires. More properly, an incendiarist.

incendiary^2 (Milit) A flammable material or device used to set a fire, such as a flame thrower or firebomb. See: Molotov cocktail.

incendive (Fire Pr) Capable of starting a fire or likely to do so.

incentive wage system (Man Sci) A method of wage payment that relates earnings of workers to their actual production, individually or as a group.

inch^1 (Metrol) A unit of length equal to 1/12 of a foot or 2.54 cm.
NOTE: The inch was established in 1305 in the English system as a length equivalent to three grains of dry round barley laid end-to-end.

inch^2 (Techn) a. An intermittant motion of a machine by momentary operation. b. To move or be moved in small increments.

in charge (Fire Pr) 1. To be in command of a unit or over other officers and firefighters. 2. Referring to the ranking officer actively directing fire-fighting operations.

inch of mercury (Metrol) A unit of pressure equal to the pressure exerted by a column of mercury one inch high at a standard temperature. NOTE: 2.02 inches of mercury = 1 psi.

inch of water (Metrol) A unit of pressure equal to the pressure exerted by a column of water 1 inch high at a temperature of 4° C or 39.2° F.

incidence[1] (Fire Pr) Occurrence of something, esp with reference to its rate and locality.

incidence[2] (Phys) Arrival at a surface.

incident history (Fire Pr) A count of the fires that have occurred in a given area in a given period of time.

incinerate (Comb) To reduce to ashes by burning; to cremate.

incineration (Comb) The controlled process by which solid, liquid, or gaseous combustible wastes are burned and changed into gases, leaving little or no combustible material in the residue.

incinerator (Comb; Techn) A furnace-like apparatus used to burn waste substances.

incipient (Fire Pr) Referring to a small fire or one in its initial stages.

incision (Med; Surg) A smooth cut or wound made by a sharp instrument; esp one made in a surgical procedure.

inclinometer[1] (Fire Pr) An instrument mounted on an aerial ladder to indicate the angle of elevation.

inclinometer[2] (Geophys) A dip needle for measuring the geomagnetic field.

incombustible = noncombustible

incompressible flow (Fl Mech) The motion of a fluid without change in its density.

increaser (Fire Pr) An adapter used on a hose, pump, or fitting to permit connection of a large-size hose. Ant: reducer.

incubation (Bact) Holding cultures of microorganisms under conditions favorable to their growth.

incubation time (Med) The elapsed time between exposure to infection and the appearance of disease symptoms, or the time period during which microorganisms grow.

incus (Meteor) An anvil-shaped cloud. See: cloud.

independent action (For Serv) Fire suppression activities by other than regular fire control organizations or a fire cooperator. Also called: self-help.

independent failure (Ind Eng) A failure that occurs without being related to the failure of associated items.

independent gate (Fire Pr) A gate valve that controls the flow from a single outlet. The valve may be built-in or attached just before opening the main valve on a hydrant or pump, thereby permitting connection of a hose without having to shut down lines already in service.

independent suspension (Transp) A system of arms, springs, wheels, etc., for elastically supporting the sprung mass of a vehicle, which permits the deflection of any of the supporting wheels without substantially changing the load or position of the remaining wheels (as distinguished from solid axle or bogie suspension systems).

independent variable (Psych; Statist) The variable controlled by an experimenter and applied to determine its effect on a dependent variable.

index[1] (Math) An exponent that shows the power or root of a number.

index[2] (Info Sci) A list of references to a collection of data on information, such as a catalog, bibliography, and the like; usu a numerical or alphabetical arrangement of topics, part numbers, authors names, etc.

index of refraction (Opt) The ratio

of the velocity of light in a vacuum to the velocity in a given medium. Also expressed as the ratio of the sines of the angles of incidence and refraction relative to a normal to the interface for a ray passing obliquely from a vacuum into the medium.

Indian pump (Fire Pr) A trade name for a back-pack fire extinguisher with a capacity of 2.5 to 5 gallons of water and equipped with a hand-operated pump.

indicator[1] (Chem) A substance that changes color at a definite hydrogen ion (or other specific ion) concentration; e.g., litmus, starch, etc.

indicator[2] (Fire Pr) A recording instrument for the pressure-volume relationship in an engine.

indicator[3] (Metrol) a. Any meter or instrument that shows a physical value, such as temperature, pressure, etc. b. (Brit) A device that visually signals a fire alarm or other warning.

indictment (Law) A formal statement presented by a grand jury naming and accusing an individual of committing a specified crime.

indirect application (Fire Pr) A technique of throwing a water stream into a confined space to generate steam, distribute unvaporized droplets of water, and thereby cool hot materials beyond the direct reach of a fire stream.

indirect cause (Safety) A causal factor associated with an occupational accident other than medical costs and workmen's compensation payments.

indirect cost (Insur; Safety) Monetary loss, other than medical expenses and workmen's compensation payments, resulting from an accident. See: accident costs.

indirect damage (Insur; Surg) Loss resulting indirectly from a hazardous condition or incident.

indirect method (For Serv) Combatting a forest fire by establishing a control line along natural firebreaks or one remote from the fire and backfiring the fuel in the path of the fire. See: direct method.

indirect recirculating oven (Techn) An oven equipped with a gas-tight duct system, a furnace, and a circulating fan.
NOTE: Combustion gases are circulated through this closed system and mixed with fresh combustion gases generated by the burner in the combustion chamber. A vent or overflow removes a portion of the gases to allow for the fresh gases added by the burner.

individual assignment method (For Serv) A method of combatting a fire in which each crewman is assigned a specific section of a control line and is made responsible for all fire suppression tasks on that section, from hotspotting to mop-up.

individual rate (Man Sci) 1. A rate paid to a worker in an establishment without a standard wage-rate system. 2. A rate paid to an individual worker, as distinguished from the standard job rate.

individual-rung ladder (Techn) A fixed ladder, each rung of which is individually attached to a structure, building, or equipment.

indoctrination (Man Sci) 1. Instruction in the principles or methods of a job, task, or activity. 2. An educational program aimed at instilling an attitude.

induced current (Electr) An electrical current in a conductor produced by varying a magnetic field across it.

induced radioactivity (Nucl Phys) Radioactivity that is created when substances are bombarded with neutrons, for example, from a nuclear explosion or in a reactor, or with charged particles produced by an accelerator. See: activation.

inductance (Electr) A property of an electrical circuit by which an electromotive force is induced by a variation of current, either in the circuit itself or in a neighboring circuit.

induction[1] (Electr) The act or process of producing voltage by the variation of a magnetic field across a conductor.

induction² (Psych) The process of reasoning from the particular to the general, or from a part to the whole.

induction heating (Metall) Heating caused by the dissipation of high-frequency radiation in solids.
  NOTE: The method is used to heat metals by alternating currents that can be induced in conductors by an alternating magnetic field. Furnaces using this principle are used for preparing special alloys.

induction period (Comb) The period of time required by certain combustibles (such as wood, paper, linseed oil, etc.) before oxidation and burning can proceed independently of heat or energy input.
  NOTE: Depending on materials and conditions, this may vary from several milliseconds to weeks or months.

inductor (Electr) A circuit element, such as a coil, in which electromotive force is induced or affected by the magnetic field of a varying current flow.

industrial chemicals (Chem) Inorganic and organic chemicals produced primarily for use in chemical manufacture or in chemical processes.
  NOTE: Industrial chemicals are generally sold to other manufacturers or formulators. They range from relatively simple substances, such as chlorine, caustic soda, and sulfuric acid, tc complex dyes and bulk medicinals. Other industrial chemicals are used as solvents, as intermediates in further chemical manufacture, and as the basis for making synthetic rubber, plastics, and man-made fibers.

industrial engineer (Ind Eng) A management engineer who plans and oversees the use of production facilities and personnel, largely for purposes of production quality, control, efficiency, economy, and safety.

industrial fire brigade (Fire Pr; Brit) A private fire suppression organization that provides fire protection to a factory or industrial plant. Syn: works fire brigade.

industrial-health engineer (Ind Hyg) A trained individual who plans and coordinates private or government industrial health programs, with particular emphasis on hazards and diseases.

industrial hygiene (Ind Hyg) The science and practice of controlling environmental factors in a workplace that may cause sickness, impair health, or induce discomfort and inefficiency among workers.

industrial occupancy (Fire Pr) A factory or plant manufacturing one or more products or a property devoted to processing, assembling, mixing, packaging, finishing or decorating, repairing, and similar operations.
  NOTE: General industrial occupancy is one in which all manufacturing, except high-hazard operations, are conducted in conventional buildings suitable for the type of manufacture housed. High-hazard industrial occupancy is a factory, plant, or property having contents which are liable to burn with extreme rapidity or which may generate poisonous fumes, gases, or explode in the event of fire. Special-purpose industrial occupancy includes all buildings, except high-hazard occupancies, that are designed and suitable only for particular types of operations, characterized by a relatively low density of employee population with much of the area occupied by machinery or equipment.

industrial research (Ind Eng) Systematic search for new knowledge undertaken by an industrial enterprise to expand its own opportunities or performed for an outside agency to find solutions to specified problems.

industrial structures (Ind Eng) Platforms used for necessary access to operations conducted in the open air, such as in oil refining and chemical processing plants.
  NOTE: The platforms sometimes have roofs or canopies to provide some shelter, but no walls.

industrial truck (Mat Hand) A heavy-duty vehicle used to carry, push,

pull, lift, stack, or tier material.
NOTE: Powered industrial trucks may be classified by power source, operator position, and means of engaging the load. Power sources are usu electric motors energized by storage batteries and engines using gasoline, LP-gas, diesel fuel, or a combination of gas or diesel and electric.

inequality (Math) A relation in which one quantity is less than or greater than another, written respectively as a < b or b > a.

inert (Fire Pr; Chem) 1. To make a substance nonreactive. 2. The state of being nonreactive.

inert element (Chem) An element of the zero group in the periodic table. The elements in this group have no chemical reaction properties.

inert gas (Chem; Fire Pr) 1. A gas, such as CO_2 or N_2, that excludes oxygen or dilutes it below the concentration necessary for combustion. 2. The gases in the family of elements in the zero valence column of the periodic table, including helium, neon, argon, krypton, xenon, and radon.
NOTE: Although they are normally considered inert, a few compounds of krypton and xenon with fluorine have recently been discovered.

inert-gas welding (Techn) An electric welding operation using an inert gas such as helium to flush away the air to prevent oxidation of the metal being welded.

inertia (Phys) The tendency of a body at rest to remain at rest, or of a body in motion to resist a change in speed or direction. See: **Newton's laws**.

inerting (Fire Pr; Comb) Exclusion of oxygen from an enclosure or other space by displacing it with an inert gas so that ignition or combustion cannot occur.

infarction (Med) A localized region of tissue death resulting from a vascular obstruction.
NOTE: An infarction may be gradually absorbed, leaving a fibrous scar, as in the case of a heart muscle after a coronary occlusion.

infection (Med) An invasion of the body tissues by pathogenic organisms. When these organisms become established, some of them produce toxins that damage the tissues.

inference The drawing of a conclusion based on one or more previous conclusions.

inferential statistics (Statist) A set of techniques for inferring something about information obtained from a representative sample. Methods of drawing conclusions from sample data about a larger body of data.
NOTE: Inferential statistics fall into two main categories: those that give the margin of error in predicting or estimating some population measure, such as a mean, and those that permit the testing of hypotheses about populations.

inferior (Anat) Pertaining to a lower part of the body.

infestation (Med) 1. The invasion of the body surface by parasites. See: **infection**. 2. An affliction caused by parasites.

infiltration (Techn) Air flowing inward as through a wall, crack, etc.

infinite (Math) Extending indefinitely; having innumerable parts, capable of endless division within itself.

inflammable = **flammable**

inflammation[1] (Comb; Brit) Ignition of a gas or vapor.

inflammation[2] (Med) A local tenderness; usu a painful response of the tissues to injury, infection, or irritation, characterized by heat, swelling, reddening, and tenderness (rubor, tumor, calor, et dolor).

information (Info Sci) Processed data. The meaning assigned to data, or a description, extension, or elaboration of data.
NOTE: Important qualities of information to a given receiver are novelty and utility.

information analysis center (Info Sci) A service directed toward collecting technical information in a specific area of effort and its evaluation and filtering into the form of condensed data, summaries, or state-of-the-art reports, as an aid to information users.

information officer (For Serv) A staff officer of a fire suppression organization responsible for releasing information on a fire situation to the news media and the public.

information retrieval (Info Sci) Recovery of data from a collection for the purpose of obtaining information. Retrieval includes all the procedures used to identify, search, find, and recover specific information or data from storage.

information science (Info Sci) The study of the processes of identifying, generating, acquiring, storing, retrieving, disseminating, and using information; the study of the properties, structure, and transmission of information; and the development of methods for the useful organization of data and dissemination of information.

information system (Info Sci) A network of information services providing facilities by which information and data are gathered, processed, and transmitted from originator to user.

infrared (Opt; Abb: ir) Invisible heat rays beyond the long wavelength end of the visible spectrum (beyond 7600A). Wavelengths of the electromagnetic spectrum that are longer than those of visible light and shorter than radio waves (10^{-4} cm to 10^{-1} cm wavelength).
NOTE: All warm bodies radiate infrared rays. A candle flame emits ir radiation below 4 microns; oxyacetylene above 15 microns.

infrared detector (Fire Pr) A device that detects fires by sensing infrared rays emitted by the fire.

infrared heater (Techn) A device that generates and radiates infrared heat rays.
NOTE: Commonly used in restaurants to keep food warm.

infusion (Med) Administration of a fluid into a vein. Injection of a saline or sugar solution into the body intravenously or subcutaneously when liquids or nourishment cannot be given by mouth.

infusomanometer (Med) An intra-arterial blood-pressure recording apparatus that measures the back resistance to a slow intra-arterial infusion.

ingestion[1] (Physiol) The act or process of taking food, drugs, etc. into the body.

ingestion[2] (Biol) The act of engulfing or taking up bacteria or other foreign matter by cells.

ingot (Metall) A block of iron or steel cast in a mold for ease in handling prior to processing.

ingress (Build) 1. Access to the interior, such as a building. 2. The act of entering.

inhabited building (Build) A building or structure regularly used in whole or part as a place of human habitation.
NOTE: Included is any church, school, store, railway passenger station, airport terminal for passengers, and other building or structure in which people are accustomed to congregate or assemble but excluding any building or structure occupied in connection with the manufacture, transportation, storage, and use of explosives.

inhalation (Med) 1. Aspiration; breathing in. 2. Drawing in air, vapor, fumes, or other substances in the act of breathing.

inhalation injury (Med) Damage to the respiratory system caused by breathing in irritating substances.

inhalation valve (Safety) A device that allows respirable air to enter the facepiece of a breathing apparatus and prevents exhaled air from leaving the facepiece through the intake opening.

inhalator (Med) A type of breathing device that supplies a constant flow of oxygen to assist respiration.

NOTE: Used after artificial respiration has revived a victim of respiratory failure.

inherent reliability (Ind Eng) The theoretically achievable maximum reliability of a design, assuming operation in an ideal, standard, or theoretical environment (for example, a standard summer day or an ideal plant environment).

inhibition[1] (Fire Pr; Comb) The reduction or moderation of a fire or flame by chemical species that interfere with the flame reactions.
NOTE: Inhibition implies reduction in burning intensity; extinction refers to carrying inhibition to completion.

inhibition[2] (Bact) Prevention of growth or multiplication of microorganisms.

inhibitor[1] (Chem) A substance that retards or prevents normal chemical change from taking place. Syn: negative catalyst.
NOTE: Inhibitors are used to moderate reactions or to minimize undesirable ones. Antioxidants; corrosion, foam, and ultraviolet inhibitors; preservatives; fuel oil additives; and rust preventors are typical examples.

inhibitor[2] (Comb) A chemical compound that reduces the degree of intensity of burning, or one that makes ignition more difficult. See: retardant.
NOTE: Common examples are salts of alkali metals, such as sodium bicarbonate, and violatile halogenated hydrocarbons, such as trichlorobromomethane.

initial alarm (Fire Pr) The first notification received by a fire department that a fire or other emergency exists.

initial attack (Fire Pr) The first activity performed in suppressing a fire, usu aimed at preventing fire extension and safeguarding life while additional hose lines are laid and the main attack force is brought into position. Syn: first attack.

initial shift = first work period

initiation (Civ Eng) 1. The beginning of the deflagration or detonation of an explosive. 2. The first action in a fuze when it is set off. 3. In a time fuze, the starting of the action that explodes the fuzed munition.

injunction (Law) A writ issued by a court prohibiting a named individual from doing a particular thing or continuing a specified activity.

injury[1] (First Aid; Med; Safety) Physical harm or damage to a person resulting in the marring of appearance, pain or discomfort, infection, or physical impairment.
NOTE: Sometimes injury includes illness. Some particular usages: Bed Disabling Injury: one that confines a person to bed for more than half the daylight hours either on the day of the accident or on some following day. Restricted Activity Injury: one that causes a person to reduce his usual activities for a whole day. Disabling Injury: one that prevents a person from performing a regular job that is open and available to him for a full day beyond the day of the accident that produced the injury. Work-Related Injury: any injury arising in the course of employment that involves medical treatment; loss of consciousness; restriction of work or motion; or transfer to another job. These categories are further subdivided for reporting and recording purposes.

injury[2] (Law) Loss or damage suffered or a violation of rights for which compensation may be claimed under law.

inlet (Fire Pr) A water intake connection, usu the suction-hose connection of a fire department pumper.

inlet gage (Fire Pr) A gage that shows the pressure at the intake side of a pump. An inlet gage is usu a compound gage, because intake pressure may be above or below atmospheric pressure.

in-line (Fire Pr) Referring to a device or fitting inserted in series with (or parallel to the flow in) a hose line.

innocuous (Med) Harmless; not likely to injure.

inoculation (Med) Introduction of microorganisms or antigens into a living organism for the purpose of stimulating the production of antibodies.

inoculum (Med) The material or serum used for inoculation.

inorganic (Chem) Composed of matter that does not contain carbon. Of mineral origin. Referring to substances other than vegetable or animal.
NOTE: Although containing carbon, carbon monoxide and carbon dioxide are considered to be inorganic.

inorganic chemistry (Chem) The field of chemistry, divided into the branches of synthetic, theoretical, and industrial chemistry, concerned with the study of all substances other than organic compounds of carbon and hydrogen and their derivatives.

in phase (Phys) The state of coincidence of two waves of the same frequency passing through their maximum and minimum values of like polarity at the same instant.

input¹ (Techn) Energy or effort introduced into a system.

input² (Electr) The signal fed into a circuit. Ant: output.

inquest (Law) An official inquiry or judicial proceeding, esp before a jury, held to ascertain the facts related to a given matter, such as a corner's inquest to determine the circumstances of a death.

inrunning nip point (Techn) The point at which a rotating mechanism can seize and wind up loose clothing, hair, fingers, and the like; the point where two or more adjacent or contacting surfaces rotate toward each other.

in service (Fire Pr) 1. A company or other fire-fighting unit that is ready and able to respond to its assignments. Also called: on the line or on the track when the apparatus is in quarters. 2. A firefighting unit engaged in active firefighting operations at a fire as compared to one that is standing by.

in-service inspection (Fire Pr) Inspection work performed by fire companies that are in radio contact with headquarters and therefore are able to respond to fire alarms.

in-service training (Educ) A program of instruction provided by an employer to a worker during working hours. Syn: on-the-job training.

inside diameter (Abb: ID) The interior diameter of any circular opening, such as the inside of a pipe, fitting, or other hollow circular object.

inside fire fighting (Fire Pr) Fire fighting in which firefighters enter a burning building and attack the fire at close range. Ant: outside fire fighting.

insolation (Meteor) The quantity of solar energy reaching a unit area of the earth's surface.

insoluble (Chem) A substance which is incapable of being dissolved.

inspect (Ind Eng) To view closely and critically; to examine; to scrutinize and evaluate.

inspection (Fire Pr) An on-site survey of buildings, structures, and neighboring features conducted by fire department personnel in their areas of jurisdiction to familiarize firefighters with occupancies and locations in which fires could occur.
NOTE: Inspection is carried out to identify and eliminate both common and special fire hazards and to enforce fire prevention ordinances and regulations.

inspection bureau (Insur; Fire Pr) A state or regional fire insurance office concerned with community and property fire protection problems. Inspection bureaus and fire departments cooperate in maintaining approved fire prevention standards. Also see: fire prevention bureau.

inspection district (Fire Pr) An area assigned to a fire company or other unit to carry out routine fire department inspections.

inspiration (Physiol) The process of drawing air into the lungs.

instantaneous radiation (Nucl Phys) The radiation emitted during the fission process.
NOTE: Instantaneous radiations are frequently called prompt gamma rays or prompt neutrons. Most of the fission products continue to emit radiation after the fission process.

instantaneous value (Techn) The magnitude at any particular instant when a value is continually varying with respect to time.

instinct (Psych) An inborn characteristic tendency toward a certain behavior.

institutional occupancy (Fire Pr; Insur) A building or part of one used for medical care or other treatment of persons suffering from physical or mental illness, disease, or infirmity; for the care of infants, orphans, convalescents, or aged persons; or for detention of persons for penal or corrective purposes. See also: occupancy.

instruction¹ (Comput) A coded word or part of a word used in a computer to cause some operation to be performed.

instruction² (Educ) The practice of teaching.

instructor (Fire Pr) A fire department training officer, usu certified by an educational institution to give training in fire-related subjects.

instrument (Techn) 1. A device for measuring or recording data. 2. A tool or other means for accomplishing a task, esp one that requires precision.

instrumental analysis (Chem) Chemical analysis and physical measurements made by optical, electrical, thermal, or nuclear instruments.

instrument shelter (For Serv) A structure built to protect meteorological instruments from direct sunlight and precipitation. Syn: thermoscreen.

insulating board (Build) A low-density thermal insulating building material, such as Celotex (trade name), usu made of wood or sugarcane fibers.

insulation¹ (Electr) A nonconducting (dielectric) material used to isolate electrical conductors to prevent short-circuits, grounding, and injury. Also called: electrical insulation.

insulation² (Thermodyn) A nonconductor of heat used as a barrier to deter heat from escaping from a high-temperature system or flowing into a low-temperature system. Also called: thermal insulation.

insulin (Med; Pharm) 1. A hormone secreted by the pancreas, necessary for carbohydrate metabolism. 2. A medication used to treat and control diabetes.

insurable interest (Insur) Ownership, financial investment, or other legal interest that establishes a basis for insurability.

insurance (Insur) 1. Financial indemnity for a specified loss or injury to life and body, property, health, and the like, due to a given contingency or peril, or protection against liability for negligence and similar adverse legal judgements. 2. The business of insuring people or property. 3. The amount of money for which something is insured.

insured (Insur) The person protected by an insurance policy.

insured costs (Insur) Accident losses that are covered by workmen's compensation, medical, or other insurance programs.

intact stability (Transp) Stability of an operable, roadworthy vehicle.

intake screen (Fire Pr) A mesh filter used to prevent foreign material or objects from entering a pump.

integer (Math) Any whole number, such as 0, 1, 2, 5, etc., or negative thereof.

integral (Math) Pertaining to integers.

integrate (Math) 1. To bring together. 2. To perform the function of integration.

integration¹ (Math) Addition of infinitesimally small parts together by the methods of integral calculus.

integration² (Fire Pr) See: consolidation.

integrity (Build) The quality or state of wholeness; of being complete and unimpaired. Solidity.
NOTE: Integrity fails when cracks or fissures appear, as in a wall, allowing flame or hot gases to pass through to ignite materials on the other side.

integument (Anat) The skin, hair, and nails that comprise the outer covering of the body.

intelligence (Educ; Psych) The ability to reason, grasp relationships, to learn quickly, to adapt effectively to new situations, and to solve abstract problems.
NOTE: Often related to scholastic aptitude.

intelligence officer (For Serv) A staff officer of a fire suppression organization responsible for gathering and compiling all information required to develop the suppression plan for controlling a fire.

intensity¹ (Phys) The quantity of energy or the number of photons or particles of any radiation incident upon a unit area or flowing through a unit of solid material per unit of time. In connection with radioactivity, the number of atoms disintegrating per unit of time. See: flux.

intensity² (For Serv) The rate of heat release per unit length of fire front.

intercept (Math) The measured distance between the point where a graph crosses one of the coordinate axes and the origin of a coordinate system.

intercom = intercommunication system

intercommunication system (Commun) A two-way communication system with a microphone and speaker at each station.

intercostal (Anat) Referring to the space between adjacent ribs.

intercostal muscle (Anat) Muscles running from rib to rib.

interface¹ (Chem; Phys) The surface of contact between two immiscible phases.
NOTE: The phases may be solid-solid, liquid-liquid, liquid-solid, solid-gas, or liquid-gas. Gas-gas interfaces exist only temporarily.

interface² (Comput; Electr) The point at which two units or systems are joined so as to interact or communicate with each other; for example, the connection or coupling of a device or machine to a computer.

interference¹ (Opt; Acoust) The additive and subtractive effects between two superimposed sound or light waves.

interference² (Commun) Extraneous signals or noise that masks or drowns desired signals.

interference³ (Educ; Psych) A conflict between incompatible associations in learning or memory.

interferogram (Opt) A photograph of an optical-interference pattern.

interferometry (Opt) The technique of splitting a light beam, passing the parts along different paths, and reuniting them to produce a pattern of light and dark called interference fringes. Since they travel different paths, the components of the reunited beam are alternately in and out of phase, which additively and subtractively produces the fringes.
NOTE: The principle is used to determine exact distances, compare wavelengths, study density gradients, measure fluid flow rates, and investigate many other phenomena.

interior court See: court

interior cushioning (Transp) Animal and vegetable fibers, corrugated board, excelsior, glass wool, expanded mica, diatomaceous earth, shredded paper, foam rubber, foam plastics, creped cellulose wadding, indented chipboard, sawdust, or shavings used to protect goods from shocks.

interior finish (Build) The material of walls, fixed or movable partitions, ceilings, and other exposed interior surfaces of buildings, including plaster, paneling, wood, paint, wallpaper, floor and ceiling tiles, and the like. Interior finish implies more or less permanent attachment, in contrast to decorations, such as hanging pictures, rugs, draperies, and furnishings.
NOTE: Interior finishes are classified with respect to their surface burning characteristics as tested in the 25-ft tunnel fire test. Class A: Any material classified at 25 or less. Class B: Any material classified at more than 25 but not more than 75. Class C: Any material classified at more than 75 but not more than 200. Class D: Any material classified at more than 200 but not more than 500. Class E: Any material classified at over 500. See: flame spread classification.

interior hung scaffold (Build) A scaffold suspended from a ceiling or roof structure.

interlock (Safety) 1. A device actuated by the operation of some other device to govern succeeding operations. 2. A device that prevents the operation of a machine while a cover or door of the machine is open or unlocked, and which will also hold the cover or door closed and locked while the machine is in operation. An elevator, for example, will not move until the door is closed. 3. A device that interrupts power until certain predetermined conditions are met.

intermediate (Chem) A chemical formed as a middle step in a series of chemical reactions, esp in the manufacture of organic dyes and pigments.
NOTE: In some cases the intermediate may be isolated and used to form a variety of desired products. In other cases, the intermediate may be unstable or may be consumed in the reaction.

intermediate package (Transp) An interior container that holds two or more unit packages of identical items.

intermediate sill (Transp) The main longitudinal members of the car underframe between the side sills and the center sills.

intermittent flame-exposure test (Comb) A flame test in which specified gas flames are applied to a roof specimen from 3 to 15 times, depending on the classification of the roofing material.

intermolecular forces (Phys) Forces of mutual attraction or repulsion acting between molecules. Separation and other conditions cause deviations from the ideal gas law and affect transport properties.

internal energy (Thermodyn; Symb: u or E) The energy attributed to a given state of a system, excluding kinetic energy and the energy of external fields.

internal friction (Soil Mech) The portion of shear strength of soil that is proportional to the normal stress on the shearing surface.

internal motion resistance (Transp) Resistance to motion of a vehicle due to forces acting within and upon the vehicle, such as friction between moving parts, hysteresis, inertia, vibrations, etc.

internal respiration (Physiol) Gas exchange across cell walls and capillary membranes. Oxygen and nutrients diffuse into the cells from the capillary blood while carbon dioxide diffuses out into the capillaries.

interpolation (Math) A procedure used with tabulated functions, such as logarithms and trigonometric functions, to find intermediate values between those given in the table. Tables of values often contain columns showing proportional parts for use in interpolation. Ant: extrapolation.

interstice (Med; Phys) A small space within a phase or between parts and particles. A small interval or gap within a structure or tissue.

interstitial (Biol) 1. Pertaining to the small spaces between cells or structures. 2. Occupying the interstices of a tissue or organ. 3. Designating connective tissue occupying spaces between the functional units of an organ or a structure.

interstitial implants (Med) Solid or encapsulated radiation sources, made in the form of seeds, wire, or other shapes, to be inserted directly into tissue that is to be irradiated.

interstitial space (Fire Pr; Build) Concealed interconnecting passages capable of conducting heat, smoke, and flame from one part of a building to another.

interval[1] The time between two events or the distance between two objects.

interval[2] (Statist) A class interval in a frequency distribution.

intortus (Meteor) A twisted cirrus cloud type. See: cloud.

intoxication (Med; Toxicol) Poisoning, for example, by a drug, alcohol, or other toxic substance. Esp drunkenness produced by over-indulgence of alcohol. See: blood alcohol.

intracranial (Med) Situated in the interior of the skull.

intraperitoneal (Med) Inside the membrane (the peritoneum) that lines the interior wall of the abdomen.

intravenous (Med) Into or inside the vein.

intubation (Med) Insertion of a plastic or rubber tube into the larynx, trachea, or through the nose or mouth to permit breathing when the larynx swells shut. This could occur from an inhalation injury. Cf: tracheotomy.

intumescent paint (Build) A type of coating applied as a paint to a surface to protect it from flame or heat. Such paints may contain sodium silicate or other materials that expand in thickness when exposed to high temperature and thus tend to insulate and protect the surface underneath. A coating that produces an insulating, flame-resistant foam upon exposure to heat.

invasive infection (Med) Infection due to invading organisms. Beginning or initial attack by a disease.

inventory management (Man Sci) The activity involved in determining levels of materials and supplies and ensuring that they are readily available as needed to support the work and productivity of an organization without delay or interruption.

inverse proportion (Math) A relationship between two variables such that an increase in one is accompanied by an equivalent decrease in the other, and the product is therefore always constant.

inverse square law (Phys) A law that describes the decrease in radiation intensity over distance. When energy is propagating through space, its intensity is inversely proportional to the square of the distance it has traveled. An object 3 feet away from an energy source receives 1/9 the energy of that received by an identical object 1 foot away; 4 feet away, 1/16; and so on.

inverse thermal diffusion (Phys) The transport of energy in a system due to a concentration gradient. Also called: Soret effect. See: transport processes.

inversion[1] (Meteor) A stable layer of air in which temperature abnormally increases with height.
NOTE: A coastal inversion or marine inversion occurs when cool moist air from the ocean spreads over low coastal lands, forcing the dryer, warmer land air upward. Air cooled at night at the earth's surface causes a surface inversion or night inversion.

inversion[2] (Math) 1. A transposition of numbers in a series. 2. A change in the direction of a curve.

inverted siphon (Techn) A curved pipe used to transfer liquids from a higher to a lower elevation. The pipe can be used to convey a liquid from a source, pass it underneath an obstruction, and deliver it on the other side.
NOTE: Such a siphon will operate as long as the outlet is below the surface level of the liquid in the

source. In contrast to a regular siphon, an inverted siphon does not require priming.

investigation (Fire Pr) 1. Inquiry into the presence or suspicion of smoke, heat, odor, or other indication of a possible fire by a fire department officer or detail. 2. Examination of a fire reported as extinguished or found extinguished on arrival. 3. Inquiry into a fire of suspicious origin to determine cause and circumstances, or to verify the possibility of arson.

investigator (Law; Safety) An individual employed to verify the circumstances of an event, to verify the truth or accuracy of statements, to gather evidence and information, or to determine compliance with laws or regulations.

involved (Fire Pr) 1. The state of being enveloped in flame and smoke. 2. Referring to the part or whole of a building or structure on fire.

ion (Chem; Phys) An atom or molecularly bound group of atoms that has gained or lost one or more electrons and is thus positively or negatively charged. Often occurring in solutions of electrolytes.
 NOTE: Examples are an alpha particle, which is a helium atom minus two electrons, and a proton, which is a hydrogen atom minus one electron.

ion exchange (Chem) A chemical process involving the reversible interchange of various ions between a solution and a solid material, usu a plastic or a resin.
 NOTE: The process is used to separate and purify chemicals, such as fission products, rare earths, etc., in solutions.

ion-exchange resin (Chem Eng) Synthetic resins containing active groups that give the resin the property of combining with or exchanging ions with a solution.

ionization (Chem; Phys) The process in which a neutral atom or molecule gains or loses a charge, usu by absorption of radiation, through solution in a liquid, by collisions, by electrical discharges, or by heating.

ionization chamber (Nucl Phys) An instrument that detects ionizing radiation by measuring the electrical current that flows when radiation ionizes the gas in the chamber and makes it electrically conductive. Cf: chemical dosimeter; film badge; Geiger counter.

ionization constant (Chem) The product of the concentration of the ions in a solution divided by the concentration of the unionized molecules of the solute.
 NOTE: For practical purposes it is a constant under given conditions, and its measured values can be used to calculate concentrations.

ionization detector (Fire Pr) A device that detects combustion products and actuates an alarm and/or an extinguishing system. More correctly, an ionization chamber smoke detector.
 NOTE: Typically an alpha-ray detector with radium^{226} or americium^{241} as the ionizing source. Combustion products entering the chamber of the detector lower the current flow between two plates, and associated circuitry generates a signal when the current flow drops to a predetermined value.

ionization potential (Chem) The energy per unit charge for a given kind of atom required to remove an electron from the atom to an infinite distance. The potential is expressed in electron volts.

ionizing radiation (Nucl Safety) Any radiation that displaces electrons from atoms or molecules and produces ions.
 NOTE: Alpha, beta, gamma radiation, x-ray, neutrons, and short-wave ultraviolet light are included in ionizing radiation. Ionizing radiation may produce severe skin or tissue damage.

ionosphere (Meteor) A layer of ionized air in the upper atmosphere. The ionosphere is the reflecting surface for radio waves and is therefore important in communications. See: atmosphere. Syn: Kennelly-Heaviside layer.

IRFAA = International Rescue and First Aid Association

iris (Anat) The pigmented annular muscle around the pupil that controls the quantity of light entering the eye.

iron oxide fume (Metall; Toxicol) A noxious emission from cutting and welding of ferrous metals.

iron pipe thread standard (Techn; Abb: IPS) A thread standard for rigid pipes.
  NOTE: IPS threads are finer and more difficult to connect than American National Standard threads. Appliances that have American iron pipe thread require adapters to make connections with appliances that have standard threads.

irradiation (Phys) Exposure to radiation.

irrespirable (Physiol) Unfit for breathing safely; referring to a noxious or toxic atmosphere.

irritant (Med; Toxicol) A substance or stimulus that causes tissue inflammation or similar adverse reaction.
  NOTE: A primary irritant is one that has been found to produce an irritating effect at the area of the skin contact. Although irritants affect everyone, they do not produce the same degree of irritation.

IR window (Opt) A transparency in a medium to a band of infrared frequencies.
  NOTE: Important windows exist in the atmosphere that pass wavelengths between 3.6 and 4.7 microns and 8 to 13 microns. Water vapor and carbon dioxide are relatively opaque to infrared.

ischemia (Med) A decrease in blood flow to an organ or tissue due to a constriction or occlusion in an artery.
  NOTE: A condition that leads to hypoxia and possible necrosis.

isentropic flow (Fl Mech) Flow of gases in which the entropy of the fluid remains constant.

isentropic process (Phys) A process that takes place without change in entropy.

ISFSI = International Society of Fire Service Instructors

island (For Serv) An unburned area within a burning forest.

iso- A prefix meaning equal.
  NOTE: For example, an isobar is a line connecting points of equal pressure on a chart; isotherm, points of equal temperature; etc.

ISO = International Standardization Organization

isobar[1] (Nucl Phys) One of two or more nuclides having about the same atomic mass but different atomic numbers, hence different chemical properties. Cf: isomer; isotope.

isobar[2] (Meteor) A line representing constant pressure on a meteorological chart or map.

isobaric process (Thermodyn) A process that takes place without change in pressure.

isodose curves (Nucl Safety) Curves or lines drawn to connect points where identical radiation intensities reach a certain depth in tissue.

isogenic (Med) 1. Related to a group of individuals or strain of animals genetically alike with respect to gene pairs. 2. From oneself, such as skin grafts taken from one part of the body for application to another part. Ant: nonisogenic.

isointensity contours (Nucl Safety) Imaginary lines on the surface of the ground or water, or lines drawn on a map, joining points in a radiation field that have the same radiation intensity at a given time.

isolated Not readily accessible; without special means of access.

isolated pitting (Metall) Small defects, usu caused by corrosion, that do not effectively weaken a metallic plate but are indicative of possible penetration. Since the pitting is isolated, the original metal is essentially intact.

isolation (Mech Eng) Shock and vibration decoupling of a component or assembly from its mounting by

239

means of a resilient mounting system.

isomer (Chem) Either of two substances that have the same number and kinds of atoms but in which the atoms are arranged differently within the respective molecules, giving the substances different physical and chemical properties.
 NOTE: A nuclide in the excited state and a similar nuclide in the ground state are isomers. Cf: isotope.

isomerism (Chem) The existence of two or more compounds with the same molecular composition (molecules containing the same number and kinds of atoms) but different molecular structure.

isometric (Phys) Having or pertaining to equality of measure.

isomorphism (Chem) The relation of two or more different substances having closely similar crystal structures, molecular spacing, and chemical composition.

isoteniscope (Chem; Phys) An instrument for measuring vapor pressure.

isotherm (Meteor) A line on a chart joining points of equal temperature.

isothermal atmosphere (Meteor) A mass of air in which temperature does not vary with altitude.

isothermal process (Thermodyn) A process in which temperature remains constant.

isotope (Nucl Phys) Either of two or more atoms with the same atomic number but with different atomic weights; that is, the nuclei of each have the same number of protons but different numbers of neutrons. Isotopes often have nearly the same chemical properties but somewhat different physical properties. Cf: allotrope.

isotope separation (Nucl Phys) The process of separating isotopes from one another, or changing their relative abundances, for example, by gaseous diffusion or electromagnetic separation. All separation systems are based on the mass differences of the isotopes.

isotropic (Phys) Having the same properties, such as conductivity, elasticity, etc., regardless of the direction of measurement.

isotropy (Phys) The state of having one or more properties identical in all directions.

J

j (Phys) Abbrevation for joule.

J (Chem) Symbol for iodine (Europe).

jack¹ (Techn) A mechanical device used for lifting heavy objects.

jack² (Fire Pr) A ground jack or outrigger jack extended from the side of an aerial ladder or aerial platform truck to brace and stabilize the chassis when the ladder or platform is raised.

jack³ (Commun) A connector plug on a communications switchboard.

jacket See: hose jacket

jackknife¹ a. A folding pocket knife. b. A folding action of something upon itself.

jackknife² (Fire Pr) To position an aerial ladder truck tractor at an angle to the trailer to increase its stability when the ladder is raised, esp when the ladder is used in cantilever position (fly unsupported). See also: jack; wheel chocks.

jackknife³ (Transp) Accidental doubling of a tractor and trailer truck resulting from a skid.

jack pad (Techn) A foot plate for a jack, used to spread the load on the ground.

Jacob's ladder (Navig) A flexible marine ladder of rope or chain with wooden or metal rungs.

jamb (Build) The vertical side framing element of a wall opening.

jaundice (Med) A serious symptom of disease that causes the skin, the whites of the eyes, and sometimes the mucous membranes to turn yellow. Syn: icterus.

J-bolt (Techn) An open-end fitting with threaded ends, used to secure wire or band ties, or as a direct tie-down securement.

JCNFSO = Joint Council of National Fire Service Organizations

jet (Aerodyn) A high-velocity gas or liquid stream moving in a relatively stationary fluid medium.
NOTE: Jets are formed by fluids under pressure passing through nozzles and orifices. Shear forces acting on the surface of a jet cause turbulence, which eventually breaks up the jet.

Jet-Axe (Fire Pr) A trade name for an explosive forcible entry tool used to cut openings in doors, walls, and roofs.

jetsam See: jettison

jettison (Insur) To throw part of the cargo or gear of a vessel overboard to lighten the load and save the vessel.
NOTE: The loss is usu apportioned. The owner of the jettisoned goods is entitled to a general average, i.e., his loss is shared by the owners of the vessel and the owners of the cargo that was not thrown over. Sacrificed cargo that sinks is called jetsam; that which floats and which may be cast up on shore is called flotsam.

jib (Mat Hand) An extension attached to a boom to provide added length for lifting loads.

jig¹ (Techn) A machine for dyeing piece goods.
NOTE: The cloth, at full width, passes from a roller through the dye solution in an open vat and is then wound on another roller. The operation is repeated until the desired shade is obtained.

jig² (Mech Eng) A device used for accurately guiding and locating tools during machining operations.
NOTE: In contrast, a fixture holds the work but does not guide the tool.

job analysis (Man Sci) Systematic study of a job to discover its specifications, its mental, physical, and skill requirements, its relation to other jobs, etc., usu for employee selection, wage setting, or job simplification purposes.

job classification (Man Sci) Grouping of tasks in an establishment or industry into job or occupation categories, rated in terms of skill, responsibility, experience, training, and similar considerations, usu for wage-setting purposes. Job class may also be used in reference to a cluster of jobs of approx equal worth.

job description (Man Sci) A written statement listing the elements of a particular job or occupation; e.g., purpose, duties, equipment used, qualifications, training, physical and mental demands, working conditions, etc. Also called: position description.

job evaluation (Man Sci; Safety) Determination of the relative importance or ranking of jobs in an establishment for wage-setting purposes by systematically rating them on the basis of hazards involved and the requirements or qualifications of those who are to perform the jobs.

job mobility (Man Sci) Pertaining to the ability or freedom to change from one job to another. Syn: occupational mobility.

job placement (Man Sci) The assignment of an employee to a job on the basis of ability, education, experience, interests and the like, as well as the requirements of the job.

job safety training (Safety) Training associated with or emphasizing the safety aspects of a job and the hazards of tasks and their interrelationships within a job.

joint[1] (Mech Eng) The point or surface of contact between mechanical parts.
NOTE: Joints may be temporary (e.g., interlocking, snapped, screwed, bolted) or permanent (e.g., glued, mortared, welded, riveted). Friction joints are held by nails, clips, etc., and shrink or press fits. Articulated joints are variations of the hinge. See also: universal joint.

joint[2] (Anat) The point of union between two or more bones.

joint compound (Techn) A nonhardening substance used to seal pipe connections.

joint probability (Statist) The likelihood that two random events or variables will occur or assume specific values within a given discrete sample space.

joist (Build) A structural member, often spanning beams or walls, used to support a floor or ceiling. A timber two to five inches thick and four or more inches wide. A heavy joist is eight or more inches wide.

joker (Fire Pr; Colloq) A small fire alarm bell in a fire station watch room. Syn: tapper.

Jones snap coupling (Fire Pr) A trade name for a nonthreaded snap-type fire hose coupling.

joule (Phys; Symb: j) A unit of energy or work equal to the force of 1 newton acting through a distance of 1 meter. One joule = 10^7 ergs = 1 watt/sec.
NOTE: In laser technology, joule per square centimeter is a unit of energy density used to measure the amount of energy per area of absorbing surface or per cross section area of a laser beam. It is a unit for predicting the damage potential of a laser beam.

Joule's law (Phys) A heat-power relationship in which the internal energy of a given quantity of gas depends only on temperature and not on volume or pressure.

jounce (Transp) The downward movement of the sprung mass of a vehicle toward the unsprung mass in response to suspension system disturbances. Ant: heave.

jounce distance (Transp) The maximum downward travel of the sprung mass of a vehicle toward the unsprung mass, measured from the free standing position, before downward deflections of the suspension mechanism are rigidly restrained.

journal[1] (Mech Eng) The part of a shaft in contact with a bearing.

journal[2] (Fire Pr) A day book or diary-type record kept at the watch desk for noting events and activities of a fire-department company.

journal³ (Info Sci) A periodical publication, esp by a learned or professional society. See: literature, technical.

jumping sheet (Brit; Obsolete) = life net

jumping cushion (Fire Pr; Brit) An inflatable mattress used to catch a person jumping from a height.

jump spot (For Serv) A designated landing area for smoke jumpers assigned to fight a forest fire.

junior aerial (Fire Pr) 1. A small, 55-foot aerial ladder or 65-foot service aerial. 2. The truck carrying such ladders. See: aerial ladder.
NOTE: In contrast to large, tractor-trailer, aerial ladder trucks, a junior aerial is not articulated and does not require a tillerman.

junior captain (Fire Pr) A fire company platoon officer equivalent to lieutenant. See: ranks and titles.

junior officer (Fire Pr) A company officer rank, including lieutenants and sergeants, but sometimes also chiefs with little seniority in grade. See: ranks and titles.

junk (Techn) Unprocessed materials suitable for reuse or recycling.

jurisdiction (Law) The bounds or limits of the exercise of authority or responsibility.
NOTE: Jurisdictions may be delimited by geographic or political boundaries, by type, monetary values, physical characteristics, such as size or capacity, nature of usage, or by any of a variety of other criteria.

just cause (Man Sci) Fair and adequate reasons for discipline.
NOTE: Commonly included in agreement provisions safeguarding workers from unjustified discharge or unfair punishment.

jute (Techn) A natural fiber obtained from an Asiatic plant, used extensively for coarse fabrics (e.g., burlap) and twine.
NOTE: The fiber is not particularly strong, and deteriorates rapidly when wet.

juvenile court (Law) A court that deals with offenses committed by minors; one having special jurisdiction over delinquent and neglected children.

K

__k__ Abbreviation for kilogram, knot.

__K¹__ (Chem; Phys) Symbol for potassium, Kelvin, thermal conductivity.

__K²__ (Comput) A unit of computer storage capacity equal to 1024 bytes.

__Kalamein__ See: __metal-clad fire door__

__kapok__ (Techn) A light, silky, water-resistant fiber obtained from silk tree pods, used as stuffing in mattresses, pillows, furniture, cushions, and life jackets.

__Karman vortex street__ (Fl Mech) A double row of eddies in the wake of a cylinder at right angles to a fluid stream.
 NOTE: The eddies shed alternately from one side of the cylinder and then the other and can be observed when steady winds blow past smoke stacks, bridges, and other objects. Vortex streets produce singing wires (__aeolean tones__) when blowing past power transmission lines.

__katabatic wind__ (Meteor) A drainage wind caused by gravitation of cold air off high ground. See: __wind__.

__katherometer__ (Brit) = __caleometer__

__keel__ (Navig; Transp) Principal fore-and-aft member of the vehicle framing, extending along the bottom centerline.

__Kelly day__ (Fire Pr) A rotating off-duty period in addition to the normal off-duty schedule of a fire department platoon.

__Kelly tool__ (Fire Pr) A claw-type forcible-entry tool with a chisel at one end and an adz blade at the other. See: __forcible entry tool__.

__keloid__ (Med) Thick, dense, scar formation produced as a result of skin damage in certain individuals; tending to regrow after removal.

__Kelvin temperature scale__ (Abb: °K) A scale suggested by Lord Kelvin (William Thompson) in 1892. He calculated that molecular motion ceases at −273.16°C. He set this temperature at 0°K. A Kelvin degree is equal to a Celsius degree. Also called: __Celsius absolute__. See: __absolute zero__; __temperature scale__.

__Kennelly-Heaviside layer__ See: __ionosphere__

__keratin__ (Physiol) Sulfur-containing, fibrous proteins that form the chemical basis for epidermal tissues; found in nails, hair, feathers.

__keratitis__ (Med) Inflammation of the cornea of the eye.

__keratogenic agent__ (Med) A substance or factor that stimulates abnormal growth in the horny outer layer of the skin and may cause tumors.

__kerosine__; __kerosene__ (Chem; Techn) A petroleum distillate, bp 350 to 550°F, mp −25°F, used as a fuel for illumination, cooking, heating, and internal combustion engines (including jet aircraft power plants). Syn: __coal oil__; __paraffin oil__ (Brit; obsolete).

__ketone__ (Chem) An organic compound in which two organic radicals are bonded to a CO group in the molecule. Cf: __acetone__.
 NOTE: An example is methyl ethyl ketone, $CH_3COC_2H_5$.

__kev__ (Phys) Abbreviation for kilo electron volts. A unit of energy equal to 1000 electron volts.

__key event__ (Safety) The event in a series of events in an accident that determines the exact time, place, type, and extent of consequence of an accident.

__K-factor__ (Thermodyn) The thermal conductivity of a material expressed at Btu per sq ft per hour per degree Fahrenheit per inch. See: __thermal conductivity__.

__kickback__ (Safety) A sudden, forceful reaction of a piece of material (usu wood) being cut by a circular saw that is rotating in the opposite direction from the material as it is fed into the saw.
 NOTE: Since the saw blade is rotating at high speed, the force that can be exerted on the material is substantial and can cause serious

injury to the operator. A kickback occurs when the saw blade is dull, the material binds behind the cut, and the tangential force of the saw is greater than the operator can exert on the material.

kickout (Civ Eng) Accidental release or failure of a shore or brace.

kilo (Metrol) Short for **kilogram**.

kilo- A prefix meaning one thousand; e.g., a kilogram equals one thousand grams.

kilocurie (Nucl Phys) A unit of radioactivity equal to 1000 curies.

kilogram (Symb: kg) A metric unit of mass equal to 1000 grams. A standard unit of mass in the SI system equivalent to 2.20462 lb.
NOTE: The prototype kilogram is a platinum-iridium cylinder that is kept in a vault at Sevres, France, by the International Bureau of Weights and Measures.

kilometer (Metrol) A distance of 1000 meters; approx 0.62 or 5/8 of a mile.
NOTE: A speed of 35 mph is about 55 kmph.

kiloton energy (Nucl Phys) The energy of a nuclear explosion which is equivalent to that of an explosion of 1000 tons of TNT.

kilowatt hour (Electr; Abb: kWh) The energy in a current of 1000 amperes at 1 volt in one hour. A standard commercial unit of electricity.

kindle (Comb) To start a fire, esp by using easily ignited materials to set fire to heavier fuels.

kindling Wooden sticks of small cross section that require a greater ignition source than a match but that can be ignited easily with paper or other combustible material. Burning kindling develops sufficient heat to ignite larger pieces of wood or other solid fuel.

kindling temperature (Comb) The lowest temperature at which a substance ignites. Kindling temperature often depends on the physical state of the fuel. A finely divided material usu ignites more readily than a massive form of the same fuel.

kinematics (Mech) The branch of mechanics concerned with the possible motion of mechanical parts or points, disregarding the forces acting upon them.

kinesthesis (Physiol) The sense by which muscular motion, weight, and positions of the limbs and body are perceived; a sense mediated by the muscles, tendons, joints, and the inner ear.

kinetic energy (Phys) The energy of motion, equal to one half the mass times the velocity squared.

kinetics[1] (Chem) The branch of physical chemistry dealing with chemical reaction rates, mechanisms, and activation energies under nonequilibrium conditions.

kinetics[2] (Mech) The branch of mechanics based on **Newton's laws** that deals with the motions of bodies and the forces acting upon them.

kink (Fire Pr) 1. To bend hose back upon itself to temporarily block the flow of water. 2. A chance bend or fold in a hose that restricts the flow of water.

kip (Metrol) A unit of load on a structure, equal to 1000 pounds.

kit (Fire Pr) 1. Referring to fire department apparatus and appliances as a whole. 2. The particular apparatus to which a firefighter is assigned. 3. A set of personal firefighting tools and equipment.

knapsack tank (Fire Pr; Brit) A small, portable tank of up to 4 gal capacity, carried on the back, and used to feed foam compound to a small foam-making branch. Cf: **back pack**.

knee jerk (Med) An involuntary kick reflex that occurs when the tendon below the knee cap is struck.

knock down (Fire Pr) 1. To reduce flame and dissipate heat for the purpose of preventing fire spread. 2. The initial phase of a firefighting operation in which flame and heat are markedly reduced in order to bring the fire under control.

knot¹ (Fire Pr) A tie or fastening formed by interweaving flexible ropes or cords for the purpose of lifting, binding, pulling, or securing objects.
  NOTE: Knots are used to make a loop, noose, knob, tie a bundle, or join one line to another. There is a large variety of knots, some made for purely decorative purposes, but the most important practical kinds are hitches, reefs, bends, bights, and splices. See: line.

knot² (Navig) A unit of ship speed. One knot = 1 nautical mile (6076.115 feet) per hour.

Knudsen number (Fl Mech) The ratio of the mean free path of a molecule to the characteristic length of the structure in a fluid stream. When the Knudsen number is ≤0.01, the flow can be treated by gas-dynamic methods; when ≥1, it conforms to the laws of rarefied gas dynamics.

Kortick tool (For Serv) A combination hoe and rake tool, similar to a McLeod tool, but lighter in weight.

kurtosis (Statist) The degree of flatness or peakedness of a frequency distribution in the region of the mean, as compared with a normal distribution.

L

<u>l</u> (Phys) Abbreviation for lambert, liter.

L (Electr) Symbol for inductance.

<u>LA</u>^{50} (Med) The extent of burn that results in death 50% of the time.

<u>label</u> (Nucl Phys) An isotopic tracer.

<u>labeled</u> (Fire Pr) Carrying the identification of a recognized testing laboratory, indicating that the tagged item or material complies with accepted standards, that production is inspected periodically, or that the results of appropriate fire tests conducted on essentially identical materials or construction were satisfactory. See also: <u>approved</u>.

<u>labia</u> (Anat) Lips or lip-like organs.

<u>labor</u> (Med) The childbirth process, including contractions, dilation, delivery of the infant, and delivery of the placenta.

<u>laborer</u> (Techn) A worker who performs manual labor or is engaged in an occupation that requires physical strength.

<u>labor intensive</u> (Techn) Characteristic of a production process in which manual labor predominates, as in certain kinds of parts assembly work, wiring, logging, small-scale farming, and the like. See: <u>machine intensive</u>.

<u>labor pool</u> (Man Sci) A centrally controlled group of workers who are assigned to particular jobs or areas when needed.

<u>labyrinth</u> (Anat) A system of interconnecting canals in the ear containing the receptors for hearing and concerned with equilibrium.

<u>laceration</u> (Med) A tear or jagged wound in the skin or body tissue.

<u>lacquer</u> (Techn) A decorative and protective coating made of nitrocellulose, resins, plasticizers, and pigments; used on furniture, automobiles, paper, fabrics, and plastics.

<u>lacrimator</u> (Toxicol) A gas or vapor, such as chloroacetophenone ($C^6H^5COCH^2Cl$) that is intensely irritating to the eyes, causing them to weep. Syn: <u>tear gas</u>. See: <u>blepharospasm</u>.

<u>ladder</u>^{1} (Fire Pr) A climbing implement consisting of two parallel side-pieces called beams, joined at intervals by cross-pieces called rungs. See also: <u>aerial ladder</u>; <u>Bangor ladder</u>; <u>attic ladder</u>; <u>folding ladder</u>; <u>pompier ladder</u>; <u>trussed ladder</u>.
NOTE: Depending on size, use, and design, ladders are made of wood, aluminum, steel, or fiberglass. The principal types of ladder: a) <u>extension ladder</u> - A 30- to 45-foot ladder with three sections that can be extended by means of a hand line or a winch and cable, fitted with jacks, handling poles, and plumbing gear. b) <u>hook ladder</u> - A lightweight ladder fitted with a toothed hook that is used to suspend the ladder from a window sill or the like. c) <u>scaling ladder</u> - A sectional ladder so designed that a number of sections can be joined together. d) <u>short extension</u> - A light-weight ladder having two sections extending to about 13 feet. e) <u>first-floor ladder</u> - A single-section ladder, about 15 feet long, used for gaining access to buildings at first-floor level. Also used for additional height from an escape.

<u>ladder</u>^{2} (Fire Pr) A fire department ladder truck; loosely, a ladder company.

<u>ladder bed</u> (Fire Pr) A rack installed on ladder trucks for carrying ladders.

<u>ladder block</u> (Fire Pr) A wooden wedge used under a ladder beam to level an erected ladder. Cf: <u>ladder spur</u>.

<u>ladder butt</u> (Fire Pr) The bottom end of a ladder. Syn: <u>heel</u>.

<u>ladder carry</u> (Fire Pr) Any of several methods of carrying heavy fire department ladders, such as a four-man carry.

<u>ladder company</u> (Fire Pr) A fire company equipped with a ladder truck and trained in ladder work, ventilation, rescue, forcible entry, and

salvage work.

ladder dog chain See: **dog chain**

ladder fly See: **fly**

ladder-jack scaffold (Build) A light duty scaffold supported by brackets attached to ladders.

ladder locks (Fire Pr) Pawls that engage ladder rungs to lock fly sections of a ladder in position. The locks are operated through cables and pulleys by a lever at the base of the ladder. See also: **dog**; **pawl**; **leg lock**.

ladderman (Fire Pr) A firefighter who is assigned to a ladder company, or one handling ladders at a fire. Also called: **trucker**; **truckman**.

ladder pipe (Fire Pr) A master stream nozzle that can be attached to the rungs of an aerial ladder and controlled by ropes from the ground. The pipe is usu fed through a siamese connection at ground level.

ladder raise (Fire Pr) 1. To erect, lift, or extend a ladder. 2. One of several methods developed for erecting ladders suitable for specific locations and situations, as an **underwire raise**, **auditorium raise**, etc.

ladder safety device (Safety) Any device, other than a cage or well, designed to eliminate or reduce the possibility of accidental falls. It may incorporate such features as life belts, friction brakes, and sliding attachments.

ladder shoe (Fire Pr) A device with a rippled sole plate that can be attached to the heel of a ladder to prevent it from slipping.

ladder spur (Fire Pr) A leveling device attached to a ladder heel. Cf: **ladder block**.

ladder stand (Techn) A mobile, fixed-size, self-supporting ladder consisting of a wide, flat tread ladder in the form of stairs, sometimes with handrails.

ladder the building (Fire Pr) A command to place ladders at various points around a burning building for rescue work, to provide access to the roof for ventilation, and to advance hose lines to appropriate vantage points.

ladder tie (Fire Pr) A knot used to secure the free end of the rope used to extend the fly sections of a ladder.

ladder truck (Fire Pr) A large fire truck carrying ladders and other equipment; either a pumper-ladder or aerial ladder truck.

ladies' auxiliary (Fire Pr) A women's group made up of wives and girl friends of fire-department members, organized for social and civic activities.

lading clearance (Transp) The limitations for height and width of cars and loads, as published in the "Railway Line Clearances."

lag (Electr) a. The tardiness of one wave with respect to another. b. The distance that one wave is behind another in time, expressed in electrical degrees.

lagging (Mining) Planking or poles used for flooring, platforms, and walls in a mine.

lambert (Opt) A unit of luminance equal to 1 lumen/cm². Cf: **foot-lambert**.

laminar flow (Fl Mech) The motion of a fluid in layers without turbulence. Smooth flow.

laminated construction (Build; Techn) A structure consisting of sheets or sections of material bonded together with adhesives.

landing¹ (Build) A level space at the top, bottom, or between sections of a staircase.
  NOTE: A landing at a point where a flight of stairs reverses direction between floors is called a half-space landing.

landing² (Transp) A platform for loading and discharging passengers or cargo.

land occupancy fire (For Serv) A fire started in order to clear land for some useful purpose, such as agriculture, industry, construction, road building, and the like.

language (Comput) A set of codes or symbols and the rules used to instruct a computer to perform desired operations.

lantern light (Brit) = skylight

lanyard (Safety) A rope suitable for supporting one person. One end is fastened to a safety belt or harness and the other end is secured to a substantial object or a safety line.

lapping (Mech Eng) A fine abrasion process used to obtain precise dimensions and smooth surfaces on machine parts. See also: abrasives.

lapse rate (Meteor) 1. The rate of temperature decrease with altitude. 2. The rate of cooling of a rising or warming of a falling parcel of air.
NOTE: Unsaturated rising air has a dry adiabatic lapse rate of 5.5°F per 1000 feet. The process is reversed for a dry falling parcel. The measured rate of change in temperature with change in height is called the environmental lapse rate. If the environmental lapse rate exceeds the adiabatic rate, the air becomes unstable. If the parcel of air is saturated with moisture, temperature changes associated with vertical movement follow the moist adiabatic lapse rate, which is less than the dry rate (2 to 5°F per 1000 feet) due to the latent heat of vaporization released by condensation. For dry air, the lapse rate equation is $dT/dz = g/C_p$, where dT is the change in temperature, dz is the change in height, g is the acceleration of gravity, and C_p is the heat capacity of the air at constant pressure.

larceny (Law) Theft, involving trespass and the taking of another's personal property without consent and with the intent to steal.

large-loss fire (Fire Pr) A fire causing property loss of $250,000.00 or more.

large-lot (Storage) Four or more stacks of supplies stored to maximum height.

laryngitis (Med) Inflammation in the larynx.

larynx (Anat) The voice box in the upper anterior part of the neck, connecting with the pharynx, above, and the trachea, below; part of the airway containing the vocal chords.

laser (Phys) An acronym for Light Amplification by Stimulated Emission of Radiation.
NOTE: Laser radiation can be stimulated in solids (such as ruby crystals), gases (neon-helium, for example), or even in liquids and plasma. The radiation is emitted in a very narrow wavelength band. By proper selection of lasing material and design, many wavelengths in the visible, uv, and ir ranges are now available. Beams from laser devices can be used to detect fire in large enclosed spaces owing to the bending of the beam as it passes through heated air. Laser beams can be extremely intense and can produce serious burns in sensitive tissues, esp the retina of the eye.

last back (Fire Pr) A signal indicating that the last company has returned to quarters from a fire.

last due (Fire Pr) The last engine or truck companies to arrive in response to a given alarm; the third engine at a three-company alarm and the third and fourth engines at a four-company alarm.

latent heat (Phys) The heat absorbed or liberated when a substance changes from one state to another at a fixed temperature. Latent heat is involved in fusion, vaporization, and sublimation.
NOTE: The latent heat of fusion of water at 32°F is 143.4 Btu/lb. The latent heat of vaporization of water at 212°F is 970.3 Btu/lb.

latent period (Dosim) The time that elapses between exposure to radiation and the first manifestation of damage.

lateral (Mining) A side tunnel or secondary drift running parallel to a main drift.

latex[1] (Chem Eng) A milky extract from the rubber tree containing about 35 percent rubber hydrocarbon;

the remainder is water, proteins, and sugars.

latex² (Techn) A water emulsion of synthetic rubbers or resins. The film-forming resin in emulsion paints.

lath (Build) 1. A thin, narrow strip of wood nailed to studding, joists, or rafters to serve as a base for plaster. 2. Sheets of building material or expanded metal (mesh) used for the same purpose.

lathe (Techn) A machine tool used to shape wood or metal by the rotation of the work piece against a cutting edge.

lattice (Phys) An orderly array or pattern of atoms in a crystal.

laundry An establishment in which clothes are washed, ironed, and finished; excluding dry cleaning and dyeing.

law (Law) A body of principles and regulations enacted by legislation or deriving from government, applied to a people, and enforced by judicial decision.

lay¹ (Fire Pr) The method or sequence of connecting and stretching hose between the hydrant, pumper, and fire. See: hose lay.

lay² (Techn) The manner in which strands of rope are twisted together. In a right-hand lay, the strands twist toward the right; the opposite holds for a left-hand lay. See also: hawser; rope.

layering¹ (Techn) Stratification, usu due to density differences.

layering² (Build; Fire Pr) The partial separation of a one- to two-inch layer of concrete from a heavy concrete structural assembly due to severe fire exposure.

lay in (Fire Pr) A command to lay out or stretch hose while the truck is still moving toward the fire, usu in a forward lay, from the hydrant toward the fire.

lay out (Fire Pr) a. An order to stretch hose lines. Syn: lay in. b. To stretch hose lines from a pumper or hose truck.

layout (Fire Pr) The distribution of hose lines at a fire.

LC_{50} = lethal concentration$_{50}$

LD_{50} = lethal dose$_{50}$

leaching (Chem) Removal of a soluble material from an insoluble solid by a solvent.
  NOTE: Fire retardants, for example, are leached from treated material by water, thereby causing them to lose fire resistance. Leaching is also an important industrial process, used for extracting oil from shale, metals from ores, etc.

lead (Fire Pr) A hose line stretching from a pump or hose bed toward a fire.

lead (Symb: Pb) A heavy, metallic element, mp 327.4°C, bp 1740°C, used extensively in storage batteries, pigments, tetraethyl lead, solder, bearings, etc. The metal and its compounds, esp in the form of vapor or dust, are toxic. Some lead compounds can be absorbed through the skin.

leader (Man Sci) One who guides, directs and controls the actions of others by the exercise of authority.

leader line (Fire Pr) A hose extending from a pump to a nozzle, as distinguished from a feeder line, which supplies a pump.

lead plane (For Serv) An aircraft used to direct air tankers in combating a forest fire. Syn: bird dog.

lead poisoning (Toxicol) A state or condition of being poisoned by lead fumes, lead dust, or lead compounds. Also called: lead intoxication; plumbism.
  NOTE: Inorganic lead compounds commonly cause symptoms of lead colic (i.e., abdominal pains) and lead anemia. Organic lead compounds can attack the nervous system, producing mental symptoms.

leaf (Build) The hinged, swinging portion of a door.

leaf spring (Mech Eng) A flat bar spring that is relatively thin in proportion to its length and width, designed to be loaded in bending. In vehicle suspensions, leaf springs usu are a lamination of several leaves of unequal lengths.

leakage[1] (Electr) The electrical loss due to poor insulation.

leakage[2] (Nucl Phys) The escape of neutrons from a radioactive source.

leakage flux (Testing) Magnetic lines of force that leave and enter the surface of a part, because of a flaw or other discontinuity in the material making up the part.

leakage rate (Build) The quantity of air, in cubic feet per minute, passing through cracks, pores, and other openings in a building.

leak detector (Techn) An instrument used to locate the point from which gas is escaping from a pipe, container, or system.

leaker[1] (Transp) A shipping container that has been damaged and from which the contents are escaping.

leaker[2] (Fire Pr) A perforated hose.

lean-to collapse See: floor collapse

learning (Educ) 1. The process of acquiring a change in behavior as a result of practice or experience. 2. Acquisition of knowledge, understanding, or skill through study or instruction.

learning curve (Educ) A graphic plot of performance with respect to the number of trials in a learning task.
NOTE: Learning rate generally follows one of three basic patterns, depending on the individual, the task, and other factors: 1) rapid initial rise in learning followed by gradual decline; 2) gradual and steady acceleration in learning as practice continues; 3) gradual initial increase, followed by a dramatic spurt of accelerated learning, followed in turn by gradual deceleration.

least-cost theory See: economic theory

least-square law (Statist) The assumption that the best estimate of a value is that for which the sum of the squares of the deviations is minimum.

leave (Fire Pr) Off-duty time, such as annual leave, sick leave, or compensatory time off.

leave of absence (Man Sci) Excused but unpaid time away from work without loss of job or seniority.

LeChatelier's law (Chem; Phys) A system in equilibrium disturbed by an external factor, such as temperature or pressure, will adjust itself to minimize the effect of the disturbance.

ledger (Build) A horizontal scaffold member that extends from post to post and supports putlogs or bearers, forming a tie between the posts. Also called: stringer.

legal aid (Law) Legal advice, counseling, and representation provided by lawyers at nominal or no cost to indigent clients.

legal chemistry See: forensic chemistry

legal liability (Law) Liability imposed by law, as opposed to liability arising from an agreement or contract.

legitimate smoke (Fire Pr) Smoke from an intentional fire, such as that from an industrial process or from authorized burning of waste.

leg lock (Fire Pr) A method of securing oneself to a ladder by placing one leg between the rungs and hooking the foot over the next rung below or around the beam in order to free both hands.

length[1] (Metrol) A measure of spatial extent; one of the three fundamental physical quantities, the others being time and mass. The standard unit is the international meter.
NOTE: In the English system of measures, the foot, equal to 0.3048 meter, is the standard.

length[2] (Fire Pr) A standard 50-foot section of hose.

lens (Opt) A symmetrical, curved, disk-shaped or cylindrical piece of

lesion (Med) 1. An injury, damage, or abnormal change in a tissue or organ, esp one that impairs the function of the part involved or, even if it causes no impairment of function, expresses a symptom or sign of a disease. 2. A wound or localized patch of disease on the skin or other tissue.

lesser circulation (Physiol; Rare) Blood circulation in the lungs Cf: greater circulation.

lethal (Med; Toxicol) Capable of causing death.

lethal concentration^{50} (Toxicol; Abb: LC^{50}) A concentration of a substance which, if inhaled, swallowed, absorbed through the skin, or ingested, kills 50 percent of a population of experimental animals during an exposure of a pre-established duration (usu 4 hrs). See: approximate lethal concentration.

lethal dose (Dosim) A dose of ionizing radiation sufficient to cause death.

lethal dose^{50} (Toxicol; Abb: LD^{50}) The median quantity of a substance that is fatal to 50% of test animals within a specified time period after ingestion or exposure. Syn: median lethal dose.
NOTE: The x-ray LD-50/30 for man (i.e., the dose of x-rays that within 30 days kills 50% of humans exposed) is about 400-450 roentgens.

lethality (Med) Killing power; the ratio of deaths from a given disease or condition to the existing cases of that disease or condition. The ratio of the total bacteria killed by a substance to the bacteria present.

leukemia (Med) A blood disease distinguished by overproduction of white blood cells. It may result from overexposure to radiation or it may generate spontaneously. It is almost always fatal. The disease is attended by progressive anemia and exhaustion.

leukocyte (Physiol) A white blood cell.

leukocytosis (Med) An increase in the number of white blood cells, which occurs naturally during digestion and pregnancy and abnormally in the presence of infection.

leukopenia (Med) A serious reduction in the number of white blood cells.

level (Mining) 1. A horizontal tunnel used for working and hauling in a mine. 2. The horizon at which an orebody is opened for working. 3. All of the horizontal workings on the same horizon.

level gage (Techn) An instrument for determining the quantity of liquid in a container by measuring the distance between its surface and a known reference point.
NOTE: The simplest level gage is a graduated stick that is dipped manually into the liquid; the wetted portion indicating the depth. Gage glasses mounted on the side of a liquid container show the level of the liquid directly. Float gages show liquid levels on dials. Other gages measure differential pressure, hydrostatic head, etc., in different ways to determine liquid levels. See also: fluid gage.

Lewis acid See: acid

Lewis number (Phys; Abb: Le) The ratio of the energy transported by conduction to that transported by diffusion. See: dimensionless numbers.

liability insurance (Insur) 1. Insurance that obligates an insurance company to pay any liability for which the insured may be covered and to defend the policyholder in any damage suits. 2. Insurance that reimburses the policyholder for sums paid to others as the result of his negligence.

liability limit (Insur) A specified amount beyond which an insurance company is not liable to protect the policy holder.

license (Fire Pr) Written permission granted by a fire department or other authority to a citizen or corporation to conduct business or engage in some activity, usu for a prescribed period of time, in accordance with law or regulations. See: permit.

licensed material (Nucl Phys) Radioactive source material received, stored, used, or transferred under a general or special license.

licensing (Law) Official authorization or permission to engage in some activity, usu under specified conditions and for a stated period of time.

lieutenant (Fire Pr) A fire company officer rank between captain and sergeant, second in command to the captain and usu in charge of a platoon or working shift. See: ranks and titles.

life belt (Fire Pr) A wide belt with a snap hook that can be fastened to a ladder to secure the wearer. Syn: ladder belt; pompier belt.

life gun (Fire Pr) A gun that fires a cord attached to a projectile to persons trapped in inaccessible places. The cord is used to pull in a life line for rescue.

life hazard (Fire Pr) A condition or material that is dangerous to life, for example, fire, smoke, fumes, toxic products, panic, and the like.

life line (Fire Pr) 1. a. A rope secured to a firefighter and leading to a safe exit, used to enable a firefighter to find his way out of a smoke-filled building. Syn: guide line; escape line. b. A rope, usu attached to a belt and secured to a structure to prevent or limit the fall of a person working aloft. c. A hand rope, usu with an eye spliced at one end, used in fire-department rescue work. See: line. 2. (Brit) A 150- to 220-foot rope used in rescue work with turntable ladders.

life net (Fire Pr) 1. A circular canvas tarpaulin stretched on a folding metal hoop or frame with springs, held by six to eight firemen, used to catch and cushion the impact of persons jumping from burning buildings. 2. = (Brit) jumping sheet.
NOTE: Life nets were originally made of rope. Nets are usu considered practical for jumps of up to four stories. Although successful jumps have been made from up to eight stories, the probability of injury increases above the fourth floor.

life safety (Fire Pr) The preservation and protection of life from the hazards of flame, heat, smoke, etc.

life saving (Rescue) Techniques of rescuing persons from drowning or other hazards. See: first aid.

life support (Med) The science and technology concerned with health, safety, protection, sustenance, escape, survival, and recovery of personnel.

lift[1] (Fire Pr) The vertical distance between the surface of a static water source and a fire department pumper intake.

lift[2] (Brit) = elevator

lift pump (Fire Pr) A pump with a hollow cylinder that has a gate valve through which water can pass only in one direction.

lift truck (Mat Hand) An industrial truck used for lateral transportation and equipped with a power-operated lifting device, usu in the form of forks, for piling or unpiling lumber units or packages.

ligament (Anat) Fibrous tissue that connects bone to bone. A fibrous band supporting a body organ.

light (Opt) Radiant electromagnetic energy in the visible spectrum that can be optically diffracted, refracted, reflected, and polarized.
NOTE: In vacuum, light travels in straight lines at the speed of 299,792.8 km/sec, or about 186,000 miles/sec.

light adaptation (Physiol) The adjustment of the eye to bright light. The iris contracts to narrow the pupil, thus reducing the quantity of light entering the eye. Cf: dark adaptation.

light burning (For Serv) Periodic broadcast burning of light fuels to prevent accumulations in quantities that would cause extensive damage or

253

present difficulties in suppression in the event of fire.

light end (Chem Eng) The portion of a mixture of liquids, usu from petroleum, that distills at a lower boiling point than other portions of the mixture.

lighter 1. An electrical or mechanical device for lighting cigarettes. 2. An electrical heater for lighting a charcoal grill.

light fuel (For Serv) Fast-burning fuel such as dry leaves, dry grass, and evergreen needles.

light hazard (Fire Pr) Referring to occupancies with low fuel loads such as offices, schoolrooms, churches, assembly halls, etc. Cf: low hazard.

light hydrogen (Chem) Ordinary hydrogen.

lighting outlet (Build) An electrical connector or receptacle for a lampholder, light fixture, or hanging lamp cord.

lighting unit (Fire Pr) A fire department apparatus equipped with an array of floodlights to provide illumination for firefighting and rescue operations at night. Also called: floodlight unit; light plant; searchlight unit.
 NOTE: The apparatus usu also carries generators, communications equipment, and various electrically driven power tools.

lightning (Meteor) A naturally occurring, intensely luminous, high-current electrical discharge between clouds and the earth or from cloud to cloud, usu occurring during turbulent atmospheric conditions. See also: ball lightning.

lightning activity level (For Serv) The relative frequency of lightning strikes expressed on a scale of 1 through 5, on which 2 represents twice the frequency of 1, 3 twice the frequency of 2, and so on.

lightning arrester (Techn) A device used to protect an electrical appliance from damage by lightning-induced power line surges.

lightning fire (For Serv) A fire caused in any way by lightning.

lightning risk (For Serv) A measure of the number of lightning strikes expected during the rating day.

lightning rod (Build) A metal pole erected on the highest point of a building and connected to ground to safely discharge high electrical potentials or lightning strikes near the building.

light water (Nucl Phys) Ordinary water (H_2O), as distinguished from heavy water (D_2O).

Light Water foam (Fire Pr) A froth or foam formed by discharging water containing a fluorocarbon surfactant through a foam maker. See: foam. Syn: aqueous film-forming foam.

lightweight concrete (Build) Concrete made with perlite or vermiculite aggregate.

lignin (For Serv) A major polymeric component of wood, composed of phenyl propane, and amounting to 20 to 30% of dry wood by weight. See: wood.

lignite A type of fuel intermediate between peat and coal.

limit control (Techn) A device that turns on, shuts off, or throttles the fuel supply to an appliance in response to changes in pressure, temperature, or liquid level.

limited combustibility (Build; Testing) The tendency or disposition of a material to burn that lies between noncombustibility and combustibility.

limiting oxygen index (Comb; Abb: LOI) The lowest oxygen concentration in an oxygen-nitrogen mixture, expressed in percent, that will support the combustion of a fuel specimen.
 NOTE: The index, used extensively to rate the flammability of plastics, is defined in one specific test as $I=O_2/(O_2+N_2)$, where I is the index and O_2 and N_2 are the concentrations of oxygen and nitrogen, respectively, in consistent units. The lowest value of I for a given specimen is its LOI.

limit switch (Safety) A switch that cuts off the power supply when a moving part reaches a specified limit.

line[1] (Fire Pr) One or more lengths of connected hose. A <u>fire line</u>.

line[2] A length of rope.
  NOTE: Rope is used extensively in the fire service for a variety of purposes: 1) <u>belt line</u> = <u>pocket line</u> (See: <u>pouch line</u>, below). 2) <u>bobbin line</u> - A line wound on a bobbin and carried in a pouch on a hook belt and used for hauling small gear or as a guy or guide line. Cf: <u>pouch line</u>. 3) <u>escape line</u> - a) (Brit) A 15- to 20-foot rope secured to the top round of an escape or extension ladder. b) (US) (1) A light line used to assist firefighters in finding their way back from smoke-filled rooms. (2) A line used to secure the heel of a first-floor ladder when it is used to extend the height or to secure the head of an extension ladder. 4) <u>grass line</u> - A line made of coir and used as an endless whip on fireboats. 5) <u>ground control line</u> - A line used to control a turntable ladder and monitor from ground level. 6) <u>guide line</u> - A rope of up to 300 feet in length used to guide men in and out of buildings when breathing apparatus is worn. 7) <u>guy line</u> - a) Usu a rope 130 feet long attached to a turntable ladder to stabilize it in a high wind. b) A line bent on another line to keep an object that is being lowered clear of obstructions on its way down. 8) <u>long line</u> - A 100-foot, 2-inch manila line. 9) <u>lowering line</u> - A two-inch, 130-foot line of Italian hemp or terylene, with two legs spliced at one end with a running noose, used for rescue work. 10) <u>personal line</u> - A 20-foot line secured to a man at one end and fitted with a snap hook at the other, used for hooking on to a guide line when breathing apparatus is worn. 11) <u>pocket line</u>; <u>pouch line</u> - Similar to a bobbin line, but made up on a special former. Carried in a pocket, it is used to lash branches, etc. When carried on the belt is is called a belt line. 12) <u>rescue line</u> - A special, two-inch rope, usu 220 feet long, used for rescue work with turntable ladders. 13) <u>short line</u> - A fifty-foot, two-inch rope. 14) <u>tail line</u> - A line of up to 20 feet permanently attached to a turntable ladder rescue sling and used to prevent a person being lowered from swinging when the turntable ladder is trained away from a building. See also: <u>halyard</u>; <u>hand line</u>; <u>hawser</u>; <u>life line</u>; <u>fire line</u>.

line[3] (Electr) An electrical conductor, such as a <u>transmission line</u>.

linear (Math; Phys) 1. Descriptive of a steady progression or a straight line. 2. Continuous, as opposed to discrete or discontinuous, esp with respect to mathematical functions.
  NOTE: A linear system, for example, is one in which the output varies directly with the variations of the input.

linear accelerator (Nucl Phys) A device that increases the speed of charged particles. Particles passing through receive successive increments of energy.
  NOTE: A linear accelerator differs from other accelerators in that the particles move in a straight line at all times instead of in circles or spirals.

linear energy transfer (Abb: LET) A measure of the ability of biological material to absorb ionizing radiation; the radiation energy lost per unit length of path through a biological material. In general, the higher the LET value, the greater the relative biological effectiveness of the radiation in that material. See: <u>biological dose</u>; <u>relative biological effectiveness</u>.

linear programming (Math) A mathematical method for finding the maximum or minimum of a linear function.
  NOTE: Linear programming is used to solve agricultural and transportation problems; e.g., to determine the least cost of shipping goods to various customers from warehouses at different geographic locations.

linear relation (Math) A relationship between values of measurement bearing a straight line proportionality to each other.

line boss (For Serv) A supervisory officer in a fire-suppression organization responsible for carrying out the fire-suppression plan

adopted by the fire boss.
  NOTE: The line boss may assist in organizing large fire attack forces, coordinate the work of two or more divisions, or supervise three or four sector bosses if divisions have not been established.

line corrosion (Metall) Pits that are connected or nearly connected to others in a narrow band or line. Also called: crevice corrosion.

lined hose (Fire Pr) Fire hose consisting of a rubber tube covered with one or two woven jackets.

line firing (For Serv) Setting fire to the fuel at the edge of a control line. See: strip firing.

line function (Man Sci) Management and control activity directly related to the production operations of an organization.
  NOTE: Line functions include the responsibility and the authority for accomplishing the mission or objectives of an organizationn.

line item (Man Sci) A separate, identified item of supply on a transaction document.

line locator (For Serv) A person responsible for determining where a control line is to be established to contain a forest fire.

linen hose (Fire Pr; For Serv) A fire hose made of linen or flax fabric and having no rubber lining, used with first-aid standpipes and in the forest service. Syn: unlined fire hose. See: hose.

line scout (For Serv) An individual in a fire suppression organization assigned to reconnaissance duties on a fire line. See also: scout.

lineup (Fire Pr) Assembly of fire-company or -unit personnel for inspection, roll call, or to honor official visitors.

lintel (Build) A horizontal building member supporting the weight above an opening, such as a window or door.

linters (Techn) Short cotton fibers.

liquefaction (Chem Eng) The process of converting a gas or vapor into a liquid by cooling below its critical temperature, usu accompanied by application of pressure.

liquefied compressed gas (Techn) A gas that is partially liquefied and stored in this state in a pressure vessel.

liquefied petroleum gas (Abb: LPG) 1. A mixture chiefly of butane and propane gas maintained in liquid state by pressure, stored in cylinders or bottles and used for heating, lighting, and as an internal combustion engine fuel. 2. = (Brit) liquid fuel gas.

liquid^1 (Phys; Chem) One of the states of matter. A relatively incompressible, structureless fluid that flows under low shear forces, conforms to the shape of a containing vessel, and presents an upper surface. See fluid.
  NOTE: Under suitable conditions of temperature and pressure, metals, gases, and various other substances have a liquid phase.

liquid^2 (Fire Pr) Any material, including combustible and flammable liquids, that has a fluidity greater than 300 penetration asphalt. See: combustible liquid; flammable liquid.
  NOTE: An unstable liquid is one that polymerizes, decomposes, or becomes self-reactive under conditions of shock, pressure, or temperature.

liquor (Chem Eng) A chemical liquid used to reduce wood chips to pulp and remove impurities.
  NOTE: Used also in many other applications, such as tanning liquor.

listed (Fire Pr) Referring to a device or product that has been tested, passed, and certified by a nationally recognized testing laboratory.

liter (Abb: l) A metric unit of volume equal to 1000 cubic centimeters or 0.2642 US gallon (1.1 quarts). Spelled litre in Great Britain and France.
  NOTE: A liter of water at 4°C and standard pressure (760 mm Hg) weighs

one kilogram, or 2.2 pounds.

literacy test (Educ; Physiol) A test that measures the ability to read and write.

literature, technical (Info Sci) The principal medium of disseminating technical information to scientists, engineers, practitioners, and researchers in printed form.
  NOTE: Primary sources of technical information include periodicals, monographs, research reports, patents, and manufacturer's publications. Secondary sources include lists of and guides to literature, largely through subject access points. These include abstract journals, indexes, bibliographies, and library catalogs. Tertiary sources are educational in nature, including reviews, advances, progress series, treatises, surveys, and text books on subjects in particular fields by authoritative experts. A quaternary source is the reference work: encyclopedias, dictionaries, handbooks, tables, and directories.

litter[1] (For Serv) The top layer of a forest floor consisting of loose debris, dead sticks, branches, twigs, leaves, dry needles, and the like. See also: duff.

litter[2] (Env Prot) Carelessly discarded material.

litter[3] (First Aid) A stretcher for carrying a sick or injured person.

live burning (For Serv) Progressive burning of green slash as it is cut.

live edge (Comb) The boundary of glowing combustion.

live line (Fire Pr) 1. A hose line or reel with all connections made to a pump and ready for immediate use. 2. A line containing water under pressure. Syn: charged line.

live load[1] (Transp) The cargo or payload of a vehicle, contrasted with a dead load, which must be transported but cannot be used.
  NOTE: The contents of a water tank is a live load, but the tank itself is a dead load.

live load[2] (Build) The weight of the nonintegral and nonstructural parts and contents of a building, or portions of a building, such as furnishings, movable equipment, occupants, etc., usu expressed in terms of pounds per square foot.

liver (Anat) The largest gland or organ in the body, situated on the right side of the upper part of the abdomen.
  NOTE: The liver performs many important functions: regulating the amino acids in the blood; storing sugar, iron, and copper for the body; forming and secreting bile, which aids in the absorption and digestion of fats; transforming glucose into glycogen and vice versa; disposing of worn-out blood cells.

live room (Acoust) A reverberant room that is characterized by an unusually small amount of sound absorption.

live steam (Techn) Steam issuing directly from a boiler.

living fuel (For Serv) Living plants or parts of plants, such as grass, leaves, needles, and twigs, that are small enough to be consumed in an advancing fire.

living room (Build) A room in a residence used primarily for entertaining and other general and social activities. Cf: family room.

living unit (Build) An apartment or house used by one family.

load factor (Techn) The ratio of average load carried by an electric power plant or system during a specific period to its peak load during that period. Cf: plant factor.

load hoist (Mat Hand) A hoist drum and rope reeving system used for hoisting and lowering loads.

loading (Fire Pr) The accumulation of foreign material on a sprinkler head, which may impair or prevent its operation.

loading platform (Transp) A flat surface to facilitate loading or unloading, usu erected alongside a warehouse at the approximate level of a rail car or truck floor.

load limit (Build; Transp) The upper weight limit capable of safe support by a structural element, floor, or vehicle.

load weight (Transp) The heaviest weight that can be carried by a vehicle safely. Also called: allowable load.

lobby (Build) 1. A public passageway or waiting room adjoining a series of rooms or offices. 2. A large hall serving as a foyer in a theater or hotel.

local alarm (Fire Pr) An alarm given only to the company nearest a small fire, such as burning leaves, rubbish, or automobile.

localized (Med) Restricted to one spot or area in the body and not spread all through it. Cf: systemic.

location (Fire Pr) A designated rendezvous point for fire apparatus in response to an alarm. The location may be an address, a street intersection, or coordinates on a map. Fire units report to the location to receive their fireground assignments. Similar in purpose to a staging area.

location marker (Storage) Numbers or letters applied to and used to identify a specific space for storage of supplies.

lock See: air lock; fox lock; ladder lock; leg lock; safety lock; vapor lock.

locomotive crane (Mat Hand) A rotating superstructure with a powerplant, operating machinery, and boom, mounted on a base or car equipped for travel on railroad track; used to hoist and swing loads.

lode (Mining) An ore deposit, esp a rich one.

lodged tree (For Serv) A tree that has not fallen to the ground after being partly or wholly separated from its stump or otherwise displaced from its natural position.

loft (Build) 1. A room or similar space under a sloping roof. 2. An upper story of a warehouse or mercantile building.

log (Fire Pr; Navig) A diary of events, such as a ship's log, sometimes used to refer to a company journal or the record of the progress of an extensive fire.

logarithm (Math; Symb: log or ln) The power to which a base must be raised to give a desired number. For example, if the base is ten, the logarithm of 100 is 2; i.e., 10^2 = 100. Logarithms can be used to multiply numbers by addition, to divide by subtraction, and to extract roots by division. A logarithm is usu given as a whole number, called the characteristic, plus a decimal fraction, called the mantissa. Tables of logarithms are available to six or more significant figures so that any required accuracy may be obtained in numerical calculations.
NOTE: Two systems of logarithms are in general use: common (also called Briggs logarithms) and natural (also called Napierian or hyperbolic). Common logarithms have the base 10, and the notation for the logarithm of a number N is log N. Natural logarithms have the base e = 2.71828, and the usual notation, esp in scientific work, is ln N.

log carriage (For Serv) A framework mounted on wheels which run on tracks or in grooves in a direction parallel to the face of the saw, and which is equipped with dogs, clamps, or other devices to hold a log securely and advance it toward the saw.

logdeck (For Serv) A platform in the sawmill on which the logs are stored prior to sawing.

logged (Brit) = charged

logger (Commun; Fire Pr) A communications recording device that automatically records telephone and radio calls and provides a time check.

log haul (For Serv) A conveyor for transferring logs to a mill.

logic The branch of philosophy that is concerned with principles of reasoning, rules for correct thinking, and deductive and inductive mental processes.

logistics (Milit) The science of planning for and handling personnel, supplies, transportation, and facilities to accomplish a required task effectively and economically.

LOI = limiting oxygen index

long day (Fire Pr) A 24-hour tour of duty used to rotate a man or platoon from a day to night shift.

longevity (Fire Pr) The length of time served in a fire department. Syn: seniority.

longitudinal study (Psych) A study design in which the same subject or population is observed over a long period of time. Cf: cross-sectional study.

longshoring (Mat Hand) The loading, unloading, moving, or handling of ship cargo, ship's stores, gear, etc.

lookout (For Serv) 1. An observer assigned to a vantage point to detect and report fires. 2. A lookout station or tower.

lookout dispatcher (For Serv) An individual acting both as a lookout and as a dispatcher.

lookout fireman (For Serv) An individual acting both as a lookout man and as a fireman.

lookout house (For Serv) An improved structure with living quarters and large windows built on a tower or natural elevation to permit unobstructed viewing of a large forest area.

lookout man (For Serv) An individual manning a lookout station to detect and report forest fires. Also called: tower man. See: secondary lookout man.

lookout observatory (For Serv) A lookout house which may lack living quarters.

lookout observer = lookout man

lookout patrolman (For Serv) A roving lookout who travels ridges and other vantage points to detect, report, and suppress fires, and often to carry out prevention activities.

lookout point (For Serv) An observation point used for fire detection.

lookout station (For Serv) An observation point with structures used for fire detection.

lookout tower (For Serv) A structure built to elevate an observer above nearby obstructions and carrying a lookout house or observatory; used to detect signs of fire and to fix its location. Also called: fire tower.

loop^1 (Fire Pr) A water main laid in a loop so that water can be supplied to a given location from two directions, thereby minimizing the possibility of the water supply being interrupted.

loop^2 (Testing) A closed circuit of pipe in which materials and components may be placed for testing under different conditions of temperature, irradiation, etc.

Loschmidt number (Phys) The number of molecules in a cubic centimeter of an ideal gas at 0°C and 1 atmosphere of pressure, equal to 2.6868×10^{21}. Syn: Avogadro's constant [in Europe].

loss See: fire loss

loss control (Safety) A program to minimize financial losses, based on detailed analysis of both indirect and direct accident costs, including property damage as well as injurious and potentially injurious accidents.

loss of pressure (Fire Pr) A sudden drop in pressure for fire streams due to hose failure, pump failure, water system failure, or depletion of the supply because of the number of pumpers drawing water.

loss prevention (Safety) A program designed to identify and correct potential accidents before they occur.

loss ratio (Insur) The amount of losses divided by the cost of premiums, expressed as a percent of the premiums.

<u>loss reserve</u> (Insur) An estimate of the amount an insurer expects to pay for losses incurred but not yet due for payment.

<u>lost line</u> (For Serv) The part of a fire control line that has been crossed by a wildland fire.

<u>lost-time accident</u> See: <u>lost time injury</u>

<u>lost-time injury</u> (Safety) A work injury which results in death or disability and in which the injured person is unable to report for duty on his next regularly scheduled shift. See: <u>injury</u>.

<u>loudness</u> (Acoust) The intensity of sound as it affects the organs of hearing.
NOTE: Loudness varies with sound wave amplitude and with frequency. At equal intensity, sound in the middle range of 1000 to 5000 cps appears to be louder than at higher or lower frequencies. The unit of sound intensity or pressure level is the <u>decibel</u> (0.1 bel), which is the intensity ratio of a given sound to a reference sound. See: <u>sense organs</u>.

<u>loudness level</u> (Acoust) A sound, in phons, numerically equal to the median sound pressure level, in decibels, relative to 0.0002 microbars of a free progressive wave of 1000 Hz presented to listeners facing the source, which in a number of trials is judged by the listeners to be equally loud.

<u>louver</u> (Build) A ventilation opening, such as one provided by a slatted panel.

<u>low density</u> (Build) Descriptive of materials that are exceptionally light in weight and usu deficient in strength properties.

<u>lower explosive limit</u> (Abb: LEL) 1. The minimum concentration of combustible gas or vapor in air below which flame does not propagate away from an ignition source. 2. The lower limit of flammability of a gas or vapor at ordinary ambient temperatures expressed in precent of the gas or vapor in air by volume.
NOTE: This limit is assumed constant for temperatures up to 250° F. Above this it should be decreased by a factor of 0.7 because explosibility increases with higher temperatures.

<u>lower flammability limit</u> (Comb) The lowest percent concentration by volume of a flammable vapor or gas mixed with air that will ignite and burn with a flame. See: <u>flammability limit</u>.

<u>low-expansion foam</u> See: <u>foam expansion</u>

<u>low explosive</u> (Abb: LE) An explosive thatdeflagrates or burns rather than detonates; that is, the rate of advance of the reaction zone into the unreacted material is less than the velocity of sound in the unreacted material. Low explosives include propellants, certain primer mixtures, black powder, photoflash powders, and delay compositions.

<u>low hazard</u> (Fire Pr; Insur) Having such low combustibility that no self-propagating fire can occur and the only probable danger would be from panic, fumes, smoke, or fire from some external source.

<u>low-level analysis</u> (Nucl Safety) A procedure to measure the radioactive content of materials with very low levels of activity, using sensitive detecting instruments with good shielding to eliminate the effects of background radiation. Also called: <u>low-level counting</u>.

<u>low population zone</u> (Nucl Safety) An area of low population density sometimes required around a nuclear installation.
NOTE: The number and density of residents is of concern in providing, with reasonable probability, that effective protection measures can be taken if a serious accident should occur. See: <u>exclusion area</u>.

<u>low pressure</u> (Fire Pr) Erroneous for inadequate water supply.

<u>low pressure tank</u> (Techn) A storage tank designed for service pressures between 0.5 to 15 psig.

lowrise (Build) A multistory building the top floor of which is within reach of fire department aerial equipment.

Lowry hydrant (Fire Pr) A trade name for a flush-type hydrant that requires a portable hydrant chuck.

low service (Fire Pr) A public water system generally supplying city districts that do not require high pressures. Fire department pumpers are needed to supply pressure for hose streams. See: high service.

LPG = liquefied petroleum gas

L/T ratio (Transp) A steering ratio in which L represents the length of track in contact with the ground and T represents the lateral distance between the centerlines of the tracks.

lugging (Transp) A condition in which an engine labors due to overload.
NOTE: Lugging can be corrected by shifting to a lower gear when climbing a steep hill, or by adjusting engine speed in pumping operations.

lumbar (Anat) Pertaining to the part of the back between the chest and the pelvis.

lumber (For Serv) Timber sawn into smaller, usu uniform, standard sizes for use in building.

lumbering fire (For Serv) Any fire, except one caused by a smoker, resulting from lumbering operations, including harvesting, transporting, and processing of wood.

lumen[1] (Anat) The internal cavity or bore of a hollow tubular organ, such as a blood vessel, the intestine, etc. Cf: alveolus.

lumen[2] (Opt) A unit of luminous flux equal to that emitted through a unit solid angle by a point source of one candela.

luminance (Opt; Symb: B) The brightness of an illuminated surface, typically expressed in candelas per square foot.
NOTE: Since surfaces do not reflect perfectly, the reflection factor must be taken into account. Thus, luminance = illuminance x reflection factor. This is expressed in foot lamberts (1/4 candelas per square foot); luminance is equal to KE, where K is the reflection factor and E is footcandles incident.

luminescence (Phys) Emission of light produced by the action of biological or chemical processes, by radiation, or any other stimulation other than high temperature, which produces incandescence. Cf: fluorescence.
NOTE: The emitted light is usu of a longer wavelength than the exciting light.

luminous efficiency (Opt) The ratio of the luminous flux to the radiant flux, usu expressed in lumens per watt of radiant flux.
NOTE: For energy radiated at a single wavelength, luminous efficiency is synonymous with luminosity factor. The reciprocal of the luminous efficiency of radiant energy is sometimes called the mechanical equivalent of light.

luminous flux (Opt) The rate of flow of light, measured in lumens, indicating the intensity of the source.
NOTE: The standard light source is the candela, which emits 4π lumens and has the intensity of 1 candlepower.

luminous intensity unit See: candle.

lung irritant (Med; Toxicol) A vapor, gas, or aerosol that irritates the lungs and causes pulmonary edema. Lung irritants can be effectively screened out by gas masks.

lymph (Anat; Physiol) A clear, coagulable fluid that flows through the lymphatic vessels into the blood stream and serves as an intermediary between the blood and the body tissues.
NOTE: The lymphatic system supplements the venous system, picking up excess fluid, bacteria, and debris. It contains nodes or way stations at that bacteria are attacked.

lymphatic system See: circulation system

lysis (Med) 1. Gradual decline of disease symptoms and abatement of fever. 2. Dissolution of cells, such as bacteria and red corpuscles into

a structurless liquid.
  NOTE: Some bacteria, notably the streptococci, release materials which destroy the white blood cell walls that develop to seal off the bacteria from the rest of the body.

M

m (Phys) Abbreviation for meter, mile, milli-.

MAC = **maximum allowable concentration**

machine guarding (Safety) The installation of equipment or devices on machines to eliminate hazards created by operation of the machines.

machine intensive (Ind Eng) Characteristic of a production process in which machines predominate over manual labor, such as in an automated petroleum refinery. See: **labor intensive**.

mackerel sky (Meteor) Uniform bands of altocumulus and cirrocumulus clouds. See: **cloud**.

macroscopic (Opt) Large enough to be seen by the eye. Syn: **megascopic**.

magazine (Safety) Any building, compartment, or structure reserved for the storage of explosives.

magnaflux (Testing) A test in which particles of iron are applied to the surfaces of a magnetized specimen; the particle pattern indicates surface or near-surface flaws or irregularities.

magnesium (Metall; Symb: Mg) A silvery, ductile metal used in lightweight alloys and chemical processes.
NOTE: Magnesium burns with an intense white light and is often used in pyrotechnics.

magnetic circuit (Electr) The complete path of magnetic lines of force.

magnetic field (Phys; Symb: H) The space in which a magnetic force exists.

magnetic particle inspection See: **magnaflux**.

magnification (Opt) The number of times the apparent size of an object has been increased by the lens system of a microscope.

main^1 (Electr) A large electrical conductor used for transmitting or distributing power to service lines.

main^2 (Hydraul) A large water conduit that supplies and distributes water to smaller branches.

main aisle See: **aisle**

main section (Fire Pr) The bottom section of an extension ladder. Syn: **bed section**.

maintainability (Ind Eng) A characteristic of design and installation which is expressed as the probability that an item will be retained in or restored to a specified condition within a given period of time, when the maintenance is performed in accordance with prescribed procedures and resources.

maintenance (Ind Eng) The activity involved in keeping plant and equipment in serviceable condition or to repair and restore them to serviceability.
NOTE: Maintenance includes inspecting, testing, servicing, repairing, overhauling, modifying, modernizing, and rebuilding. **Preventive maintenance** is a system of scheduled overhaul and replacement of key parts to forestall breakdown. **Routine breakdown maintenance** involves the restoration of facilities and equipment by making needed repairs on a preset schedule.

maintenance engineering (Ind Eng) The function of providing policy guidance to maintenance activities for the purpose of exercising technical and management review for effective maintenance programs.

major injury (Safety) An injury involving a loss of time to the injured person and a medical expense.

major medical (Insur) Referring to a plan designed to insure workers against the costly medical expenses resulting from catastrophic or prolonged illness or injury.

make coupling (Fire Pr) To join two lengths of hose by connecting the couplings.

make up (Fire Pr) 1. To pick up hose and equipment and prepare to return to service. See: pick-up. 2. (Brit) = To reinforce.

makeup air (Ind Hyg) Clean outdoor air supplied to a work space to replace air removed by exhaust ventilation or by an industrial process.

malaise (Med) A general feeling of discomfort and illness.

male coupling (Fire Pr) 1. A threaded hose nipple that mates with the bowl of a female swivel coupling having the same thread and diameter. 2. A coupling to which nozzles and other similar appliances can be attached.

malice (Physiol) A disposition to wish or cause harm to others.

malicious false alarm See: false alarm

malignant (Med) 1. Cancerous, either locally invasive or capable of undergoing spread to remote parts of the body (metastasis). Ant: benign. 2. Virulent; tending to go from bad to worse, such as malignant high blood pressure.

malingerer (Psychol) One who feins illness or other abnormalities. Referring to one who is unable to function productively.

malleability (Metall) The property of a substance, usu metallic, that permits it to be rolled or hammered into sheets.

malpractice (Insur; Law) Misconduct, negligence, or lack of proper care or skill on the part of any professional, such as a doctor, dentist, attorney, or engineer.

Maltese cross (Fire Pr) A badge worn by firefighters.
  NOTE: The Maltese cross is an eight-pointed figure, the heraldic insignia of the crusader Knights of Malta. The four arms of the Maltese Cross have deep triangular indentations, which make it an eight-pointed figure. It is being replaced by the cross patee, which has blunt arms. The latter is similar to the Imperial German Iron Cross.

mamma (Meteor) An udder-shaped cloud type. See: cloud.

management (Man Sci) The decision-making, directing, controlling, and performance review process of an organization's administration.

management control system (Man Sci) An orderly procedure, generally documented, assisting managers in defining or stating policy, objectives and requirements and assigning responsibility; achieving efficient and effective use of resources; periodically measuring performance; comparing that performance against stated objectives and requirements and and taking appropriate action.

manager (Man Sci) An individual with line responsibility and authority to direct the operations of an enterprise, and who is held accountable for the effective performance of some phase of operations.

man-caused risk (For Serv; Abb: MCR) A measure of the number of firebrands originating from human activities during the rating day for a given protection unit.

mandatory Required; obligatory

mandrel (Mech Eng) 1. A rod used to align holes in drilled materials. 2. A rod that forms a hole in tubular materials, such as pipes during manufacture.

maneuverability (Transp) Property of a vehicle relating to ability to control, change, or retain direction of motion and speed.

man-function (Ind Eng) The function allocated to the human component of a system.

manganese (Chem; Symb: Mn) A silvery gray metal; hazardous if fumes or dusts are inhaled.

man-hours (Man Sci) The hours worked by one or more employees, usu with regard to a specific task or project. A man-hour is the equivalent of one man working for one hour. Total man-hours is the sum of hours worked by a given population in a specified period.

manifold (Fire Pr; Techn) A liquid or gas distribution device that has a large inlet and several smaller

outlet connections, or several small inlets and one large outlet.
  NOTE: An exhaust manifold collects and channels combustion products from an engine into a muffler and the exhaust pipe. A pumper manifold is part of the internal plumbing of a pumper. A portable manifold is used in the same way as a siamese.

manifold truck = distributor truck

manila hemp A plant with large leaves, related to the banana tree, extensively cultivated in the Philippine Islands for its fibers, which are used to make rope and other products.

manipulative dexterity (Ind Eng) The degree of skill a person has in using his fingers, hands, and wrists for fine tasks. See also: manual dexterity.

manipulator (Ind Hyg; Nucl Safety) A mechanical device, commonly in the form of claws or clamps mounted on an articulated arm, used for safe handling of hazardous materials. The arms and clamps are usu operated remotely from behind a protective shield. See: hot cell. Cf: glove box.

manlift (Build) A device consisting of a power-driven endless belt moving in one direction only and provided with steps or platforms and handholds attached to it for the transportation of personnel from floor to floor. Syn: escalator.

manometer (Metrol) An instrument for determining the pressure of gases, vapors, or liquids, both above and below atmospheric pressure. Cf: absolute manometer. See also: gage.
  NOTE: The most common form is a U-shaped tube partially filled with liquid, one end open to the atmosphere and the other attached to the object in which pressure is to be measured. The levels of the liquid in the legs of the tube indicate the differential pressure.

man-passing-man method (For Serv) A system used in fire suppression in which each crewman is assigned a specific task, such as clearing a specified section of a control line. When the assigned task is completed, he passes other workers in moving to a new assignment, which may be the identical task but in a new location.

mansard roof (Build) A roof with a double pitch on both sides. Cf: gabled roof.

mantissa See: logarithm

manual¹ By hand.

manual² (Educ) A handbook.

manual³ (Fire Pr) An alarm initiated by hand to a central station, in contrast to an automatic signal.

manual control switch (Techn) 1. An auxiliary device for manual operation of or intervention with an automatic controller. 2. Any control switch that is actuated by hand.

manual dexterity (Ind Eng) The ease with which a person uses his hands, including arm-hand, as well as wrist-finger movements. See also: manipulative dexterity.

manual rate (Insur) The compensation insurance premium in dollars per hundred dollars of payroll for a specified classification of operation or risk as listed in an official manual for a given state or jurisdiction.

manway (Mining) A shaft with a ladder used by miners to move from one level in a mine to another. Also a narrow communicating tunnel.

map (Cartog) A drawing representing a portion of the earth's surface.

maps and records officer (For Serv) A staff officer in a fire suppression organization responsible for preparing and maintaining administrative records at a fire, such as the disposition of men and equipment, instructions issued, progress maps, and organization charts.

map symbols (Cartog) A system of marks, characters, lines, abbreviations, and drawing conventions used to identify features represented on maps.

margin of safety (Safety) The limit beyond which a particular behavior, condition, or situation becomes

265

hazardous or unsafe.

marine chemist (Fire Pr; Safety) A chemist who inspects ships for hazardous conditions on board prior to repair work.

marine fire signal (Navig) A fire signal sounded by a ship in port by blowing its whistle five times with blasts of about 5-sec duration. The signal is repeated until a fireboat responds.

marine inversion = coastal inversion. See: inversion.

marine service station See: service station

marking (Storage; Transp) Numbers, nomenclature, or symbols stamped, painted, or otherwise affixed to items or containers.

marlinspike (Techn) A pointed metal or wooden tool used to separate strands of rope for splicing.

marsh gas = methane

maser (Electr) = Microwave Amplification by Stimulated Emission of Radiation. See: laser.
  NOTE: For optical masers, it is often interpreted as molecular amplification by stimulated emission of radiation.

mask (Fire Pr; Safety) A face mask designed to filter out noxious fumes, toxic gases, and lung irritants and to provide breathable air during firefighting operations. See: air mask; airline respirator; all-service mask; demand-type mask; oxygen mask; breathing apparatus.

mask service unit (Fire Pr) A fire department unit that inspects, repairs, recharges, and services breathing apparatus and related equipment.
  NOTE: The Mask Service Unit of the New York Fire Department also maintains and services acetylene cutting equipment.

masonry (Build) A structure consisting of individual building units laid in and bound together with mortar. The material used is commonly brick, stone, concrete block, glass block, and tile.

mass^1 (Phys) The quantity of matter in a body and a measure of its inertia; the property of a body that determines the acceleration it will acquire when acted upon by a given force. Often used as a synonym for weight, which, strictly speaking, is the force exerted by a body under the influence of gravity. See: inertia; weight.
  NOTE: Mass is one of the fundamental quantities basic to all physical measurements, the others being length and time.

mass^2 (Mining) A large ore body of irregular shape.

mass action law (Chem) The rate of a chemical change is proportional to the concentration of the reacting substances.

mass burning rate (Comb) The mass of material burned per unit time.

mass fire (Fire Pr) A fire involving many buildings, structures, or a large forest area.
  NOTE: Mass fire is a relative term, including group fire, conflagration, and fire storm. For example, what would be called a group fire in a large city might be called a mass fire or conflagration in a small community.

mass median drop size (Phys) The mean size of a drop in a spray; a drop of such a size that half the drops in the spray are smaller.

mass moment of inertia (Mech) The second moment of mass; for a discrete mass, the mass moment of inertia is the mass multiplied by the square of the distance from the reference axis to the center of mass, slug-ft^2.

mass number (Nucl Phys; Symb: A) The sum of the neutrons and protons in a nucleus. It is the nearest whole number to an atom's atomic weight. For instance, the mass number of uranium-235 is 235. Cf: atomic number.

mass separation (Nucl Phys) The spacing between fissionable material which controls the rate of fission.

mass spectrograph (Spectr) An instru-

ment for detecting and analyzing the relative masses of isotopes in an element. See: spectrum.
  NOTE: Nuclei that have different charge-to-mass ratios are separated by passing them through electrical and magnetic fields.

mass spectrometry (Chem; Spectr) A technique of qualitative and quantitative chemical analysis in which gaseous ions of differing mass and charge are separated by electrical and magnetic fields. Mass measurements of the individual species are made with an instrument called a spectrograph, while the relative abundances of different ions are determined by a spectrometer.

mass transfer (Phys) The transport of material under a force gradient. Mass transport in a pressure gradient is convection, transport under a concentration gradient is called diffusion. Mass transport in a temperature gradient is called thermal diffusion. Heat and mass transfer are major processes along with chemical reaction in fires.

master planning (Fire Pr) A systematic process for determining how much risk a community is willing to assume with respect to fire protection.

master stream (Fire Pr) A large-volume fire stream developed by siamesing two or more large hose lines into one. Syn: heavy stream.
  NOTE: Master streams are usu discharged at 80 to 100 psi nozzle pressure at the rate of 400 to 1200 gpm. Master streams are produced by monitors, deluge sets, ladder pipes, wagon batteries, and turret pipes.

mastic (Build) One of a variety of bituminous-like materials used as sealers, adhesives, and waterproofing materials in building construction.

mastoid process (Anat) The bony part of the skull protruding below and behind the ear; part of the temporal bones that are often tender when the middle ear is infected.

match A wooden splint with a head consisting of a combustible material and an eye that ignites when it is rubbed against a rough surface. The eye is made of a red phosphorus compound. So-called safety matches and book matches have heads containing an oxidizing compound, and the red phosphorus compound is attached in a strip to the book or box.

materials handling (Mat Hand) The activity involved in lifting, transporting, and depositing material.
  NOTE: In manual handling of materials, a variety of hand or hand-operated accessories may be used, such as hooks, bars, jacks, handtrucks, dollies, and wheelbarrows. Mechanical materials handling involves mechanized devices or equipment, such as hoisting apparatus, conveyors, elevators, railways, powered industrial trucks, as well as ropes, slings, and chains.

materials handling equipment (Mat Hand) Any devices, mechanically or manually operated, used for movement or handling of supplies.

materials lock (Techn) A chamber through which materials and equipment pass from one air pressure environment into another.

mathematical statistics (Statist) The branch of statistics that is concerned with the mathematical proofs and theorems used in statistics.

mathematics (Abb: Math) The branch of science that deals with quantities, their measurement, and their interrelationships.

mating (Mech Eng) Fitting together of interconnecting parts. If mating involves insertion of one part into another, the receiving element is called a female connector and the other a male connector. See: coupling.

matrix[1] (Metall) A mold, as one used in a casting.

matrix[2] (Math) A set of related elements arranged in rows and columns. A matrix of one row or one column is called a vector.

matter (Phys) An aggregate of particles that occupies space and has mass. See: atom; element; molecule.

mattock (For Serv) A hand tool with a pick or chopping blade at one side

267

of the head and a hoeing blade at the other, used for digging and hoeing, esp when constructing fire control lines.

maul and bar (Fire Pr; Salv) A heavy sledge hammer used with a crow bar to break sewer pipes during salvage operations.

maximum allowable concentration (Toxicol; Abb: MAC) The quantity of dust, vapor, or fumes, measured in parts per million, which if exceeded would be detrimental to health.

maximum boiling point (Chem) The boiling point of a multicomponent liquid system such that all other proportions and the components individually have lower boiling points.

maximum credible accident (Nucl Safety) The most serious nuclear reactor accident that can be imagined, within reason, from any combination of equipment malfunction, operating errors, or other foreseeable causes.

maximum evaporative capacity (Ind Hyg) The quantity of evaporating sweat from a human being that the humidity will permit.

maximum freezing point (Chem) The freezing point of a multicomponent liquid system such that all other proportions and the components individually have lower freezing points.

maximum permissible concentration (Nucl Safety; Abb: MPC) The quantity of radioactive material in air, water, or food which might be expected to result in a maximum permissible dose to persons inhaling or ingesting them at a standard rate of intake. See: radiation protection guide; radioactivity concentration guide.
NOTE: Maximum average concentrations of radionuclides to which a worker may be exposed 8 hours a day, 5 days a week, and 50 weeks a year have been recommended by the National Committee on Radiation Protection.

maximum permissible dose (Nucl Safety; Abb: MPD) The dose of ionizing radiation below which there is no reasonable expectation of risk to human health, and which is somewhat below the lowest level at which a hazard is believed to exist. See: radiation protection guide.

maximum permissible exposure See: maximum permissible dose.

maximum permissible level (Nucl Safety) The largest flux of nuclear radiation to which humans can be safely exposed. Also called: maximum permissible limit.

maximum rated load (Ind Eng; Transp) The total of all loads, including the working load, the weight of the carrier or vehicle, and any other load that may be expected.

MCA - Manufacturing Chemists Association

McLeod gage (Phys) A manometer used for determining low gas pressures by compressing a sample until a measurable pressure is reached.

McLeod tool (For Serv) A combination hoe and rake tool used in control line construction. Similar to a Kortick tool, but heavier.

mean (Math; Statist) 1. Usually the arithmetic mean or average, obtained by adding a set of values and dividing the sum by the number of values in the set. 2. Sometimes refers to the geometric mean, which is obtained by multiplying the individual values together, then extracting the n-th root of the product, where n is the number of values.

mean deviation (Statist) A number that indicates the variability of a set of measurements, numbers, or values. Obtained by finding the difference between each value and the mean, and then averaging the differences, ignoring the signs. See also: standard deviation.

mean effective pressure (Mech Eng) The average pressure exerted by the working fluid on the piston of an engine or other piston mechanism during one stroke.

mean free path[1] (Acoust) The average distance traveled by a sound wave between successive reflections in an enclosure.

mean free path² (Phys) The average distance traveled by a particle, such as a molecule or electron, between successive collisions with other particles.

mean life (Phys) The average time during which an atom, an excited nucleus, a radionuclide, or a particle exists in a particular form. See: scattering.

mean maintenance time (Ind Eng) The total preventive and corrective maintenance time divided by the total number of preventive and corrective maintenance actions during a specified period of time.

mean sea level pressure (Meteor) The pressure of a standard atmosphere: 29.92 inches of mercury or 1013.2 millibars. See: pressure measurement.

means of egress (Build; Fire Pr) A route for exiting a building, preferably a continuous and unobstructed path from any point in a building or structure to a public way.
NOTE: A means of egress consists of three separate and distinct parts: the way of exit access, the exit, and the way of exit discharge. A means of egress comprises the vertical and horizontal paths of travel, including intervening room spaces, doorways, hallways, corridors, passageways, balconies, ramps, stairs, enclosures, lobbies, escalators, horizontal exits, courts, and yards.

mean time between failures (Ind Eng; Abb: MTBF) For a particular interval, the total functioning life of a population of an item divided by the total number of failures within the population during the measurement interval. The definition holds for time, cycles, miles, events, or other measure of life units.

mean time to repair (Ind Eng; Abb: MTTR) The total corrective maintenance time divided by the total number of malfunctions during a given period of time.

mechanic (Techn) A worker skilled in the use of tools.

mechanical ability (Psych) An aptitude for dealing with mechanisms and machines.

mechanical brakeman (Transp) A slotted metal plate nailed to the floor of a car, used to retard movement of lading.

mechanical equivalent of heat (Phys; Thermodyn) The number of units of work equivalent to one unit of heat; 778 foot pounds = 1 Btu, or 1 calorie = 4.18 joules = 41,800 ergs.

mechanical foam See: foam

mechanical foam generator See: foam generator

mechanical hazard (Safety) Any unsafe condition involving machinery, equipment, tools, etc.

mechanical impedance (Mech Eng) The ratio of a force-like quantity to a velocity-like quantity when the arguments of the real (or imaginary) parts of the quantities increase linearly with time. Impedance is the reciprocal of mobility.

mechanical property (Techn) Any of several characteristics relating to elastic and inelastic reactions of a material when a force is applied to it, or that involve a stress-strain relationship, as for example the modulus of elasticity, tensile strength, or fatigue limit. Cf: physical property.

mechanical shock (Techn) A nonperiodic excitation, e.g., a motion of the foundation or an applied force of a mechanical system, that is characterized by suddenness and severity and usu causes significant relative displacements in the system.

mechanics (Abb: Mech) The science that deals with the effects of forces upon bodies at rest and in motion, divided into the study of fluids (fluid mechanics) and of solids (mechanics of solids). Mechanics is also divided into statics and dynamics, with the latter subdivided into kinematics and kinetics. Statics deals with bodies at rest or in equilibrium; kinematics with abstract motion; and kinetics with the effect of forces or moment on the motion of material bodies.

medial (Anat) Referring to the midline of the body, or toward the midline.

median (Math; Statist) The middle value in a distribution; the 50th percentile; the point that divides a group of numbers in a distribution into two equal parts.

median interval (Statist) The class interval that contains the median.

median lethal dose See: lethal dose.

medical benefit (Insur) Financial assistance provided to employees and often to their dependents, as well, for specified medical care, doctor's visits, and sometimes prescription drugs. Generally a part of a health insurance program.

medicine (Med) 1. The art or science of healing diseases. 2. Any drug or remedy.

medium (Phys) 1. An intervening solid, fluid, or empty expanse through which a force acts or energy is transmitted, such as sound through air, solar energy through space, or a magnetic field through an iron bar. 2. A material that contains or holds physical particles, such as the oil that holds pigments in suspension in paint.

medium-expansion foam See: foam expansion

medium lot storage (Storage) A quantity of supplies, arranged in one to three stacks, stored to a maximum height.

mega (Symb: M) A prefix meaning one million; e.g., a megawatt is one million watts. Sometimes Meg, as in Megohm.

megaton energy (Nucl Phys) The energy of a nuclear explosion which is equivalent to that of an explosion of one million tons of TNT.

melt (Techn) The total batch of ingredients introduced into a pot or furnace at one time in glassmaking or metallurgy.

melting point (Chem; Phys; Abb: mp) The temperature at which the solid and liquid phases of a substance are in equilibrium; usu used in reference to temperatures above 0°C.
NOTE: The heat absorbed by a melting substance is called the latent heat of melting. The melting point is usu affected by pressure. In some cases the melting point is raised and in others lowered when pressure is increased, depending on whether the liquid or solid phase has the greater density. See: freezing point.

membrane (Anat) A thin, pliable layer or sheet of animal tissue that serves to cover a surface, line the interior of a cavity or organ, or divide a space, often permeable to fluids and their dissolved constituents.

membrane barrier (Techn) A thin layer of material impermeable to the flow of gas or water.

memory[1] (Physiol) 1. The store of past experience that can be recalled. 2. The capacity to store experiences.

memory[2] (Comput) The component or medium used for storing data in electrical or magnetic form in a computer.

membrane barrier (Techn) A thin layer of material impermeable to the flow of gas or water.

meniscus[1] (Phys) The curved surface of a liquid in a tube.

meniscus[2] (Opt) A lens that is convex on one side and concave on the other.

mensuration (Metrol) The process of measurement.

mental disorder (Physiol) Any serious mental disturbance or maladjustment, including psychoses and neuroses.

mephitic (Toxicol) Noxious; poisonous.

mercantile occupancy (Fire Pr) A building or part of one that houses a store, market, or other merchandise outlet.

mercury (Chem; Symb: Hg) A silvery-white liquid metallic element, density 13.6 g/cm² (0.49 lb/in³), mp -39°C, bp 375°C, used in thermometers, barometers, and manometers. Syn: quicksilver.
 NOTE: Mercury is 13.546 times heavier than water.

merit rating (Man Sci) An evaluation or appraisal of performance on a particular job. An efficiency rating.

MESA = Mine Enforcement and Safety Administration

meson (Phys) One of a number of short-lived, unstable particles carrying a positive, negative, or no charge and having a mass intermediate between the electron and proton.

mesosphere (Meteor) A layer in the atmosphere above the stratosphere, extending to approx 50 miles altitude and characterized by gradual increase in temperature to about 30 miles and then gradual decrease in temperature up to the mesopause. See: atmosphere; pause.

message (Fire Pr) A fire department communication consisting of a call, response, text, and acknowledgement.

metabolic shock (Med) A state resulting from the loss of body fluids.

metabolism (Physiol) The sum of material interchanges between a living organism and the environment by which tissue changes occur and energy is obtained for vital processes.
 NOTE: Metabolism consists of anabolism, energy consumption processes, and catabolism, energy production processes.

metacenter (Navig) The intersection of a line drawn vertically through the heeled center of buoyancy and the ship centerline plane. The heeled center of buoyancy is not on the centerline.

metacentric height (Navig) Distance from the metacenter to the center of gravity of the vehicle, symbolically GM.

metacentric radius (Navig) Distance from the center of buoyancy to the metacenter.

metal (Techn) An electropositive element, usu a solid with high strength and plastic properties, that is a good conductor of heat and electricity, reflects light, has luster, and forms basic hydroxides and oxides.
 NOTE: Metallic sodium, potassium, and lithium burn vigorously in air and react violently with water. Magnesium will burn and cannot be extinguished by carbon dioxide, vaporizing liquid extinguishers, or many of the common chemical extinguishing agents. Burning magnesium combines with atmospheric air to form magnesium nitride. Molten magnesium reacts with water, liberating heat and hydrogen gas, which can lead to an explosion hazard. Powdered iron, titanium, zirconium, and steel wool (esp if oiled) present combustion and explosion hazards. Magnesium and aluminum are extremely dangerous in powdered form. Powdered aluminum and iron oxide form a high-temperature combustible material called thermite. Copper can burn in a chlorine atmosphere. Other combustible metals and dust-explosion hazards are germanium, hafnium, plutonium, and uranium.

metal-clad fire door (Build) A flush or panel door consisting of metal-covered wood cores or stiles and rails and insulated panels covered with 24-gauge or lighter steel. Also called: Kalamein.

metalizing (Techn) An operation involving the melting of wire by means of a flame in a special device which sprays the atomized metal onto a surface to be coated. The metal may be steel, lead, or other metal or alloy.

metalloid (Chem) A semimetal that is neither very electronegative or electropositive and the oxides of which do not form strong acids or bases. Some examples are arsenic, carbon, silicon, germanium, selenium, antimony, and tellurium. Also called: semiconductor. Cf: nonmetal.

metallurgy (Metall) The science of extracting metals from their ores and processing them in pure states or converting them into alloys.

metal poisoning (Toxicol) Illness or death resulting from contact with, ingestion, or absorbtion of metallic substances; frequently resulting from inhaling metal dust or fumes.

metastasis (Med) Spread of malignancy from the site of primary cancer to secondary sites due to transfer through the lymphatic or blood system.

meteorology (Abb: Meteor) The study of the state of the atmosphere, the processes occurring therein, and the weather.

meter[1] (Metrol) A measuring instrument. Unless otherwise specified, usu a device that integrates the value of a quantity being measured with respect to time.
NOTE: A gas meter, for example, measures the volume of gas that passes through it.

meter (Metrol; Abb: m) A metric unit of length equal to 39.37 inches and consisting of 100 centimeters, or 1000 millimeters.
NOTE: Originally supposed to be 1/10,000,000 of the earth's meridian quadrant at sea level but now based on the red line of cadmium (1,553,164.13 wavelengths at standard conditions.) In the international (SI) system, a meter is exactly 1,650,763.73 wavelengths of the orange-red line of Krypton 86 under given conditions.

methane (Chem) A gaseous combustible hydrocarbon, CH_4, of the paraffin series, bp -161°C, frequently found in mines, sewers, and oil fields. The gas is colorless, odorless, half as dense as air, slightly soluble in water, and forms explosive mixtures with air. Syn: marsh gas; firedamp.

methanamine (Testing) The chemical substance of the pill used in the pill test for floor coverings (carpets and rugs).

methanol (Chem) A liquid organic compound, CH_4O, obtained by synthesis or by destructive distillation of wood. Syn: methyl alcohol; wood alcohol.
NOTE: Used in chemical synthesis, as a denaturant of ethyl alcohol, as a solvent, and as an antifreeze. Highly toxic by ingestion or inhalation.

methemoglobin (Med) Hemoglobin in which oxygen is tightly bound with iron; it is therefore unable to combine reversibly with oxygen as normal hemoglobin does in normally oxygenated blood. The condition is caused by poisoning, as by potassium chlorate.

method (Man Sci) A systematic procedure or a series of calculated steps to carry out a task effectively or to reach a desired goal.

methyl alcohol = methanol

methyl bromide (Chem; Fire Pr) A vaporizing liquid insecticide and extinguishing agent, CH_3Br, used on electrical fires. See: Halon.
NOTE: The compound itself is toxic and produces toxic substances when applied to a fire. From 1922 to 1940 it was marketed as a fire extinguishing agent.

methyl iodide (Chem) An iodine substituted, methane compound, CH_3I, that is an effective flame suppressant but is dangerously toxic. See: Halon.

metric system (Metrol) A decimal system of measurement based on the meter as the primary standard. The secondary standard of weight, the gram, is derived from a cubic centimeter of water. See: standard international units; exponential notation.
NOTE: Common prefixes used in the metric system are:

```
deci  =         one tenth = 10^-1
centi =     one hundredth = 10^-2
milli =    one thousandth = 10^-3
micro =    one millionth  = 10^-6
nano  =    one billionth  = 10^-9

deka  =          ten times = 10^1
hecto =  one hundred times = 10^2
kilo  = one thousand times = 10^3
mega  = one million times  = 10^6
pico  = one billion times  = 10^9
```

Mev (Phys) One million (10^6) electron volts. Also written as MeV.

mezzanine (Build) An intermediate or fractional story between a floor and ceiling of a building.

mgd = million gallons per day

mica (Mineral) One of a large group of silicates of varying composition, but similar in physical properties. All have excellent cleavage and can be split into very thin sheets. Used in electrical insulation.

micro- (Symb: u or μ) A prefix meaning extremely small, or one millionth; e.g., one microsecond is one millionth of a second, or 10^{-6} sec.

microbe (Bact) A microscopic organism belonging to either the plant or animal kingdom.

Microdon (Med) A trade name for a nonadherent burn dressing, consisting of porous polypropylene backed by absorbent cellulose acetate batting.

microform (Info Sci) A sheet of microfilm, often 4x6 in., carrying images of document pages.

micron[1] (Metrol) A unit of length equal to one millionth of a meter, or 10^{-6} m. Syn: micrometer.

micron[2] (Metrol) A unit of pressure equal to 10^{-3} mm Hg, or about one millitorr.

microorganism (Bact) A minute organism; microbes, bacteria, cocci, viruses, molds, etc., are microorganisms.

microphone (Acoust) An electroacoustic transducer that responds to sound waves and delivers essentially equivalent electric waves.

microscopic reversibility (Chem) The principle in molecular reaction kinetics which asserts that every process and its exact opposite occur at equal rates at equilibrium.

microstructure (Metall) The fine texture of a surface, as of a polished and etched metal specimen observed through a microscope.

microwave radiation (Commun) Electromagtic radiation with extremely short waves, particularly those which are less than one meter long.

middle ear (Anat) The part of the ear between the eardrum and the cochlea.

middle management (Man Sci) The management level between top management and supervisory management responsible for the successful operation of divisions or departments.

midship (Navig; Transp) Midway between the ends of a vehicle; for a vehicle with twenty-one stations the midship is station 10.

midship pump (Fire Pr) A fire pump mounted on a truck chassis behind the driver's seat.

mil (Metrol) A unit of measure equal to one one-thousandth of an inch. Also called: thou.

mile (Metrol) A distance of 5,280 feet, 1760 yards, or 1609 meters. Also called: statute mile; English mile. Cf: nautical mile.
NOTE: Originally supposed to be the equivalent of one thousand paces of a Roman legionnaire.

mileage death rate (Safety) The number of deaths from vehicle accidents per 100,000,000 miles of vehicle travel in a given geographical area.

mill (Techn) A machine consisting of two adjacent metal rolls, set horizontally, which revolve in opposite directions; used for the mechanical working of plastic materials.

milled refuse (Env Prot; Techn) Solid waste that has been mechanically reduced in size.

milli- (Metrol; Symb: m) A prefix meaning one thousandth; e.g., one milligram is one thousandth of a gram, or 10^{-3} g.

milliampere (Electr; Abb: ma) One-thousandth of an ampere.

millicurie (Nucl Phys) One-thousandth of a curie.

milligram (Metrol; Abb: mg) One thousandth of a gram.

milligrams per liter (Env Prot; Toxicol) A ratio used to express concentration; e.g., concentration of salts in the blood serum or contaminant concentrations in gases.
NOTE: One pound equals 454 grams or 454,000 milligrams; one cubic foot equals 28.32 liters; hence 1 mg/liter is approximately equivalent to 62.4 pounds per million cubic feet.

millimeter (Metrol; Abb: mm) A unit of length equal to one thousandth of a meter, or one-tenth of a centimeter.
NOTE: The diameter of the wire of an ordinary paper clip is approx 1 mm.

millimeter of mercury (Phys; Abb: mmHg) A unit of pressure equal to the pressure exerted by a column of liquid mercury one millimeter high at a standard temperature. Syn: **torr**.

millimicron (Metrol) A unit of length equal to 0.001 micron or one thousandth of a micron.

million gallons per day (Abb: mgd) A unit of measure used in reference to water supply rates.

milliroentgen (Nucl Phys) One one-thousandth of a roentgen.

mill-type construction (Build; Fire Pr) A structure with masonry walls, heavy plank floors, and heavy supporting timbers. Syn: **heavy timber construction**.

millwork (Build) Wooden building materials produced by millwork plants and planing mills, such as sashes, doors, frames, panels, and other items of interior and exterior trim.

mine (Mining) An opening in the earth's crust from which valuable materials may be extracted.

mine gas (Mining) A mixture of methane and nitrogen given off by coal in coal mines.

mine inspector (Mining; Safety) An individual who makes periodic examinations of mines to determine if operators are complying with state laws and regulations.

mineral (Mineral) An inorganic substance of definite chemical composition found in nature, such as granite or quartz.

mineral pitch (Techn) Tar from petroleum or coal in distinction to wood tar.

mine safety engineer (Mining) A specialist who inspects underground or open-pit mine workings and trains mine personnel in safety matters to insure compliance with state and federal laws and accepted mining practices. He also investigates explosions, fires, and accidents and reports causes; recommends remedial action to insurance companies, mine management, and state authorities; and maintains rescue equipment.

minimum acceptable reliability (Techn) The reliability that must be achieved before material can be approved for use.

minimum boiling point (Phys) The boiling point of a multicomponent system in such proportions that all other proportions and the components individually have higher boiling points.
NOTE: Solder, a mixture of lead and tin, has a much lower boiling point than either metal in its pure state.

minimum damage theory (For Serv) A theory which states that the objective of forest fire control is minimum fire damage. See: **economic theory**.

minimum maintained velocity (Ind Hyg) The velocity of air movement that must be maintained in order to meet minimum specified requirements for health and safety.

mining (Mining) The process of extracting useful minerals from the earth's crust, both from the surface and underground.

minor injury (Safety) An injury involving no lost time or appreciable medical treatment or costs.

miscellaneous cause (Fire Pr) A fire cause that does not fit properly into any of the standard categories listed in fire-cause statistics.

miscellaneous fire (For Serv) A fire that cannot be classified under any of the other seven fire causes. See: causes of fire.

miscibility (Chem) The ability of two or more substances to mix and form a homogeneous phase.
NOTE: Two liquids may form a conjugate solution: two layers of liquid, each saturated with the other.

miscible (Phys) Mixable; referring to two or more substances that can be mixed together and remain as a mixture without tending to separate. Ant: immiscible.

misfire (Civ Eng; Mining) An explosive charge which failed to detonate.

mission time The period of time in which an item must perform a specified mission.

mist¹ (Meteor) A thin suspension of very small droplets in the air. Also called: damp haze. See: fog; precipitation.

mist² (Techn) An aerosol of liquid droplets formed by condensation from the gaseous to the liquid state or by breaking up a liquid into finely divided state, such as by atomization.
NOTE: Mists are often formed in industrial processes such as spray painting, grinding, cutting, electroplating, and pickling.

mixmaster (For Serv) An individual responsible for providing fire retardants to airborne tankers engaged in forest-fire fighting.

mixture (Chem) A combination of substances in any proportion that can be separated into two or more constituents by physical methods, as opposed to a compound, which is a chemical entity.

MM = mouth-to-mouth resuscitation. See: resuscitation.

mnemonic (Physiol) A cue for remembering something, usu an association made between something familiar with what needs to be remembered.

mob (Physiol) A crowd of individuals acting together as a group under strong emotional feelings.

mobile home (Build) 1. A living unit mounted on a wheeled chassis, towed as a trailer by a road vehicle, and used as a movable residence. 2. = (Brit) caravan.
NOTE: Usu parked in a trailer court or trailer park, connected to utilities, and used as a long-term residence. Two units parked side-by-side and interconnected are called a double-wide.

mobile unit (Fire Pr) A traveling fire department radio communications unit, as distinguished from a base station.

mobile work platform (Ind Eng) A platform, mounted on casters or wheels, that can be moved from place to place and used to reach various heights above floor level.

mobility (Transp) The ease with which a vehicle can move from one point to another under its own power.

mobility index (Transp) A dimensionless number obtained from a mathematical interrelationship of certain characteristics of a vehicle.

mode (Statist) A local maximum in a frequency distribution for a given population or sample. The value or score that occurs most frequently. See: mean.

model¹ An ideal or standard to be followed or emulated.

model² (Ind Eng; Techn) A mathematical, graphic, or physical representation of a real object or system, the behavior of which can be studied, analyzed, and tested.
NOTE: Used to predict the characteristics and performance of the real object or system.

model arson law (Fire Pr) A model law for arson recommended by the Fire Marshals' Association of North America and adopted into law by many states. See: arson.

model fireworks law (Fire Pr) A model law for regulating the use of fireworks recommended by the Fire Marshals' Association of North

America and adopted into law by many states. See: <u>fireworks</u>.

<u>modeling</u> (Techn) The art or practice of predicting the results of an experiment by analogy with another, usu simpler and smaller experiment. Modeling is accomplished by expressing physical processes in numerical form and varying the values to observe the behavior of other values.
  NOTE: Models are small in scale and mathematical and are much cheaper than full-scale experiments. Successful modeling often improves the physical insight into a system. Modeling may be complete or partial, depending on whether a single parameter or all important physical parameters are modeled. Where a single or dominant parameter is important, such as a linear dimension, the process is called <u>scaling</u>.

<u>modem</u> = <u>modulator-demodulator</u>

<u>moderator</u> (Nucl Phys) A material, such as ordinary water, heavy water, or graphite, used in a reactor to decelerate high-velocity neutrons, thus increasing the likelihood of further fission.

<u>modular</u> (Build) Having standard dimensions, shape and composition; interchangeable with others similar to it. Replaceable with off-the-shelf items or easily available substitutes.

<u>modulation</u> (Electr; Commun) The purposeful variation of the frequency or amplitude of a given wave by impressing another wave upon it.

<u>module</u> (Techn) A functional and structural assembly or subassembly so designed that, if it becomes unserviceable, it may be readily replaced by a similar standard unit.

<u>modulus of elasticity</u> (Mech Eng) The ratio of the stress (load) to the strain (deformation) of a structural material. Also called: <u>coefficient of elasticity</u>. See also: <u>Young's modulus</u>.

<u>moiety</u> (Chem) One of the components into which something is divided.

<u>moisture content</u> See: <u>humidity</u>

<u>mol</u> = <u>mole</u>

<u>molal solution</u> (Chem) A solution that contains one mole of solute per kilogram of solvent.

<u>molar concentration</u> (Chem) The number of gram molecules (moles) of a substance per liter.

<u>molar solution</u> (Chem) A solution that contains one mole of solute per liter of solution.

<u>mold</u>^{1} (Bact) A growth of fungi forming a furry patch, as on stale bread or cheese. See: <u>spore</u>.

<u>mold</u>^{2} (Metall) A hollow form or matrix into which molten material is poured to produce a casting.

<u>mole</u> (Chem) The mass of a substance in grams equal numerically to its molecular weight. Syn: <u>gram molecule</u>. Also written: <u>mol</u>. See: <u>Avogadro's number</u>.
  NOTE: One mole of any substance contains 6.023×10^{23} molecules.

<u>molecular volume</u> (Chem) The volume occupied by one mole of a solid, liquid, or gas, found by dividing the molecular weight by the density.
  NOTE: The volume occupied by one mole of any gas at 0°C and 760 mm Hg pressure is 22.4 liters.

<u>molecular weight</u> (Chem) The sum of the atomic weights of the atoms in a molecule. Atomic weights are so scaled that carbon-12 has the atomic weight of 12. See: <u>atomic weight unit</u>.

<u>molecule</u> (Chem; Phys) 1. The smallest physical unit of a substance, consisting of atoms held together by chemical forces and possessing the properties of the substance in quantity. 2. The smallest unit of a substance capable of independent existence. Cf: <u>atom</u>; <u>ion</u>.
  NOTE: A molecule can consist of identical atoms, such as oxygen (O_2) or different atoms, such as carbon dioxide (CO_2). In a few gases and metals the molecule and the atom are identical.

<u>Mollier diagram</u> (Phys) The plot of the properties of a vapor on a plane whose coordinates are <u>enthalpy</u> (total heat) and <u>entropy</u> (unavail-

able energy in the system). The diagram is used in determining actual and theoretical expansions of vapor, as well as pressure, superheat, and temperature.

Molotov cocktail (Milit) A crude incendiary device consisting of a bottle filled with combustible or flammable liquid, such as kerosine or gasoline, fitted with a wick or saturated rag that is ignited just prior to hurling. Syn: (Brit) petrol bomb. Cf: fire bomb.

moment (Mech; Phys) 1. The product of a quantity, such as a force, and a distance to a given point. 2. The turning effect of a force about a point. Also called: torque.

momentum (Phys) The product of the mass and velocity of a body in motion. See: inertia.

monaural hearing (Physiol) Hearing with one ear only.

mond gas (Brit) 1. A fuel gas produced by air and steam acting on hot coal and slack. 2. = (US) producer gas.

monitor[1] (Safety) To observe an area closely to see whether it is safe for workers.

monitor[2] (Fire Pr) A master-stream device usu fed by several hose lines through a siamese connection, operated by handwheel and gears, and used to throw a heavy fire stream or fog spray.
NOTE: The device is usu mounted permanently on a fire apparatus, fireboat, or elevating platform, but some models are designed to operate on the ground unattended.

monitor[3] (Fire Pr) A firefighter trained and equipped to detect radiation hazards and to deal with them in emergency situations.

monitor[4] (Dosim) An instrument that measures the level of ionizing radiation.

monitor[5] (Build) A vent extending above the roof line, opening on at least two opposite sides.
NOTE: The openings are normally closed by louvered panels, metal doors, or glass panes. Monitor vents are often designed to open automatically.

monitoring[1] (Commun) Listening for and recording radio signals and messages.

monitoring[2] (Fire Pr) Guarding against life hazards such as radiation and toxic substances by early detection and warning.

monitor wagon (Fire Pr) A fire truck equipped with a heavy stream device and large hose.

Monnex (Fire Pr) A trade name for a dry-powder extinguishing agent made of urea and alkali bicarbonate, with noncaking characteristics and greater effectiveness than sodium bicarbonate or potassium bicarbonate.

monobasic acid (Chem) An acid having one replaceable hydrogen atom per molecule.

monochromatic (Opt) Having a single fixed wavelength.

monochromatic light (Opt) Light of a single wavelength or of one color of the spectrum.

monolayer = monomolecular layer

monomer (Chem) A compound of relatively low molecular weight which, under certain conditions, either alone or with another monomer, forms various types and lengths of molecular chains called polymers or copolymers of high molecular weight. Example: Styrene is a monomer that polymerizes readily to form polystyrene. See also: polymer.

monomolecular layer (Phys; Chem) A film one molecule thick. Such films are used in lubrication to reduce the evaporation of water, to reduce rusting, and to stabilize solid dispersions, emulsions, and foams. Also called: monolayer.

monthly average (For Serv) The average number of man-made fires in a fire-protected unit during a given month.

moonlighting (Fire Pr) Work performed by firefighters on their off-duty time for employers other than the fire department. Colloq: C-shift.

MOP = man on probation

mop-up (Fire Pr) 1. The final stage of fire fighting, in which remaining hot spots are quenched with small quantities of water and hazardous conditions are corrected or made safe. Syn: overhaul. 2. = (Brit) clearing up.

mop-up time (For Serv) The elapsed time between control of a fire and completion of organized mop-up.

morbidity (Med) 1. Sickness; a diseased state. 2. The ratio of sick to well persons with respect to a given disease in a community or population.

mordant (Chem) A substance that combines with a stain or dye to make it insoluble.

morphine (Pharm) A bitter, crystalline alkaloid, $C_{17}H_{19}NO_3$, obtained from opium. An addictive narcotic used as a pain reliever and sedative.

morphology (Biol) The branch of biological science that deals with the study of the structure and form of living organisms.

mortality (Med) 1. The state of having a finite life span. 2. Death rate; the number of deaths in a given population per unit time, such as deaths per year per 100,000 population.

motion (Phys) The change in position of matter in space.

motivation (Psychol) An internal state or condition that impels one to respond toward some object or situation in his environment with goal-directed activity or behavior.
  NOTE: Motivation differs considerably in different individuals, depending on previous experience, personal values, and other factors. A high level of motivation is often called a drive.

motive (Psych) Inducement or reason for thought or action.

motor[1] (Electr) A machine that converts electrical energy to mechanical energy. See: engine.

motor[2] (Mining) An electrically powered transporter or locomotive in a mine.

motor[3] (Phys) Referring to muscular activity.

motor set (Phys) Positioning of the body and preparation of the muscles to perform a particular action. A condition of readiness to do something.

motor spirit (Brit) = gasoline

motor vehicle (Transp) Any self-propelled vehicle, such as an automobile, bus, truck, tractor, or truck and trailer used for the transportation of persons or freight over public highways.

motor vehicle accident (Safety) Any accident involving a motor vehicle in motion that results in death, injury, or property damage.

mouth-to-mouth resuscitation (First Aid; Abb: MM) Reestablishment of normal respiration by blowing air into the mouth of a person who has stopped breathing. Various devices are available to avoid direct mouth-to-mouth contact. Also called: artificial respiration; resuscitation.
  NOTE: General rules for the procedure are to: 1) clear the victim's mouth of obstructions, 2) start artificial resuscitation immediately, 3) don't give up - continue, 4) summon medical assistance as quickly as possible.

move up (Fire Pr) Transfer of fire companies from home stations to vacated stations nearer to a major fire to give coverage to districts that have been stripped of normal protection. Some move-ups are automatic, following additional alarms, while others are ordered by fire alarm dispatchers. Syn: relocate; change of quarters.

move-up method = progressive method

moving parts (Mech Eng) Gears, sprockets, revolving shafts, clutches, belts, pulleys, or other revolving or reciprocating parts that are attached to or form an integral part of a machine.

MPC = maximum permissible concentration

muck (Mining) Broken ore.

mucking (Mining) Breaking ore into small pieces for hauling out of a mine.

mucous membrane (Anat) The internal lining of the body passages, principally the respiratory, alimentary, and genitourinary tracts, that secretes a viscous lubricating liquid called mucus.

muffler (Transp) A special duct or pipe that impedes the transmission of sound by reducing the velocity of the air or gas flow (dispersive type), by absorbing sound (dissipative type), or by reflecting the sound back toward the source through a series of cavities and side chambers (reactive type).

mullion (Build) A vertical dividing member in a window or between adjacent doors.

multiple alarm (Fire Pr) 1. A second or greater alarm. Also called: additional alarm. 2. Calls for additional assistance from several alarm boxes for the same fire.

multiple death fire (Fire Pr) A fire in which three or more fatalities have occurred.

multiple line insurance (Insur) A policy that combines coverage for several perils. See: homeowner's policy.

multiple peril policy (Insur) A package form of commercial insurance that includes a wide range of coverages.

multiple stage pump (Techn) A centrifugal pump with two or more impellers.

multiplex (Commun) The communications mode in which signals or messages may be transmitted and received simultaneously.

multiplication (Math) A mathematical operation in which one number (the multiplicand) is added to itself as many times as indicated by another number (the multiplier), resulting in a number called the product.

multiversal (Fire Pr) Any one of the class of adjustable master-stream devices such as ladder pipes, monitors, deck pipes, and deluge sets.

municipal Pertaining to an urban political unit, usu a city or large town.

murder (Law) Taking the life of another unlawfully, esp with malice aforethought.

mushroom cloud See: atomic cloud

mushroom effect (Fire Pr; Brit) The horizontal spread of hot gases and smoke at ceiling and roof levels.

mushrooming (Fire Pr) Rapid upward and lateral extension of a fire through upper floors, caused by the build-up of heat from below and lack of venting for heat release.

muster (Fire Pr) 1. A recreational gathering of volunteer firefighters at which competitions are held in firemanship skills. 2. A roll-call for recording attendance.

mutation (Biol) A permanent transmissible change in the characteristics of an offspring from those of its parents.

mutual aid (Fire Pr) Prearranged assistance rendered to each other by two or more fire departments in emergencies and for joint response to fire alarms near jurisdictional boundaries.

myocardial infarction (Med) Necrosis (death) of a portion of the heart muscle that occurs when its blood supply is blocked.

myocardium (Anat) The heart muscle.

N

<u>n</u> (Phys) Symbol for refractive index.

<u>N</u> (Chem) Symbol for nitrogen.

<u>NACA</u> = <u>National Academy of Code Administrators</u>

<u>NAFO</u> (Brit) = <u>National Association of Fire Officers</u>

<u>NAFSA</u> = National Association of Fire Science and Administration

<u>nano-</u> (Electr; Phys) A prefix meaning one billionth (10^{-9}); e.g., a nanosecond is one billionth of a second.

<u>nanometer</u> (Metrol) A unit of length equal to one billionth of a meter or 10^{-7} centimeter.

<u>naphthalene</u> (Chem) A white crystalline hydrocarbon $C_{10}H_8$, obtained from coal tar or petroleum fractions. Used as a moth repellent and a basic material in the manufacture of dyestuffs, synthetic resins, lubricants, and other products.

<u>narcosis</u> (Med) Numbness, stupor, or general anesthesia produced by a drug or other chemical substance.
   NOTE: The same state induced by electrical means in called <u>electronarcosis</u>.

<u>narcotic</u> (Pharm) A drug that relieves pain, induces sleep, or produces coma. A central nervous system depressant, usu addictive.

<u>nascent</u> (Biol; Chem) Just forming, as from a chemical or biological reaction.

<u>nasotracheal tube</u> (Med) A tube inserted through a nostril and trachea to supply air to the lungs.

<u>National Automatic Sprinkler and Fire Control Association</u> A national trade association of manufacturers and installers of atuomatic sprinkler compcnents and systems.

<u>National Board of Fire Underwriters</u> (Abb: NBFU) See: <u>American Insurance Association</u>.

<u>National Building Code</u> (Build) The building code maintained and published by the American Insurance Association.

<u>National Committee on Radiation Protection</u> (Nucl Safety) An advisory group of scientists and professional people that makes recommendations for radiation protection in the United States.

<u>National Concensus Standard</u> A standard that has been adopted by a nationally recognized standards-producing organization under procedures whereby persons interested in or affected by the standard have reached substantial agreement on its adoption and diverse views have been considered.

<u>National Electrical Code</u> (Fire Pr) A code of basic minimum provisions for safeguarding life and property from electrical hazards.
   NOTE: The code is prepared by the NFPA National Electrical Code Committee, which reviews and revises it every three years. The code is intended to be advisory, but is widely used as a basis for community ordinances, regulations, and electrical standards.

<u>National Fire Codes</u> (Fire Pr) The collected technical fire protection standards prepared by various committees of the National Fire Protection Association and published annually in 15 volumes.

<u>National Fire Danger Rating System</u> (For Serv; Abb: NFDR) A uniform rating system adopted in the US in 1973 and operated by the National Fire Weather Service that describes the cumulative effects of weather on wildland fire behavior.

<u>National Fire Protection Association</u> (Abb: NFPA) A nonprofit educational and technical association, founded in 1896, devoted to the protection of life and property from fire.

<u>National Safety Council</u> A private, nonprofit organization dedicated to public service in safety matters.

<u>National Standard Thread</u> See: <u>standard thread</u>

<u>natural gas</u> (Techn) A mixture of light, combustible hydrocarbon gases

obtained from pockets in the earth and used as fuel for heating, cooking, and various industrial processes, such as the manufacture of carbon black and ammonia. Typically consisting of about 85% methane (CH_4), bp -254°F; 10% ethane (C_2H_6), bp -128°F; 3% propane (C_3H_8), bp -44°F; and smaller quantities of butane (C_4H_{10}), pentane (C_5H_{12}), hexane, heptane, and octane. Syn: (Brit) casinghead gas.

natural philosophy See: physics

natural radiation = background radiation

natural radioactivity (Geophys) The background radioactivity of natural elements.

nautical mile (Metrol) A distance of about 1.13 statute miles, intended to be one minute of arc on a great circle of the earth.
 NOTE: Since the earth is not a perfect sphere, some confusion has resulted. The US standard is 6076.115 feet or 1852 meters. The British unit, also called the Admiralty mile, is 6080 feet.

NBBPVI = National Board of Boiler and Pressure Vessel Inspectors

NBS = National Bureau of Standards

NCRP = National Committee on Radiation Protection

NCSBCS = National Conference of States on Building Codes and Standards

near accident (Safety) An accident in which no injury or property damage occurred. Syn: noninjury accident.

nebulizer (Med) An aerosol dispenser of medication that is intended to be inhaled.

NEC = National Electrical Code

necropsy (Med) = autopsy; a postmortem examination.

necrosis (Med) Death of one or more cells or, more often, a circumscribed portion of tissue or organ. See also: infarction.

needless alarm (Fire Pr) A fire alarm given in good faith but which proves to be unnecessary. Also called a good intent false alarm. See: false alarm[1].

negative[1] (Math) Minus, or in the minus direction. Less than zero.

negative[2] (Commun) = no in voice communication.

negative[3] (Photog) An exposed and developed transparency of an image on film in which lighter objects are registered in grays shading into black and darker objects are recorded in grays shading into clear. Ant: positive.

negative catalyst (Chem) A catalyst that retards the rate of a chemical reaction.

negative feedback (Techn) Return of energy or a signal from the output to the input of a system to modify, reduce, or shut off the output.

negative pressure (Phys) Pressure below atmospheric. A partial vacuum indicated on a compound gage. See: pressure measurement.

negatron = beta particle

negligence (Law) Failure to exercise due care and reasonable prudence in any act or circumstance.

neoplasm (Med) A cellular outgrowth characterized by rapid cell multiplication. It may be benign (semi-controlled and restricted) or malignant.

neoprene (Chem) A kind of rubber made of polychloroprene, used in the manufacture of automotive inner tubes, characterized by low combustibility.

neper (Metrol) A number used to express the ratio of two powers as a natural logarithm. A neper is equal to 8.686 decibels.

nephritis (Med) Inflammation in the kidneys.

nervous system (Anat) A complex system of receptors and effectors interconnected by nerves in the body. Environmental changes stimu-

late responses in the receptors, which transmit excitations through the nerves to the effectors, which in turn act to adapt the organism to the changes in the environment detected by the receptors.
   NOTE: The nervous system includes the brain, spinal cord, peripheral nerves, the various sense organs, and numerous subsystems of nerves.

nested ladders (Fire Pr) Several ladders stored in the same bed on a ladder truck, the smaller ladders resting between the spars and on the rungs of larger ladders below.

net See: life net

net fire effect (Fire Pr) The sum of all effects, both good and bad, resulting from a fire.

net pump pressure (Fire Pr) The pressure developed in a pump, obtained by subtracting the intake pressure from the discharge pressure.

net storage (Storage) Floor area upon which bins or racks are erected, plus the floor area upon which material can be stored.

net weight (Transp) Weight of a load or commodity exclusive of its container.

neural (Physiol) Pertaining to a nerve or the nervous system.

neuritis (Med) Inflammation of a nerve.

neurodermatitis (Med) A chronic skin ailment of uncertain origin, possibly neurotic; a scaly itching skin lesion.

neurogenic shock (Med) A state of impaired blood circulation caused by the failure of the nervous system to control the diameter of the blood vessels, resulting in sudden puddling of blood with insufficient circulation. See: shock.

neutralization (Chem) 1. The counteraction of a chemical property, for example, combining an acid and a base to produce a salt and water solution that is neither acidic nor basic. 2. The union of hydrogen ions of an acid with the hydroxyl ions of a base to form water.

neutral pressure plane (Build) A level, esp in a tall building, at which the interior pressure is equal to the outside pressure.

neutron (Phys; Symb: n) An uncharged elementary particle with a mass slightly greater than that of the proton, found in the nucleus of every atom heavier than hydrogen. A free neutron is unstable and decays with a half-life of about 13 minutes into an electron, proton, and neutrino. Neutrons are effective atomic projectiles for the bombardment of nuclei and are used to sustain the fission chain reaction in a nuclear reactor. See: fast neutron; intermediate neutron; thermal neutron.

neutron absorber (Nucl Phys) Material with which neutrons interact significantly by reactions resulting in their disappearance as free particles.

neutron capture (Nucl Phys) The process in which an atomic nucleus absorbs or captures a neutron. The probability that a given material will capture neutrons is measured by its neutron capture cross section, which depends on the energy of the neutrons and on the nature of the material.

newton (Phys; Symb: N) A unit of force that imparts an acceleration of 1 m/sec^2 to a mass of one kilogram; equal to 10^5 dynes.
   NOTE: A newton is equal to approx 0.225 pounds of force. The unit for pressure, expressed in newtons/meter squared, is the pascal.

Newtonian fluid (Phys) A fluid in which the stress at any point is a linear function of the time rate of strain at the same point.

Newton's laws of motion (Phys) The three fundamental principles of classical mechanics: 1) A body at rest or in motion remains at rest or in motion at constant speed in a straight line unless acted upon by an external force (also called: Galileo's law of inertia). 2) The acceleration of a body is directly proportional to the force acting on the body and inversely proportional to the mass of the body. 3) A force exerted by one body on another is equal and opposite to the force exerted by the second body on the first (also called: law of action and reaction).

NFC = National Fire Codes

NFPA = National Fire Protection Association

NFPCA = National Fire Prevention and Control Administration

NHTSA = National Highway Traffic Safety Administration

nidus (Med) 1. The central point of origin or nucleus of a nerve. 2. A focus of lodgement and development of a pathogenic organism.

night blindness (Med) The condition of subnormal vision in dim light.

night hitch (Fire Pr) Bunker clothing placed at a sleeping firefighter's bunkside, arranged with boots tucked inside trousers or cover-alls in order to permit the firefighter to dress quickly when an alarm is sounded. Also called: quick hitch. See also: bunker clothing.

night shift (Fire Pr) A nighttime tour of duty.

night vision (Physiol) The capacity to see at night or under conditions of extremely dim light.

night watch (Fire Pr) 1. A nighttime tour of duty at a watch desk while other firefighters are sleeping. 2. A nighttime fire patrol in a hazardous district.

nimbostratus (Meteor) A cloud, classification Ns. See: cloud.

nimbus (Meteor) A type of rain cloud. See: cloud.

NIOSH = National Institute for Occupational Safety and Health

nip point (Techn) The point of intersection or contact between two rotating circular surfaces, or between a plane and a circular surface.

niter (Chem) A white salt, KNO_3, widely distributed in nature and formed in soils from nitrogenous organic bodies by the action of bacteria. Also called: potassium nitrate.
NOTE: Used in making fertilizer, medicinals, and other products.

nitrate (Chem) A salt of nitric acid containing the characteristic NO_3^- group, used extensively in the manufacture of fertilizers, explosives, and photographic films and papers.
NOTE: Nitrates in general are oxidizers and can react with combustible substances. Ammonium nitrate explodes when subjected to shock.

nitric acid (Chem) A strong mineral acid, HNO_3, colorless in pure form, fp $-47°C$. A strong oxidizer, the acid reacts with most metals. Hot, concentrated nitric acid oxidizes carbon to CO_2, sulfur to sulfuric acid, and phosphorus to phosphoric acid.
NOTE: Nitric acid and ammonium chloride or hydrochloric acid (which yields chlorine in reaction with nitric acid) dissolve gold and even platinum. A mixture of 1:4 nitric and hydrochloric acids is called aqua regia (royal water) because of its ability to dissolve noble metals. Nitric acid was once called aqua fortis (strong water).

nitrocellulose (Chem) A highly explosive product of nitric acid and cotton. Syn: guncotton.

nitrogen (Chem; Symb: N) A colorless, odorless, tasteless, diatomic gaseous element that forms approx 78% of the air in the earth's atmosphere, bp $-185.8°C$, fp $-209.8°C$. Nitrogen is extremely inert, but it is an essential element in all plant and animal life and is important in the production of nitric acid, ammonia, cyanides, and as a cryogenic when liquefied.
NOTE: Calcium, magnesium, and aluminum will "burn" in a nitrogen atmosphere without oxygen.

nitrogen dioxide (Chem; Env Prot) An extremely toxic gaseous compound, NO_2, formed from atmospheric nitrogen and oxygen when combustion takes place under conditions of high temperature and high pressure, as in internal combustion engines, or from

burning cellulose nitrate.
NOTE: Nitrogen oxides are considered major air pollutants.

nitrogen fixation (Chem) Chemical combination or fixation of atmospheric nitrogen with hydrogen, as in the synthesis of ammonia.
NOTE: Nitrogen is fixed in soil by bacteria, providing an industrial and agricultural source of nitrogen compounds.

nitrogen oxides (Chem; Env Prot) Gaseous compounds of nitrogen and oxygen, often formed from atmospheric nitrogen and oxygen when combustion takes place under conditions of high temperature and high pressure, as in internal combustion engines. Nitrogen oxides are considered major air pollutants.

nitroglycerine (Med; Pharm) A chemical compound for the relief of pain (as a vasodilator) in the treatment of angina pectoris.

nitrous fumes (Fire Pr; Toxicol) A general term referring to a variety of toxic oxides of nitrogen. Also called: **nitrogen oxides**.

nitrous oxide (Chem) A colorless gas, N_2O, of sweetish odor and taste; used as a general anesthetic. Also called: **laughing gas**.

NLPGA = **National Liquid Petroleum Gas Association**

node[1] (Electr) A point, line, or surface in a standing wave where some characteristic of the wave field has essentially zero amplitude.

node[2] (Med) A small, round or oval mass of tissue; a collection of cells; one of several constrictions occurring at regular intervals in a structure.
NOTE: A lymph node is a gland-like structure along a lymphatic passage that tends to eliminate bacteria.

node[3] (Fire Pr) A point on a map, usu a street intersection, that is assigned a unique number for computer processing.
NOTE: Nodes are used to develop travel route networks and to calculate travel times of fire apparatus.

nodule (Med) A small mass of rounded or irregularly shaped cells or tissue; a small node.

noise[1] (Acoust) Unwanted sound, often loud and unpleasant.

noise[2] (Electr) An unwanted stray signal in an electronic system, such as radio static, which can interfere with normal operation.

noise level (Acoust) Sound pressure expressed as the logarithm of the ratio of the measured sound intensity to a reference quantity of the same kind.

nomenclature A name or a naming system used to describe and identify items, e.g., chemical compounds.

nomograph (Math) A chart with three (or more) scales, so graduated and positioned that any straight line intersecting them indicates values satisfying an equation, relationship, or given set of conditions. Syn: **alignment chart**; **line coordinate chart**.

nonadiabatic = **diabatic**

nonburning (Comb; Testing) Having the property of not igniting and combusting when exposed to fire.
NOTE: The term is meaningful only when identified with a particular test. Some tests involve small or very small flames that are applied briefly to a material.

noncombustibility (Build; Comb) The property of a material to withstand high temperature without ignition.

noncombustible (Build; Comb) Resistant to burning under normal conditions. Also called: **incombustible** (erroneous). Ant: **combustible**. See: **fire resistant**.
NOTE: Standards and codes as a rule specify degrees or conditions for a noncombustible with reference made to specific tests. For example, a material that has a flame spread rating not exceeding 25 and will not support combustion independently regardless of how it is cut, or a material which, in the form in which it is used and under the conditions anticipated, will not ignite, burn, support combustion, or release flam-

mable vapors when subjected to fire or heat.

noncombustible construction (Build; Fire Pr) A structure made of material that does not ignite or spread flame readily when exposed to fire.

nondegree course (Educ) A course given by an educational institution that is not acceptable for credit toward a degree.

nondestructive testing (Testing) Testing of objects or materials to detect internal defects without damaging the specimen.
  NOTE: Common techniques include x-rays, isotope radiation, and ultrasonics.

nonelectric delay blasting cap (Civ Eng) A blasting cap with an integral delay element in conjunction with and capable of detonation by a detonation impulse or signal from a miniaturized detonating cord.

nonferrous metal (Metall) A metal or alloy, such as nickel, brass, or bronze, that does not include any appreciable quantity of iron.

nonflammable (Fire Pr) 1. Not readily capable of burning with a flame. 2. Not liable to ignite and burn when exposed to flame. Ant: flammable.

nonhygroscopic (Phys) Not absorbing moisture from the air.

noninductive circuit (Electr) A circuit in which inductance is reduced to a minimum or negligible value.

noninjury accident = near accident

nonmetal (Chem) An element, such as a gas (oxygen, hydrogen, nitrogen), liquid (bromine), or solid (boron, carbon, sulfur) that is a poor conductor of heat and electricity, has no characteristic metallic luster, and forms acidic oxides and hydroxides.
  NOTE: The term nonmetal is not used for all substances that are not metals, such as plastics or wood, which are properly called nonmetallic materials. Tellurium, antimony, and one form of tin have nonmetal properties but are called metalloids.

nonperishable items (Transp) Items that do not require refrigeration during transportation and storage.

nonpersistent (Chem; Toxicol) Not lasting; dissipating or weakening in effect relatively quickly. Often used in reference to toxic and noxious substances that volatilize and disperse or decompose readily into harmless materials. Ant: persistent.

nonpolar compound (Chem) An organic material or solvent having an electrically and magnetically balanced molecular structure.
  NOTE: Hexane, for example, is a balanced nonreactive flammable liquid that is a weaker solvent than the polar compound methyl ethyl ketone, which has a nonuniformly distributed internal electric charge and therefore a dipole moment. The latter is capable of breaking down fire-fighting foams. See: polar compound.

nonpolar solvent See: solvent

nonreturn valve (Mech Eng) A valve that passes a fluid flow in one direction only. A type of gate valve.

nonstandard (Techn) 1. Deviating from a norm. 2. Not usual or customary. 3. Not meeting established standards.

nonstatistical fire (For Serv) Any fire that does not qualify as reportable in the national statistical fire summaries. Such fires include those that burn themselves out; legitimate slash burning; fires contained in fireplaces, stoves, or campfires that cannot spread; several fires so close together that they are reported as one fire; fires confined to private lands; etc.

nontoxic (Toxicol) A substance that does not cause physiologic, morphologic, or functional changes adversely affecting the health of man or animal according to reasonable and appropriate tests.

nonvesicant (Toxicol; Med) A substance that may be otherwise harmful or toxic but does not produce blisters or eruptions on the skin.

Ant: <u>vesicant</u>.

<u>norepinephrine</u> (Med; Pharm) A hormone-like, crystalline compound, $C_6H_{11}NO_3$, that has a strong vasoconstricting effect and is used to increase blood pressure and cardiac output. One of the catecholamines, which include epinephrine and dopamine.

<u>norm</u> (Techn) 1. Typical, or average. 2. A principle, standard, or guide for acceptable behavior, achievement, or performance.

<u>normal</u> (Geom; Opt) At right angles, or perpendicular.

<u>normal distribution</u> (Statist) An expected or predictable spread of values, such as test scores or the diameters of ball bearings, represented by a bell-shaped curve when plotted on a graph. Syn: <u>Gaussian distribution</u>.
 NOTE: The curve is centered on the mean of the distribution, and the spread is determined by the variance.

<u>normal fire season</u> = <u>fire season</u>

<u>normal salt</u> (Chem) A salt containing neither replaceable hydrogen nor hydroxyl.

<u>normal solution</u> (Chem; Symb: N) A solution that contains one gram of replaceable hydrogen or its equivalent per liter. In general, a solution that contains one equivalent weight of an ion, element, or compound dissolved in enough water to make one liter.

<u>nose</u> (Build) The portion of a tread projecting beyond the face of the riser immediately below. Also called: <u>nosing</u>.

<u>noxious gas</u> (Toxicol) A toxic compound that is a gas under normal conditions and in limited concentrations is capable of causing physical distress, injury, or death. See: <u>toxic substances</u>.

<u>nozzle</u> (Fire Pr) A tubular, metallic, constricting attachment fitted to a hose to increase fluid velocity and form a jet. Nozzles are often adjustable to provide a solid stream, a spray, or fog. Many special-purpose nozzles have been designed to produce particular kinds of stream or spray patterns. See: <u>diffuser nozzle</u>; <u>spray nozzle</u>.

<u>nozzle foam maker</u> See: <u>aspirator foam nozzle</u>

<u>nozzleman</u> (Fire Pr) 1. A firefighter assigned to handle the nozzle of a fire hose. 2. = (Brit) <u>branchman</u>.

<u>nozzle reaction</u> (Fire Pr; Abb: NR) The force acting in the opposite direction to the water stream leaving the nozzle. The magnitude of this force depends upon the size of the nozzle tip and the water pressure and is due to the acceleration of the water in the nozzle.
 NOTE: The formula for calculating nozzle reaction is NR = gpm x P x 0.505, where gpm is gallons per minute and P is psi pressure at the base of the nozzle.

<u>nozzle spanner</u> (Fire Pr) A tool used to tighten or loosen the joint between a nozzle and a playpipe.

<u>nozzle tip</u> (Fire Pr) An interchangeable tip for a nozzle used to alter the pattern of water discharge or size of the water jet. The term is used as a synonym for <u>nozzle</u>. See: <u>tip</u>.

<u>NSC</u> = <u>National Safety Council</u>

<u>NSMS</u> = <u>National Safety Management Society</u>

<u>NSPE</u> = <u>National Society of Professional Engineers</u>

<u>NTP</u> = <u>normal temperature and pressure</u>. See: <u>standard conditions</u>.

<u>nuclear battery</u> (Nucl Phys) A radioisotopic generator of electricity. A device in which the energy emitted by decay of a radioisotope is converted first to heat and then to electricity.

<u>nuclear bombardment</u> (Nucl Phys) Exposure of atomic nuclei to a stream of high-energy particles produced in an accelerator, usu in an attempt to split the atom or to form a new element.

<u>nuclear energy</u> (Nucl Phys) The energy released when a neutron splits the

nucleus of an atom (fission) or when two nuclei are joined under extremely high heat (fusion). Also called: atomic energy.

nuclear excursion (Nucl Phys) A technical euphemism for a runaway chain reaction.

nuclear explosion (Nucl Phys) The rapid fissioning of a quantity of fissionable material, which produces an intense heat and light flash, a powerful blast, and a quantity of radioactive fission products. These may be attached to dust and debris, forming fallout. Nuclear explosions also result from nuclear fusion.

nuclear explosive (Milit) An explosive based on fission or fusion of atomic nuclei.

nuclear power plant (Nucl Phys) Any device, machine, or assembly that converts nuclear energy into some form of useful power, such as mechanical or electrical power. In a nuclear electric power plant, heat produced by a reactor is generally used to make steam to drive a turbine that in turn drives an electric generator.

nuclear reaction (Nucl Phys) A reaction involving a change in an atomic nucleus, such as fission, fusion, neutron capture, or radioactive decay, as distinct from a chemical reaction, which is limited to changes in the electron structure surrounding the nucleus. Cf: thermonuclear reaction.

nuclear reactor (Nucl Phys) An apparatus in which nuclear fission may be sustained in a self-supporting chain reaction.
  NOTE: A reactor contains fissionable material (fuel), usu moderating material (see: moderator), a reflector to conserve escaping neutrons, provision for heat removal, and measuring and control elements.

nuclear weapons (Milit) A collective term for atomic bombs and hydrogen bombs. Any weapon based on a nuclear explosive.

nucleon (Nucl Phys) A constituent of an atomic nucleus; a proton or a neutron.

nucleonics (Nucl Phys) The science and technology of nuclear energy and its applications.

nucleus (Phys) The small, positively charged core of an atom. It is only about 1/10,000 the diameter of the atom but contains nearly all the atom's mass. All nuclei contain both protons and neutrons, except the nucleus of ordinary hydrogen, which consists of a single proton.

nuclide (Nucl Phys) An atomic form of an element. Not a synonym for isotope. Whereas isotopes are the various forms of a single element (hence are a family of nuclides) and all have the same atomic number (number of protons), nuclides comprise all the isotopic forms of all the elements. Nuclides are distinguished by their atomic number, atomic mass, and energy state. See: element.

nuisance dust (Ind Hyg) Generally innocuous dust, not recognized as the direct cause of a serious pathological condition.

number system (Math) An orderly method of showing numerical quantities. The most common systems are the duodecimal, based on 12 (which has the advantage of being divisible by 1, 2, 3, 4, and 6); the decimal, based on 10; the binary, based on 2; and the octal, based on 8. The last two are widely used in computer technology. See: base^2, binary notation; octal notation; biquinary notation.

numerical control system (Comput) A system in which actions are controlled by the insertion of numerical data at some point. The system must automatically interpret at least some portion of the data.

numerical data (Comput; Math; Statist) Facts expressed by numbers, or by symbols that assume discrete values.

n-unit (Nucl Safety) A measure of radiation dose due to fast neutrons.

nurse (Med) A trained specialist in tending sick and injured persons or one engaged in preventive medicine, usu under the direction of a physician.

nurse tanker (Fire Pr) A water tank truck used to supply water to a pumper.

nursing home (Build; Fire Pr) A building or part of one used for continuous lodging, boarding, and nursing care of 4 or more persons who, because of mental or physical incapacity, may be unable to provide for their own needs and safety without the assistance of another person.

Nusselt number See: **dimensionless number**

nutrient (Biol) A substance that can be used for food.

O

O (Chem) Symbol for oxygen.

oakum (Techn) Loosely twisted hemp or jute fiber treated and used to pack joints and boat seams before sealing.

obligated storage (Storage) The portion of net usable storage area reserved for the storage of incoming supplies.

observation time (For Serv) The time required to take fire danger observations.

obstruction to distribution (Fire Pr) Any barrier or interference to the application of an extinguishing agent on a fire, such as stock, displays, merchandise, partitions, and the like.

occlusion[1] (Metall) The absorption or trapping of a gas by a solid.

occlusion[2] (Med) The obstruction of a passageway, as a blood vessel by a clot or thrombus.

occlusion[3] (Meteor) A complex interface that occurs when a cold front overtakes a warm front. Also, a cyclone having such a front.

occult blood (Med) 1. Concealed bleeding. 2. Minute traces of blood in the stool.

occupancy (Fire Pr) 1. The purpose for which a building, floor, or other part of a building is used or intended to be used. 2. Any stationary structure or enclosure in which one can normally expect to find one or more persons. Syn: (Brit) occupation.
NOTE: The principal types of occupancy are residential, institutional, educational, storage, shop and department store, office, assembly, and factory.

occupant load (Build; Fire Pr) The largest number of persons that may occupy a building or portion of it at any given time.

occupational disease (Ind Hyg; Safety) Any abnormal physical condition or disorder resulting from exposure to factors associated with employment, including illness caused by inhalation, absorption, ingestion, or direct contact with toxic substances or irritants. Syn: occupational illness.

occupational injury (Safety) Any injury such as a cut, fracture, sprain, amputation, etc., that occurs in the course of employment. See: injury.

occupational safety (Safety) A program of practical and reasonable precautions taken to ensure that employees' health and well being are protected on the job from foreseeable dangers and hazards.

occurrence (Safety) Any notable incident or event; an instance of something happening.

occurrence index (For Serv) A measure of the potential fire incidence within a given fire protection unit.

OCD = Office of Civil Defense

octal notation (Math) A number system, based on powers of 8, widely used in computer technology. For example, the decimal number $10 = 8^1 + 2$, so in octal notation the decimal number 10 is written 12 (one eight and two ones). Decimal 25 = octal 31 $((3 \cdot 8^1) + 1)$; decimal 148 = octal 224 $((2 \cdot 8^2) + (2 \cdot 8^1) + 4)$; and so forth. See also: number systems.

octane (Chem) An alkane, C_8H_{18}, occurring in 18 possible isomers, bp 99.3°C to 125°C, depending on its isomeric form. Octane is used as an antiknock additive to gasoline.

octane number (Techn) A measure of the antiknock quality of gasoline, determined by comparing the detonation of a gasoline sample with that of a mixture of 2,2,4-trimethylpentane (often called isooctane), rated at 100, and normal heptane, rated at 0. Hence, the percentage of isooctane by volume in the matching mixture is the octane number of the gasoline being tested. Also called: octane rating. Cf: cetane number.

octave (Acoust) The interval between two sounds in which the higher is twice the frequency of the lower.

octave band analyzer (Audiom) An instrument for measuring the intensities of specific octave bands, usu beginning at 75 Hz and continuing through 10,000 Hz.

octet (Phys) A group of eight electrons in the highest energy levels of an atom.

ocular (Anat) Pertaining to the eye.

OD = outside diameter

ODM = outside diameter of male thread

odometer (Transp) A device on a vehicle for measuring the distance traveled, usu as a cumulative total but sometimes for individual trips, with an indicator on the instrument panel, where it is often combined with a speedometer, or mounted in the hub of a wheel on some trucks.

odor (Physiol; Psych) A chemical stimulus affecting the olfactory nerve cells (the organs of smell). See: chemical senses; smell.
  NOTE: Many burning materials have distinctive odors that can be readily identified: e.g., rubber, cloth, pine, feathers, etc. Dense smoke and strong odors overpower the sense of smell and the ability to recognize specific odors decreases rapidly with continued exposure. One scheme for classifying odors is based on a prism, the paired vertices of which are labeled fragrant, spicy; putrid, burned; ethereal, resinous.

office occupancy (Fire Pr) Any building or structure or part of one used to transact business or to render or receive professional services.

officer (Fire Pr) A member of a fire department with supervisory responsibilities. See: ranks and titles.

officer in charge (Fire Pr) The highest ranking officer at a fire or commanding an on-duty fire department unit.
  NOTE: The senior officer of a given rank is usu in command, unless another officer of equal rank is specifically assigned to command. In a mutual aid situation, the highest ranking officer of the local fire department is in command, regardless of the ranks of the officers of assisting units.

officer training (Fire Pr) Special training programs or courses given to fire department members in leadership and administrative subjects and skills.

official action guide (Fire Pr) An official fire department handbook of accepted practices and procedures.

offline operation (Comput) Operation of peripheral equipment, such as data channels or printers, independent of the central processor of a computer system.

off shift (Fire Pr) Off duty.
  NOTE: Off-shift units may be subject to recall to duty in case of emergency.

off the air (Commun) 1. An interruption in radio communication due to equipment breakdown or other malfunction. 2. Termination of radio contact at completion of communication, as during return to quarters, where normal alarm and telephone communication is resumed.

off-the-job safety (Safety) Accident prevention or safety programs associated with activities not related to a job.

ohm (Electr) A unit of electrical resistance across which there is a potential drop of one volt when the current flow is one ampere.

Ohm's law (Electr) The principle that relates voltage, current, and resistance: in an electrical circuit the current (I) varies directly with voltage (E) and inversely with resistance (R); that is, $I=E/R$.

oil vapor (Ind Hyg) Oil in a gaseous state.
  NOTE: Oil vapor content is usu stated as parts per million of gas by weight (PPM/w).

olefin (Chem) An unsaturated hydrocarbon, sometimes referred to as an alkene, containing one or more double bonds between carbon atoms. Olefins contain twice as many hydrogen atoms as carbon atoms and are of the general formula $C(n)H(2n)$ for

singly unsaturated (one double bond) compounds; for example, butene (CH_2:$CHCH_2CH_3$) ethylene (H_2C:CH_2), and propylene (CH_3CH:CH_2).
NOTE: Olefins, obtained from petroleum and natural gas, are characterized by relatively high chemical activity. Used in making textile fibers.

olfaction (Physiol) The sense of smell.
NOTE: The stimulus for smell is a substance in the gaseous state.

one and one (Fire Pr; Colloq) A small fire that is controlled by one engine company and one ladder or truck company.

one-brick wall (Build; Brit) A brick wall 9 inches thick, or equal to the length of one brick.

one-lick method = progressive method

one-line fire (Fire Pr) A small fire that can be extinguished by a single hose line.

one-way radio (Commun) A radio communication system between fire alarm headquarters and fire department stations and mobile units. See also: two-way radio; three-way radio.

on the air (Commun; Fire Pr) The state of being in radio contact.

on the line (Fire Pr) 1. The state of readiness of apparatus to respond to an alarm. 2. Notification that a fire company unit is back in quarters and ready to resume its normal assignments. Also called on the track. Syn: in service.
NOTE: The term arose when horse-drawn apparatus was guided into its berth by steel rails set in the floor.

open burning (Env Prot) Uncontrolled burning of wastes in the open or in an open dump.

open center system (Hydraul) A hydraulic system equipped with a relief valve that allows the fluid from the pump to empty into the reservoir when it is no longer required to run a power take-off.

open circuit[1] (Electr) A discontinuity in an electrical circuit in which the path for the flow of electrical current between two points is broken.

open circuit[2] (Commun; Fire Pr) A signal circuit in which current flows only when a signal is being transmitted.

open fracture (First Aid; Med) A compound fracture. A fracture in which the skin is broken in the area of the broken bone. See: fracture.

open-head system (Fire Pr) A sprinkler system in which the sprinkler heads are open, and water is turned on either manually or automatically by a thermostat. See: sprinkler system.

opening (Build) Any aperture or passageway through a wall, floor, ceiling, roof, or partition.
NOTE: A Class A opening is a doorway or other opening through a wall separating buildings or dividing a single building into fire areas; closures for these openings normally require a fire endurance rating of 3 hours. Class B: A doorway or other opening in enclosures of vertical communication through buildings (stairs, elevators, etc.); doors for these openings normally require a fire endurance rating of 1 or 1-1/2 hours. Class C: A doorway in a corridor or in room partitions; doors for these openings normally require a fire endurance rating of 3/4 hour. Class D: A doorway or other opening in an exterior wall that may be subjected to severe fire exposure from outside of the building; doors and shutters for such openings normally require a fire endurance rating of 1-1/2 hours. Classes E and F: Doorways or other openings in exterior walls that are subject to moderate or light fire exposure, respectively, from outside of the building; doors, shutters or windows for such openings normally require a fire endurance rating of 3/4 hour.

open-loop system (Techn) A control system without a means for comparing the output with the input.
NOTE: An example is a switch (controlling input), light bulb (controlled object), and light (output). The essential characteristic is that the quality and magnitude of the output are independent of the con-

trolling input. See: closed-loop system.

open plan school (Build) A building for educational purposes, the main part of which is not subdivided into classrooms, study areas, and the like by interior walls.

open riser (Build) The vertical space between the treads of a stairway without full-width upright members (risers).

open roof (Build; Brit) A roof with exposed structural members. A roof without a ceiling.

open storage (Storage) An outdoor area designated for storage of material.
NOTE: Types of open storage: Improved - An area that has been graded and hard-surfaced or prepared with topping of some suitable material to permit effective material handling operations. Unimproved - An area that has not been surfaced but is used for storage purposes. Unimproved Wet Water - An area specifically alloted for the storage of floating equipment.

open up (Fire Pr) 1. To ventilate a building and permit accumulated smoke and heat to escape. 2. To force entry into a closed burning building. See: forcible entry.

operant learning (Educ; Psych) A method of learning in which the learner is rewarded for a desired response in order to reinforce that response in a subsequent trial.

operating device (Techn) A pushbutton, lever, or other manual device used to actuate a control.

operating suction lift (Fire Pr) The vertical distance between the center line of a pump and a static water level.

operations floor (Fire Pr) The story immediately below the fire floor, commonly used as the base for launching fire-fighting operations to the fire floor.

operations research (Man Sci) The systematic study of complex problems involved in the interrelated activities of organizations of people and resources to provide rational bases for setting goals, assessing risks, and making decisions. See also: systems analysis.

operator[1] (Fire Pr) A dispatcher in a fire alarm center.

operator[2] (Fire Pr) A firefighter operating an apparatus or pump. Also called: engineer.

ophthalmologist (Med) A physician who specializes in the structure, function, and diseases of the eye.

ophthalmology (Med) The branch of medicine that deals with diseases of the eyes, their treatment, and correction of vision.

optical density (Opt; Abb: OD) A logarithmic expression of light attenuation by a filter. See: smoke density.

optician (Med) A specialist in grinding lenses.

optics (Phys) The science of light and the phenomena associated with it, such as refraction, reflection, and interference.
NOTE: Optics encompasses electromagnetic radiation above and below the visible range.

optimum The best or most favorable degree, condition, amount, or the like.

optometry (Med) The science of measuring the refractive and focussing power of the eye and the practice of fitting glasses to improve vision.

oral fluids (Med) Fluids taken by mouth.

orbit[1] (Phys) The region occupied by an electron as it moves about the nucleus of an atom. See: shell.

orbit[2] (Acoust) The bony socket of the eye.

order of magnitude (Math) A range or interval extending from a value to ten times the value, such as from 1 to 10, from 10 to 100, etc.

orders (Build) The five orders of classical architecture: Tuscan,

Doric, Ionic, Corinthian, and Composite.

ordinary combustibles (Fire Pr) Commodities, packaging, or storage aids which have a heat of combustion similar to wood, cloth, or paper and which produce fires that may be extinguished by the quenching and cooling effect of water.

ordinary construction (Build) Construction in which exterior bearing walls or bearing portions of exterior walls are of noncombustible materials and have a minimum fire endurance of 2 hours. Nonbearing exterior walls are of noncombustible construction, and roofs, floors, and interior framing are wholly or partly of wood (or other combustible material) of smaller dimensions than required for heavy timber construction. See: building construction.

ordinary hazard (Fire Pr) A situation where the amount of combustibles or flammable liquids present is such that fires of moderate size may be expected. These may include mercantile storage and display, auto showrooms, parking garages, light manufacturing, warehouses not classified as extra hazard, schools, shop areas, etc.

ordinary hazard contents (Build; Fire Pr; Insur) Building contents that are liable to burn with moderate rapidity and to give off a considerable volume of smoke, but from which neither poisonous fumes nor explosions are likely to occur in case of fire.

ordinate (Geom; Math) The vertical axis of a cartesian coordinate system; usu labeled the y axis.

ore (Mining) A naturally occurring mineral from which a metal or nonmetal can be extracted.

organ (Anat) An organized collection of tissues that performs a special function in the body.

organic[1] (Chem) Referring to substances that contain carbon-to-carbon bonds, which are characteristic of compounds originating in living matter.
NOTE: To date nearly one million organic compounds have been synthesized or isolated. Many occur in nature; others are produced by chemical synthesis.

organic[2] (Med) a. Referring to an organ of the body. b. Organized into a structure.

organic chemical (Chem) Designation of any chemical compound containing carbon (some of the simple compounds of carbon, such as carbon dioxide, are frequently classified as inorganic compounds). To date nearly one million organic compounds have been synthesized or isolated. Many occur in nature; others are produced by chemical synthesis. In medicine, producing or involving alteration in the structure of an organ; opposed to functional.

organic chemistry (Chem) The branch of chemistry that deals with carbon compounds.
NOTE: Carbon itself and a few simple carbon compounds, such as carbon dioxide, are considered inorganic. Carbon combines most commonly with hydrogen, oxygen, nitrogen, sulfur, and phosphorus, in decreasing order of occurrence.

organic disease (Med) Disease in which some change in body tissue structure can either be seen or positively inferred from indirect evidence.

organic matter (Chem) A compound containing carbon.

organism (Biol) A being possessing life functions, as a human, an animal, germ, plant, etc., esp one consisting of several parts, each specializing in a particular function. Life functions include metabolism, respiration, digestion, excretion, and reproduction.
NOTE: A microorganism is any living thing that is microscopic or submicroscopic in size.

organization (Man Sci) The association of people with specified functions and responsibilities for pursuing an agreed purpose.

orientation (Physiol; Psych) Familiarization with and adaptation to a situation or environment; interpretation of the environment as to time, space, objects, and persons.

orifice¹ (Anat) An opening that serves as an entrance or outlet of a body cavity or organ, esp the opening of a canal or a passage.

orifice² (Hydraul) a. An opening intended for the passage of a fluid, often used to meter the rate of fluid flow. b. A discharge opening, such as in a nozzle.

oropharyngeal airway (First Aid; Med) A curved breathing tube that holds the tongue forward and allows the patient to breathe.

orotracheal tube (Med) A tube inserted through the mouth and trachea to supply air to the lungs.

oscillation (Phys) A variation, usu with time, of the magnitude of a quantity with respect to a specific reference when the magnitude is alternately greater and smaller than the reference.

oscilloscope (Electr) An electronic instrument for showing graphical representations of the waveforms encountered in electrical circuits.

OSHA = Occupational Safety and Health Administration

osmosis (Chem; Phys) The flow of solvent through a semipermeable membrane that separates solutions of different concentrations and that does not allow the dissolved material to pass through. The net flow dilutes the more concentrated solution.

osmotic pressure (Phys) The pressure differential across a semipermeable membrane that counterbalances the flow of osmosis.

osseous (Anat) Pertaining to bone; bony, or bone-like.

ossicle (Anat) Any one of the three small interconnected sound-conducting bones (maleus, incus, or stapes) between the outer membrane of the tympanum (eardrum) and the membrane covering the oval window of the inner ear.

otitis (Med) An inflammation and infection of the ear.

otologist (Med) A physician who specializes in diagnosis, treatment, and diseases of the ear.

otology (Med) The science of hearing and of the ear, including its anatomy, physiology, and disorders.

otosclerosis (Med) Formation of bony tissue about the foot plate of the stapes and the oval window of the inner ear. Hearing is gradually lost; however, the condition can often be corrected by surgery.

ototoxic (Toxicol) Having a toxic effect on the ear.

outdoor air Air taken from outdoors and therefore not previously circulated through an air-handling system.

outlet¹ (Fire Pr) An opening provided for the discharge of a fluid, usu water.

outlet² (Electr) An electrical receptacle, usu a female wall plug.

out of service (Fire Pr) A fire-fighting unit that is not available to answer assigned calls because of accident, mechanical failure, or because it is already engaged in fire-fighting or emergency operations. Syn: tied up.

out on arrival (Fire Pr) A fire that has been extinguished before the arrival of firefighting personnel and fire department equipment.

outpatient (Med) A patient who visits a clinic or hospital for treatment but is not hospitalized.

outrigger jack (Fire Pr) A support that can be extended from an aerial ladder or aerial platform apparatus to stabilize the truck chassis. Also called: ground jack; jack; outrigger; stabilizer.

outside aid (Fire Pr) Fire-fighting assistance given to adjacent areas and nearby communities by contract or other agreement that cover conditions and payment for assistance rendered and services performed. Contrasted to mutual aid, in which neighboring fire-fighting organizations assist each other without charge.

outside diameter (Mech Eng) The distance from the outside surface through the center to the opposite outside surface of a circular, annular, tubular, or cylindrical object.

outside fire fighting (Fire Pr) Fire fighting conducted outside a burning structure, either because the fire has gained too much headway, the structure has become weakened, or for other reasons that make it inadvisable to send men and equipment into the building. Cf: inside fire fighting.

outside sprinklers (Fire Pr) An open-head sprinkler system installed on the outside of a building to protect it from severe exposure hazards.

outside standpipe (Fire Pr) A standpipe on the exterior of a building with a siamese connection for a fire department pumper.

overbreathing = hyperventilation

overcome (Med) Physically incapacitated by exposure to a fire environment. The state of unconsciousness or helplessness due to heat, smoke, toxic gases, exhaustion, or combination of factors.
NOTE: The condition may develop in varying degree from a number of causes, including oxygen deprivation, carbon monoxide poisoning, breathing poisonous gases, and ingesting particulate materials contaminated with toxic substances. Withdrawal from reality due to psychologic stress may also occur.

overexposure (First Aid; Med) Exposure beyond tolerable limits.

overhanging load (Transp) The portion of lading overhanging one or both ends of a car.

overhauling (Fire Pr; Salv) The terminal phase of fire fighting, in which the fire area is carefully examined for embers or other traces of fire. In addition, steps are taken to protect remaining property from further damage. Syn: mopping up.

overload (Fire Pr; Transp) Gross vehicle weight in excess of that specified by the chassis manufacturer, by axle load ratings, or wheel and rim load ratings.

overload test (Fire Pr) A test required of a new pumping apparatus to demonstrate its ability to pump 10% above its rated capacity for a brief period. Syn: spurt test.

overpressure (Comb) The transient pressure above atmospheric pressure caused by a shock wave from an explosion. See: shock wave.

overstory (For Serv) The foliage of a forest canopy.

overt Not hidden. Open to public observation. Ant: covert.

over the air (Commun) By radio.

oxidant (Comb) An oxidizing agent. Syn: oxidizer.

oxidation (Chem) 1. A process in which oxygen combines with another substance. 2. A reaction in which electrons are transferred to or from an atom or group of atoms. Ant: reduction.

oxidation number (Chem) The number of electrons that must be added to or subtracted from an atom in combined state to convert it to its elemental state.

oxidation rate (Chem; Comb) Oxidation as a function of time. The process is accompanied by the evolution of heat, which is extremely slow in the case of rusting iron or rotting wood, rapid in the case of combustion, and more or less instantaneous in the case of detonation and explosion.

oxide (Chem) A compound of oxygen and another element. Four classes of oxide are distinguished: 1) acidic (generally nonmetallic) as SO_2, P_2O_5, CO_2, etc.; 2) basic (generally metallic) as CuO, PbO, Fe_2O_3, etc.; 3) peroxide (generally metallic), containing more oxygen than the common oxides, as for example PbO_2, BaO_2, Na_2O_2, etc.; and 4) neutral oxide, as CO, N_2O, and H_2O.

oxidizer (Chem) 1. A substance that readily gives up oxygen without requiring an equivalent of another

element in return. Also called: oxidant; oxidizing agent. Ant: reducer. 2. A substance that contains an atom or atomic group that gains electrons.
  NOTE: Common oxidizers are oxygen, ozone, chlorine, hydrogen peroxide, nitric acid, metal oxides, and the chlorates and permanganates.

oxygen (Chem; Symb: O) A gas, bp -182.9°C, fp -218.4°C, the most abundant element on earth, making up approx 21% of the atmosphere and occurring in water and metallic oxides. Oxygen is essential to the life cycle of plants and animals, as well as to many chemical processes, including combustion. See also: ozone.

oxygen balance (Comb) Ratio of self-contained oxygen to fuel in propellants, explosives, and pyrotechnics. Gives the extent that an explosive is deficient or rich in oxygen compared to the quantity required for complete combustion.

oxygen deficiency[1] (Med) Insufficiency of oxygen. When the oxygen content of the air falls below 16%, thinking and coordination become difficult and there is danger of asphyxiation; below 6% breathing ceases. See: anoxia; asphyxiation.

oxygen deficiency[2] (Comb) Insufficiency of oxygen to support combustion.

oxygen index (Comb) An oxygen concentration in an oxygen-nitrogen mixture that will support combustion of a plastic material or gaseous fuel; used extensively to rate plastic flammability.
  NOTE: The lowest oxygen concentration that will support combustion is called the limiting oxygen index.

oxygen mask (Fire Pr) A self-contained breathing unit with a mask that provides air to the wearer either by chemical reaction or from an oxygen cylinder. See: breathing apparatus.

ozone (Chem; Symb: O_3) A triatomic molecule of oxygen, bp -111.9°C, mp -192.5°C, that occurs in small concentrations in the atmosphere and is produced by electric sparks in air. Ozone is a highly active element and is a powerful oxidizer. Prolonged exposure to concentrations of 0.1 to 1 ppm are considered hazardous.

P

pacemaker (Med) A battery-operated electronic pulse generator that is connected to the heart muscle and stimulates the heart to beat regularly.
NOTE: Pacemakers are commonly implanted in the body and recharged periodically by an induction coil from outside the body.

package (For Serv) A unit of lumber.

packaging[1] (Storage; Transp) Any wrapping, cushioning, or container for a commodity.

packaging[2] (First Aid; Med) Application of splints and dressings, stabilization of impaled objects, and similar procedures preparatory to removing a casualty or accident victim and transporting to the hospital.

packing (Transp) Assembly and placement of goods or packaged items in exterior shipping containers or other shipping media for protection and ease of handling during transportation.

packing gland See: gland

padlock remover (Fire Pr) A lever-type forcible-entry tool with slots used to twist off padlock hasps to gain entry. See: forcible-entry tool.

paging system (Commun) A communications system for alerting people, broadcasting messages, or contacting individuals, usu by loudspeakers located at strategic points.
NOTE: A pocket pager is a small electronic device that emits a tone, which signals the wearer to call in by telephone.

paid absence allowance (Man Sci) Payment for lost working time available to workers for various types of leave not otherwise compensated for, such as excused personal leave.

paid fire department See: fire department

paid firefighter (Fire Pr) A full-time, paid firefighter, as distinguished from a volunteer. Also called: paid man.

pain (Physiol; Psych) An unpleasant physical sensation induced by an injury or intense stimulus, such as excessive light or noise, extremes of heat and cold, irritating chemicals, electric shock, etc.
NOTE: Bruises, swellings, and severe exposure to heat or sun produce hyperalgesia, a heightened sensitivity of the skin to pain. The heat pain threshold of the skin ranges from 210 to 480 millicalories (0.21 to 0.48 g-cal/sec/cm^2), with over 20 discriminable levels of pain intensity. A temperature of 52°C (126°F) is sufficient to produce thermal pain and tissue injury. Sound pressure above 120 dBA is usu sufficient to cause pain in the human ear.

pair production (Phys) The conversion of a gamma ray into a pair of particles - an electron and a positron.
NOTE: This is an example of direct conversion of energy into matter, which follows Einstein's formula for their equivalence: $E=mc^2$; (energy) = (mass) x (velocity of light)^2.

pall See: smoke pall

palladium chloride (Chem) A chemical compound that blackens upon exposure to carbon monoxide.

pallet (Storage) A low wooden platform used for storing materials a few inches above the floor level as protection from dampness and to permit forklift blades to slide underneath. Also called: skid.

palliative treatment (Med) Treatment intended to make a patient comfortable.

palpitation (Med) Rapid, forceful beating of the heart; often occurring during stress and characteristic of anxiety.

pancake collapse See: floor collapse

panelling (Build) Decorative interior finish in the form of thin laminated or pressed sheets, usu 4x8 feet, either painted on one side or faced with plastic or a quality wood veneer.
NOTE: Often applied with adhesive to old walls.

panel walls (Build; Brit) Nonload-bearing walls made of panels fitted into steel or reinforced concrete framing members of a structure.

panic (Psych) A sudden onset of demoralizing fear or terror, often affecting a group of persons and characterized by seemingly irrational behavior.

panic hardware (Build) A horizontal bar or panel across an exit door, at about waist height, used in many public buildings, that operates the door latch.
  NOTE: A body pushed or falling against the bar will unlatch the door and allow it to swing open.

panoramic photograph (For Serv) A photographic picture, taken from a lookout point, showing azimuth and elevation scales to help locate fires with a firefinder.

paper-and-pencil test (Educ) An examination in which answers are written.

papilloma (Med) A small growth or tumor of the skin or mucous membrane. Commonly a wart or polyp.

parabola (Geom) A curve formed by the intersection of a plane parallel to a generator of a right circular cone with the cone. It is the locus of points at the same distance from a point (the focus) and a straight line (the directrix).
  NOTE: A body moving in a parallel gravity field and unaffected by any other force follows a parabolic path.

paracargo (For Serv) Equipment, tools, supplies, or other items dropped or scheduled for dropping from an aircraft by parachute or free fall.

paraffin^1 (Chem) The class of saturated hydrocarbons, including methane (CH_4), ethane (C_2H_6), propane (C_3H_{10}), n-butane (C_4H_{10}), components of gasoline, kerosine, and various lubricating and fuel oils. The lower paraffins are gases, those up to n-hexadecane ($C_{16}H_{34}$) are liquids, and those with more than 16 carbon atoms are waxy solids at room temperature.
  NOTE: Paraffins are straight or branched chain hydrocarbon molecules with the carbon atoms attached to each other by single bonds (saturated). Generalized formula: $C(n)H(2n+2)$. Also called: paraffin series.

paraffin^2 (Brit) = kerosine

parallel^1 a. Anything similar or resembling another thing in all essential particulars. b. A counterpart. c. Side-by-side.

parallel^2 (Geom) The orientation of lines or planes such that they have common perpendiculars and never meet no matter how far they are extended.

parallel^3 (Navig) An imaginary circle on the earth's surface formed by the intersection of a plane parallel to the equator and used to determine the distance in degrees latitude north or south of the equator.

parallel entry (Comput) A display in which all characters are presented simultaneously, as opposed to serial or one by one.

parallel method (For Serv) A method of fire suppression in which a fireline is constructed parallel to a fire edge and sufficiently ahead of it to permit men and equipment to work safely. Fuel remaining between the fire edge and the fire control line is usu burned out as construction of the control line proceeds.

parallel operation (Fire Pr) Operation of a multistage centrifugal pump with the impellers producing volume rather than pressure. Also called: volume operation.

paralysis (Med) Partial or complete loss of voluntary motion or sensation.
  NOTE: Paralysis may be either spastic or flaccid. Spastic results from damage to the motor neurons from the brain. Flaccid results from damage to the motor neurons between the spinal cord and the muscles.

paramedic (Fire Pr; Med) An individual trained to render medical assistance to a sick or injured person under the direction of a physician. See also: emergency medical technician.

parameter^1 (Math) An arbitrary constant whose value distinguishes different cases, as the members of a family of curves.

parameter² (Phys) A physical property that determines the characteristics or behavior of a particular system.

parapet (Build) A wall extending above the roof line to provide a fire barrier.

Para rubber (Techn) A natural crude rubber obtained from rubber trees in South America.

parasite (Biol) An organism that derives its nourishment from a living plant or animal host.

parent (Nucl Phys) A radioactive nucleus that disintegrates to form a radioactive product or daughter. Also called: precursor.

parenteral (Med) Entering the system by a route other than the gastrointestinal tract.

paresthesia (Med) Tingling or pricking sensations in the skin.

parietal (Anat) Pertaining to the top central portion of the skull.

parity check (Comput) An error check on transmitted data, made by determining whether a data character has an odd or an even set of marks to describe it. Code characters are all formed from an odd or even number of marks, so that an odd count with an even code, for instance, indicates an error in transmission.

parking structure (Build) A deck used for parking or storing automobiles. Also called: parking deck.

paroxysm (Med) A convulsion or seizure, such as a fit of coughing or sudden pronounced recurrence of disease symptoms. A spasm.

partial body irradiation (Dosim) Exposure of part of the body to incident electromagnetic energy.

partial equilibrium (Comb) An approximate relation applied to sets of reactions which can be divided into fast and slow classes. The fast reactions attain an approximate equilibrium that can be calculated thermodynamically, neglecting the slow reactions.

partial pressure (Phys) The fraction of the total pressure contributed by each of two or more components of a gaseous or vapor mixture.
NOTE: Dalton's law states that total pressure is the sum of the partial pressures: $P[total] = P1 + P2 + P3 + \ldots$. For example, the oxygen in the atmosphere accounts for 21% of the atmospheric pressure, or 3.09 psia (21% of 14.7 psia) = 160 mm Hg.

particle¹ (Phys) A point mass used as an idealized model for various physical phenomena.

particle² (Phys) An actual physical mass constituent; e.g., proton, electron, atom, molecule, etc. See: elementary particle.

particulars (Fire Pr) Data required for accurate and thorough reporting of a fire or other emergency incident.

particulate (Phys) Consisting of or existing in the form of particles, which may be liquid or solid.

particulate-filter respirator (Fire Pr) An air-purifying respirator that removes most of the dust, smoke particles, or fumes from air passing through the apparatus. Also called: dust respirator; fume respirator.

particulate matter (Phys) Finely divided solid or liquid particles, such as dust, mist, fog, and smoke, capable of floating in the air. A particulate suspension is commonly called an aerosol if in air or a gas, and a hydrosol if in a liquid.

partition (Build) An interior space divider in a building, esp one that has no structural function. A temporary wall.

part-paid firefighter (Fire Pr) A standby firefighter who fills in part-time during absences of regular firefighters and is usu paid for the number of hours actually spent on duty. Syn: substitute. Cf: call firefighter.

parts per million (Chem; Toxicol; Abb: ppm) The fractional quantity of one material in another by weight or volume, used in reference to the

299

presence of vapor, gas, or other contaminant in air or in highly dilute solutions.
NOTE: The ppm proportion of impurities in solids and liquids is usu by weight; in gases it is often by volume. Confusion is avoided by specifically stating weight, volume, or molar units.

parturition (Biol; Med) The act or process of giving birth.

party wall (Build) A wall dividing two adjoining properties. Syn: separating wall (Brit).

pascal (Metrol; Pa) A unit of pressure equal to one newton per square meter. One pascal is equal to 10 microbars.

Pascal's law (Phys) The principle that a confined fluid transmits externally applied pressure equally in all directions.

passageway (Build) A corridor, hallway, passage, tunnel, and the like for pedestrian traffic.

passive learning (Educ) Learning without incentive or with little intention to master the subject. Ant: active learning.

passive vocabulary (Educ) The set of words an individual understands, in contrast to active vocabularly, which is the set of words the individual uses.

patchy continuity See: fuel continuity

patella (Anat) The kneecap.

pathogen (Biol; Med) A disease-producing microorganism, such as bacteria, viruses, fungi, protozoa.

pathogenesis (Med) The onset and development of a disease. Descriptive of how a disease invades the body, spreads, and its consequences.

pathological (Med) Pertaining to a disease state. Literally, pertaining to the study of disease conditions.

pathology (Med) The study of causes and mechanisms of diseases.

patrol (For Serv) 1. To inspect along a given route to prevent, detect, and suppress fires. 2. To travel back and forth along a control line watching for any sign of breakover, spot fire, or hot spot. See: brand patrol.

patrol desk (Fire Pr) The watch desk in a fire station at which fire-alarm and communications equipment is located and where the company journal is kept. Syn: watch desk.

patrol time (For Serv) The elapsed time between completion of organized mop-up and the time a fire is officially declared out.

pattern¹ a. A systematic organization or orderly arrangement of related elements. b. A model or sample to be followed.

pattern² (Fire Pr) Referring to the shape of a stream from a nozzle.

pause (Atm Phys) A thermal transition zone in the atmosphere.
NOTE: The tropopause lies at approx 8 miles altitude, between the troposphere and the stratosphere; the stratopause at approx 15 miles altitude, between the stratosphere and the mesosphere; and the mesopause at approx 50 miles altitude, between the mesosphere and the thermosphere. Heights vary with the seasons, distance from the poles, and other factors.

pawl² (Fire Pr) A part of a locking mechanism that secures an extension ladder in the desired extended position by engaging the beams of the fly ladder to the rungs of the bed ladder. See also: dog.

pawl¹ (Mech Eng) A mechanism or catch that prevents a wheel from turning in a backward direction by engaging a rachet tooth or by cam friction on a smooth wheel.
NOTE: A pawl can also be used to advance a rachet wheel forward in increments.

pay out (Fire Pr) To feed out or pull off a hose line, rope, or electrical wire steadily from a bed, reel, or coil.

peak-to-peak value (Phys) The algebraic difference between the extremes of an oscillating quantity.

peak value[1] (Phys) The maximum value of a quantity.

peak value[2] (Fire Pr; Brit) The minimum concentration of extinguishing-agent vapor, expressed in percent by volume or weight, that prevents flame propagation in any combination of fuel, extinguishing agent, and oxidizer.

peat A low-grade fuel consisting of decayed vegetation in the first stages of conversion to coal.

peavy (For Serv) A stout wooden handle fitted with a spike and hook, used for rolling logs.

Péclet number (Thermodyn; Symb: Pe) The ratio of heat transfer by convection to heat transfer by conduction. See: dimensionless numbers.

pectoral (Anat) Pertaining to the chest; esp the chest muscles.

pedestal (Fire Pr) The platform and control console for a turntable aerial ladder operator.

pedestrian (Safety) A person traveling on foot.
  NOTE: For purposes of accident reporting, any person who at the time of the accident was not in a motor vehicle or on a bicycle in normal vehicular traffic flow.

pedestrian traffic Numbers of people moving on foot in more-or-less definable patterns along sidewalks, through building corridors, and the like.

peel strength (Techn) The force with which adhesives resist being peeled apart from a surface.

peening (Techn) To draw, bend, or flatten by hammering or by blasting with steel shot.

peer (Psych) An equal; one with equal standing. One who has equality with other members of a group in terms of age, status, experience, or the like.

pelleting (Techn) The compression of powder into pellets or small briquettes.

pendant sprinkler (Fire Pr) An automatic sprinkler head that hangs downward from piping in the ceiling, as contrasted to an upright sprinkler head, which points upward. See: sprinkler system.

penetrant See: wetting agent

penetrometer[1] (Phys) A device for measuring the penetrating power of a beam of x-rays or other penetrating radiation by comparing transmission through various absorbers. See: absorber.

penetrometer[2] (Soil Mech) A probe advanced into the ground to obtain a measure of soil strength.

penology (Law) The science and practice of punishment of criminal offenders.

Pensky-Martens tester (Testing) An apparatus used to measure the flash points of liquids.

percent (Math) One part in a hundred. One part of any whole, or number, that has been divided into 100 equal parts.

percent explosivity See: flammability index

percentile (Statist) A point in a distribution marking off the percent of cases indicated by the given percentile. Thus, the 50th precentile divides a distribution into two equal parts; 50% fall below and 50% above.

percent of slope (Civ Eng) The number of units a slope rises or falls vertically in a horizontal distance of 100 units.

perception (Psych) Awareness gathered through the medium of the senses. The meaning or interpretation given to stimuli received through the senses.

perchlorate (Chem) A compound, derived from perchloric acid, $HClO_4$, containing chlorine in the 7+ oxidation state.
  NOTE: Perchlorates are good oxidizing agents and, on heating, liberate oxygen and react with combustibles.

percolating hose (Fire Pr) A hose with porous walls that permit water to seep through.
NOTE: Linen hose used for first-aid fire fighting is usu of this type. Not having a liner, it can be stored in a small cabinet or compartment. When wet, the threads swell to make a relatively watertight conduit.

percolation (Hydrol) The downward movement or drainage of water through soil, solid waste, or other porous media.

perfect combustion (Comb; Brit) Total combustion in which the exact quantities of air and fuel are present to react completely in a combustion process. See: stoichiometry.

perfect gas law = ideal gas law

performance test See: acceptance test

peribronchial (Anat) Around the bronchi of the lungs.

perimeter[1] (Geom) The boundary enclosing a given area. The length of a closed boundary.

perimeter[2] (For Serv) The boundary of a wildland fire.

period (Phys) The time interval in which a periodically varying quantity repeats itself.
NOTE: The period is the reciprocal of frequency.

periodic law (Chem) The principle, proposed by the Russian chemist D. I. Mendeleyev, that the elements, when arranged in the order of their atomic numbers, show regular variations in most of their properties. They can thus be classified into families that behave similarly chemically. See: periodic table.

periodic table (Chem) A chart showing the chemical elements arranged in horizontal rows (periods) in the order of their atomic numbers or weights and in vertical columns (groups) according to similarities in properties.
NOTE: The table is based on the periodic law that the physical and chemical properties of the elements vary periodically with their atomic weights.

periodic vibration (Phys) A vibration, all of whose characteristics recur at fixed time intervals.

peripheral blood (Med) Blood that is circulating through the less central portions of the body. See: circulation system.

peripheral nervous system See: nervous system

peripheral vision (Physiol) Seeing with the outer portions of the retina.

perishable items (Transp) Items requiring refrigeration during transportation and storage.

perivascular (Med) Around the blood vessels.

perlite (Build) Crushed and heat-expanded volcanic rock used as an aggregate in lightweight concrete.

permanent disability (Safety) Permanent physical impairment, regardless of how incapacitating.

permanent firefighter (Fire Pr) A full-fledged member of a fire department. A firefighter who has successfully passed his probationary period.

permanent impairment = permanent disability

permanent partial disability (Insur) Any injury other than permanent total disability which results in the loss of use of any part or all of a member of the body or any permanent impairment of functions of any part of the body.

permanent total disability (Insur) Any injury that permanently and totally incapacitates a person from following a gainful occupation, or that results in the loss of use of both eyes; one eye and one hand, arm, leg, or foot; any two of the following not on the same member: hand, arm, foot, or leg.

permeability (Phys) The property of a material that allows gases or liquids to pass through it.

permeability coefficient (Hydrol) The

rate of discharge of water under laminar flow conditions through a unit cross-sectional area of a porous medium under a unit hydraulic gradient and standard temperature.

permeable (Phys) Capable of passing fluids; allowing liquids or gases to penetrate or pass through.

permissible dose See: maximum permissible dose

permit (Fire Pr) An official document issued by a fire department or other agency authorizing the use of fire or other hazardous process for destroying waste, for manufacturing purposes such as welding, and many other applications.

permutation (Statist) The number of ways that a set of objects can be arranged in a different order. See also: binomial coefficient; combination.
NOTE: A number of objects n can be arranged in n! (n factorial) different ways. If n = 5, n! = 5 • 4 • 3 • 2 • 1 = 120 ways.

peroxide (Chem) A compound characterized by an -O-O- peroxy group, as hydrogen peroxide (HOOH).
NOTE: Peroxides are extremely reactive and most peroxide vapors are irritating to the mucous tissues and cause headaches.

persistence The quality or tendency of continuing.

persistent (Toxicol) Lasting or enduring for an extended time. Often used in reference to noxious or toxic products that have low volatility or high vapor density and therefor do not disperse readily. Ant: nonpersistent.

person (Law) Any individual, partnership, corporation, agency, association, or organization.

personal factor (Safety) A mental or physical characteristic which permits or occasions an act contributing to an accident.

personal liability insurance (Insur) Insurance that reimburses the insured if he becomes liable to pay damages for injury to others.

personal protective equipment (Safety) Any material or device worn to protect the worker from exposure to or contact with any harmful material or force.

personnel monitoring (Dosim) Determination of the amount of ionizing radiation to which an individual has been exposed, e.g., by measuring the darkening of a film badge or performing a radon breath analysis. Cf: radiation monitoring. See: hand and foot counter.

persuader (Fire Pr) A metal bar used for prying, smashing, hammering, forcing entry, and, in general, overcoming obstinate mechanical frustrations in firefighting. See: forcible-entry tools.

pesticide (Chem) Any of a large group of chemicals used to control or kill rodents, insects, fungi, bacteria, weeds, etc., that cause or carry disease or destroy agricultural products.
NOTE: Specific types are called insecticides, herbicides, fungicides, rodenticides, miticides, nematocides, fumigants, and repellents.

petri dish (Bact) A flat, usu round, glass, plastic, or ceramic dish with a cover, used to culture microorganisms.

petrochemical (Chem) Any of a great variety of chemicals produced from petroleum or natural gas and used for chemical synthesis rather than for hydrocarbon fuels or lubricants.

petroleum (Chem) A mixture of solid, liquid, and gaseous hydrocarbons found in natural deposits in the earth and used to produce gasoline and other fuels, lubricants, solvents, and many other products.

pH (Chem) A measure of the degree of acidity or alkalinity of a solution. It is the negative logarithm of the hydrogen ion concentration.
NOTE: A pH of 1 to <7 is acid, 7 is neutral, and >7 to 14 is alkaline.

phagocyte (Physiol) A cell in the body that characteristically engulfs foreign material and consumes debris and invading microorganisms.

303

phantom (Dosim) A quantity of material closely approximating the density and effective atomic number of living tissue used in biological experiments involving radiation.

phantom box (Fire Pr) A fire alarm code number for a location at which there is no fire alarm signal box. Also called: imaginary box.

phantom box number (Fire Pr) A number representing an area of 250,000 sq ft assigned to a block on a grid map for pattern recognition purposes. The number may be considered as a street fire alarm box number.

pharmaceutical (Pharm) 1. Any of a large group of drugs and related chemicals intended for medical use and distributed through drug suppliers. Pharmaceuticals include not only medicinals such as aspirin and antibiotics, but also nutriments, such as vitamins and amino acids for both human and animal use. 2. Pertaining to medicines and drugs; esp to their physiological action.

pharmacology (Pharm) The science and practice of the physiological action of drugs.
NOTE: Distinguished from pharmacy, which deals with the combination, blending, and preparation of drugs for medicinal use.

pharynx (Anat) The chamber at the back of the mouth leading to the esophagus.

phase difference (Electr) The angle, or portion of a cycle, by which one wave leads or lags another.

phase equilibrium (Chem) The condition in which two or more phases (gas, liquid, solid) coexist in thermodynamic balance.

phase of fire (Fire Pr) A defined stage in the history of a fire.
NOTE: Three phases of fire development and progress are said to occur in enclosed spaces. Phase 1 is a smoldering interior fire involving ordinary combustibles, and the oxygen content of the air is approx normal. Phase 2 is the period during which flames are consuming ordinary combustibles, and the oxygen content of the air ranges down from 21% to 15%. Phase 3 is a hazardous smoldering period in which the oxygen content of the air has been depleted below 15%, and a smoke explosion (back draft) becomes possible if outside air is suddenly admitted. See: time-temperature curve.

phenol (Chem) An organic acid compound, C_6H_5OH, containing a hydroxyl group attached to a benzene ring. Commonly called: carbolic acid. A chemical intermediate and base for plastics, pharmaceuticals, explosives, and other products.
NOTE: Phenol is toxic to all living cells and in diluted form is used as an antiseptic.

phlebitis (Med) Inflammation of the walls of a vein, often associated with blood clot formation. See: thrombus.

phon (Acoust) A unit of loudness level equal to the sound pressure in decibels of a 1000-Hz reference tone judged by listeners to be equally loud.
NOTE: The basic unit of loudness is the sone. Sones and phons are related as follows: 1, 2, 4, 8, 16, ... sones are equal, respectively, to 40, 50, 60, 70, 80, ... phons.

phosgene (Chem; Toxicol) A colorless, tasteless gas with a disagreeable odor, $COCl_2$, formed from carbon monoxide and chlorine in the presence of hot charcoal. Syn: carbonyl chloride.
NOTE: Phosgene forms also when carbon tetrachloride is applied to burning materials in the presence of carbon monoxide. A strong, delayed-action irritant; 2.5 ppm is deadly. When inhaled, the gas dissolves in the mucus of the respiratory tract and forms hydrochloric acid.

phosphor (Chem) A substance that emits light when stimulated by radiation. See: scintillation.
NOTE: A phosphor absorbs radiant energy, such as x-rays, cathode rays, or ultraviolet light, and then emits a portion of the energy in the ultraviolet, visible, or infrared region of the electromagnetic spectrum. Phosphors are classed as fluorescent or luminescent substances.

phot (Optics) A unit of illumination

equal to the density of one <u>lumen</u> per square centimeter.

<u>photochemical process</u> (Chem) A chemical change brought about by radiation acting on a chemical system.
NOTE: Sunlight acting on various pollutants in the atmosphere produces photochemical smog.

<u>photochemistry</u> (Chem) The branch of chemistry dealing with chemical effects induced in matter by electromagnetic radiation, generally in the ultraviolet or optical regions of the spectrum.

<u>photoelectric effect</u> (Phys) The production of electric current when electrons are knocked out of the atoms of a conductor by light rays.
NOTE: The effect is used in a photoelectric cell or electric eye.

<u>photon</u> (Phys) A quantum of electromagnetic energy having an effective momentum but no mass or electrical charge.

<u>photosynthesis</u> (Biol) The food-producing process by which plants synthesize carbohydrates from carbon dioxide and water when the green chlorophyll in plant tissues is exposed to sunlight.
NOTE: In reducing the carbon dioxide, plants release oxygen.

<u>physical change</u> (Phys) A change in state or form that does not involve a chemical change.

<u>physical chemistry</u> (Chem) The branch of chemistry, lying between physics and chemistry, that deals with the physical properties of substances and their chemical compositions and transformations.
NOTE: Physical chemistry includes the thermodynamics of matter, kinetics of chemical changes, and determination of molecular structure.

<u>physical factors</u> (Safety) Environmental factors conducive to accident occurrence. Physical, material, or environmental hazards.

<u>physical property</u> (Phys) Any of a number of descriptive characteristics of matter, such as density, electrical and thermal conductivity, expansion, and specific heat. Cf: <u>mechanical property</u>.

<u>physics</u> (Phys) The study of the interaction of matter and energy and of the origin, nature, and properties of gravitational, electromagnetic, nuclear, and other force fields for the purpose of explaining and understanding natural phenomena.
NOTE: Formerly, physics was called <u>natural philosophy</u> and included classical mechanics, heat and thermodynamics, kinetics of gases and statistical mechanics, optics, acoustics, electricity, and magnetism. Newer branches of the science include the physics of high-energy particles; nuclear, atomic, and molecular physics; the physics of solids, liquids, gases, and plasmas; cryogenics (low-temperature physics); mathematical physics, biophysics, and a number of other specialized branches. See: <u>science</u>.

<u>physiological effects</u> (Physiol) The effects of environmental and other factors on the organic functions of living organisms.

<u>physiology</u> (Physiol) The branch of biology that deals with the physics and chemistry of living matter, body processes, and cell and tissue functions of living organisms.

<u>pi</u> (Geom; Symb: π) The symbol for the ratio of the circumference of a circle to its diameter.
NOTE: The ratio, to eight decimal places, is 3.14159265, commonly rounded to 3.1416 in practical computations.

<u>pickaroon</u> (For Serv) A device with a head similar to an axe but with a point rather than a blade mounted on the end of a handle which is used to assist in the lifting and placement of bolts of wood.

<u>pickling</u> (Techn) The process of removing a coating of scale, oxide, or tarnish from metals by immersing them in an acid bath to obtain a chemically clean surface.
NOTE: For leather, pickling refers to the treatment of light skins with a dilute solution of sulfuric acid and salt.

<u>pick up</u> (Fire Pr) To gather up hose and other fire-fighting equipment

after a fire and stow it on the fire apparatus for return to quarters.

pick-up tube (Fire Pr) A tube through which foam making concentrate or liquid is fed into foam-making equipment.

pico A prefix meaning one trillionth (10^{-12}). Syn: micromicro.

piece¹ (Fire Pr; Colloq) a. A length of fire hose. b. A fire apparatus.

piece² (Police; Slang) A handgun.

pier (Transp) A platform-like structure projecting into a waterway to allow vessels to moor alongside to load and unload cargo.

piercing pole (Salv) A long wooden or fiberglass pole with a sharp metal point used to puncture holes in ceilings to drain water during salvage operations. See: salvage.

piezometer tube (Fire Pr) A tube inserted into a fluid stream to measure the flow pressure. See: pitot tube.

pig¹ (Metall) A small ingot from the casting of blast furnace metal.

pig² (Nucl Phys; Slang) A heavily shielded container (usually lead) used to ship or store radioactive materials. Syn: cask; coffin. Cf: rabbit.

pigtail¹ (Electr) A short length of electrical cord with adapters to match different types of electrical connectors.
NOTE: Used to connect electrical appliances and equipment to power sources.

pigtail² (Fire Pr) A short extension of hose.

pike pole (Fire Pr) A long wooden or fiberglass pole with a sharp metal point and a spur-like hook, used to pull down plaster, esp from ceilings, to expose hidden fire. Syn: ceiling hook.
NOTE: Pike poles may also be used to construct water chutes, to advance hose vertically, and in other innovative ways.

pilaster (Build) A small pillar projecting from the face of a wall.

pile¹ (Storage; Transp) A load or portion of a load composed of two or more units which may be placed either side by side or one on top of another, or secured as a single unit.

pile² (Nucl Phys) The core of a nuclear reactor, consisting of numerous blocks of graphite and slugs of uranium. See: reactor.

piling and burning (For Serv) Collecting logging slash and other forest debris into piles and destroying them by burning.

pillar¹ (Build) An upright column used as a loadbearing support in a building.

pillar² (Mining) A column of undisturbed material, roughly circular in cross section and splayed at the top and bottom, left at intervals in a mine as a roof support.

pillar³ (Fire Pr; Brit) = street box

pillar hydrant (Fire Pr; Brit) = post hydrant

pill test (Build; Testing) A test method for floor coverings to determine whether sustained burning occurs when a small pill placed on the material is ignited.

pilot¹ (Comb) A small flame used to ignite gaseous fuel issuing from a large burner. Also called: pilot light.

pilot² (Fire Pr; Brit) An individual assigned to direct fire department units as they arrive at the fireground.

piloted ignition (Comb; Testing) Ignition initiated by a pilot flame.
NOTE: Used in flammability testing.

pilot plant (Techn) Small-scale operation preliminary to major enterprises. Common in the chemical industry.

pilot study A preliminary simplified and limited version of a full-scale study to determine feasibility, to verify the study design, and to see that the results are the ones sought in the detailed study.

pin lug coupling (Fire Pr) A hose coupling with pin-shaped lugs or projections for use in tightening or loosening the coupling with a wrench or spanner.

pinch point (Safety) Any point other than the point of operation at which it is possible for a part of the body to be caught between the moving parts or between moving and stationary parts of a machine. Cf: nip point.

pinna (Anat) Ear flap; the part of the ear that projects from the head. Also called: auricle.

pipe[1] (Techn) A tubular structure used to convey fluids. A conduit.

pipe[2] (Mining) An approx cylindrical vertical geological formation.

pipe flow (Fl Mech) The motion of fluids in closed conduits.
  NOTE: In a round, smooth pipe, the flow is laminar when the product of the flow velocity and the pipe diameter divided by the kinematic viscosity of the fluid is less than 2000. Above 2000, flow becomes turbulent. In laminar flow, the velocity distribution across the pipe diameter is parabolic. Flow velocity is maximum along the axis of the pipe and approaches zero velocity at the walls. See: Reynold's number.

pipeman (Fire Pr) A firefighter who directs the stream from a nozzle and operates the nozzle shut-off or control valve. Syn: nozzleman.

piping system (Techn) A complex of interconnecting pipe, tubing, connections, valves, the pressure-containing parts of other components such as expansion joints and strainers, and devices which serve such purposes as mixing, dividing, metering, or controlling flow.

piston pump (Hydraul) A positive-displacement pump, usu with 2, 4, or 6 reciprocating pistons. See: pump.
  NOTE: Piston pumps can be single- or double-action. In the single-acting, water is delivered only on the positive stroke; double-acting pumps deliver water on both strokes.

piston tool (Techn) A hand tool in which the impact of a piston is used to drive a stud, pin, or other fastner into a workpiece or surface.

pitch[1] The degree of steepness, as the slope of a roof.

pitch[2] (Fire Pr) The included angle between the horizontal and a leaning ladder, measured on the side opposite the climbing side.

pitch[3] (Chem Eng) A dark-colored residue left after distillation of tar.

pitch[4] (Techn) The distance between consecutive threads of a bolt or coupling, or the distance between the centers of two adjacent teeth of a gear wheel.

pitch[5] (Navig) Motion about the transverse axis of a vessel or aircraft.

pitch[6] (Acoust) The quality of highness or lowness of sound, determined by its frequency.
  NOTE: In the audible range, 16 Hz is the lowest and 20,000 Hz is the highest sound that can be heard by the normal human ear.

pitch pocket (For Serv) An opening in wood, e.g., a timber, extending parallel to the annual growth rings and containing either solid or liquid pitch.

pitot gage = pitot tube

pitot tube (Aerodyn) An instrument used to measure the stagnation (total or impact) pressure of a flowing fluid.
  NOTE: In most common use, the tube, bent to a right angle, is inserted into the center of a flow with the open end pointing upstream and aligned with the flow. A gage attached to the other end of the tube indicates the pressure. Velocity measurement errors can arise if the tube is used incorrectly. In fire practice the tube is inserted into a hose stream at a distance of half the nozzle diameter away from the orifice, and flow rates are determined with the aid of published tables. Also called: impact tube; pitot gage. See: piezometer tube.

placard (Fire Pr) 1. A sign or notice, esp one that identifies a hazardous material and indicates appropriate precautions. Syn: label. 2. A sign with instructions posted above a local fire alarm station.

placarding (Fire Pr) 1. Signs or notices posted to warn of possible hazards. 2. Affixing and displaying of such signs. Syn: labeling.

placenta (Physiol) A vascular membrane that unites a fetus to its mother's uterus and provides for metabolic exchanges.

place of assembly (Fire Pr) A building or portion of one used by a large number of people for deliberations, worship, entertainment, dining, awaiting transportation, etc.
 NOTE: Places of assembly are classified according to capacity: Class A, capacity of 1,000 persons or more; Class B, capacity of 300 to 1,000 persons; Class C, capacity of 100 to 300 persons.

Planck's constant (Phys; Symb: h) A fundamental physical constant equal to 6.6256×10^{-27} erg • sec, used to represent an elementary quantum of action or the product of energy and time.

plank (Build) A piece of lumber 2 to 4 in. thick and 8 or more in. wide.

planograph (Storage) A scale drawing of a storage area showing the approved layout.

plans chief (For Serv) A fire suppression staff officer responsible for compiling and analyzing data required for the development of fire suppression plans.

plant[1] (Fire Pr) 1. An arrangement of incendiary materials made by an arsonist to set fire to a structure. Syn: set.
 NOTE: The plant may involve flammable liquids, trailers, scattered combustibles, and may include various time-delay or remote-control ignition devices.

plant[2] (Build; Brit) The materials, tools, machines, scaffolding, ladders, and other equipment used by a builder in the construction of a building.

plant factor (Electr) The ratio of the average power load of an electric power plant to its rated capacity. Also called: capacity factor.

plant protection (Law) Protection of property and control of access to a business or industrial establishment.

plasma[1] (Med) The liquid part of the lymph, or of the blood, in which the blood cells are suspended. See: blood plasma.

plasma[2] (Phys) An electrically neutral, gas-like mixture of positive and negative ions within a metal, or one that can be contained by strong magnetic and electrical fields.
 NOTE: Sometimes called the fourth state of matter, since it is unlike a solid, liquid, or gas. In contrast to a gas, plasma is a good conductor of electricity. High-temperature plasmas are used in controlled fusion experiments.

plaster board = gypsum board

plaster hook (Fire Pr) A barbed hook attached to a long handle, used to break through and pull down ceilings and wall boards to expose and give access to concealed fire. Syn: ceiling hook; pike pole.

plastic[1] (Metall; Phys) The property of being easily deformable.

plastic[2] (Chem) Any of a wide range of natural or synthetic organic materials of high molecular weight that can be formed by pressure, heat, extrusion, and other methods into desired shapes. Plastics are usu made from resins, polymers, cellulose derivatives, caseins, and proteins. The principal types are thermosetting (irreversibly rigid) and thermoplastic (reversibly rigid).
 NOTE: Plastics may be grouped by physical characteristics into five categories: film, formed, filled, foam, and fiber or filament. Films include sheets, coatings, and adhesives; formed plastics include rods, blocks, castings, extrusions,

and moldings; filled forms include reinforced plastics in which another material is distributed throughout the plastic; foamed plastics include cellular and honeycomb structures; fibers and filaments are essentially plastic threads.

plastic deformation (Metall) Permanent change in the shape of metals produced when crystal fragments slide over one another under pressure, as by rolling, molding, or extrusion. Ductility, malleability, and creep properties of metals determine the extent to which they can be shaped.

plasticity (Soil Mech) The property of a soil that allows it to be deformed beyond the point of recovery without cracking or appreciable volume change.

plasticizer (Chem) An organic chemical used in modifying plastics, synthetic rubber, and similar materials to facilitate compounding and processing, and to impart flexibility to the end product.

plate (Build) A structural element that forms the top or bottom of a framed wall.

plateau (Psych) A level segment on a learning curve, indicating that learning has ceased.
  NOTE: A temporary halt in learning may be due to fatigue, loss of motivation, etc.

platelet (Med) A disk-shaped body in the blood that helps to form clots.

platform¹ (Techn) A working space for persons that is elevated above the surrounding floor or ground, such as a balcony, for the operation of machinery and equipment.

platform² (Build) An extended step or landing breaking a continuous run of stairs.

platform ladder (Ind Eng) A self-supporting ladder of fixed size with a platform provided at the working level. The size is determined by the distance from the platform to the base of the ladder.

platoon (Fire Pr) A group of firefighters working the same tour of duty.

play (Fire Pr) To aim and apply a firestream on a fire.

playpipe (Fire Pr) 1. A rigid, tapered extension pipe between a hose and a nozzle, fitted with handles to assist in controlling the firestream. 2 = (Brit) branchpipe.

plea (Law) The response of a defendant to a charge in court.

plenum¹ (Techn) A pressure-equalizing chamber.

plenum² (Build) The space above a hanging ceiling used as part of the air-handling system in a building.

pleura (Anat) The membrane surrounding the lungs and lining the walls of the thoracic cavity.

pleurisy (Med) An inflammation of the pleura, usu accompanied by irritation, pain, and breathing difficulty.

plot 1. A set of data points on a scatter diagram. 2. To enter data points on a graph.

plowline (For Serv) A firebreak constructed by a tractor-drawn plow.

plug (Fire Pr; Colloq) A fire hydrant.
  NOTE: The term derives from the time when water-main openings were stoppered by wooden plugs. To tap the main, the plug had to be removed and a tapered standpipe driven into its place.

plug pressure (Fire Pr) 1. The water pressure in a hydrant. 2. Pertaining to a hose stream delivered at hydrant pressure without pumping.

plumb (Techn) A device used to find the vertical. Used to measure the angle of incline of aerial ladders.

plumbago = graphite powder

plumbism = lead poisoning

plume (Fire Pr) A column of smoke or visible particles from a fire or a smoke stack.

plume trap (Phys) An exhaust ventilation hood designed to capture and remove the plume given off the target upon impact of a laser beam.

plutonium (Nucl Phys; Symb: Pu) A heavy, radioactive, man-made, silvery metallic element with atomic number 94, mp 639.5°C.
NOTE: Its most important isotope is fissionable plutonium-239, produced by neutron irradiation of uranium-238, which is used as reactor fuel and in weapons. In finely divided form it can ignite spontaneously in moist air. Plutonium is a radiological poison, being specifically absorbed by the bone marrow.

pluviograph (Meteor) A recording rain gage.

plywood (Build) A sandwich construction of three or more layers of wood veneer, usu laid cross-grain to each other for added strength, and bonded by an adhesive. One of the facing plies is often of a better grade of wood than the other.

PMMA = polymethylmethacrylate

pneolator (First Aid) A positive pressure resuscitator that delivers oxygen at a preset pressure into the lungs of a person who has stopped breathing. Exhalation occurs without assistance by the natural contraction of the chest and diaphragm. See: breathing apparatus.

pneumatic barrier See: barrier

pneumatic spring (Techn) A self-contained spring assembly that derives its spring action from the compressibility and elasticity of an enclosed gas.

pneumococcus (Med) A microorganism that is the major cause of lobar pneumonia.

pneumoconiosis (Med) A lung condition resulting from prolonged inhalation of dust or other particulate matter. Syn: black lung.
NOTE: A form called Shaver's disease is found in workers exposed to aluminum oxide fumes and to silica particles in bauxite smelting and corundum manufacturing.

pneumonia (Med) Inflammation of the lungs caused by microbial, physical, or chemical agents, often accompanied by fluids filling the lung cavities. A common occurrence in burn patients with thermal injuries to the respiratory tract.

pneumonitis (Med) A mild form of pneumonia.

pneumothorax (Med) The presence of gas or air in the pleural cavity outside the lungs.

pocket[1] (For Serv) An angular unignited area between advancing fingers of fire.

pocket[2] (Mining) a. A small cavity containing a deposit of ore. b. A holding area in a mine used to collect broken ore for subsequent transport to the surface.

pocket rot (For Serv) Advanced decay which appears in the form of a hole, pocket, or area of soft rot usu surrounded by apparently sound wood.

point of perception (Safety) The time and place at which an individual first saw, heard, smelled, or felt a hazard; the first sign that an accident was about to occur.

point of possible perception (Safety) The place and time at which an unusual or unexpected movement or condition could have been observed by a normal person. This point always comes at or before the point of actual perception.

point operation (Techn) The part of a machine where cutting, shearing, squeezing, drawing, or manipulating is performed.

point-to-point communication (Commun) Transmission of messages between two fixed stations.

poise (Phys; Symb: P) The cgs unit of absolute viscosity, named after Poiseuille, discoverer of the laws of flow.
NOTE: Water at 20°C has a viscosity of one centipoise. A centipoise is one-hundredth of a poise. See: viscosity.

poison[1] (Toxicol) A substance that chemically or physicochemically

causes serious tissue damage or organic malfunction from small doses.
  NOTE: Four types of poison are distinguished: a) <u>asphyxiants and depressants</u>, such as carbon monoxide, cyanide, and certain hydrocarbons that cause anoxia and circulatory collapse; b) <u>surface agents</u>, such as acids and alkalis; c) <u>systemic poisons</u> such as arsenic, lead, mercury, and cadmium compounds that damage specific body organs or tissues and cause local irritation at the point of entry; and d) <u>systemic poisons</u> such as blood poisons and nerve gases that do not cause local irritation at the point of entry but attack specific portions of the body after absorption. Poisons are also described as corrosives, irritants, asphyxiants, enzyme inhibitors, neurotoxins (nerve gases), and hemotoxins (blood poisons). See: <u>first aid</u>.

<u>poison</u>² (Nucl Phys) A neutron-absorbing material, such as boron, used in control rods to absorb neutrons in nuclear reactors to decrease reactivity.

<u>poisoning</u> (Med) The effects, esp systemic, of toxic materials.

<u>Poisson distribution</u> (Math) A distribution whose frequency function is of the form:

$$P(x) = (m(x)/x!) d(-m),$$

the probability that x events will occur; where m is a parameter that is both the mean and variance, the mean and variance of the Poisson distribution being equal.
  NOTE: This distribution often appears in observed events which are very improbable compared to all possible events, but which occur occasionally since so many trials occur, e.g., traffic deaths, industrial accidents, and radioactive emissions.

<u>pokethrough</u> (Build; Colloq) An opening made in a building wall or other fire barrier during construction or alteration for a utility conduit, wiring, and the like, that is not sealed or otherwise firestopped after the service element has been installed. A utility hole.

<u>polar compound</u> (Chem) An organic, nonionic material or liquid solvent having an electrically unbalanced molecular structure, a local separation of electric charge, that results in magnetic polar susceptibility.
  NOTE: Some polar compounds are capable of dispersing fire-fighting foams. Methyl ethyl ketone and other ketone solvents, aldehydes, and ethers are common organic polar solvents. See: <u>nonpolar compound</u>; <u>solvent</u>.

<u>polarization</u> (Phys; Opt) A condition of light or radiant heat in which the transverse vibrations of the rays assume different forms in different planes.

<u>pole</u> (Fire Pr) 1. A sliding pole used by firefighters to descend from upstairs sleeping quarters to the apparatus floor. 2. = <u>tormentor pole</u>.

<u>pole ladder</u> (Fire Pr; Brit) = <u>folding ladder</u>

<u>pole man</u> (Fire Pr) A firefighter handling a <u>tormentor pole</u> during ladder raising and lowering operations.

<u>pole shutoff</u> (Fire Pr) A pole, varying in length from 6 to 20 feet, with a metal wedge at one end, used to close fused sprinkler heads.

<u>policy</u>¹ (Man Sci) A basic statement that guides administrative action and defines authority.

<u>policy</u>² (Insur) A written contract which guarantees responsibility or indemnification.

<u>policyholder</u> (Insur) An individual with whom an insurance company has contracted to make certain losses good if they occur.

<u>pollution</u> (Man Sci) Contamination of the environment, including the soil, water, and atmosphere, particularly resulting from human activity. See: <u>air pollution</u>.

<u>polyamide</u> (Chem) A high molecular weight polymer, such as nylon, containing amide ($CONH_2$) groups.

polybutadiene (Chem) A kind of rubber blended with natural rubber and used in the manufacture of automobile tires.

polychromatic light (Phys) White light. Light composed of many wavelengths.

polyethylene (Chem) A macromolecular form of ethylene, $CH_2=CH_2$, used in packaging, linings, squeeze bottles, and the like.

polyethylene oxide (Fire Pr) A polymer of ethylene oxide, trade name "Polyox" and others, with an average molecular weight of several thousand.
NOTE: The compound is usu supplied in the form of a white powder which, added in small quantities to water, reduces the frictional flow forces in hoses and pumps. This allows significant increase in water discharge rates and distance. Common formulations are called "Slippery Water," "Rapid Water," etc.

polyisoprene (Chem) A synthetic analog of natural rubber.

polymer (Chem) A high-molecular-weight organic or inorganic material consisting of long chains of one or more kinds of simple molecules called monomers.
NOTE: There may be hundreds or even thousands of the original molecules linked end to end and often cross-linked. Rubber and cellulose are naturally occurring polymers. Most resins are chemically produced polymers. Polymers are used in the manufacture of plastics, rubbers, fibers, coatings, films, and the like.

polymerization (Chem) A chemical reaction that produces very large molecules by a process of repetitive addition of simple ones.

polymorphism (Chem; Phys) The ability of a substance to exist in two or more crystalline forms. See also: **allotropy**.

Polyox (Techn) A trade name for **polyethylene oxide**.

polystyrene (Chem) A thermoplastic polymer that has a self-ignition temperature of 910°F and burns with a slow sooty flame. Impact polystyrene is used extensively in refrigerator doors, packaging, cabinets and enclosures, sheet rock, and furniture. Expanded polystyrene is used for insulated containers and in buoyancy and flotation applications.

polytropic atmosphere (Meteor) A parcel of air that is changing in volume but has insufficient heat entering or leaving the system to maintain a constant temperature.

polyurethane (Chem) A resin containing the urethane group -NHCOO- used in adhesives, coatings, castings, fibers, and both rigid and elastic foam. The latter is widely used in upholstery, rug backing, and insulation.

polyvinyl chloride (Chem; Abb: PVC) A thermoplastic material that has high resistance to weathering, water, and chemicals and has self-extinguishing properties. Widely used in interior finishing, floor coverings, rainwear, upholstery, siding, roofing, phonograph records, and for chemical and electrical insulation.

pompier (French) = **firefighter**

pompier belt (Fire Pr) An adjustable belt with a snap hook used to secure oneself to the spar of a **pompier ladder** or a ladder rung and thus free the hands.
NOTE: A pompier belt may also be used for **rappelling**.

pompier chain (Fire Pr) A series of pompier ladders suspended from several windows, one above the other, to provide a means for scaling a building.

pompier ladder (Fire Pr) A single-spar ladder with rungs projecting from both sides and a long gooseneck at the top for hooking over window-sills and other openings. On reaching a window, the ladder can be hauled up and hooked over another sill above, thereby providing access to the next floor. See: **ladder**.

pool fire (Comb; Fire Pr) A free-burning fire above a pool of combustible liquid or a large area of molten solid.
NOTE: Pool fires are often used as test fires in experimental studies.

population (Statist) All members of a defined set of objects, values, measurements, or data, as, for example, all items in a given production lot. The set from which a sample is drawn.

porch (Build) A covered platform attached to a building; often open, but sometimes enclosed by walls, windows, or screens.

porphyrin (Chem) One of a group of complex chemical substances that form the basis of the respiratory pigments of animals and plants, as hemoglobin and chlorophyll.

port[1] An opening in a research apparatus through which objects are inserted for testing and other purposes.

port[2] (Navig) The left side of a ship or vehicle, when looking from aft to forward.

portable director (Fire Pr; Brit) A holder for a standard branch that can be left unattended to direct a stream of moderate size. Syn: portable monitor.

portable generator (Electr) A small electric generator carried on a fire truck to provide power for lights and electrically powered tools.

portable monitor (Fire Pr) A heavy stream device that can be dismounted from an apparatus and set up to operate in a position that cannot be reached by wagon batteries or ladder pipes.

portable pump (Fire Pr) A small electric or gasoline-driven pump carried on fire apparatus and capable of supplying small hose at a typical rate of 50 gpm at 90 psi or relaying water to a fire department pumper.

portable tank (Transp) 1. A closed container having a liquid capacity of more than 60 US gallons and not intended for fixed installation. 2. Any compressed gas container that can be attached temporarily to a motor vehicle, railroad car other than a tank car, or marine vessel, and equipped with skids, mountings, or accessories to facilitate handling of the container by mechanical means.

Portland cement See: cement

position (Fire Pr) An area of operation assigned to a fire company or an area of command assigned to a chief officer at a fire.

position sensor (Commun) A device for measuring a position and for converting this measurement into a form convenient for transmission, as a source of feedback. See: transducer.

positioning time (Techn) The time required to traverse a tool from one cutting operation to another.

positive-displacement pump (Hydraul) A pump that moves a liquid by alternately filling and emptying an enclosed volume. Varieties of such pumps include reciprocating, direct-acting, rotary, vacuum, and air-lift types.

positron (Phys) An elementary particle with the same mass as an electron and with a positive charge equal in magnitude to the negative charge of an electron.
NOTE: The positron reacts readily with an electron to yield two gamma rays.

post (Mining) A vertical supporting timber in a mine.

postburn (Med) Referring to a time following a burn injury.

posterior (Anat) Toward the back, or to the rear. Ant: anterior.

post hydrant (Fire Pr) 1. The common street fire hydrant. 2. = (Brit) pillar hydrant.

post indicator valve (Fire Pr; Abb: PIV) A valve used in sprinkler systems that indicates open and shut positions.

postnatal (Med) Occurring after birth.

post partum (Med) Following birth; referring to conditions following childbirth.

posttension concrete (Build) One of several ways of preparing prestressed building elements in which plastic-sheathed cables are laid loosely in forms before concrete is poured. These cables, called <u>tendons</u>, are jacked, or tensioned, several days later, after the concrete has set-up.

<u>postulate</u> (Math) An assumption taken as true in a line of reasoning. Cf: <u>hypothesis</u>.

<u>potable</u> Suitable for drinking.

<u>potassium bicarbonate</u> (Chem; Fire Pr) A dry chemical agent, KCO, that is compatible with aqueous film-forming foam (Light Water). The two together are used as twinned extinguishing agents, applied from separate nozzles, to put out liquid fuel fires.

<u>potato roll</u> (Fire Pr) Hose rolled up in the manner of a ball of string.

<u>potential difference</u> See: <u>electromotive force</u>

<u>potential energy</u> (Phys) Energy possessed by a body by virtue of its position rather than its motion.
  NOTE: For an elevated body, this is equal to its weight multiplied by the distance it has been raised. For a spring, potential energy is increased due to compression.

<u>potential hazard</u> (Safety) A person, situation, thing, or event that exhibits behavior or characteristics conducive to an accident.

<u>potential heat value</u> = <u>calorific potential</u>

<u>potentiometer</u> (Electr) A variable voltage divider; a resistor which has a variable contact arm so that any portion of the potential applied between its ends may be selected.

<u>pound</u> (Metrol; Abb: lb) A unit of mass in the English absolute system and unit of force in the English gravitational system of units. The US (avoirdupois) pound is equal to 0.454 kg. A one-pound force is equal to 4.4482 newtons and will impart an acceleration of 32.174 ft/sec^2 to a one-pound mass.

<u>poundal</u> (Phys; Symb: pdl) A force capable of accelerating a one-pound mass one-foot per second. A one pound weight exerts a force of 32 poundals. A poundal is equal to 0.138255 newton.

<u>powder-actuated tool</u> (Techn) Descriptive of a tool or machine which drives a stud, pin, or fastener by means of an explosive charge.

<u>powderman</u> (Civ Eng; Mining) In mining, quarrying, road building, etc., the specialist responsible for explosives. At times the powderman may be in charge of all blasting work on a project.

<u>powder metallurgy</u> (Metall) Fabrication of objects from pure or alloyed metal powders under pressure and heat.

<u>power</u>^{1} (Phys) The time rate at which work is done, usu expressed as watts or horsepower. See: <u>joule</u>; <u>horsepower</u>; <u>watt</u>.

<u>power</u>^{2} (Math) The number of times a number is multiplied by itself, commonly indicated by a superscript called the <u>exponent</u>.

<u>power</u>^{3} (Opt) Referring to the magnification strength of a lens or optical system.

<u>power density</u> (Phys) The intensity of electromagnetic radiation per unit area expressed as watts/cm^2.

<u>power factor</u> (Electr) The ratio of the actual power of an alternating or pulsating current, as measured by a wattmeter, to the apparent power, as indicated by ammeter and voltmeter readings. The power factor of an inductor, capacitor, or insulator is an expression of the losses.

<u>power function</u> (Statist) A measure of whether a hypothesis should be rejected at a certain level of risk.

<u>power level</u> (Acoust) The ratio of a given power to a reference power, expressed in decibels, as 10 times the logarithm to the base 10.
  NOTE: To be meaningful, the reference power must be indicated.

<u>power reactor</u> (Nucl Phys) A reactor designed to produce useful nuclear power, as distinguished from reactors used primarily for research or for producing radiation or fission-

able materials.

<u>powersaw boss</u> = <u>saw boss</u>

<u>power take-off</u> (Mech Eng; Abb: pto) A mechanism used to couple various tools, devices, or machines to a primary mechanical power source.

<u>power train</u> (Mech Eng) A series of power-transmitting components between an engine and a power-consuming machine, such as a pump.

<u>power transmission</u> Equipment such as shafting, gears, belts, pulleys, or other parts used for transmitting power to a machine.

<u>PPBS</u> = <u>planning-programming-budgeting system</u>

<u>ppm</u> = <u>parts per million</u>

<u>practice</u>[1] (Fire Pr) a. A drill session. b. Repetition for the purpose of perfecting a skill.

<u>practice</u>[2] (Man Sci) An approved procedure or method for accomplishing a given task, or one that has been proved convenient and workable and is generally accepted and followed.

<u>prairie fire</u> (Fire Pr) A dry grass fire on a prairie.

<u>Prandtl number</u> (Fl Mech) A <u>dimensionless number</u>, the product of the coefficient of viscosity and the specific heat at constant pressure divided by the thermal conductivity of a fluid. It is used in the analysis of heat transfer in the forced flow of fluid and characterizes the fluid while the <u>Reynolds number</u> characterizes the flow.

<u>preaction system</u> (Fire Pr) A dry sprinkler system in which automatic devices charge the pipes before the individual heads open and begin to discharge the extinguishing agent.

<u>preburn time</u> (Fire Pr) A timed period after ignition of a combustible fuel and before extinguishment is started. Usu used in testing or training exercises to allow the fire to reach the desired stage of development.

<u>precautionary</u> (Fire Pr) Referring to an action taken in advance of need, esp when the nature or extent of the need appears serious but cannot be evaluated immediately.

<u>precipitate</u> (Chem) 1. To form a separable, insoluble solid in a liquid medium, for example, silver chloride when table salt is added to a silver nitrate solution. 2. The solid material formed by precipitation.

<u>precipitation</u> (Meteor) Any form of water condensing in the atmosphere and falling to earth or condensing directly on objects on the ground. In general, the fallout can be categorized into five types: liquid, mixed liquid-solid, solid, subliming, and suspended forms.
1) <u>Liquid</u>: The most common liquid form is rain, consisting of water droplets ranging from 0.5 to 5 mm in diameter falling from cumulus clouds. Drops larger than 5 mm tend to break up. <u>Drizzle</u> is a gentle sprinkle consisting of droplets finer than rain and falling in showers from fog or thick stratus clouds. Also called: <u>mist</u>. <u>Freezing rain</u> is rain or drizzle that freezes to form a sheet of ice when it reaches the ground. Also called: <u>freezing drizzle</u>; <u>glazed frost</u>; <u>ice storm</u>; <u>glaze storm</u>; <u>silver thaw</u>; or <u>sleet storm</u>.
2) <u>Solid</u>: The most common solid precipitation is <u>snow</u>, which falls in the form of symmetric hexagonal shapes or six-pointed spangles, sometimes agglomerating into large irregular flakes. <u>Snow pellets</u>, also called <u>graupel</u> or <u>soft hail</u>, are soft opaque ice particles, irregular and sometimes scalloped at the edges with protruding crystals, typically 2 to 5 mm in diameter and falling in showers. <u>Snow grains</u> are similar to snow pellets but are smaller and flatter (1 mm diameter or less), falling from stratus clouds or fog as the solid analog of <u>drizzle</u>. <u>Ice crystals</u> are small rods or platelets falling in a steady shower in cold stable air, often without visible clouds. <u>Ice needles</u> are irregular slivers of ice, often coated with <u>rime</u>. The crystals are narrow and pointed, 1 to 3 mm long and 0.25 mm in diameter. <u>Ice pellets</u> are spherical or irregular, hard, clear or opaque ice particles formed from frozen rain or drizzle. <u>Hail</u> is formed of layers of rime and ice,

typically 1 cm in diameter but sometimes reaching 10 cm in size, falling in showers during thunderstorms from extremely thick clouds. Small hailstones are sometimes called pellets, prisms, or sleet.

3) Mixed forms: In US usage sleet is partly frozen rain or the frozen coating that forms on trees, wires, and other objects by freezing rain. In British usage, sleet is a mixture of falling rain and snow. See: rime, below. Glaze is formed by cold raindrops that freeze when they strike an object or the ground. Also called: freezing rain.

4) Subliming forms: Dew is moisture condensing in droplets from the air on solid objects on the ground, esp at night. Frost is formed by small, needle-shaped ice crystals subliming directly from the vapor phase on objects that have cooled below the dew point when the dew point is below the freezing point. Also called: white frost; hoarfrost. Rime is a white icy coating formed on trees and shrubs directly from vapor or fog.

5) Suspended or aerial forms: Fog is an opaque, hanging or drifting cloud or layer of small water droplets in the air near the earth's surface. Mist is a thin fog in which horizontal visibility is >1 km. Also called: drizzle. Haze is a thin mist. See: cloud; water.

NOTE: Any precipitation, including clouds, blowing snow, spray, etc. is called a hydrometeor.

precipitator (Techn) A device, usu electrostatic, that precipitates solid or liquid particles suspended in a gas by means of a unidirectional electric field.
NOTE: The precipitated particles are attracted to and collected on the positive electrode.

preconnected line (Fire Pr) Referring to a hose carried connected to the pump to save time at a fire.

pre-employment examination (Med) A physical examination of a job applicant conducted prior to his employment.

prefire planning (Fire Pr) Advance planning for fire-fighting operations based on surveys and inspections of identified hazards and taking into account physical layouts of structures, location of fire-fighting equipment and hydrants, quantities of water available, and other factors.

prefire survey (Fire Pr) An inspection of a property carried out to gather data and information for prefire planning.

pregnancy (Med) The state of carrying a fetus.

preheater (Comb) A device that heats air or fuel prior to its entry into a combustion chamber.

preheat fire region (Comb) A region ahead of a spreading or propagating fire that is being heated by radiation, conduction, and hot gases from the fire but has not yet become involved in flames.

preheat zone (Comb) The region in a flame front in which temperature and composition changes take place through diffusion and thermal conduction but reaction has not yet begun.

premature (Med) Pertaining to a newborn infant weighing less than five pounds or one that is less than 270 days of conceptual age at delivery.

premeditation (Law) Advance planning and deliberate intent to commit a crime.

premise A proposition used to support a conclusion.

premixed flame (Comb) A flame for which the fuel and oxidizer are mixed prior to combustion, as in a laboratory Bunsen burner or a gas cooking range. Propagation of the flame is governed by the interaction between flow rate, transport processes, and chemical reaction.

preparedness (Fire Pr) 1. The condition, degree, or readiness to cope with a fire situation. 2. Alertness and readiness to recognize fire danger in order to respond promptly and correctly to existing conditions.

preponderance of evidence (Safety) A substantial accumulation of evidence (data, information, past experience) that supports one judgment or decision as opposed to another. Descriptive of the basis upon which a decision is made, such as determining if an injury is occupational or nonoccupational.

presbycusis (Med) A hearing loss due to aging; believed to be caused by the degeneration of the nerve cells.

prescribed burning (For Serv) Planned destruction of natural wildland fuel by burning for the benefit of silviculture, wildlife management, grazing, reducing fire hazards, and other useful purposes. Syn: controlled burning.

preservation (Storage; Transp) Application or use of measures to prevent deterioration resulting from exposure to atmospheric conditions during shipment and storage.

preservative (For Serv) Any substance applied to wood that makes it resistant to wood-destructive fungi, wood borers, and other destructive organisms for an extended period of time.

presignal delay (Fire Pr) A delay in transmitting an alarm of fire to the fire department while local authorities verify the local alarm as real.

pressure (Phys) The force exerted at right angles to a surface, for example, by a fluid on a wall, and defined as the ratio of force to area, expressed in pounds per square inch (psi) or any other convenient term. See: absolute pressure; atmospheric pressure; gage pressure; hydraulic pressure; pressure measurement; standard temperature and pressure; static pressure; vapor pressure.

pressure control device (Fire Pr) A pressure-regulating valve or governor installed in a pumping apparatus to stabilize discharge pressure. Syn: pressure regulator.

pressure gage (Techn) A device for measuring fluid pressure (force per unit area). See: gage.
NOTE: The four most common types of pressure-measuring instruments are liquid column, bell instruments, expanding-element gages, and electrical pressure transducers.

pressure loss (Hydraul) A decrease in water pressure due to friction in conduits, fluid viscosity, lift, and other factors. See: friction loss.

pressure measurement (Metrol) The determination of the force applied to a unit area and expressed as gage pressure (psig), the difference between atmospheric pressure and the pressure being measured; as absolute pressure (psia), the total pressure including atmospheric; or as vacuum pressure, when the pressure is less than atmospheric. Some common pressure equivalents are given in the table below.

Unit	atm	kg/cm^2	psi	in.Hg	in.H$_2$O	mmHg	mb
atm	1	1.033	14.7	29.92	406.8	760	1013
kg/cm^2	0.9678	1	14.22	28.96	393.7	736.6	980.7
psi	0.0680	0.0703	1	2.036	27.68	57.71	68.95
in.Hg	0.0334	0.0345	0.491	1	13.60	15.40	33.86
in.H$_2$O	0.00246	0.00254	0.0361	0.0736	1	1.868	2.486
mmHg	0.00132	0.00136	0.0193	0.0394	0.535	1	1.333
mb	1013	0.00102	0.0145	0.0295	0.401	0.75	1

Note:
The table is read from left to right, assuming "one" for each entry in the units column: e.g., one atm = 1.033 kg/cm^2 = 14.7 psi = 20.92 in.Hg . . .; one psi = 0.0680 atm = 0.0703 kg/cm^2 . . . etc.

pressure operation (Hydraul) The operating mode of a centrifugal pump in which successive impellers increase the pressure of the fluid being passed. Also called: series operation.

pressure regulator = pressure control device

pressure regulator service (Techn) A fuel gas supply protected by a device that reduces or limits the pressure of the supply system.

pressure release coupling (Fire Pr; Brit) A type of coupling that can be disconnected while the hose is still charged with water.

pressure relief valve (Hydraul) A valve design in which hydraulic force is used to open a bypass to feed water from the discharge side of a pump back to the intake side when a predetermined pressure is exceeded.

pressure-sensitive adhesive (Techn) An adhesive that requires only briefly applied pressure at room temperature to make it stick to a surface.

pressure surge (Hydraul) A sudden jump or fall in pressure when a valve or nozzle is suddenly opened or closed. A pressure wave tends to travel rapidly and cause damage in long hose lines. See: water hammer.

pressure tank (Hydraul) A fire protection water storage tank pressurized with air to provide the required pressure head. Cf: gravity tank.

pressure vessel[1] (Techn) A storage tank or vessel that has been designed to operate at pressures above 15 psig.

pressure vessel[2] (Techn) A fired or nonfired device designed to contain gas or vapor (such as steam) under pressure. A steam boiler is an example.

pressurization (Build) Increasing the pressure above atmospheric in stairwells, elevator shafts, and lobbies to keep escape routes in highrise buildings clear of smoke.

prestressed concrete (Build) Concrete in which embedded cables or rods have been tensioned either before or after the concrete has set.
NOTE: If tensioning occurs after the concrete has set it is called posttensioned; if before, pretensioned.

presuppression activity (For Serv; Fire Pr) Measures taken beforehand to reduce fire incidence and to improve fire protection. See: fire prevention; fire suppression.

prevention (Safety) Reduction or elimination of hazards, usu as part of a planned program, to lessen the probability of a harmful or dangerous event. See also: fire prevention.

prevention guard (For Serv) A person assigned to fire prevention duties, such as advising campers and other visitors in fire safety, inspecting fire prevention measures and equipment of commercial forest operations, etc. Also called: prevention patrolman.

preventive maintenance See: maintenance

primary air (Comb) Air premixed into a fuel gas prior to burning. Also called: primary combustion air.

primary alarm circuit (Fire Pr) A circuit over which an initial alarm is transmitted from fire alarm headquarters to fire stations and other units. A primary circuit usu rings a bell (tapper) and operates a punch register. In some departments the primary alarm is transmitted by voice. See: secondary alarm circuit.

primary battery See: battery

primary blasting (Civ Eng) A blasting operation in which an original rock formation is dislodged from its natural location.

primary circuit (Electr) The driving circuit of two or more coupled circuits in which a change in current induces a voltage in the other or secondary circuits such as the primary winding of a transformer.

primary high explosive (Comb) An explosive that is sensitive to heat and shock and is normally used to initiate a secondary high explosive. Cf: secondary high explosive.
  NOTE: A primary explosive is capable of building up from a deflagration to detonation in an extremely short distance and time interval; it can also propagate a detonation wave in a small-diameter column.

primary lookout (For Serv) 1. A lookout point manned continuously to assure adequate observation of a given forest area, esp during the normal fire season. 2. The person manning such a lookout point; more properly called a primary lookout man or primary observer.

primary observer (For Serv) A person manning a primary lookout point.

prime mover (Techn; Transp) 1. An original source of power, steam, gas, oil, and air engines or motors, and steam and hydraulic turbines. 2. A powerful truck or tractor, esp one with all-wheel drive.

prime number (Math) A number that is divisible a whole number of times only by itself and 1.

primer[1] See: priming pump

primer[2] (Civ Eng; Mining) A relatively small and sensitive device actuated by friction, precussion, heat, pressure, or electricity, used to initiate an explosive, igniter, train, or pyrotechnic device.

primer[3] (Build) A base or groundwork, such as an undercoat for paint.

priming (Hydraul) Filling a pump, esp a centrifugal pump, with water to enable it to start operating. See: pump.

priming composition (Comb) A physical mixture of materials that is sensitive to impact or percussion and, when exploded, undergoes rapid autocombustion.
  NOTE: Priming compositions are used to ignite primary high explosives, black powder igniter charges, propellants in small arms ammunition, and so on.

priming pump (Hydraul) A small positive-displacement pump used to expel air from a pump or displace the air with water to enable the pump to start operating.
  NOTE: Centrifugal pumps must be primed before they can draft water. A pumper engine exhaust can be used for pump priming. A valve diverts the engine exhaust through a venturi eductor, which lowers the pressure in the centrifugal pump chamber, allowing atmospheric pressure to force water into the pump and thus prime it.

printer (Comput) An output mechanism which prints or typewrites computer results.

prism[1] (Opt) A transparent, wedge-shaped solid body used to decompose light into its colors or to refract or reflect light beams.

prism[2] (Geom) A body having parallelogram-shaped faces and congruent, parallel bases.

private (Fire Pr) A firefighter, such as one handling a hose or a ladder.

private box (Fire Pr) 1. A fire alarm box located on private property, in schools, institutions, and the like, not accessible to the general public but tied in to the public fire alarm system. 2. A fire alarm box in a proprietary fire alarm system.

private fire brigade See: fire brigade

private hydrant (Fire Pr) A hydrant in a private water system for the protection of private property. See also: reverse hydrant.

probability (Statist) The likelihood or chance that an event or result will occur, expressed as a number between 0 and 1, 0 meaning that the event is certain not to occur and 1 meaning that the event is certain to occur.
  NOTE: An unbiased coin, for example, has the probability of 1/2 or 0.5 of turning up heads on any flip.

probability theory (Math; Statist) The branch of mathematics that deals with chance occurrences.

probable error (Statist) An error range r around a value such that a measurement of the value is equally likely to be within or outside the range of ±r.

probang (First Aid; Med) A slender, supple rod with a sponge at one end, used to remove obstructions lodged in the esophagus.

probationer (Fire Pr) A recruit firefighter receiving indoctrination and training, which upon satisfactory completion qualifies the recruit for permanent appointment. Also called: proby; rookie.

probenicid (Med) A medicinal preparation used to combat infection and septicemia in burn patients.

proby = probationer

procedure (Man Sci) A series of prescribed steps establishing management policy by which recurring business action is initiated, performed, and controlled.

proceedings (Info Sci) The formal published record of a meeting of a professional society or other organization, including reports and papers presented.

processing (Techn) Any method, system, or other treatment designed to change physical form or chemical content.

processing error (Comput) An error that occurs during recording, manipulation, or analysis of data.

producer gas (Techn) A fuel gas rich in carbon monoxide and nitrogen, formed by passing air and steam over a bed of hot coke. Syn: mond gas (Brit). See: water gas.

product (Math) The result of multiplication.

product liability (Insur) The liability a merchant or a manufacturer may incur as the result of some defect in the product he has sold or manufactured, or the liability a contractor may incur from improperly performed work.

productivity (Man Sci) The ratio of units of output to units of input, e.g., as 10 units produced per man-hour. Also called: output per man-hour.

products of combustion See: combustion products

profession A vocation or occupation that requires advanced training in some liberal art or science, and usu involving highly specialized work (e.g., teaching, engineering, medicine, law, etc.).

profile (Psych) A graphic or written representation of test results that permits identification of areas of strength or weakness.

prognosis (Med) The duration, severity, and prospect of future outcome of a disease or other disorder as estimated from a patient's condition and response to therapy.

program (Comput) A stored set of instructions to define, select, and control a sequence of computer operations.

programming (Comput) The preparation of a detailed sequence of operating instructions for a particular problem.

progressive burning (For Serv) Disposal of slash by burning it as it is gathered into a pile. Also called: forced burning; swamper burning.

progressive hose lay (For Serv) A hose lay with double-gated wyes inserted at intervals in the main hose line, permitting continuous application of water on a fire while lateral lines are laid to the fire edge and the main line is being extended.

progressive method (For Serv) A fireline construction method in which crewmen work in single file, moving forward as the line is constructed. Also called: bump up; functional line construction; move up; step-up.
NOTE: In one variation, crewmen

begin work at about 5-yard intervals, and when one man overtakes another, he bypasses him, and all those ahead of him also move forward. In another variation, each man makes one or more strokes with his tool and then moves forward a specified distance.

project (Man Sci) A planned undertaking of something to be accomplished, produced, or constructed, having a finite beginning and a finite ending.

project fire (For Serv) A fire requiring manpower and equipment beyond the resources of the local fire protection units.

projection booth (Build) An enclosure at the upper rear of a theater, housing the lighting controls and projection equipment and used to project pictures on the screen at the front of the theater.

prolapsed cord (Med) Presentation of the umbillical cord first in childbirth. A case of abnormal delivery.

promoter (Chem) A substance that increases the activity or efficiency of a catalyst. Also called: activator. Ant: poison; inhibitor.

propagation (Comb) The spread of a combustion process through a combustible medium, or the spread of a fire from one combustible to another. Any movement of flame or of a combustion reaction through a combustible solid, gas, or vapor. See also: deflagration; detonation; flame propagation. See also: flame spread.

propagation of flame (Comb) The spread of flame through the entire volume of the flammable vapor-air mixture from a single source of ignition.
 NOTE: A vapor-air mixture below the lower flammable limit may burn at the point of ignition without propagating (spreading away) from the ignition source.

propane (Chem) An alkane hydrocarbon, $CH_3CH_2CH_3$, mp -187.7°C, bp -42.1°C, present in quantities of 3 to 18% in natural gas. Used as a domestic fuel in liquefied mixtures with butane. See: liquefied petroleum gas.
 NOTE: Burning of one pound of propane releases 21,646 Btu, which is sufficient heat to raise the temperature of 1000 pounds of water 21.6°F under standard conditions.

propellant (Comb) A solid or liquid explosive material that has a sufficiently low rate of combustion and other properties suitable to permit its use as a propelling charge.

propellant-actuated device (Techn) Any tool or special mechanized device or gas generator system that is actuated by a smokeless propellant charge. Cf: powder-actuated tool

propeller pitch (Transp) The theoretical distance that a point on the surface of a propeller blade will advance in one revolution.

proper fraction (Math) Any fraction less than unity.

property[1] (Chem; Phys) A characteristic by which a substance is identified; e.g., color, odor, taste, solubility, hardness, density, and so on.

property[2] (Fire Pr) Any physical object valued by its owner, including goods, personal items, real estate, etc.

property and liability insurance (Insur) Financial protection against loss or damage to property from fire, wind, hail, explosion, riot, aircraft, motor vehicles, vandalism, and smoke. The liability portion protects against loss resulting from injuries to persons or damage to their property.

property damage accident (Insur; Safety) An accident in which property is damaged or destroyed.
 NOTE: Accidents involving loss of human life or personal injury are not included, even though property has been damaged.

prophylactic (Med) Preventing or protecting against disease.

prophylaxis (Med) The art of guarding against disease, including preventive treatment.

proportioner (Fire Pr) A device or system that meters liquid concentrates into water in proportion to the flow rate for the purpose of making water "wetter" or to produce fire-fighting foam. See: **foam proportioner**.

proprietary alarm system (Fire Pr) A private emergency, fire alarm, and communications system connected to a central guard room and not to a public or commercial station.

proscenium (Build; Theater) 1. The wall that separates a stage or enclosed platform from the spectators' area of an auditorium or theater and forms the arch for the stage. 2. The portion of a stage wall in front of the curtain.

prospecting (Mining) Searching for ore.

prostatic agent (Electr) A substance that increases the charging tendency of a liquid fuel without increasing its conductivity.

prosthesis (Med) An artificial substitute for a body part, such as an artificial limb, eye, denture, etc.

prostration (Med) Extreme exhaustion.

protected noncombustible construction (Build) Noncombustible construction in which exterior or interior bearing walls or bearing portions of walls are required to have certain minimum **fire endurance** ratings.
 NOTE: For example, bearing walls having 2-hour fire endurance; roofs and floors and their supports, 1 hour; enclosures for stairways and other openings, 1 hour.

protected opening (Build) An opening in a wall or partition fitted with a door, window, or shutter that has a fire endurance rating required for the opening. See: **opening**.

protected ordinary construction (Build) Construction in which roof and floor assemblies and their supports have 1-hour fire endurance, stairways and other openings through floors are enclosed with partitions having 1-hour fire endurance, and the building as a whole meets all the requirements of ordinary construction. See: **building construction**.

protected wood frame construction (Build) Construction meeting all the requirements of wood frame construction, and in which roof and floor assemblies and their supports have 1-hour fire endurance, and stairways and other openings through floors are enclosed with partitions having 1-hour fire endurance. See: **building construction**.

protection boundary (For Serv) The perimeter of a fire protection area covered by a given fire protection organization.

protection plate[1] (Fire Pr) A metal shield provided on a ladder to protect wear points. See: **ladder**.

protection plate[2] (Transp) A metal guard under a wire or band securement of a shipping package to protect it from sharp edges.

protection unit (For Serv) A geographical area for which fire suppression plans are formally drawn up.

protective action guide (Dosim; Abb: PAG) The dose of ionizing radiation absorbed by individuals in the general population which would warrant protective action following a contaminating event, such as a nuclear explosion. See: **radiation protection guide**.

protective clothing (Fire Pr) Fire-fighting clothing, including coats, boots, turnout pants, and gloves, specially designed to afford protection in a hazardous environment. Any special clothing worn to protect the body or personal clothing from contamination or other hazard.

protective coating (Build) A thin layer of metal or organic material, such as paint, applied to a surface primarily to protect it from oxidation, weathering, and/or corrosion.

protective hand cream (Techn) A cream designed to protect the hands and other parts of the skin from exposure to harmful substances.

protective shield (Safety) A device or guard attached to the muzzle end of a tool for the purpose of confining flying particles. Also called: protective guard.

protective survey (Dosim) An evaluation of the radiation hazards incidental to the production, use, or storage of radioactive materials or other sources of radiation under a specific set of conditions.

protein (Chem) A high-molecular-weight compound consisting mainly of amino acids, important to all biological systems.
  NOTE: Proteins are primary constituents of body tissues, hormones, hemoglobin, chromosomes, enzymes, viruses, etc.

protein foam (Fire Pr) A type of fire-fighting foam produced by proportioning a liquid concentrate containing hydrolyzed protein material into water. The resulting 3 or 6% solution (by volume) is aerated mechanically and applied by various means to flammable liquid fires. The concentrate is made from animal hooves, horns, blood, and other protein materials. See: foam.

protocol (Med) An official medical report, such as one describing an autopsy.

proton (Phys) A fundamental unit of positive electrical charge equal in magnitude to that of an electron but opposite in sign (4.803 • 10^{-10} esu). A proton has a mass of 1.6724 • 10^{-24} g, which is equal to that of 1,836.75 electrons.
  NOTE: The nucleus of a hydrogen atom has one proton. The number of protons in the nucleus of a given atom is equal to the atomic number of the atom.

protoplasm (Biol) The basic material from which all living tissue is made. Physically it is a viscous, translucent, semifluid colloid, composed mainly of proteins, carbohydrates, fats, salts, and water.

prototype (Techn) A primitive form or early model of something.

protractor (Geom) A semicircular instrument graduated in degrees, used to draw and measure angles.

proximal (Anat) 1. Nearby; touching. 2. Close to the central axis of the body.

proximate analysis (Comb) Analysis of a solid fuel to determine (in percent) how much moisture, volatile matter, fixed carbon, and ash the sample contains; usu the heat value of the fuel is also determined.

proximate cause (Safety) The cause that directly produces the effect without the intervention of any other cause.

pseudomonas aeruginosa (Med) A microorganism that frequently infects burn victims, producing a type of pneumonia. Also called: blue pus organism.

psi = pounds per square inch

psia = pounds per square inch absolute

psig = pounds per square inch gage

PSO = public safety officer

PSTN = Pesticide Safety Team Network

psychogenic deafness (Psych) Loss of hearing originating from the mental reaction of an individual to his physical or social environment. Also called: functional deafness; feigned deafness.

psychogenic shock (Med) A condition of faintness resulting from the reduction of blood flow to the brain, caused by sudden dilation of the blood vessels.

psychophysical characteristics (Psych) A combination of human mental and physical qualities, such as visual acuity, reaction time, hearing, depth perception, peripheral vision, manipulative dexterity, color vision, etc.

psychophysical measurement (Psych) Measurements of reaction times and various kinds of sensorimotor and psychomotor coordinations.

psychosocial factors (Psych) Social influences that are related to or affect psychological factors of human behavior and attitudes.

psychrometer (Meteor) A humidity-measuring instrument consisting of two identical mercury thermometers, one of which has moistened wicking around the bulb. Also called: <u>hygrometer</u>; <u>sling psychrometer</u>.
   NOTE: The instrument is whirled rapidly in the air, and the two temperatures indicated can be used to determine the relative humidity.

PTFE = <u>polytetraflouroethylene</u>

pto = <u>power take-off</u>

PTO = <u>planning, training, and operations</u>

public (Fire Pr) Belonging or accessible to the general citizenry, in contrast to private ownership or access.

public conveyance (Transp) Any railroad car, streetcar, ferry, cab, bus, airplane, or other vehicle that operates as a common carrier.

public information officer (Fire Pr) A fire department officer responsible for releasing fire and fire department information to the public and the news media.

public relations (Fire Pr) Efforts undertaken to keep the public informed of aims and activities in order to build confidence, good will, and support for the organization in the community.

public safety department (Fire Pr) A municipal government department headed by a public safety director or commissioner to whom the fire department administrator reports instead of to a mayor or city manager.

public safety officer (Fire Pr; Police; Abb: PSO) A city employee primarily assigned to police duty and usu reporting to the police or public safety department, who also serves as a call firefighter.

public way (Transp) A thoroughfare accessible for travel to the general public.

Pulaski tool (For Serv) An ax with a grub hoe opposite the blade.

pull a ceiling (Fire Pr) To remove a ceiling, as with a <u>plaster hook</u> or <u>ceiling hook</u>, to expose a concealed fire.

pull-down hook (Fire Pr) A heavy-gage steel hook attached to a heavy chain and used to pull down obstructions or remove hazardous debris at a fire.

pull-station (Fire Pr; Colloq) A pull-type fire alarm box.

pull the box (Fire Pr) An order to sound a box alarm.

pulley (Fire Pr) A small grooved wheel for guiding the <u>halyard</u> used to elevate the fly sections of extension ladders.

pulmonary (Med) Pertaining to the lungs.

pulsation See: <u>hose pulsation</u>

pulse^1 (Techn) A brief excursion in pressure, flow rate of a fluid, or variation of a quantity, such as voltage or current, consisting of an abrupt change from one level to another, followed by an abrupt change to the original level.

pulse^2 (Physiol) The regular throbbing of the arteries caused by the contractions of the heart.

pulse echo technique (Techn) An ultrasonic technique in which the presence of a flaw in a material is indicated by the amplitude and time delay of a reflected energy pulse.

pulverization (Techn) The crushing or grinding of material into very small particles or powder.

pumice (Techn) A natural silicate of volcanic ash or lava used as an abrasive.

pump (Hydraul) A device, operating on one of several principles, used to move fluids.
   NOTE: Positive-displacement pumps accelerate fluid by moving a solid boundary. These are of two types: <u>piston pumps</u> and rotary <u>gear pumps</u>. Pumps of the <u>aspirator</u> or <u>eductor</u> type move fluid by entraining it in a driving jet. In <u>turbine pumps</u> the driving force is transferred by

viscous interaction of the fluid with rotating elements called impellers. Many specialized pumps have been developed for fire fighting. For example, see: centrifugal pump; piston pump; gear pump; stirrup pump; suction pump; turbine pump; etc.

pump chance See: chance

pump drive (Fire Pr) The power train that transmits the mechanical energy to run a pump. Also called: pump transmission.

pumper (Fire Pr) A fire department pumping engine equipped with a 500 gpm or larger pump, hose, ladders, and other fire-fighting gear.
 NOTE: Fire department pumpers are generally available in five sizes, according to their water discharge ratings: 500, 750, 1000, 1250, and 1500 gpm. A Class A pumper is capable of discharging its rated capacity at 150 psi, as compared to a Class B pumper, which produces its rated capacity at 120 psi. Class B pumpers are considered obsolete. Class A pumpers must be capable of delivering 70% of their rated capacity at 200 psi and 50% at 250 psi.

pumper-ladder (Fire Pr) 1. A combination fire apparatus carrying both ladder truck and pumper equipment. Sometimes called a quadruple combination (or quad), because it combines a water tank, a pumping engine, hose, and ladders into one apparatus. See: apparatus. 2. = (Brit) pump escape.

pumper pit (Fire Pr) A large enclosed cistern used for testing pumps at draft and for training pump operators. Also called: drafting pit.

pump escape (Fire Pr; Brit) A self-propelled apparatus with a built-in 500 gpm (or larger) pump and carrying a wheeled escape ladder, hose reel equipment, and a minimum of 100 gallons of water. Also called: dual purpose machine.

pump intake (Hydraul) The entrance for water into a pump.

pump operator (Fire Pr) A firefighter trained and assigned to operate a pumper. Also called: engineer.

pump panel (Fire Pr) The console of instruments, gages, and controls used by a pump operator.

pump pressure (Fire Pr) The water pressure produced by a fire department pumper, as contrasted with pressure available from a hydrant or from a gravity tank.

pump school (Fire Pr) A fire department school or course of instruction for training pump operators.

pump slippage (Fire Pr) Leakage between the internal parts of a pump, such as through worn valves and past defective rings. The amount of slip is expressed as the percent difference between calculated discharge and actual discharge. A pump in good condition may have as much as 5% slip.

pump suction (Hydraul) An arrangement of large-diameter, hard hose and water source for drafting water at low or negative pressures to the intake of a pump.

pump tank (Fire Pr) A small fire extinguisher with a tank of 2.5 to 5 gallons capacity, a built-in hand pump, and a hose and nozzle. Also called: Indian pump (trade name).

pump transmission = pump drive

punch card (Comput) A card of specific size and shape, suitable for punching in a meaningful pattern for mechanical handling. The punched holes are usu sensed electrically or mechanically.

punch tape (Comput) Paper tape into which a pattern of holes is punched to record information for later processing or other use.

punishment (Fire Pr) The physical discomfort endured by firefighters from heat, smoke, gases, weather conditions, and overexertion.

punk (Fire Pr) Charred or partly decayed wood or other material that tends to smolder.

pup boat (Fire Pr) A small fireboat used to fight fires under piers and in other restricted spaces that are inaccessible to large fireboats.

purchase order (Man Sci; Abb: PO) A document authorizing acquisition of goods or services, stating the quantities or kind and delivery dates. Also called: purchase request.

purge (Techn) To clear or flush a gas or other unwanted substance or impurity from a conduit or container.

purlin (Build) A roof beam that joins rafters or trusses at right angles.

purple-K-powder (Fire Pr; Abb: P-K-P) A type of dry chemical extinguishing agent consisting of potassium bicarbonate.
NOTE: Developed in 1957, it is twice as effective as the older sodium bicarbonate dry chemical agent. It is usu dyed purple to differentiate it from other dry chemical agents. The symbol K stands for potassium.

purpura (Med) Bleeding into the skin tissues, mucous membranes, or internal organs.

push stick (Safety) A narrow strip of wood or other soft material with a notch cut into one end, used to push short pieces of material through saws so that hands can be kept at a safe distance from the blades.

putrefaction (Bact) The decomposition of organic matter by microorganisms and oxidation. Syn: decay; rot.

PVC = polyvinylchloride

pycnometer (Phys; Chem) A specific gravity bottle for measuring the density of liquids.

pyro = pyromaniac

pyrolysis (Chem) The transformation of a compound into one or more other substances by heat alone. Often the initial step in the combustion of solids. Syn: thermal decomposition.

pyromaniac (Fire Pr; Law) One driven by an uncontrollable urge to set fires. A psychopathological firesetter, as contrasted to an arsonist, who starts fires to defraud or conceal a crime, or an incendiarist, who sets fires in rebellion against the established social or political order.

pyrometer (Metrol) An instrument for measuring or recording temperatures beyond the range of a normal thermometer by measuring electrical resistance, optical intensity, etc. The chief types are radiation and optical pyrometers. See also: bolometer. thermocouple; thermometer.

pyrometric cone equivalent (Techn; Abb: PCE) An index for the heat resistance of a material obtained by shaping it into a cone and comparing its behavior under heat with that of a graded series of standard materials shaped into similar cones.

pyrophobia (Psych) A pathological fear of fire.

pyrophoric material (Comb) A material that ignites easily or quickly.

pyrophoric metal (Comb) A metal, usu in powder form, that ignites spontaneously when exposed to air.

pyrotechnic (Fire Pr) Pertaining to fireworks and similar devices, including warning flares and smoke candles. Pyrotechnics are almost always balanced compositions containing oxidizing agents in sufficient quantities to react completely with the fuel or reducing agents in their composition.

pyroxylin (Chem) An extremely flammable nitrocellulosic plastic invented to replace ivory in billiard balls. The plastic can burn without external oxygen and produces toxic fumes.

Q

Q¹ (Electr) The quality factor of an electronic circuit.

Q² (Phys) A symbol used to express very large energy figures. One Q equals 10^{18} (1 billion billion) British thermal units.

Q³ (Fire Pr) 1. A symbol representing hydraulic friction in fire stream equations. Q^2 = friction loss. See: <u>friction loss</u>. 2. A symbol representing flow rate, as gallons per minute.

<u>quad</u> = <u>pumper-ladder</u>

<u>quadrilateral</u> (Geom) A four-sided figure.
  NOTE: A quadrilateral with one pair of sides parallel is a <u>trapezoid</u>; if the four sides are equal it is a <u>rhombus</u>; if two pairs of adjacent sides are equal it is a <u>kite</u>; and if two pairs of sides are parallel it is a <u>parallelogram</u>.

<u>qualified</u> Descriptive of an individual who holds a recognized degree, certificate, or professional standing, or who by knowledge, training, and experience is competent and able to perform in a given field.

<u>qualitative</u> Referring to kind, or nature.

<u>qualitative analysis</u> (Chem) Identification of the elements or components of a substance or mixture. Cf: <u>quantitative analysis</u>.

<u>quality</u> A characteristic (physical or nonphysical, individual or typical) that constitutes the basic nature of a thing or is one of its distinguishing features. The relative goodness or excellence of something.

<u>quality assurance</u> (Techn) A planned and systematic pattern of all sections necessary to provide adequate confidence that material conforms to established technical requirements and will achieve satisfactory performance in actual use.

<u>quality factor</u> (Dosim) The factor by which an absorbed dose is to be multiplied to obtain a quantity that expresses on a common scale, for all ionizing radiations, the irradiation absorbed by exposed persons. See: <u>dose equivalent</u>; <u>relative biological effectiveness</u>.

<u>quantal</u> (Phys) Referring to a variable that changes in discrete steps.

<u>quantitative</u> (Chem; Phys) The property of anything that can be determined by measurements. The property of being measurable in dimensions, amounts, etc., or in extensions of these that can be expressed in numbers or symbols.

<u>quantitative analysis</u> (Chem) Determination of the specific amounts or proportions of the components present in a substance or mixture.

<u>quantum</u> (Phys) A unit quantity of energy according to the <u>quantum theory</u>.
  NOTE: The photon, for example, is one quantum of electromagnetic energy.

<u>quantum theory</u> (Phys) The investigation of energy changes on an atomic level.
  NOTE: Max Planck, a German physicist, stated that the energy of radiation emitted or absorbed by atoms does not have continuous values but is directly proportional to its frequency and is concentrated in units, or quanta, equal to h times the frequency, where h is <u>Planck's constant</u>.

<u>quarters</u> (Fire Pr) The fire station to which a given fire company or individual is assigned.

<u>quartile</u> (Statist) Any of the three points that divide a distribution into four equal groups. The lower quartile, or 25th percentile, sets off the lowest fourth of the group; the second quartile is the same as the 50th percentile, or median.

<u>quartz</u> (Mineral) A brilliant, crystalline mineral, silicon dioxide, SiO^2, occurring most often in a colorless, transparent form. Vitreous, hard, chemically resistant free silica, the most common form in nature. The main constituent in sandstone, igneous rocks, and common sands.

quench¹ (Electr) To limit or stop an electrical discharge in an ionization detector.

quench² (Fire Pr) To extinguish a fire by soaking the fuel with water.

quenching (Metall) Quick cooling of heated materials, achieved by plunging them into liquids or subjecting them to air blasts for hardening and tempering purposes.

quenching crack (Metall) A rupture that occurs during the hardening of steel when one portion cools and contracts more rapidly than the remainder of the piece.

quenching distance (Comb) The orifice diameter, wall separation, or mesh spacing just sufficient to prevent propagation of flame.
NOTE: An important parameter in flame trap designs. Quenching distance can be related to flame front thickness and minimum ignition energy.

queueing theory (Math) The study of the formation and behavior of waiting lines.
NOTE: The theory is used extensively in operations research to study traffic congestion, storage and inventory problems, fluctuating demands and service times, etc.

quick burner (Fire Pr) A structure that would burn rapidly in case of fire because of poor construction, combustible contents, vertical openings, etc. Syn: firetrap; taxpayer.

quick-disconnect device (Fire Pr) Any of a variety of connectors or couplings that can be easily separated by hand. Also, an automatic or hand-operated shut-off.

quick hitch = night hitch

quicklime (Chem) A caustic, white compound, CaO, obtained by calcination of calcium carbonate.
NOTE: Quicklime evolves considerable heat in reaction with water and forms slaked lime or hydrated lime, which is calcium hydroxide, $Ca(OH)_2$, used in plaster and cement.

quick-release knife (Fire Pr; Brit) A heavy-duty sheath knife with a razor edge on one side and a rounded edge on the other, used for cutting openings in crashed aircraft.

quint (Fire Pr) A pumper-ladder carrying a mechanical aerial ladder in addition to the standard pumper and ladder truck equipment.

Q-value (Nucl Phys) The energy liberated or absorbed in a nuclear reactor.

R

R (Chem; Phys) Abbreviation for radical, Rankine, resistance.

rabbit (Nucl Safety) A device used to move a radioactive sample rapidly from one place to another. Often a small cylinder of aluminum or plastic moved by air pressure through a pneumatic conveyor system.

raceway (Electr) A pipe or other channel for electrical wiring.

rack space (Storage) Area in which storage racks are located.

rad (Dosim) Radiation Absorption Dose; a unit of absorbed ionizing radiation equal to 100 ergs per gram of irradiated material. Cf: roentgen unit.

radial branch See: branch

radian (Geom; Abb: rad) An arc of a circle equal in length to the radius; equal to 57.2958 degrees. Frequently used as the unit of angular measurement. There are 2π radians in a circle.

radiant energy (Phys) Electromagnetic energy radiating from a source, as heat, light, X-rays, gamma rays, etc.

radiant heat (Thermodyn) Heat energy carried by electromagnetic waves longer than light waves and shorter than radio waves. Radiant heat passes through gases without warming them appreciably, but increases the sensible temperature of solid and opaque objects.

radiant heater (Build) A heating device in which hot elements radiate heat to the surroundings.

radiant panel test (Testing) A popular name for the ASTM "Standard Method of Test for Surface Flammability of Materials Using a Radiant Heat Energy Source," E162. An alternate to the eight-foot tunnel test.
  NOTE: The test apparatus, developed by the National Bureau of Standards, consists of a vertical gas-fired panel. Specimens 6x18-in. are exposed to the radiant panel and are tested for flame spread, rate of heat release, and smoke generation.

radiation (Phys; Electr; Thermodyn) Transfer of energy, including heat, by electromagnetic waves ranging from very long radio waves through visible light, to x-rays and cosmic rays. In contrast to conduction and convection, radiation does not require a conducting medium. See: electromagnetic radiation; heat transfer; ionizing radiation; quantum.
  NOTE: The term has been extended to include streams of fast-moving particles (alpha and beta particles, free neutrons, cosmic radiation, etc.). Nuclear radiation is that emitted from atomic nuclei in various nuclear reactions, including alpha, beta, and gamma radiation, as well as neutrons.

radiation accidents (Safety) Accidents resulting in the spill or scatter of radioactive material or in the unintentional exposure of individuals to radiation.

radiation area (Safety) Any area accessible to personnel in which a major portion of the body can receive a dose of more than 5 millirem in any 1 hour, or a cumulative dose exceeding 100 millirem in any 5 consecutive days. See: absorbed dose.

radiation burn (Med) Damage to the skin caused by radiation. Beta burns result from skin contact with or exposure to emitters of beta particles. Flash burns result from sudden thermal radiation. See: beta particles; flash burn; ionizing radiation; thermal burn.

radiation calorimeter (Metrol) An instrument used to measure the intensity of thermal radiation.

radiation chemistry (Chem) The branch of chemistry that is concerned with the chemical effects, including decomposition, of energetic radiation or particles on matter. Cf: radiochemistry.

radiation damage (Nucl Safety) Any harmful effects of radiation on matter.

radiation detection instrument

(Dosim) A device used to detect and record the characteristics of ionizing radiation. See: counter; dosimeter; monitor.

radiation dosimetry (Dosim) The measurement of the amount of radiation in a specific place or the quantity of radiation that has been absorbed. See: dosimeter.

radiation effects (Med; Dosim) Harmful biological effects resulting from exposure to radioactive substances.

radiation illness (Med) An acute organic disorder that follows exposure to relatively severe doses of ionizing radiation, characterized by nausea, vomiting, diarrhea, blood cell changes, and in later stages by hemorrhage and loss of hair. See: ionizing radiation.

radiation monitoring (Dosim) Continuous or periodic determination of the level of radiation present in a given area. See: monitor.

radiation protection (Law; Nucl Safety) 1. Legislation and regulations to protect the public and laboratory or industrial workers against radiation. 2. Measures to reduce exposure to radiation. See: radiation standards.

Radiation Protection Guide (Safety) The officially determined schedule of radiation doses which should not be exceeded. See: Radioactivity Concentration Guide.
NOTE: These standards, established by the Federal Radiation Council, are equivalent to what was formerly called the maximum permissible dose or maximum permissible exposure. For normal environmental conditions and for incident electromagnetic energy of 10 to 100 GHz, the radiation guide is 10 mW/cm² averaged over any 0.1-hour period.

radiation resistance (Electr) A fictitious resistance that may be considered to dissipate the energy radiated from an antenna.

radiation shielding (Safety) Reduction of radiation by interposing a shield of absorbing material between any radioactive source and a person, laboratory area, or radiation-sensitive device.

radiation source (Med; Techn) Usually a man-made, sealed source of radioactivity used in teletherapy and radiography, as a power source for batteries, or in various types of industrial gauges.
NOTE: Machines such as accelerators, radioisotopic generators, and natural radionuclides may also be considered as sources.

radiation standards (Nucl Safety) Exposure standards, permissible concentrations, rules for safe handling, regulations for transportation, regulations for industrial control of radiation, and control of radiation exposure by legislative means. See: radiation protection; Radiation Protection Guide.

radiation sterilization (Med; Techn) 1. Use of radiation to cause a plant or animal to become sterile and incapable of reproduction. 2. The use of radiation to kill all forms of life (esp bacteria) in food, surgical sutures, etc.

radiation therapy (Med) Treatment of disease, such as the destruction of cancer cells, with any type of radiation. Also called: radiotherapy. See: brachytherapy.

radiation therapy (Med) The controlled use of high-energy radiation for medical purposes, such as the destruction of cancer cells.

radiation warning symbol (Nucl Safety) An official symbol, a magenta trefoil on a yellow background, which warns of an existing radiation hazard.

radiator (Techn) Any source capable of emitting energy in wave form.

radiator cooler (Fire Pr) An auxiliary cooling unit built into fire apparatus radiators to allow cool water from the pump to reduce engine temperature.

radiator fill line (Fire Pr) A small water line leading from the pump to the fire apparatus radiator, used to refill the radiator during periods of hard pumping at a fire.

radical¹ (Chem) A bound group of atoms that are a fragment of a molecule but act as a single unit in chemical reactions. See also: free radical.

radical² (Math) A mathematical symbol indicating a root.

radio (Commun) A wireless communications device using electromagnetic waves as the transmitting medium.

radio- (Phys) A prefix denoting radioactivity or a relationship to it, or a relationship to radiation.

radioactive cloud (Milit) A mass of air and vapor in the atmosphere carrying radioactive debris from a nuclear explosion. See: atomic cloud.

radioactive contamination (Env Prot) Radioactive material deposited where it may harm people, spoil experiments, or make products or equipment unsuitable or unsafe for some specific use. The presence of unwanted radioactive matter.

radioactive decay (Nucl Phys) The spontaneous transformation of one nuclide into a different nuclide or into a different energy state. The process results in a decrease of the number of the original radioactive atoms in a sample. Decay occurs through the emission from the nucleus of alpha particles, beta particles (or electrons), or gamma rays, or by fission. Syn: radioactive disintegration.

radioactive source (Nucl Phys) A radioactive material that emits radiation.

radioactive standard (Nucl Phys) A sample of radioactive material, usu with a long half-life, in which the number and type of radioactive atoms at a definite reference time is known. The standard is used in calibrating radiation measuring equipment or for comparing measurements in different laboratories.

radioactive tracer See: tracer

radioactive waste (Nucl Phys) Equipment and materials from nuclear operations that are radioactive and for which there is no further use. Cf: fission products.
NOTE: Wastes are generally classified as high-level (having radioactivity concentrations of hundreds to thousands of curies per gallon or cubic foot), low-level (in the range of 1 microcurie per gallon or cubic foot), or intermediate (between these extremes).

radioactivity (Nucl Phys) Emission of energy in the form of alpha, beta, or gamma radiation from the nucleus of an atom. See Also: radioisotope.
NOTE: Radioactivity always involves change of one kind of atom into a different kind. A few elements, such as radium, are naturally radioactive. Other radioactive forms are induced.

Radioactivity Concentration Guide (Nucl Safety) The concentration of radioactive material in an environment which would result in doses equal, over a period of time, to those in the Radiation Protection Guide. Formerly: maximum permissible concentration.

radiobiology (Biol) The study of the effects of high-energy radiation on living tissues.

radiochemistry (Chem) The body of knowledge and the study of the chemical properties and reactions of radioactive materials. Cf: radiation chemistry.

radio communication (Commun) The transmission and reception of signals and information by means of electromagnetic waves.

radiodiagnosis (Med) A method of diagnosis that involves x-ray examination.

radioecology (Nucl Safety) The body of knowledge and the study of the effects of radiation on species of plants and animals in natural communities.

radioelement (Nucl Phys) An element containing one or more radioactive isotopes; a radioactive element.

radio frequency (Commun; Abb: rf or r-f) Any frequency of electrical energy between 10^4 to almost 10^{12} Hz capable of propagating in space.
NOTE: Radio frequencies are much

higher than sound-wave frequencies and lower than infrared.

radiogenic (Nucl Phys) Of radioactive origin; produced by radioactive transformation.

radiography (Med; Techn) The use of ionizing radiation to produce shadow images on a fluorescent screen or photographic emulsion. Some of the rays (gamma rays or x-rays) pass through the object, while others are partially or completely absorbed by the more opaque parts of the object and thus cast a shadow on a photographic film. Cf: autoradiograph.
  NOTE: Used in nondestructive testing for internal examination of metal objects to detect inclusions, defects, differences in thickness, etc.

radioisotope (Nucl Phys) Any isotope of an element that exhibits radioactivity.
  NOTE: Radioisotopes are widely used in medicine, science, and technology. They are used as energy sources in chemical processes, nuclear batteries, and the like.

radioisotope scanning (Med) A method of determining the location and amount of radioactive isotopes within the body by measurements taken with instruments outside the body.
  NOTE: Usually the instrument, called a scanner, moves in a regular pattern over the area to be studied, or over the whole body, and makes a visual record. Cf: whole-body counter. See: coincidence counting.

radiological warfare (Milit) Deliberate contamination of large areas with radioactive materials.

radiology (Med) The medical science that deals with the use of ionizing radiation in the diagnosis and the treatment of disease. Cf: tracer; radiography.

radioluminescence (Phys) Visible light caused by radiations from radioactive substances; an example is the glow from luminous paint containing radium and crystals of zinc sulfide, which give off light when struck by alpha particles emitted by the radium. See: luminescence.

radiometer (Phys) An instrument for measuring the intensity of radiant energy emitted by a heat source. Cf: pyrometer.

radiomutation (Biol) A permanent, transmissible change in form, quality, or other characteristic of a cell or offspring from the characteristics of its parent, due to radiation exposure.

radionuclide (Nucl Phys) A radioactive nuclide; one that is capable of spontaneously emitting radiation.

radioresistance (Nucl Safety) The relative resistance of cells, tissues, organs, or organisms to the injurious action of radiation.

radiosensitivity (Nucl Safety) A relative susceptibility of cells, tissues, organs, or organisms to the injurious action of radiation.

radiotherapy = radiation therapy

radium (Chem; Symb: Ra) A white, highly radioactive, metallic element; atomic number 88, atomic weight 226.0254. One of the earliest known naturally radioactive elements. It is far more radioactive than uranium and is found in the same ores.

radon breath analysis (Dosim; Med) Examination of exhaled air for the presence of radon, a heavy radioactive gas, to determine the presence and quantity of radium in the human body. See: personnel monitoring.

rafter (Build) Any one of the beams supporting a roof covering.

rafter sample (Nucl Safety) A sample of settled dust taken from a rafter or other settling place, and assumed to typify dust suspended in the air.

rail ladder (Nucl Safety) A fixed ladder consisting of side rails joined at regular intervals by rungs or cleats and fastened in full length or in sections to a building, structure, or equipment.

rail storage area (Storage; Transp) Trackage allotted for the purpose of storing rolling stock.

railing (Build) A vertical barrier erected along exposed sides of stairways and platforms to prevent falls of persons. The top member of railings usu serves as a handrail.

railroad fire (For Serv) A fire resulting from railroad operations or maintenance of the right-of-way.

rails[1] (Fire Pr) The long side members of a trussed beam ladder that are separated by truss or rung blocks.

rails[2] (Fire Pr; Archaic) Steel bars that at one time were embedded in the apparatus floor of fire stations to help guide horse-drawn engines into their berths.

railway fire signal (For Serv) Whistle signals sounded by locomotive engineers when a fire is observed along the right-of-way.

rain (Meteor) Droplets of water condensing from clouds and falling to earth. Clouds consist of droplets that are so small that normal winds keep them aloft. Rain occurs when a temperature decrease makes these droplets grow to sufficient size that they fall. Syn: hydrometeor. See: cloud; precipitation.

rain gage (Meteor) An instrument for measuring the quantity of rain falling at a given location, usu by collecting rainwater in a tube.

rain shadow (Meteor) An area of reduced precipitation on the lee side of a mountain.

raise[1] (Mining) A vertical or inclined passageway connecting one working section in a mine with another at a higher level. Cf: winze.

raise[2] See: ladder raise

rake[1] (Fire Pr) To sweep a hose stream across a fire rapidly for the purpose of knocking down flames and cooling the burning fuel.

rake[2] (Fire Pr) A type of plaster hook without the pointed pike.

rale (Med) An abnormal sound heard in the chest during auscultation, such as crackling, bubbling, clicking, gurgling, etc., each sound being characteristic of more-or-less specific lung disorders.

ram's horns (Fire Pr) A connection between a siamese and a monitor nozzle consisting of two curved arms leading from the siamese and joining into one connection for the nozzle playpipe.

ramp (Build; Civ Eng) An inclined plane serving as a way between different levels.

random (Statist) Scattered; not ordered or predictable in detail. Pertaining to a statistical selection method in which every member of a population has an equal chance of being included in the sample. See: random sample.

random access (Comput) Equality of access time to all memory locations, without dependence on the location of the previous memory reference.

random failure (Techn) A breakdown or failure that is predictable in a statistical sense.

random sample (Statist) A set of items selected from a population of items where each item in the population has an equal chance of being selected.
  NOTE: A table of random numbers, which can be found in most textbooks on statistics, is often used to ensure that a sample is truly random.

range[1] (Build) A cooking stove that has an oven and a flat top with gas burners or electrical elements.

range[2] (Techn) The maximum distance of travel, operation, or throw, as of a vehicle, radio transmitter, or hose stream.
  NOTE: The maximum range of a hose stream horizontally is obtained when the water leaves the nozzle at the angle of 32 degrees to the horizontal. Cf: reach.

range[3] (Statist) The interval between the highest and lowest values in a distribution.

ranger = forest ranger

range finder (Surv; For Serv) An optical instrument used to determine distances.

range-finding tests (Toxicol; Abb: RF) A series of tests to determine the lethality of toxic substances.
  NOTE: Since toxic products can enter the system through the skin, respiratory canal, or the gastrointestinal tract, different dosage techniques must be used in each case. Often exponential dosage scales are used, such as 1, 2, 4, 8, ... mg/kg of animal body weight. Because of sensitivity it may be necessary to use a more closely spaced scale, such as 1, 1.2, 1.44, 1.73, etc. Doses are found that kill no test animals (LD^0) and that kill all animals (LD^{100}). The 50% lethal dose is derived from these two limit quantities.

range fire (For Serv) Any unwanted fire burning in rangeland.

rangeland (For Serv) Open land suitable for cattle grazing.

rank (Statist) 1. The ordinal position of a value on a scale or in a range of values. 2. To place items in a hierarchic order, as numbers or scores according to absolute value, occurrences according to frequency, and the like.

Rankine temperature scale (Phys; Symb: $°R$) A temperature scale with a degree interval identical to that of the Fahrenheit scale, but zero set at absolute zero.
  NOTE: The freezing point of water is $491.69°R$ and the boiling point $671.69°R$. Also called: **Fahrenheit absolute**. See: **temperature scale**.

rank order (Statist) The arrangement of a series of items according to magnitude.

ranks and titles (Fire Pr) Positions of responsibility and authority in the fire service management and operations hierarchy and the official designations and names for them.
  NOTE: There is no uniformity in ranks or official titles in the fire service. Depending on the size of a given fire department, nature of jurisdiction, and local custom, fire department heads are variously designated as **chief of department**, **chief engineer**, **chief fire marshal**, **fire marshal**, **director**, **general manager**, or **fire commissioner**. Subordinate titles, not necessarily in order of rank, are **district chief**, **battalion fire chief**, **fire chief**, **deputy fire chief**, **assistant fire chief**, **administrative assistant**, **captain**, **lieutenant**, **platoon commander**, and **sergeant**. In smaller departments officers may have both line and staff responsibilities. Many departments, esp the larger ones, have special titles, such as **superintendent of fire alarm**, **assistant superintendent**, as well as numerous other designations for officers of specialized fire service units, including **head of fire prevention**, **principal fire training officer**, **superintendent of apparatus and maintenance**, **superintendent of communications**, etc. See: **trumpet**.

RANN = Research Applied to National Needs

Raoult's law (Chem) The principle that vapor pressure of solutions of nonelectrolytic substances is lowered by a value equal to the product of the mole fraction of the substance in solution and the vapor pressure of the pure solvent at the same temperature.

Rapid Water (Fire Pr) Trade name for **slippery water**. See: **polyethylene oxide**.

rappel (Rescue) To descend a sheer face by means of a rope passed across the shoulder and looped under one thigh, or by means of a running line and a special safety belt.

rare earths (Chem) The elements in the periodic table with atomic numbers 57 (lanthanum) through 71 (lutetium).
  NOTE: Used in the manufacture of special steels and glasses.

rarefaction wave (Comb) A low-pressure wave that follows a shock or detonation wave.

rash (Med) Abnormal reddish coloring or blotch on some part of the skin.

ratchet (Mech Eng) A toothed wheel that operates in conjunction with a

pawl or catch so that it rotates in one direction only.

rate (Phys) The change in a variable as a function of time, expressed as a ratio.

rate of fire spread (For Serv) The increase in the perimeter, area, or advance of a fire as a function of time, usu expressed in acres or chains per hour.

rate-of-rise detector (Fire Pr) A fire detection device that is actuated by a rapid increase in temperature but not by slow, normal fluctuations.

rate-of-spread meter (For Serv) A device used to compute the probable rate of spread of a fire for various combinations of fuel moisture, weather conditions, and other factors.

rated bursting pressure (Fire Pr) The maximum pressure at which a frangible disc is designed to burst.

rated load (Techn) 1. The combined loaded weight that a working platform is designed to lift. 2. The maximum load that a machine is designed to handle safely.

rated speed (Techn) The designed operating speed of a device or machine.

rating bureau (Insur) A fire insurance agency that sets fire insurance rates for given locations.

ratio (Math) The proportional relationship between two numbers or values. As an example, two units of one substance mixed with five units of another results in a mixture having the ratio of 2 : 5 of the two initial substances.

ravelly ground (Mining) Loose and crumbled mining debris. Broken soil and rocks.

raw score (Educ; Statist) The first quantitative result obtained in scoring a test. Usually the number of right answers, the number right minus some fraction of the number wrong (to penalize guessing), the time required for performance, the number of errors, similar direct, unconverted, uninterpreted measures.

RBE = relative biological effectiveness[2].

reach (Fire Pr) The vertical height of a fire stream. Cf: range[2].
 NOTE: The maximum effective reach (vertically) is obtained when the nozzle is held at an angle of approx 70 degrees to the horizontal.

reactance (Electr; Symb: X) The opposition offered to the flow of an alternating current by the inductance, capacitance, or both, in an electrical circuit.

reactants (Chem) Substances that undergo chemical change.

reaction[1] (Phys) The force opposing a change in motion. For example, a water stream accelerating through a nozzle imparts a backward force on the nozzle equal to the thrust of the acceleration.

reaction[2] (Chem) A chemical change in which elements and compounds are combined or decomposed to produce different chemical combinations or arrangements.

reaction distance (Transp) The distance traveled between the point at which a driver perceives that braking action is required and the point at which he contacts the braking controls.

reaction rate (Chem) The rapidity of a chemical reaction as dependent on the nature of the reacting substances and their respective concentrations at given temperature and pressure conditions, expressed as a numerical constant. In many reactions the rate doubles with an increase in temperature of 10°C. See: Arrhenius equation.

reaction time (Transp) The time taken by a vehicle to traverse the reaction distance.

reactivity (Chem) The readiness with which a substance takes part in a chemical reaction.

reactor (Nucl Phys) An atomic furnace in which nuclei of a radioactive fuel undergo controlled fission.
 NOTE: Fission is initiated by

335

neutrons and in turn produces new neutrons in a chain reaction. This releases large quantities of energy, which may be used to produce electricity. The moderator for the first reactor was a pile of graphite blocks. Reactors are classified as research, test, process heat, and power, depending on their principal function.

ready line (Fire Pr) 1. A charged hose line stretched in anticipation of need. 2. A hose line with a nozzle, preconneced to a pump and water supply and ready for immediate use.

reagent (Chem) A substance used in a chemical reaction to produce another substance or to detect its composition.

real time (Comput) Descriptive of a computation related to a physical process which takes place over a time period identical to that of the process.
  NOTE: An example is a case in which the current value of an external variable is needed before the next step of a computer program can be carried out.

rear[1] (Fire Pr) The side of a burning structure opposite the main street or command position. Also called: side two.

rear[2] (For Serv) The slowest advancing part of a forest fire, or the edge of a fire diametrically opposite the head. See: fire nomenclature.

reburn (For Serv) 1. Reignition of remaining combustible fuel in a burned-over area in which a fire is thought to be extinguished. 2. An area that has burned a second time.

recall[1] (Fire Pr) 1. A signal, similar to an all-out, indicating that fire department companies have returned to quarters from a fire and are ready to resume their normal assignments. Syn: (Brit) return. 2. To call off-duty firefighters back to duty, such as in the case of a major emergency.

recall[2] (Psych) The retrieval from memory of something that has been learned.

receiver (Commun) 1. An entity that transforms a signal into a message. 2. A device that receives electronic signals from a transmitter.

receiving (Mat Hand; Storage) 1. The receipt of new stocks or supplies; including the activities of preplanning, handling, and document processing. 2. The area used for checking, inspecting, and preparing incoming materials prior to transfer to storage areas.

receptivity (Educ; Psych) The readiness or willingness to accept an idea or suggestion.

recharge (Fire Pr) To refill for future use, such as a fire extinguisher or a storage battery.

reciprocal (Math) The number by which another number must be multiplied to give unity. For example, 2 and 1/2 are reciprocals of each other because their product is 1.

reciprocating pump (Hydraul) A positive-displacement pump in which a piston moves back and forth in a cylinder, filling the chamber when stroking in one direction and discharging in the other. Reciprocating pumps can be single or double-action. Double-action pumps discharge on both strokes.

recirculation zone (Comb) A region of turbulent flame or combustion in which residence time is extended because of eddy flow.
  NOTE: The long residence time promotes combustion in this region, and the effect can act as a flame holder or igniter in high-speed flow, which would otherwise be blown off.

recitation (Educ) Oral reproduction of learned material.

reclamation (Env Prot) Restoration to a better or more useful state, such as land reclamation by sanitary landfilling or obtaining useful materials from solid waste.

recognition test (Educ) A test in which a subject is asked to select a learned item from among several items not learned, as in a multiple choice test.

recoil energy (Phys) The energy emitted during fission or radioactive decay and shared by the reaction products.

reconnaissance (Fire Pr; Brit) size up

recoverable resources (Techn) Materials that still have useful physical or chemical properties after serving a specific purpose and can, therefore, be reused or recycled for the same or other purpose.

recovery (Techn) The process of obtaining materials or energy resources from solid waste. Syn: extraction; reclamation; salvage.

recycling (Techn) The reuse of waste materials to make new products.

red line (Fire Pr) A small, 3/4- or 1-in. rubber hose. Sometimes called: booster or chemical hose.

red network (Fire Pr) A special telephone circuit used to notify volunteer firefighters or call men of fires.

red shirt (Fire Pr; Archaic) A volunteer firefighter, from the bright red shirts formerly worn as part of the uniform. Cf: blue shirt.

reduce (Med) To restore to a former state; specifically, to set a fractured bone.

reduced assignment (Fire Pr) A temporary cutback in assignments due to multiple calls or special hazards, such as during the grass and brush fire season or holidays, when alarms are unusually frequent. Last-due companies are often dropped from assignments when the reduced assignment condition is in effect.

reducer[1] (Fire Pr) a. A coupling with a female thread on one side and a smaller-diameter male thread on the other, used to attach small hose lines to large ones. b. Any adapter used to connect a smaller-diameter conduit to a larger. Ant: increaser.

reducer[2] (Chem) a. A substance that removes oxygen from another substance in a chemical reaction. Hydrogen and carbon are common reducing agents. b. A substance containing an atom that loses one or more electrons in a chemical reaction. Ant: oxidizer.
NOTE: Oxidation and reduction are necessarily simultaneous processes in a chemical reaction. As one reactant is being oxidized, another must be reduced.

reduction (Chem) 1. Addition of one or more electrons to an atom through chemical change. 2. The removal of oxygen from a compound; for example, hydrogen gas passed over copper oxide reduces the copper oxide and forms copper and water.

redundancy (Techn) The existence of duplicate means for accomplishing a certain function.
NOTE: Redundancy is used to increase reliability. If the primary means fails, the redundant means permits continued operation.

reel (Fire Pr) A spool used to carry small-diameter hose on a tank apparatus or one with electrical cord.

reeving (Mat Hand) A rope system in which the rope travels around drums and sheaves.

refinery (Techn) A plant that produces a flammable or combustible liquid on a commercial scale from crude petroleum, natural gas, or other hydrocarbon resource.

reflash (Fire Pr) Reignition of a flammable fuel by hot objects after flames have been extinguished. Cf: flashback; rekindle.

reflection (Phys) The return of electromagnetic radiation or sound from a surface.

reflective tape (Fire Pr; Safety) Strips of light-reflecting, self-adhesive tape, attached to clothing, tools, appliances, etc. to make them visible under adverse lighting conditions.

reflector (Phys) In nuclear engineering, a material surrounding the core of a reactor to reduce the escape of neutrons by scattering them so that a large fraction may return to the core, sustaining the chain reaction.

reflex (Physiol) An automatic and

subconscious response to a stimulus, such as the blink of the eye when an object approaches it suddenly.

reflux (Chem) The portion of the vapor that is condensed and returned to the still in a distillation process.

refraction (Opt) A change in the direction of light rays or other electromagnetic waves as they pass from one medium into another.
NOTE: If the velocity of propagation is less in the second medium than in the first, the direction of the incident radiation bends toward the normal. In the opposite case, the radiation bends away from the normal. The mathematical relation for refraction is known as Snell's law. Sound waves and water waves are similarly subject to refraction.

refractive index (Opt; Symb: n) A measure of the ability of a substance to bend a ray of light. The index is equal to the sine of the angle of incidence divided by the sine of the angle of refraction.
NOTE: The constancy of the ratio of the sines is known as Snell's Law.

refractory (Techn) 1. Heat-resistant. 2. A material esp resistant to heat and hence used for lining furnaces, etc., such as fire clay, magnesite, graphite, and silica.

refrigerant (Techn) The working fluid of a cooling device that produces a cooling effect from the latent heat of vaporization.
NOTE: Common refrigerants are ammonia, sulfur dioxide, ethyl chloride, and methyl chloride. Fluorinated and chlorinated hydrocarbons, such as the freons, are commonly used in domestic refrigeration and air-conditioning systems.

refuse chute (Build; Techn) A pipe, duct, or trough through which solid waste is conveyed pneumatically or by gravity to a central storage area.

regenerate (Techn) To reproduce an earlier state or condition.

regeneration (Med) Regrowth or restoration of lost or damaged tissues.

regenerative process (Med) Replacement of damaged cells by new cells.

regimen (Med) A program of rest, diet, exercise, and the like for the purpose of improving health.

register[1] (Fire Pr) A device, usu installed at a fire department watch desk, that records fire alarm signals in code on paper tape.

register[2] (Build) A grill, diffuser, or other outlet in an air-distribution system for discharging heating, cooling, or ventilating air.

regression curve (Statist) A curve fitted to the means of a set of variables.

regression rate (Comb) The speed with which a solid or liquid burns, expressed in terms of inches per second measured normal to the surface. Syn: burning rate.

regulation (Law) A rule, ordinance, law, or device by which conduct or performance is controlled. Cf: standard.

regulator (Techn) A mechanism for controlling or governing the movement of machinery, or the flow of liquids, gases, electricity, steam, etc.

rehabilitation (Med) Restoration to health and function, esp after an injury or illness.

Reid method (Testing) A method of determining the vapor pressure of a volatile hydrocarbon.

reinforced concrete (Build) Concrete in which metal rods, mesh, or the like has been embedded to give it added strength.
NOTE: Tensile stresses are carried mostly by the metal while compressive stresses are carried mostly by the concrete.

reinforcement[1] (Educ; Psych) Strengthening of a response, as by reward, recognition, and the like.

reinforcement[2] See: follow-up

reinstatement (Build; Brit) Restoration of a building after a fire by replacement or strengthening of damaged structural elements and assemblies and making other necessary repairs.

reinsurance (Insur) The assumption by an insurance company of a risk undertaken by another company to reduce or spread the liability.

rejection (Testing) Nonacceptance.

rekindle (Fire Pr) To reignite after extinguishment, such as a fire reigniting at some time after being put out. Latent heat and hidden embers may restart a fire several hours after it has been declared out if overhauling has not been sufficiently thorough.

relapse (Med) Return of disease symptoms after a period of improvement.

relative biological effectiveness (Dosim; Abb: RBE) The ratio of absorbed dose of a reference radiation to the absorbed dose of the radiation of interest that produces the same biological effect. See: absorbed dose; distribution factor; quality factor.
NOTE: The factor is used to compare the biological effectiveness of different types of ionizing radiation.

relative density (Phys) The ratio of the density of a material to that of a reference material under given standard conditions. Syn: specific gravity.
NOTE: For solids and liquids the standard reference is usu water; for gases, usu air.

relative humidity (Meteor; Phys) The ratio (in percent) of the water vapor in the air to the maximum amount of vapor it can hold at the same temperature and pressure. Cf: absolute humidity. See also: humidity.
NOTE: Relative humidity of less than 30% is considered a hazardous fire condition in the US Forest Service.

relative motion (Phys) Motion with respect to a given frame of reference.

relay (Civ Eng) An explosive train component that provides the required explosive energy to the next element in the train, such as a detonator. Especially a small charge that is initiated by a delay element.

relay pumping (Fire Pr) A method of conveying water in which two or more pumpers operate in series, with the discharge outlet of one connected to the suction inlet of the next. The method is used to pump water over long distances or heights. See: tandem.

relay station (Commun) A radio station that retransmits a signal from a nearby transmitter to one farther away, esp to one that is out of range of the originating transmitter. Cf: repeater.

reliability[1] (Educ) The characteristic of a test that it is a true measure of that which is being tested.

reliability[2] (Techn) The probability that a component, equipment, or a system will perform its purpose adequately for a given period of time or under given operating conditions.
NOTE: For a system with independent components, the overall reliability is based on the product of the individual reliabilities; e.g., three independent components with a 90% reliability each will have an overall reliability of 0.9 x 0.9 x 0.9, or 72.9%. Similarly, 100 components with a 99% reliability each will have an overall reliability of only 36.5%.

reliability assurance (Techn) All actions necessary to provide adequate confidence that material conforms to established reliability requirements.

reliability sampling (Statist) A measure of the agreement between two or more samples chosen from the same population.

relief opening (Build; Techn) An opening provided in a draft hood to permit flue gases to escape into the atmosphere in the event that normal draft is interrupted or reversed.

relief valve[1] (Fire Pr) A pressure control device on a pump to prevent excessive pressures when a nozzle is shut down.

NOTE: Relief valves automatically bypass or dump water when the desired pump pressure is exceeded.

relief valve² (Techn) A valve used to relieve excess pressure in pressure vessels.

relieve (Fire Pr) 1. To dismiss a fire company from further duty at a fire and return it to quarters. 2. To release a platoon (or other group) from duty when a shift changes. 3. To remove an individual from duty status for any reason.

relocate (Fire Pr) To order a fire company to another fire station to carry out the assignments of that company. Also called: change-of-quarters; fill-in; transfer. See: move-up.

reluctance (Electr) The opposition to magnetic flux.

rem = roentgen equivalent, man

remission (Med) Disappearance of symptoms.

remote control system (Techn) A system in which the controlled object and the controlling device are separated by an appreciable distance.

renal (Med) Pertaining to the kidney.

renal infarct (Med) An area of tissue death in the kidney resulting from obstruction of local circulation.

rendering (Techn) A process of recovering fatty substances from animal parts by heat treatment, extraction, and distillation.

rep = roentgen equivalent, physical

repeater¹ (Fire Pr) An automatic device at fire alarm headquarters that receives alarm box signals and relays the signals over the proper alarm circuits.

repeater² (Commun) A communications device that restores a signal and retransmits it. Cf: relay station.

replacement series (Chem) The arrangement of elements, esp a group of similar ones, in order of their decreasing chemical reactivity.

report time (For Serv) The time lapse between the discovery of a fire and receipt of notification of its existence and location by the first firefighter, who subsequently begins effective work on its suppression.

representative sample (Statist) A sample that corresponds to or matches the population of which it is a sample with respect to characteristics important for the purposes under investigation. See random sample.

rescue (Fire Pr) Saving of persons endangered by fire or accident. Assistance given to persons in danger who cannot help themselves.

rescue apparatus (Fire Pr) A truck-type vehicle carrying various rescue tools and equipment, and sometimes a small boat. Light rescue trucks may also double as ambulances. Heavy-duty rescue apparatus carries heavier equipment of a greater variety. See: apparatus.

rescue carry (Fire Pr) Any of several methods developed for carrying injured or unconscious persons and removing them to safety.
NOTE: If carry methods are not practical, a drag rescue technique may be used in which the victim is pulled to safety.

rescue sling (Fire Pr) A special sling used to lower persons to safety, consisting of two loops of rope or webbing connected to a steel ring. The ring may be tied to a turntable ladder rescue line.

rescue squad (Fire Pr) A fire department company specially trained in rescue techniques and the use of respiratory assistance and resuscitation equipment.

rescue tender (Brit) = rescue apparatus

resealing pressure (Hydraul) The pressure at which water stops leaking through a safety relief seal.

research Effort directed toward increased knowledge of natural phenomena and toward the solution of problems in the physical, behavioral, and social sciences, usu by means of experimental investigation.

reserve apparatus (Fire Pr) An apparatus that is used when the first-line apparatus is in repair, or one that is put into service during emergencies and is manned by off-duty firefighters.

reservoir (Hydrol) A basin for storing water, usu made by constructing a dam across a stream.

reset (Fire Pr) To restore a fire protection or detection device to its original condition after it has been operated.

residence time (Dosim) The time during which radioactive material remains in the atmosphere following the detonation of a nuclear explosive. It is usu expressed as a half-time, since the time for all material to leave the atmosphere is not well known. See: fallout.

residential-custodial care facility (Fire Pr; Insur) A building or part of one used to lodge individuals with physical infirmities or mental illness.

residential occupancy (Fire Pr) A building in which sleeping accommodations are provided for normal residential purposes, including all buildings designed to provide sleeping accommodations except those classified as institutional occupancy.

residual nuclear radiation (Nucl Phys) Lingering radiation, or radiation emitted by radioactive material remaining after a nuclear explosion. Residual radiation is arbitrarily designated as that emitted more than one minute after the explosion. Cf: fallout; initial nuclear radiation.

residual oil (Techn) Liquid or semiliquid products remaining after petroleum distillation, including asphalt, black oil, and the like. Also called: residuum.

residual pressure (Fire Pr) The pressure remaining at the inlet side of a pumper while water is being discharged from the other side. The residual pressure compared with the static pressure before pumping is started indicates the volume of water available from the water supply.
NOTE: Residual pressure is also the pressure remaining in a water main while water is being discharged from it.

residue (Techn) Anything left behind in a process, such as the remainder from evaporation.

resin (Chem) A solid, flammable, noncrystalline, organic compound without a definite melting point and no tendency to crystallize.
NOTE: Resins may be of vegetable (gum arabic), animal (shellac), or synthetic origin (celluloid). Resins are used as adhesives, in the treatment of textiles and paper, in the manufacture of plastics, and as protective coatings.

resistance[1] (Techn) Opposition to the flow of air, or other fluid, such as through a filter or orifice.

resistance[2] (Electr) The opposition to the flow of electrical current.

resistance to control (For Serv) The relative difficulty of constructing and holding a fire control line, determined by the resistance to line construction and the behavior of the fire.

resistance to line construction (For Serv) The relative difficulty of constructing a fire control line, determined by the soil, topography, fuel, and other factors and conditions.

resistor (Electr) An electrical device that is designed to restrict the flow of electrical current.

resolving power (Opt) A measure of the ability of an optical system to form distinguishable images of distinct objects.

resonance (Phys) The condition of maximum response (particularly in amplitude as a function of frequency) when a sympathetic vibration is induced by a driving system.

respirable (Fire Pr) Suitable for breathing.

respirable dust (Ind Hyg) Airborne dust in sizes capable of passing through the upper respiratory system to reach the lower lung passages.

respiration (Physiol) The process of breathing, including the supply of oxygen to cells and tissues and the removal of carbon dioxide.

respirator (Fire Pr) 1. A mechanical device designed for resuscitation or maintaining respiration artificially. 2. A mask worn over the mouth and nose to protect the respiratory tract by filtering smoke and fumes from the air. See: breathing apparatus.

respiratory disease (Med) A disease condition in the respiratory tract, e.g., pneumonitis, bronchitis, pharyngitis, rhinitis, or acute congestion, due to chemicals, dusts, gases, or fumes.

respiratory distress (Med) Difficulty in breathing. See: first aid.

respiratory irritant An irritant affecting the respiratory tract, e.g., dust, vapor, or gas.

respiratory shock (Med) A state or insufficient oxygen in the blood due to a respiratory difficulty.

respiratory system (Anat) The breathing system of the body, consisting of the nose, throat, trachea, bronchial tree, lungs, and related muscles.

respond (Fire Pr) To answer an alarm and proceed to the scene of a fire or location of the alarm box.

response (Fire Pr) 1. The act of answering an alarm. 2. The entire complement of men and equipment assigned to and answering an alarm.

response schedule See: running card

response time (Fire Pr) The total time from the moment a report of fire is received until a fire suppression unit begins fire-fighting activities, including dispatching time, turnout time, travel time, and set-up time.
NOTE: In master planning, response time may be expectation time in a response-time model.

rest period Brief interruption in the workday, usu of 5 to 15 minutes' duration, during which the worker is relieved of work duties. Also called: coffee break; break time.

restricted area (Safety) Any area access to which is controlled by the employer for purposes of protection of individuals from exposure to radiation or radioactive materials.

resuscitation (First Aid; Med) Restoration of respiration and heart beat to a person who has stopped breathing.

resuscitator (Fire Pr) A pumping device for supplying oxygen or a mixture of oxygen and carbon dioxide to a person who cannot breathe by himself. See: breathing apparatus, pnealator.
NOTE: The pump produces alternate positive and negative pressures to restore normal breathing, after which it can be used as an inhalator.

resuscitube (First Aid) A short piece of tubing with a flexible skirt that fits snugly over a victim's mouth to assist in mouth-to-mouth resuscitation.

retained firefighter (Fire Pr; Brit) 1. A part-time firefighter who may be regularly employed in another occupation but who responds to a fire when called on a personal radio receiver or telephone, or when alerted by a siren or house bell. 2. = (US) call firefighter.

retained station (Fire Pr; Brit) A fire station manned entirely by retained firefighters.

retaining wall (Civ Eng) A wall constructed to hold back earth that might otherwise slide or collapse.

retardant See: fire retardant

retention (Educ) The persistence of a learned experience without further practice.

retention curve (Educ) A graphic representation of the retention of learning over time.

retina (Anat) The light-sensitive inner surface of the eye that receives and transmits the image formed by the lens.

retirement Withdrawal, with an income, from working life or from a particular employment because of old age, disability, etc.

retraction (Physiol) The flexion of an arm or leg.

retroactive Acting backward in time, or effective as of an earlier point in time.

retrograde Moving in the backward direction.

retrograde shock amnesia (Psych) The loss or impairment of memory due to head injuries, severe loss of blood, or other shock.

return (Fire Pr; Brit) = recall

return stairs (Build) A stairway that reverses direction at a landing halfway between floors. Cf: scissors stairs.

return sweep (Educ) In reading, the shift of the eyes from the end of a line to the beginning of the next.

return wheel (Transp) One of a number of wheels that support the top run of the track between the drive sprocket and idler of a track-laying vehicle. Also called: top roller; return roller.

reuse (Techn) To reintroduce a commodity into the economic stream without any change.

reverberation (Acoust) The persistence of sound in an enclosed space as a result of multiple reflections after the sound source has stopped.

reverberatory (Techn) Descriptive of an oven kiln in which the flame is reflected back on the material being treated.

reverse hydrant (Fire Pr) A hydrant connected to an independent water system that may receive water pumped from another water supply system. A regular hydrant may be used as a reverse hydrant to augment a poor water supply or to boost pressure to pumpers working at higher elevations.

reverse lay See: hose lay

reversible process (Thermodyn) An idealized thermodynamic process, considered in computing limits, that may proceed in either direction, permitting restoration to the initial state or condition without wasteful loss, such as by friction.

reversible reaction (Chem) A chemical reaction that can proceed in either of two directions.
NOTE: When the forward reaction balances the reverse reaction, equilibrium occurs. In practice, equilibrium conditions are varied by appropriate means to force the reaction to proceed in the desired direction.

revolving cellar nozzle (Fire Pr) A special rotating casting with four (or more) nozzles at right angles to the hose line. Two of the nozzles are canted, one upward and the other downward, while the remaining two are aimed in the horizontal plane. As the casting revolves, water is distributed in all directions. See also: cellar nozzle.

revolving door (Build) A door consisting of four vanes at right angles to each other that rotate as a unit about a central axis. A person using the door pushes against a vane, rotates the assembly, and passes through.

Reynolds number (Fl Mech; Symb: Re) The ratio of the inertial force to fluid viscosity, used in fluid mechanics for modeling flow patterns and evaluating drag. The inertial force is the product of fluid density, velocity, and characteristic length. See also: dimensionless numbers; pipe flow.

rheology (Phys) The study of the flow and deformation of matter, including such subjects as the spreading of paint on a surface, the stretching of rubber, and the moldability of plastics.

rheostat (Electr) A variable resistor.

rhinitis (Med) Inflammation of the lining of the nose and nasal passages.

ride (Fire Pr) To take a fire department vehicle out from a fire station in response to an alarm, for a drill, or other official purpose. Also called: run. Cf: roll¹.

ridge (Build) The peak, or topmost part of a roof.

rig¹ (Techn) a. Rope, tackle, and similar gear. b. To equip something for a special purpose.

rig² (Fire Pr) 1. A tractor and trailer apparatus together. 2. Loosely, any piece of fire apparatus.

right-of-way (Law; Transp) The right, based on direction, speed, and proximity, of one vehicle or pedestrian to proceed in a lawful manner in preference to another vehicle or pedestrian.

rigid brace load (Transp) A load in which the lading is secured by blocking or bracing to prevent any movement of the lading in transit.

rigid wheel (Transp) A wheel that deforms a relatively negligible amount on a hard surface, such as a steel railway wheel. See: elastic wheel.

rill stope (Mining) A stope from which ore is removed in inclined slices. The longitudinal cross section of a rill stope is in the shape of an inverted V.

rime (Meteor) Layers of ice formed when droplets of water touch objects, particularly prevalent in mountainous areas. See: precipitation.

ring main (Fire Pr; Brit) A water main circling one or more buildings or a group of fire risks to supply hydrants, risers, sprinklers, etc. The ring is connected to one or more street mains.

Ringelmann charts (Fire Pr) A set of charts numbered from 0 to 5 that simulate various smoke densities by presenting different percentages of black. See: smoke.

NOTE: These charts are set up a given distance from the observer and in line with a smoke source. By visually comparing the charts and the smoke, a judgement is made on the nature of the smoke.

ringer (Fire Pr) A fire fan or buff who enjoys assisting firemen.

riprap finish (Fire Pr) The arrangement of the end of a hose line on a hose bed in which a number of accordion folds of hose are laid at right angles across the main hose load to serve as a ready supply of hose for paying out and making connections rapidly. Syn: flaked hose finish.

ripple voltage (Electr) The fluctuations in the output voltage of a rectifier, filter, or generator.

riprap (Civ Eng) Rock, metal stripping, or wooden timbers used to contain and stabilize earth embankments and fills. Cf: lagging.

rise (Civ Eng; Fire Pr) The vertical distance between two points, often expressed as a slope or grade.

riser¹ (Fire Pr) A vertical pipe used to carry water to sprinkler system piping on floors above ground level. See: standpipe.
NOTE: Sometimes used to refer to any vertical pipe.

riser² (Build) The vertical portion of a stairway between the treads.

rising main (Brit) = standpipe¹

risk¹ (For Serv) The probability or degree of threat of a fire as governed by existing conditions favoring fire and the presence of causative agents.

risk² (Fire Pr) A fire hazard.

risk³ (Insur) An insured property.

risk⁴ (Safety) a. The degree of peril; the possible harm that might occur. b. The statistical probability or quantitative estimate of the frequency or severity of injury or loss.

RMS (Electr) An abbreviation of root mean square. See: effective value.

roadability (Transp) A rating of the collective operating characteristics of an automotive vehicle that defines the quality of the vehicle's traveling performance, including ease of steering, gradeability, acceleration, road holding, suspension stiffness, rebound control, directional stability, braking characteristics, skidding characteristics, etc.

road test (Transp) A procedure for verifying the operational performance of a vehicle.
NOTE: Typically, a road test involves driving the vehicle over specified roads (e.g., paved highway, gravel or dirt roads, and sometimes cross-country) at various speeds. For emergency vehicles the test may include fording water of specified depth.

road wheel (Transp) One of a number of wheels that support the weight of a tracked vehicle and roll on the inside of the bottom run of the track. Also called: bogie wheel; bottom roller.

roasting (Metall) A metal refining process in which ore is heated to a high temperature, sometimes with catalysts, to drive off certain impurities; e.g., the roasting of iron ore to remove sulfur.

robbing (Mining) Removing ore pillars prior to abandoning a mine working and allowing it to cave.

rocker lug coupling (Fire Pr) A hose coupling with rounded lugs to keep them from catching on obstructions.

Rockwell hardness (Techn) The penetration hardness of metal determined by measuring the depth of penetration of a 1/16-in. steel ball (B-scale) or a diamond point (C-scale) under a specified applied load. See: hardness.

roentgen (Dosim; Abb: r) The international unit of x- or gamma-radiation, esp used in establishing safe levels of exposure to persons working in the presence of ionizing radiation, defined as the quantity of energy equal to 83 ergs per gram of air or one electrostatic unit of electricity in 1 cm^3 of dry air at standard temperature and pressure.

NOTE: The rad is a more convenient unit and is preferred to the roentgen unit in dosimetry. The roentgen unit is named after Wilhelm Roentgen, a German scientist, who discovered x-rays in 1895.

roentgen equivalent, man (Dosim; Abb: rem) The unit of dose of any ionizing radiation that produces the same biological effect as a unit of absorbed dose of ordinary x-rays. Cf: rad; curie; roentgen.
NOTE: The relative biological effectiveness (RBE) dose in rems = RBE x absorbed dose in rads.

roentgen equivalent, physical (Dosim; Abb: rep) The dose of ionizing radiation that produces energy absorption of 93 ergs per gram of body tissue.
NOTE: One rep is equivalent to the ionization produced in a small air cavity by one roentgen of x-rays. It has been superseded by the rad.

roentgenography = radiography

roentgen rays = x-rays

roentgen therapy = radiation therapy

Roger's rope hose tool (Fire Pr) A loop of rope and a hook used to secure hose lines, etc. See: rope hose tool.

roll[1] (Fire Pr) To get a fire department apparatus underway in response to a fire alarm.

roll[2] (Navig) Motion of a vessel or flying vehicle about its longitudinal axis.

roll call (Fire Pr) 1. An assembly of firefighters at a change in shift in which personnel line up for inspection and are counted as present. 2. An assembly of volunteer firefighters after a fire or for a company meeting or drill.

roll center (Transp) The center about which the total sprung mass of a land vehicle rotates when a side force is imposed on the vehicle.

rolling resistance (Transp) The motion-resisting force developed by the interaction of the wheels or tracks of a vehicle and the ground.

When the rolling resistance is subtracted from the gross tractive effort, the effective propelling force remains.

rollover (Storage) An overturning that occurs when a stratus of heavier liquid develops above a stratum of lighter liquid in a storage tank.

roof car (Build) A structure for the suspension of a working platform, providing for its horizontal movement to working positions.

roof cover (Salv) A salvage cover placed over holes in roofs after a fire to protect the interior from the weather.

roof coverings (Build) Materials, composite assemblies, decks, and the like used to cover buildings.
NOTE: Roof coverings are grouped into three basic classes. Class A: Roof coverings that are effective against severe fire exposures, are not readily flammable, and do not carry or communicate fire; afford a fairly high degree of fire protection to the roof deck; do not slip from position; possess no flying brand hazard; and do not require frequent repairs in order to maintain their fire-resistance properties. Class B: Those that are effective against moderate fire exposures, are not readily flammable, and do not readily carry or communicate fire; afford a moderate degree of fire protection to the roof deck; do not slip from position; possess no flying brand hazard; but may require infrequent repairs in order to maintain their fire-resistance properties. Class C: Those that are effective against light fire exposures, are not readily flammable, and do not readily carry or communicate fire; afford at least a slight degree of fire protection to the roof deck; do not slip from position; possess no flying brand hazard; and may require occasional repairs or renewals in order to maintain their fire-resistance properties.

roof cutter (Fire Pr) A tool used to cut holes in roofs for ventilation.

roofer (Build) A board that is nailed to the rafters and serves as the base for shingles or other roof covering.

roofing bracket (Build) A bracket used in sloped roof construction, having provisions for fastening to the roof or supported by ropes fastened over the ridge and secured to some suitable object. Also called: **bearer bracket**.

roof ladder (Fire Pr) A straight ladder with swivel hooks at the top end for anchoring the ladder over a roof ridge. See **ladder**.

roof layer (Mining) A layer of methane at roof level in a coal mine.

roof manifold (Build) An extension of a standpipe system to the roof of a building to provide several hose connections for fire-fighting operations on the roof.
NOTE: The manifold is usu kept dry and is supplied from a roof tank by turning on a shutoff valve.

roof-powered platform (Build) A powered platform having its raising and lowering mechanism located on a **roof car**.

roof screen (Build; Brit) A partition in the space between the top floor ceiling and the roof of a building that divides the space into fire areas, called bays, to prevent smoke and combustion gases from spreading throughout the roof space.

roof test (Build) Submission of a roof assembly to an external source of heat and flame to determine its resistance to fire penetration and surface flame spread.

roof venting (Build) A system of vents in the roof of a building that open automatically in the event of a fire to allow smoke and hot gases to escape.

rookie (Fire Pr) A novice firefighter. A **probationer**.

room heater (Build) An unvented, free-standing appliance, without ducts, used to heat its immediate surroundings. Sometimes called: **space heater**.

root mean square (Math; Abb: rms) The square root of the arithmetic mean of the squares of a set of numbers.

rope (Fire Pr) A flexible line constructed of fibers that form a strand, three of which are usu twisted together to form the rope.

rope guide signals (Fire Pr) A code of meaningful tugs on a lifeline to indicate pay out, haul in, or to call for help. See: guide line.

rope hose tool (Fire Pr) A short length of rope with a ring at one end and a hook at the other, used for securing hose, etc.

rope sling (Fire Pr) Any of several harnesses made of rope used in rescue work and to raise and lower tools and equipment.

rosin (Techn) The resin of the pine tree, obtained as a product in the manufacture of turpentine. Widely used in the manufacture of soap and soldering flux.

Rossby number See: dimensionless numbers

roster (Fire Pr) A list of personnel assigned to a given fire department unit, showing ranks, duty tours, etc.

rotary gear pump (Hydraul) A positive-displacement pump with two rotors or gears that drive water through the pump chamber. See: pump.

rote learning (Educ) Memorization; usu with little or no understanding of the meaning of the material memorized.

rotor^1 (Mech Eng; Electr) The rotating part of a device, machine, electric motor, or generator.

rotor^2 (Transp) The lifting blade assembly of a helicopter.

rouge (Techn) A finely powdered form of iron oxide used as a polishing agent.

rough (For Serv) The accumulation of organic matter and understory vegetation, such as grass, forest litter, dead evergreen needles, and underbrush that pose a fire hazard.

roughness (Mech Eng) A characteristic of surface texture of solids.
  NOTE: All solid surfaces have three-dimensional structure, which plays an important part in heat transfer, corrosion fatigue, fluid flow, contact characteristics, and problems of surface finish, application of coatings, lubrication, and the like.

rough reduction (For Serv) Action taken to decrease the fire hazard from accumulations of rough.

round^1 (Fire Pr) A single fire alarm signal. Box number alarms and additional alarms are usu repeated; i.e., sent in two or three rounds. Less urgent signals, such as return to quarters, are given in one round.

round^2 One patrol or circuit made by a plant guard or watchman. Also called: tour.

round^3 (Brit) = Rung (of a ladder).

roundwood^1 (Build) Timber used in its natural shape, without squaring it by sawing or hewing.

roundwood^2 (For Serv) Any dead forest fuel that is roughly cylindrical, such as boles, limbs, and stems.

route (Fire Pr) Designated roads to be used by fire companies responding to a fire. Route selection takes into account distances, traffic, paths of other emergency vehicles, and advantageous approaches to the fire or emergency scene.

route card (Fire Pr) An index card used by a fire alarm dispatcher or carried on rural fire apparatus, giving specific directions to individual properties, description of the property, availability of water, and other pertinent information.

routine^1 (Comput) A set of instructions, often a subsection of a program, arranged in sequence to direct a computer to perform one or more operations.

routine^2 (Fire Pr) A schedule of normal work activities.

rpm = revolutions per minute

rubber foam (Chem) A rubber product that has been fluffed in the manufacturing process so that it has a cellular structure to give it softness.
NOTE: Widely used as padding in cushions and upholstered furniture. Foam rubber is ignitable by a cigaret and can burn as a smoldering fire, in some instances with the evolution of toxic products.

rubber gloves (Fire Pr) Insulating gloves used when live electrical wires, motors, equipment, and appliances are involved.

rubber goods (Fire Pr) Waterproof clothing and boots worn by firemen.

rubbish Any solid waste, excluding food waste and ashes, discarded from residences, commercial establishments, and institutions.

rubric (Info Sci) One or more words used as a heading for purposes of classification.

rules and regulations (Fire Pr) An official fire department book of rules specifying the responsibilities of officers and men and the proper conduct of fire department operations.

run (Fire Pr) A response to a fire alarm. Cf: ride.
NOTE: When hand-drawn fire apparatus was used, firemen ran to fires with their engines, and even the earlier horse-drawn engines had no place for riders. Sometimes the officers were required to run ahead of the engine to clear the way.

rung (Fire Pr) 1. A cross-piece between the beams of a ladder, used as a footrest in climbing. See: ladder. 2. = (Brit) round.

running away from the water (Fire Pr) A condition in which the water supply is insufficient to keep the pump chamber filled, and the pump literally "runs away" from the water and cavitation occurs.

running card (Fire Pr) A card that displays fire company assignments for given alarm boxes or locations usu from first through fifth alarm. After the fifth alarm, equipment is specially called. Also called:

response schedule.

running fire (For Serv) A fire that advances rapidly in a given direction with a well-defined head. See also: creeping fire; smoldering fire; spotting.

running fit (Mech Eng) An intentional difference in the sizes of mating parts to allow them to move freely with respect to each other.

running foot (Metrol) A unit of measure equivalent to the foot, applied to materials that are made in standard widths but varying lengths, such as lumber, rope, and the like.

running gear (Transp) Wheels and other related components of a vehicle.

running line (Mat Hand) Any moving rope, as distinguished from a stationary rope such as a guyline.

running time (Fire Pr) The time elapsed between the departure of an apparatus from quarters and its arrival at the fireground.

running wrong (Fire Pr) 1. Responding to a location other than the one to which the unit was dispatched. 2. Failure to follow the prescribed route to a fire or emergency without proper authorization.

runoff (Salv) Water collecting from hose streams and draining off.
NOTE: Control of runoff to minimize water damage is a major concern of fire department salvage crews.

runway[1] (Build) A passageway for persons, elevated above the surrounding floor or ground level, such as a footwalk along shafting or a walkway between buildings.

runway[2] (Fire Pr) A water chute made by interfolding salvage covers lengthwise.

runway[3] (Transp) An aircraft landing strip.

rupture disk (Techn) A diaphragm designed to burst at a predetermined pressure difference. Syn: burst disk.

rural Places with less than 2,500 inhabitants, except those classified as urban by the US Census Bureau.

rural fire protection (Fire Pr) A system of self-help, local fire protection resources, and mutual aid agreements with neighboring volunteer or municipal fire departments to provide fire protection to sparsely settled areas.

S

S (Chem; Phys) Symbol for entropy, standard deviation, the element sulfur.

SAE numbers (Techn) An arbitrary classification of lubricating oils based on viscosity, formulated by the American Society of Automotive Engineers.

safety (Safety) Absence of conditions that can cause injury or death to personnel, damage to or loss of equipment or property.

safety belt[1] (Safety) A device, usu worn around the waist, that is attached to a lanyard and lifeline or to a structure to prevent falls.

safety belt[2] (Fire Pr) A special belt with a swivel hook for use with turntable ladders and in training-tower drills as a safety measure to prevent falling or other injury. Also called: pompier belt.
NOTE: A line is attached to the head of the ladder or to the drill tower and to the safety belt worn by the man being carried down or lowered.

safety can (Techn) A container of not more than 5 gallons capacity having a flash-arresting screen, a spring-closing lid and spout cover, and so designed that it will safely relieve internal pressure when exposed to fire.

safety coupling (Mech Eng) A friction coupling adjusted to slip at a predetermined torque to protect the rest of the system from overload.

safety curtain = fire curtain. Also see: curtain.

safety cut-out (Electr) An overload protective device in an electric circuit.

safety design (Safety) The planning of environments, structures, and equipment and the establishment of procedures for performing tasks so that human exposure to injury potential will be reduced or eliminated.

safety director (Educ; Safety) An individual who plans and administers training programs in health, hygiene, accident prevention, fire prevention and protection, and other safety procedures for employees of an industrial organization. Also called: safety supervisor.

safety education (Educ) The transmission of information concerning the safety requirements of operations, processes, environments, etc., to personnel.

safety engineer (Ind Eng; Safety) An individual who applies knowledge of industrial processes, mechanics, chemistry, psychology, and industrial health and safety laws to prevent or correct injurious environmental conditions and minimize hazards to life and property. Inspects premises for fire hazards and adequacy of fire protection and inspects fire-fighting equipment.

safety factor (Ind Eng) A design factor to allow an adequate margin of safety.

safety fuse (Civ Eng) A flexible cord containing an internal burning medium by which fire is conveyed at a continuous and uniform rate for the purpose of firing blasting caps.

safety glass (Build; Techn) Impact resistant and shatterproof glass used as eye protection and for automobile windows, large architectural windows, and doors. Also includes heat-treated glass that breaks into granules instead of sharp-edged strands.

safety guard (Techn) An enclosure designed to confine the pieces of a grinding wheel and furnish protection in the event the wheel breaks during operation.

safety hat (Build; Civ Eng) Rigid headgear of varying materials designed to protect the workman's head from falling objects. Also called: hard hat.

safety hook (Transp) A hook with a latch to prevent slings or load from accidentally slipping off.

safety inspector (Safety) Inspects machinery, equipment, and working conditions for hazards to workers to prevent accidents and fires.

safety interlock (Safety) A device that prevents the operation of a machine while a cover or door is open or unlocked and that holds the cover or door closed and locked while the machine is in operation.

safety lamp (Mining) A protected flame lamp used to test mine atmospheres, esp to detect methane-air mixtures and oxygen deficiencies. See: flame trap.

safety lock (Ind Hyg) A small chamber with double doors separating a hazardous area from the rest of the building.

safety match See: match

safety practice (Safety) A system of standards, regulations, and precautions taken to prevent accidental injury or death to workmen.

safety relief (Techn) A device, such as a blowout plug, intended to prevent rupture of a pressurized cylinder or tank.

safety relief valve (Techn) A valve that allows venting of overpressure to prevent rupture of a gas storage cylinder.

safety ring (Fire Pr) A polished steel ring on the hook anchor of a hook ladder to which the ring hook of the hook belt is attached.

safety rope See: line

safety screen (Civ Eng; Mining) An air- and water-tight diaphragm placed across the upper part of a compressed air tunnel between the face and the bulkhead in order to prevent flooding the crown of the tunnel between the safety screen and the bulkhead, thus providing a safe means of refuge and exit from a flooding or flooded tunnel.

safety shoes (Safety) Footwear providing toe protection for the wearer.
  NOTE: In addition to impact protection, safety shoes may also have conductive soles to reduce the possibility of friction sparks in environments with a fire or explosion hazard. Other safety shoes are designed to provide protection against splashes of molten metal, construction hazards such as protruding nails, contact with energized electrical equipment, wet conditions, hot surfaces, and other hazards.

safety shutoff (Techn) A device that shuts off the gas supply to a burner and sometimes also to the pilot light if the ignition source fails.

safety solvents (Techn) Solvents that are free from fire or toxicity hazards and are nondamaging to surfaces or materials being cleaned.
  NOTE: The toxicological effects alone are not adequate to assess the hazard potential of a solvent. Ignition temperature, flash point, and other factors determine the potential for fire and explosion.

safety valve See: pressure relief valve

sagging (Build) Bending downward in the middle.

Saint Florian (Fire Pr) The patron saint of firefighters.

salamander (Techn) A small furnace, usu cylindrical in shape without grates, used for heating.

sal ammoniac (Chem) A compound, NH_4Cl, commonly used in fertilizers, drycell batteries, soldering flux, and in the textile and tanning industries. On heating, the compound dissociates into ammonia and hydrogen chloride, which recombine on cooling. Also called: ammonium chloride.

saline solution (Med) An intravenous solution of sterile water containing salt. Normal saline contains 0.9% sodium chloride.

salt (Chem) A compound, such as table salt, NaCl, derived from an acid and containing an acid radical and a metal or a cation radical that replaces part or all of a replaceable hydrogen of the acid.
  NOTE: Common salt can be made by reacting sodium hydroxide with hydrochloric acid.

salvage[1] (Fire Pr) Procedures or measures for reducing damage from smoke, water, and weather during and

following a fire by the use of salvage covers, smoke ejectors, deodorants, etc.
   NOTE: In British usage salvage includes the removal of fire victims.

salvage² (Insur; Transp) a. Rescue of a ship from wreck or fire. b. Reward paid for material assistance in saving a ship or cargo. c. Goods or property saved from fire.

salvage company (Fire Pr) 1. A company or unit specializing in fire salvage or protection. Syn: (Brit) salvage corps. 2. A crew of men maintained by an insurance company to perform salvage operations.

salvage cover (Fire Pr) A waterproof tarpaulin of standard size made of cotton duck, plastic, or other material, used to protect furniture, goods, and other property from heat, smoke, and water damage during a fire-fighting operation.

salvage tender (Fire Pr; Brit) An appliance used wholly or largely for carrying special equipment, such as waterproof sheets, sawdust, etc., to prevent or minimize water damage at a fire.

sampling (Statist) The process of taking a number of individual items or a small portion from a large population or production lot of material in such a way that the items or portions selected are representative of the entire population or lot. See also: random sample; representative sample.
   NOTE: Samples are usu selected randomly. Care must be taken that the sample truly represents the total population or lot and avoids bias, which will skew measurement results and invalidate error estimates.

Sanborn map (Fire Pr; Insur; Cartog) One of a series of detailed maps of US communities of over 2000 population prepared by the Sanborn Map Company of Pelham, N. Y. The maps are scaled and show streets, buildings, types of construction, occupancies, fire resistance, heights, water supplies, and many other details that are useful to fire departments.

sand (Soil Mech) A coarse-grained soil, the greater portion of which passes through a No. 4 sieve, according to the Unified Soil Classification System.

sand filling (Mining) Using concentration tailings for filling mined-out stopes in a mine.

sand hog (Civ Eng) Any worker doing tunneling work under elevated pressure.

sandwich construction (Build; Techn) Structural panels consisting of a thick, lightweight core (often in a honeycomb configuration) faced with thin sheets of metal or wood, providing great strength and stability.

sanitary landfilling (Env Prot) An engineered method of disposing of solid waste on land in a manner that protects the environment, by spreading the waste in thin layers, compacting it to the smallest practical volume, and covering it with soil by the end of each working day.

sanitation (Env Prot) The control of factors in the physical environment that can harm physical development, health, and survival.

saponification (Chem) 1. The hydrolysis of fats or oils by an alkali, in which an ester is converted into an alcohol and a salt. 2. The process of making soap.

sarcoma (Med) A malignant tumor in connective tissue.

satellite (Fire Pr) A fire apparatus or fireboat used on roving assignments as a supporting unit. Cf: pup boat.

saturate¹ (Chem; Techn) a. To soak or impregnate thoroughly. b. To combine the greatest quantity of one substance with another at a given temperature in solution, chemical combination, or the like.

saturate² (Phys) To load an object or device completely with energy, as with electricity or magnetism.
   NOTE: Color saturation is the degree to which a color is mixed with white.

saturated air (Meteor) Moist air containing as much water vapor as it can hold.
NOTE: This occurs when the partial pressure of the water vapor is equal to the vapor pressure of water at the existing temperature.

saturated solution (Chem; Phys) A solution in which no more of a solute will dissolve in the solvent at the given temperature and pressure.

saturated vapor pressure See: vapor pressure

saturation (Chem) The state in which the maximum quantity of matter is held in a solution.

saw boss (For Serv) The supervisor of a saw crew in a fire suppression organization responsible for felling trees, cutting logs, and other sawing operations required in the construction of fire control lines or fire breaks.

SBCC = Southern Building Code Conference

SBR (Chem Eng) A kind of rubber; a copolymer of butadiene and styrene, used extensively in automotive tire manufacture. Formerly called: GR-S.

scab[1] (Med) A hard crust of blood and serum over a wound.

scab[2] = spreader[2]

scaffold (Build) Any temporary elevated platform and its supporting structure used to hold workmen.

scalar (Phys) A measure of magnitude that does not depend on direction, as constrasted with vector.

scald (First Aid; Med) 1. To burn the flesh with hot liquid, vapor, or steam. 2. An injury caused by hot liquid.
NOTE: Water scalds at 145°F and produces third degree burns at 180°F.

scaled test (Educ) A test in which the questions or tasks are arranged in increasing order of difficulty.

scaler (Dosim) An electronic instrument for rapid counting of radiation-induced pulses from Geiger counters or other radiation detectors.
NOTE: The instrument permits rapid counting by reducing the number of pulses entering the counter. See: counter; Geiger-Mueller counter.

scaling[1] (Fire Pr) Climbing, as a ladder or up a precipitous face with the aid of a rope.

scaling[2] (Techn) a. Flaking or shedding of thin chips or flakes from a surface. b. Becoming encrusted or covered with deposits, such as the inside surfaces of pipes and boilers.

scaling[3] (Math; Techn) A method of analyzing structures or phenomena by modeling an object or system with reduced dimensions to simplify handling and save costs. In the mathematical interpretation, parameters are expressed in dimensionless form by dimensionless numbers. Results of studies and analyses of the model are then extended with appropriate corrections to full size. Cf: simulation.

scaling ladder (Fire Pr) A ladder with one or more hooks that can be placed over a window sill and used to scale a building by climbing from window to window, passing the ladder upward as each successive window is reached. See: ladder.
NOTE: In US practice, often a pompier ladder. In Europe, generally a tapered two-spar ladder with brackets at the bottom for attaching additional ladders below. All such ladders are interchangeable.

scantling (Build) 1. The dimensions of structural members, including timber and stone used in buildings. 2. A small piece of lumber, such as a stud, used in framing.

scatter (Statist) The degree of dispersion of values around a central tendency.

scatter diagram (Statist) A rectangular coordinate plot of two variables. Also called: scattergram; scatter plot.

scattered radiation (Phys) Radiation that is diverted from its path by interaction with objects or within tissue.

scattering (Phys) A process in which particle pathways change.
NOTE: Scattering occurs when particles collide with atoms, nuclei, and other particles or when they interact with magnetic fields. If the internal energy of the particle, as contrasted with its kinetic energy, is unchanged by the collision, scattering is elastic; if there is a change in the internal energy, the process is called inelastic scattering.

scavenging[1] (Chem) The use of a nonspecific precipitate to remove one or more undesirable radionuclides from solution by absorption or coprecipitation.

scavenging[2] (Atm Phys) The removal of radionuclides from the atmosphere by the action of rain, snow, or dew. See: fallout.

scene of fire See: fire scene

scheduled charge (Insur) The specific charge (in days) assigned to a permanent partial, permanent total, or fatal injury.

schematic (Techn) A diagram showing the relative position and/or function of different components or elements of an object or system.

schlieren (Opt) Literally, "streaks." Interference shadow patterns caused by changes in density of air or gas, as in heated air rising from a flame, or the density jumps in shock waves. See also: streak photography.

Schmidt number (Phys) The ratio of kinematic viscosity to molecular diffusivity. See: dimensionless numbers.

scholastic aptitude (Educ) A capacity for learning or doing well in academic subjects.

science An organized body of knowledge that has been correlated and generalized into a system for the purpose of describing and explaining natural phenomena uniquely and unambiguously.
NOTE: Applied to specific disciplines, sciences are referred to as exact, such as physics and chemistry, and descriptive, such as psychology and botany.

scientific management (Man Sci) The application of scientific methods to increase return by better use of resources, improving worker efficiency, and the like.

scintillation (Phys) A flash of light produced in a phosphor by ionization. Cf: fluorescence; luminescence.

scintillation counter (Phys) An instrument that detects and measures ionizing radiation by counting the light flashes (scintillations) caused by radiation impinging on a phosphorescent material.

scissors stairs (Build) A stairway in which each flight is continuous from floor to floor, reversing direction at each floor level. Cf: return stair.

sclera (Anat) The outer coat of the eyeball.

scleroderma (Med) Hardening of the skin.

scleroscope hardness See: hardness

scoosher (Fire Pr; Brit) A combination tank truck, pumper, and water tower. The apparatus is equipped with a hydraulically operated articulated boom carrying a monitor that throws a jet or fog spray. A heat detector is used for pinpointing fire in smoke or in locations hidden to view from the ground. The apparatus also carries ladders and other standard fire-fighting equipment.

scorch (Fire Pr) 1. To burn or char a surface by heat. 2. The trace of damage on a surface caused by scorching.

scorching (Fire Pr) Pyrolysis or partial combustion of a surface, usu by exposure to a heat source.

scorchline (For Serv) The average height of foliage that has been browned and killed by fire.

scout (For Serv) A staff worker in a fire suppression organization responsible for gathering information and reporting on the location, behavior, and progress in control of a fire, as well as on conditions that must be accounted for in developing an effective fire suppression plan.

scram (Nucl Safety) 1. The sudden shutdown of a nuclear reactor, usu by rapid insertion of safety rods by the operator or automatic control equipment. 2. To shut down a reactor quickly.

scrap (Techn) Discarded or rejected material or parts of material that result from manufacturing or fabricating operations and are suitable for reprocessing.

scratch line (For Serv) A hastily constructed, incomplete emergency fire control line.

screen See: strainer

screen door (Build) An outer door usu closed automatically by a spring or pneumatic device, consisting of wire mesh or gauze in a frame; used primarily to keep insects out while allowing ventilation when the inner solid door is open. Cf: storm door.

screening (Man Sci) Segregating items or individuals, excluding some and passing others for further treatment or consideration on the basis of established criteria.

scrubbing (Chem Eng) Removal of one or more constituents of a solution by adding a compound that dissolves the constituents to be removed.
  NOTE: An example is the passing of inhibited butadiene through a caustic solution to remove the inhibitor prior to polymerizing the butadiene.

SCUBA The acronym for Self-Contained Underwater Breathing Apparatus. Syn: aqualung.
  NOTE: Exhaled air can be recycled, and closed-circuit, semiclosed circuit, and open-circuit equipment is available. Of these, the last is least efficient but cheapest and safest to use.

scupper (Build; Fire Pr) A type of drain opening near floor level for water to drain out and thereby reduce possible water damage from sprinklers and hose streams.

scuttle (Build) 1. An opening or trapdoor in a ceiling, leading to an attic. 2. An opening in a roof, fitted with a cover, providing access to the roof.

scuttle ladder (Fire Pr) A small folding or extension ladder used for crawling through restricted openings into attics and crawl spaces or through manholes. Syn: attic ladder.

sea-level pressure (Meteor) The pressure of a standard atmosphere: 29.92 in. of mercury or 1013.2 millibars. See: pressure measurement.

sear (Fire Pr; Med) To scorch or cause a surface burn.

searchlight (Fire Pr) A bright, directional light used to illuminate fire-fighting and rescue operations. See also: floodlight.

searchlight unit = lighting unit

search warrant (Law) A legal writ that authorizes a law enforcement officer to search specified premises for evidence.

seasoning (Build; For Serv) Removing moisture from green wood in order to improve its serviceability.
  NOTE: Wood may be air dried or air seasoned by exposure to the air, usu in a yard, without artificial heat, or dried in a kiln by artificial heat.

sea state (Navig) An arbitrary scale of sea conditions, based on significant wave height.

seat of fire (Fire Pr) The location of the main body of a fire; the area producing most of the heat, flames, and gases. Cf: body of fire.

sebaceous gland (Anat) An oil producing gland of the skin.

secator (Techn) A separating device that throws mixed material onto a rotating shaft; heavy and resilient materials bounce off one side of the shaft, while light and inelastic materials land on the other.

second[1] (Metrol; Abb: sec; Symb: ") A fundamental unit of time, equal to

355

1/86,400 of a mean solar day, 1/360 of an hour, and 1/60 of a minute.
  NOTE: In the SI system, a second is equivalent to 9,192,631,770 cycles of the vibration between certain unperturbed, hyperfine levels of cesium-133.

second² (Geom) A unit of angle equal to 1/60 of a minute of arc or 1/3600 of a degree.

second alarm (Fire Pr) An alarm ordered by the officer in command at a fire calling for a full second-alarm assignment. A second alarm may also bring out higher-ranking officers and special units that respond only to more serious fires.
  NOTE: A call for certain additional equipment is not considered to be a second alarm. Some cities use an intermediate signal, not a second alarm, to call for additional equipment or companies.

second-alarm assignment (Fire Pr) The complement of men and equipment designated to respond to a second alarm.

second-degree burn See: burn²

second front (Fire Pr; Colloq) Employment outside the fire department by a firefighter to supplement his income from the fire department. Syn: C-shift; moonlighting.

second-line apparatus (Fire Pr) A standby fire apparatus maintained and ready for use by off-duty firefighters in case of major emergency or by a second-line company that is assigned to answer alarms when the first-line unit is out of service or otherwise engaged.

secondary (Electr) The output coil of a transformer. See: primary circuit.

secondary blasting (Civ Eng) The reduction of oversize material, by the use of explosives, to the dimensions convenient for handling.

secondary circuit (Fire Pr) An alarm circuit that has an alternate means for sending alarms to fire companies in case the primary circuit fails.
  NOTE: The secondary circuit may actuate a gong or provide voice communications. Police and water service agencies are usu included in the secondary alarm circuit.

secondary combustion air (Comb) Air introduced at the point of combustion above a fuel bed by natural, induced, or forced draft.
  NOTE: Generally referred to as overfire air if supplied above the fuel bed through the walls of the combustion chamber.

secondary exit (Build) An alternate escape route in a building that already has adequate exits; provided as an extra precaution should an unforeseen condition block the primary exits or otherwise make them unusable or inaccessible.

secondary fire (Fire Pr; Brit) = spot fire

secondary high explosive (Comb) A high explosive which is relatively insensitive to heat and shock and is usu initiated by a primary high explosive. Also called: noninitiating high explosive.
  NOTE: This type of explosive requires a relatively long distance and time to build up from a deflagration to detonation and will not propagate in small-diameter columns. Secondary high explosives are used for boosters and bursting charges.

secondary lookout (For Serv) 1. A lookout point supplementing a primary lookout to provide observation of a blind area or coverage of areas when visibility conditions are poor or fire danger is esp severe. 2. Short for secondary lookout man.

secondary lookout man (For Serv) A member of a fire-suppression organization manning a secondary lookout point.

section¹ (Fire Pr) a. A length of hose. Sometimes called: piece. b. A segment of an aerial or other extension ladder.

section² (Med) a. To cut or separate. b. To take a thin slice of tissue for microscopic examination.

sectional ladder (Techn) A nonself-supporting portable ladder, nonadjustable in length, consisting of two or more sections that may be combined into a single ladder. Its size is designated by the overall length of the assembled sections.

sector[1] (For Serv) A segment of a fire perimeter or control line being worked by two or more crews under one sector boss.

sector[2] (Fire Pr) a. A command area or position at a fire under the direction of a subordinate chief. b. A fire department district.

sector[3] (Build; Brit) An area protected by more than one group of detectors. A signal transmitted by a detector in any group also identifies the area. Syn: zone.

sector boss (For Serv) A staff officer responsible for all fire suppression activities at a given sector of a wildland fire.

security (Law) Protection of supplies against theft, sabotage, or other malicious acts.

security doors (Build; Law) Metal doors or curtains used to cover store fronts or close off loading platforms at night. Curtain types are usu flexible and roll up like a window blind. Door types usu slide or retract upward.

security system (Build; Law) Safeguards, including locks, alarms, and other equipment, used to protect people from harm and property against trespass and theft.

sediment See: bottom settlings

sedimentation (Chem Eng) The process of settling, usu of solid particles in a liquid.

seen area = visible area

seen area map = visible area map

seepage (Hydrol) Movement of water or gas through soil without forming definite channels.

seizure (Med) A sudden, often violent attack caused by an acute illness or disease. A fit, or convulsion.

self-closing device (Build) A device intended to insure that a fire door, when opened, will return to the closed position.

self-contained breathing apparatus See: breathing apparatus

self-extinguishing (Build; Comb) The property of a material to stop burning; incapable of sustained combustion without external heat, as determined by a small-scale test procedure.

self-generating extinguisher (Fire Pr) A fire extinguisher that operates upon demand by the interaction of its components, such as the familiar soda-acid device. Most modern extinguishers operate on stored energy (pressurized gas).

self-heating (Comb) The result of exothermic reactions, occurring in some materials under certain conditions, whereby heat is liberated at a rate sufficient to raise the temperature of the material. Syn: spontaneous heating.

self-heating temperature (Chem) The lowest temperature at which an exothermic process can occur in a substance.

self-help (Fire Pr) Action taken by an individual to cope with an emergency. Syn: independent action.

self-ignition (Comb) Ignition resulting from self-heating. Syn: spontaneous ignition.

self-ignition temperature (Comb) The minimum temperature at which the self-heating properties of a material lead to ignition. Syn: autoignition temperature.
NOTE: The process is dependent on specimen size, heat-loss conditions, and other variables, such as moisture content.

self-induction (Electr) The production of a counterelectromotive force in a conductor when its own magnetic field collapses or expands with a change in current in the conductor.

self-insurance (Insur) A term used to describe the assumption of one's own financial risk.

self-powered platform (Build) A powered platform with the raising and lowering controls located on the working platform.

self-priming pump (Fire Pr) A centrifugal-type pump that contains a filled water cavity so that water floods the impeller on start-up. Positive-displacement pumps are usu self-priming.

semiautomatic feeding (Safety) Feeding wherein the material or part being processed is placed within or removed from the point of operation by an auxiliary means controlled by operator on each cycle.

semicircular canals (Anat) The organs of balance in the inner ear that are closely associated with the hearing mechanism.

semiconductive hose (Electr) A hose with an electrical resistance high enough to limit flow of stray electric currents to safe levels, yet not so high as to prevent drainage of static electric charges to ground.

semifloating load (Transp) A load where lading is secured in two units with a space at center but without space between the units and the ends of the car. This permits movement of each unit in one direction only.

semitrailer (Transp) A nonpowered vehicle having integral wheels at the rear only and designed to carry material, supplies, or equipment and to be towed by a self-propelled motor vehicle that also supports the front end by means of a fifth-wheel coupling assembly.
 NOTE: The front end can also be supported by a dolly that is provided with a fifth-wheel assembly for coupling to the semitrailer and a tongue and lunette for coupling to the prime mover.

senility (Med) The state of mental and physical infirmity due to advanced age.

senior aerial (Fire Pr) An aerial ladder truck consisting of a trailer that carries the ladder and a tractor to pull it. A senior aerial requires a tillerman to steer the rear wheels of the trailer. See also: aerial; service aerial.

seniority (Fire Pr) The length of time served in a fire department or in grade. Usually, the man with the greatest seniority in a department is in charge when several members of the same rank are present.

sense organs (Physiol) The several structures of the nervous system of the body that are influenced by the environment or given internal conditions. Sense organs are generally divided into three classes: 1) exteroceptors, which include the auditory (hearing), tactile (touch), olfactory (smell), and optic (sight) organs; 2) interoceptors, which include the sense organs of pain, hunger, nausea, and thirst; and 3) proprioceptors, which include the organs of balance and equilibrium (muscles and the semicircular canals of the inner ear). See: chemical senses; temperature senses.

sensible (Physiol) Capable of being perceived or apprehended through the senses.

sensible heat (Physiol) 1. That part of the heat flowing into a body or system that increases its temperature but produces no change in state. See: chemical senses, temperature senses. 2. The degree of human comfort as affected by the ambient temperature, humidity, air circulation, heat radiation from and to the body, the heat generated in the body from muscular activity, clothing worn, and other factors.

sensing element (Techn) The component in a system that detects a change in pressure, temperature, etc., and triggers an appropriate signal for other components of the system. Cf: detector.

sensitivity 1. (Physiol) Responsiveness to a stimulus. 2. (Med) The degree of susceptibility of bacteria to an antibiotic. 3. (Electr) The capability of producing an output for a given input.

sensitizer (Med) A chemical that, after extended or repeated exposure, produces in some individuals an allergic type of skin irritation called sensitization dermatitis.

sensorimotor ability (Psych) The ability to react physically to specified stimuli received through the senses. Related to physical coordination.

sensorineural deafness (Med) Nerve deafness or the lack of sensitivity of the auditory mechanism in the chochlea or paralysis of the acoustic nerve.

separating wall (Build; Brit) A fire wall between two adjacent properties or occupancies.

separation (Fire Pr) Space provided between buildings, stacked goods, etc., as a firebreak. See: fire separation.

separator (Techn) 1. A material separation device that relies on ballistic or gravity separation of materials having different physical characteristics. 2. A device that removes ferrous metals by means of magnets.

sepsis (Med) Infection. The presence of pus-forming bacteria, other pathogens or their toxins in the blood or tissues. See: septicemia.

septicemia (Med) Infection in the blood stream caused by pathogenic microorganisms or their toxins. Fever or chills are symptoms of the condition. Sulfonamide drugs and antibiotics are effective controlling agents. Also called: blood poisoning.

septic shock (Med) A state caused by severe infection in which toxins cause the blood vessels to dilate and lose plasma.

sequential Arranged in some predetermined logical order.

sergeant (Fire Pr) The third officer of a fire company, below captain and lieutenant. See: ranks and titles.

serial number (Fire Pr) An administrative number assigned to various units and categories of equipment by a fire department.

series operation (Fire Pr) Operation of a multistage fire pump in which the water passes through each impeller consecutively to build up pressure and provide volume output. Syn: pressure operation.

serious injury (Safety) The classification for a work injury that includes disabling work injuries and injuries requiring treatment by a physician or hospitalization, eye injuries, fractures, loss of consciousness, and the like.

serum (Med) 1. The liquid fraction of the blood; a thin, yellowish residue of the blood from which blood cells and clotting elements have been removed. 2. Blood plasma lacking fibrinogen and certain protein constituents.

service aerial (Fire Pr) An aerial ladder truck on a four-wheel chassis, not requiring a tillerman for steering. Service aerials range from 65 feet to 100 feet in length and are equipped with the same basic equipment as the tractor-drawn trailer aerials. Syn: junior aerial. See also: aerial ladder; senior aerial.

service chief (For Serv) A staff officer in a fire suppression organization responsible for obtaining, maintaining, and allocating men, equipment, and supplies, and providing facilities as required by the fire suppression plan.

Service drop (Electr) A junction box that connects the electrical power lines to the customer's service entrance conductors.

service life[1] (Ind Eng) The period of time a material, device, piece of equipment, or other item remains effective for its intended purpose.

service life[2] (Fire Pr) The period of time a gas mask may be used safely, or the duration of the air supply in a self-contained breathing apparatus.

service station (Fire Pr) Premises on which combustible or flammable motor fuels and lubricants are stored and dispensed from fixed equipment into fuel tanks, oil and grease receptacles of various types of self-propelled vehicles.
 NOTE: An automotive service station serves road vehicles and includes facilities for sale and service of parts and accessories and minor maintenance and repair work. A marine service station is one that

serves boats, ships, and other water craft from a floating dock or a pier or other shore-based structure and includes all directly related facilities.

service test (Fire Pr) A pumper test required once a year or after a major repair to determine that the pumper is in good working order and capable of delivering its rated output.

service truck (Fire Pr) An older type of ladder truck carrying minimum equipment and a small booster pump. Now replaced by service aerials or pumper-ladder combinations. Also called: city service truck.

servomechanism (Electr) A power device for directly affecting machine motion. It embodies a closed-loop system in which the controlled variable is mechanical position.
NOTE: Usually some amplification is necessary to convert the feedback signal into a command signal.

set[1] (Fire Pr) a. An incendiary fire. Syn: torch job; touch-off. b. The point of origin of an incendiary fire. c. A device or material left to ignite an incendiary fire after a time delay. See also: plant[1].

set[2] (For Serv) a. Lightning or railroad fire, esp when several are started within a short time. b. Burning material deliberately ignited for backfiring, slash burning, prescribed burning, and other purposes.

set[3] (Math) Any collection of elements without restriction as to their nature or number, for example, all real numbers, or all odd integers.
NOTE: The algebra of sets is called Boolean algebra or symbolic logic.

set[4] (Techn) A kit or complete assortment of items, such as tools, that have a common or closely related purpose.

set[5] (Mining) A frame, usu of timber, consisting of a cap, two posts, and a sill, used in a mine to brace the roof and walls of a tunnel, to support a working platform, etc.

Setaflash tester (Testing) An apparatus used to determine flash points of liquids.

set-back (Build) A recess in a wall or an offset in the upper part of a building.

set pressure (Techn) The pressure marked on a valve at which the valve is set to discharge.

settling velocity (Aerodyn) The velocity at which a given dust will fall out of dust-laden gas under the influence of gravity alone.

setup time (Fire Pr) The time between the arrival of a fire-suppression unit at the scene of a fire and the time it begins fire-fighting activities.

severity index (For Serv) A number that indicates the relative daily fire danger for an area during a given period, such as a fire season.

severity rate (Insur; Safety) The total days charged for work injuries per 1,000,000 employee-hours of exposure. Days charged include actual calendar days of disability resulting from temporary total injuries and scheduled charges for deaths and permanent disabilities. These latter charges are based on 6000 days for a death or permanent total disability, with proportionately fewer days for permanent partial disabilities varying in degrees of seriousness.

sewage sludge (Techn) A semiliquid substance consisting of settled sewage solids combined with varying amounts of water and dissolved materials.

sferics (Commun) 1. Interfering sounds or noise in electromagnetic communications due to electrical discharges in the atmosphere. Also called: atmospherics; atmospheric interference; static. 2. (Brit) statics; strays; X's.

SFPE = Society of Fire Protection Engineers

shadowgraph (Opt) A technique for visualizing disturbances in fluid flow by passing light through the fluid flow field. Density variations in the flow refract the light rays, producing a pattern of shadows. See also: schlieren.

shaft¹ (Mech Eng) A mechanical bar, rod, or cylindrical member, such as a drive shaft, for transmitting power, usu by rotation or reciprocating action.

shaft² (Build) A passageway, usu vertical, as for an elevator, stair, utility conduits, air ducts, and the like.

shaft³ (Mining) A vertical or inclined mine passageway used to find or mine ore, remove water, or to ventilate underground workings.

shake¹ (For Serv) A fissure in the annual rings of a tree or a split in a timber.

shake² (Build) A kind of wooden shingle used for roofing and siding, usu much wider than the common shingle.

shakeout (Metall) The separation of the solid, but still not cold, casting from its molding sand.

shakes (Ind Hyg) Metal fume fever.

shale oil (Chem Eng) A tarry oil obtained by distillation from bituminous shale.

shallow bag (Fire Pr) A salvage cover placed on the floor with the edges elevated or rolled to create a depression and catch dripping water. Also called: catchall. Cf: deep bag.

shank¹ (Mech Eng) The stem or butt end of a tool.

shank² (Fire Pr) An outer ring or sleeve that fits around an expander and a hose to crimp the hose to a coupling. See: coupling.

shear cutter = bolt cutter

shear strength (Soil Mech) The maximum resistance of a soil to shearing stresses.

shear stress (Mech Eng) The load applied to a structural member causing one part of the member to slide over another along a plane.

shear wave (Phys) A form of wave motion in which the particle displacement at each point in a material is at right angles to the direction of propagation. Also called: transverse wave.

shed (Build) A structure meant for storage of tools, equipment, and products, or for shelter, frequently without complete side and end walls.

sheepshank knot See: knot

sheet-metal fire door (Build) Flush, paneled, or corrugated doors of formed steel. See: fire door.

Sheetrock = asbestos cement board

shelf life = storage life

shell (Phys) One of a series of concentric spheres, or orbits, at various distances from the nucleus, in which, according to atomic theory, electrons move around the nucleus of an atom.
NOTE: The shells are designated, in the order of increasing distance from the nucleus, as the k, l, m, n, o, p, and q shells. The number of electrons which each shell can contain is limited. Electrons in each shell have the same energy level and are further grouped into subshells.

shelter (Civ Def) A prepared place of safety or refuge for people in war or disaster, such as a fallout shelter or bomb shelter.

shepherd hook (Fire Pr) A hook provided on a pike pole and used to advance hose vertically.

shield¹ (Fire Pr) A transparent screen or one with a window, used to protect the face.

shield² (Safety) A body of material used to reduce the passage of radiation. See: barricade shield; barrier shield; biological shield; radiation shielding; thermal shield.

shielding (Electr) A metallic barrier used to prevent magnetic or electrostatic coupling between adjacent circuits.

shift (Fire Pr) 1. A working tour of duty consisting of a specified number of hours. 2. A group of firefighters or platoon working the same duty hours.

shifter (Mining) A miner working a shift underground.

shimmy (Transp) The vibratory oscillation of the steerable wheels of a vehicle about the kingpins. Also called: wheel wobble.

shingle (Build) A thin, flat, wedge-shaped piece of wood or a rectangular piece of asphalt tile used in overlapping rows to cover roofs or sides of buildings.

shipping (Transp) Actions necessary to deliver material to a carrier for movement to a consignee.

shipping area (Transp) Floor space used to assemble material pending its loading for shipment.

shipping container (Transp) Any suitable exterior container used for shipment of supplies.

shipping document (Transp) Form used to authorize the shipment of property.

shock[1] (Mech Eng) An abrupt acceleration or deceleration of a body.

shock[2] (Med) A condition involving changes in blood circulation associated with bodily injury, bleeding, burns, or other traumas.
  NOTE: The mechanism is not completely understood, but is thought to be caused by local loss of blood and body fluids, release of toxic substances by the injured tissues, or changes in the tone of the blood vessels. Symptoms are apathy, clammy skin with grayish color, rapid pulse, and low blood pressure. Treatment usu includes administration of oxygen, restoration of body fluids and blood pressure by transfusion, conservation of body heat, and sedation.

shock front See: shock wave

shock isolator (Techn) A resilient support that acts to isolate a system from shock forces. Syn: shock absorber

shock spectrum (Mech) A plot of the maximum response experienced by single-degree-of-freedom systems, as functions of their natural frequencies, in response to applied shocks. The response may be expressed in terms of acceleration, velocity, or displacement.

shock wave (Aerodyn) A pressure pulse in air, water, or earth, propagated from an explosion, which has two phases: in the first, or positive phase, the pressure rises sharply to a peak, then subsides to the normal pressure of the surrounding medium; in the second, or negative phase, the pressure falls below that of the medium, then returns to equilibrium.
  NOTE: A shock wave in air is often called a blast wave. A shock wave can be induced by a spark, a rupturing diaphragm, a detonation, or other mechanism.

shoe See: ladder shoe; safety shoe.

shop (Fire Pr) A fire department maintenance and repair facility.

shop rules (Man Sci) Regulations established by an employer dealing with day-to-day conduct in plant operations, safety, hygiene, records, etc.; working rules adopted in collective bargaining agreements. Also called: working rules.

shore[1] (Civ Eng; Salv) To brace or prop temporarily with timbers and the like, such as a weakened wall or bank of earth that threatens to collapse. Also called: underpinning.

shore[2] (Civ Eng) A supporting member that resists a compressive force imposed by a load. A timber used as a brace.

short circuit (Electr) A low-impedance or zero-impedance path between two points.

shorthanded (Fire Pr) Undermanned; referring to a fire-fighting unit with fewer men on duty than called for by the table of organization or by the National Bureau of Fire Underwriters.

short line (Fire Pr; Brit) A 2-inch, 40-foot manila rope. See: line.

short-term appointment (For Serv) Employment for less than 12 months, excluding firefighters hired to fight a particular fire. The position of Fire Control Aide in the Forest Service is usu a short-term appointment.

shotcrete (Civ Eng; Mining) A fine aggregate mixture of portland cement, sand, and water that can be sprayed. Cf: gunite.

shoulder carry (Fire Pr) A method of carrying hose in which firefighters in single file drape several folded loops of hose over their shoulders. See: hose carry; ladder carry.

shoulder load (Fire Pr) Hose folded for a shoulder carry.

shower (Meteor) A brief rain that starts and stops suddenly, generally falling from cumulonimbus clouds. See: precipitation.

shrinkage stope (Mining) An overhand stope from which broken ore is removed periodically to make room for more ore to be broken. Enough broken ore is allowed to remain to furnish a footing for miners working the back, or roof, of the stope.

shut-off nozzle (Fire Pr) A common type of fire-hose nozzle equipped with a valve for shutting down the hose line.

shut-off valve (Fire Pr) A valve that can be closed to stop the flow of a liquid.

siamese (Fire Pr) A hose fitting that joins two hose lines into one. Properly a two-into-one connector, with one male outlet and two female inlets, usu larger than the male, available with or without gates or clapper valves. Syn: (Brit) two-into-one breeching. Cf: wye.

sibling (Soc Sci) One of two or more children of the same parents.

sick leave (Fire Pr) Paid off-duty time allowed for illness.

side camp (For Serv) A small, temporary camp set up to accommodate a crew working in an isolated area. Also called: spike camp.

sideflash (Build) A spark between neighboring metallic objects or from metallic objects to ground or to a lightning protection system.

siding (Build) Material consisting of tongue-and-groove boards that fit together to make a continuous surface, used to cover the sides of buildings.

side loading (Techn) A force or load applied at an angle.

side-rolling ladder (Storage) A semifixed ladder, nonadjustable in length, supported by attachments to a guide rail, which is generally fastened to shelving, the plane of the ladder being also its plane of motion.

siderosis (Med) Lung disease resulting from inhalation of iron oxide or other metallic particles.

side slope (Transp) A slope that is at right angles to the longitudinal axis of a vehicle. See: slope.

side-step ladder (Safety) A ladder from which a climber getting off at the top must step sideways in order to reach the landing.

side two (Fire Pr) The rear of a burning structure.

sidewalk elevator (Build; Mat Hand) A freight elevator, the hatch opening of which is located either partially or wholly outside the building and which has no opening into the building at its upper terminal landing.

sidewalk fire chief (Fire Pr; Colloq) A spectator at a fire who comments on how the fire should be fought.

sidewall sprinkler (Fire Pr) A sprinkler head designed to be installed in pipes along a wall instead of across a ceiling. Syn: wall sprinkler. See: sprinkler system.

sign (Med) Observable evidence of injury or disease.

signal (Fire Pr) An alarm of fire from a street fire alarm box, transmitted in code, indicating the location of the alarm box.

signaling (Fire Pr) Communication between pumpers by hand, voice, or radio.
  NOTE: If hand signaling is used, the standard signals are <u>hold the water</u>: hands overhead, spread wide, palms toward the person being signaled; <u>charge lines</u> or <u>increase pressure</u> (if lines are already charged): arm outstretched to side and waved up and down; <u>lower pressure</u>: arm waved horizontally from across chest to the side (same signal back from the operator indicates inability to increase pressure); <u>shut down but do not break up</u>: arms overhead in inverted V; <u>break up</u>: same signal as above when already shut down. Horn or bell signals are used between the driver and tillerman on tractor-trailer ladder trucks: one ring or blast: <u>stop immediately</u>; two, <u>proceed forward</u>; three, <u>reverse</u>. Signals from the tillerman for forward and reverse are repeated by the driver before he takes action.

signaling system See: <u>alarm system</u>

significance (Statist) The probability that an apparent difference between two sets of data is a real difference and not due to chance.
  NOTE: Significance tests of carefully selected hypotheses allow one to judge the acceptability of the difference as real or chance. A difference is usu significant if one is 95% or more certain that it is real.

significant figures (Math) All meaningful digits in a number, including zero, except for zeros used solely for fixing the position of a decimal point.
  NOTE: The zeros in 0.0123 are not significant, whereas in 1.2300 they are. The number 0.0123 has three significant figures, 1.2300 has five.

significant wave height (Navig) The average height from crest to trough of the 1/3 highest waves.

signs (Safety) Warnings of hazard, temporarily or permanently affixed or placed, at locations where hazards exist.

silent alarm (Fire Pr) An inaudible alarm to which response is usu limited to one or two company units. Syn: <u>still alarm</u>.

silica gel (Chem; Techn) A regenerative absorbent consisting of amorphous silica made by reacting hydrochloric acid with sodium silicate.
  NOTE: Hard, glossy, quartz-like in appearance. Used in dehydrating, in drying, and as a catalyst carrier.

silicate (Chem) A compound consisting of silicon, oxygen, and one or more metals with or without hydrogen.
  NOTE: Silicates are the most common minerals and include sand, clay, and many rocks.

silicon (Chem; Symb: Si) A nonmetallic element, atomic number 14, atomic weight 28.086. Next to oxygen, the chief elementary constituent of the earth's crust.

silicon carbide (Techn) Bluish-black, hard crystals consisting of silicon and carbon, SiC. Syn: <u>carborundum</u>.
  NOTE: Used as an abrasive and refractory material.

silicone (Chem) One of a group of compounds made by molecular combination of the element silicon or certain of its compounds with organic chemicals.
  NOTE: Produced in a variety of forms, including silicone fluids, resins, and rubber. Silicones have special properties, such as water repellency, wide temperature resistance, high durability, and high dielectric strength.

silicosis (Med) A chronic disease of the lungs caused by the continued inhalation of silica dust.
  NOTE: The disease is characterized anatomically by generalized fibrous changes and the development of small lesions in the lungs, and clinically by shortness of breath, decreased chest expansion, lessened capacity for work, presence of fever, increased susceptibility to tuberculosis, and by characteristic X-ray findings.

sill (Mining) 1. A stratum of igneous rock between beds of other rock. 2. A footing timer of a timber set.

sill seal (Fire Pr) A compressible,

continuous pad used to seal spaces between the foundation and the sill of a structure.

silt (Soil Mech) A fine-grain soil having liquid limits and plasticity indexes that plot below the "A" line on the United Soil Classification System plasticity chart.

silver solder (Techn) A solder usu containing an appreciable proportion of cadminum.

silver sulfadiazine (Med; Pharm) A topical chemotherapeutic agent used for treating burn injuries.

silviculture (Forest) The branch of forestry concerned with the theory and practice of planting and growing stands of trees of selected species for given purposes.

simple hose lay (Fire Pr) A hose line consisting of a series of hose lengths without intervening lateral connections.

simplex (Commun) A mode of two-way radio communication in which messages are transmitted in one direction at a time, switching alternately from transmit to receive.

simulation (Math) The investigation of the behavior of a complex system through a mathematical description, possibly built up from descriptions of the parts rather than physical construction of a prototype. Also called: modeling.
NOTE: Physical events may be simulated by a computer. A distribution of possible end results of a random process can be generated by many repeated trials (Monte Carlo method).

simulator (Techn) Any device in which a physical or conceptual process or a mechanical, electronic, biological, or social system is represented in such a way that the phenomenon can be imitated and studied.

simultaneous ignition (For Serv) A backfiring or broadcast burning technique in which fuel is ignited at many points simultaneously with the sets so spaced that they support each other. See: area ignition.

sine wave (Phys) The curve traced by the projection on a uniform time scale of the end of a uniformly rotating arm or vector. Also called: sinusoidal wave.

singe (First Aid) To burn lightly and superficially; to remove hair or fuzz from a surface by flame.

single dwelling (Fire Pr) A one-family house, standing apart from other structures.

single ladder (Techn) A nonself-supporting portable ladder, nonadjustable in length, consisting of one section. Its size is designated by the overall length of the side rail.

single powder See: foam powder

single-source system (Fire Pr) A fire protection system, such as automatic sprinklers, that depends on a single water supply.

sinistral (Anat) Pertaining to the left side of the body.

sintering (Techn) The process of making coherent powder of earthy substances by heating but without melting.

sinusoid (Math) A mathematical curve whose ordinates are proportional to the sines of the abscissas, as defined by the equation $y = a \sin x$.

sinusoidal motion See: harmonic motion

siphon[1] (Hydraul) An inverted U-shaped pipe used to lift a liquid above its normal surface level and discharge it at some point below.
NOTE A siphon must be primed before it can operate. The maximum theoretical lift of a water siphon is 34 feet.

siphon[2] (Fire Pr) a. A hose connection used to suck up water from basements. b. To remove excess water from a building. To dewater.

siren (Fire Pr) An emergency warning device that generates a loud continuous or varying sound wave by mechanically interrupting an air stream with a perforated rotating disk or cylinder or by amplifying an electronic signal.

NOTE: Sirens are mounted on fire department vehicles and are used to warn traffic to make way when vehicles are responding to alarms. Sirens are also used at volunteer station houses to summon volunteer firefighters.

sit-up (Educ) An exercise in which an individual lies on his back, hands clasped behind the head, elevates his upper torso to a sitting position and then returns to the original position.

size up (Fire Pr) To make a mental evaluation of a fire situation, taking into account life and property hazards, extent of the fire, fire-fighting forces available, and other factors for the purpose of developing an effective plan of attack.

sizeup (Fire Pr) The act or result of evaluating a fire situation.

skelic index (Anat) The ratio of leg length to the length of the trunk.

skewness (Statist) Lack of symmetry in the shape of a frequency curve; a lopsidedness of the curve measured by the spread between the arithmetic mean and the median.

skid¹ (For Serv) To pull or drag bolts, logs, or trees by a winch, vehicle, or animal.
NOTE: Skidding is sometimes called **yarding**.

skid² (Storage) Low platforms of wood or metal, raised on side members or legs, used to support merchandise and other stock for ease of handling and to prevent damage from water accumulations on the floor. Skids have spaces underneath for forklift blades. Syn: **pallet**.

skid hose load (Fire Pr) A number of flakes of hose resting on top of a standard hose load and so arranged that the skid load can be pulled off as a unit at a fire and placed quickly into operation as a working line.

skill (Educ) 1. A high-order ability to perform a complex task smoothly and efficiently. 2. A human action having special requirements for speed, accuracy, or coordination.

skin¹ (Build) A veneer of glass, metal, stone, or other material suspended on metal frames attached to the outside of a building.

skin² (Anat) A thin layer of tissue covering the outer surface of the body; divided into two main strata: the superficial scarfskin and the deeper true skin. The scarfskin is called the **epithelium**, **epidermis**, or **cuticle**. The true skin is known as the **corium** or the **derma**. The skin tissues contain hair follicles, sweat glands, sebaceous (oil) glands, nerve endings, tactile corpuscles (touch sensors), blood vessels, and muscle fibers.
NOTE: The average adult has approx 21.5 square feet of skin area: 9% on the head and neck, 9% each on posterior chest and lower back, 9% each on anterior chest and abdomen, 1% genital area, 9% each upper extremity, and 18% on each lower extremity. The remaining 9% is distributed between these major areas. For each year under 10, subtract 1% for each leg and add to the head area.

skin dose (Dosim; Med) A special instance of radiation dose to the tissues, referring to the dose immediately on the surface of the skin.

skin effect (Electr) The tendency of alternating currents to flow near the surface of a conductor, thus being restricted to a small part of the total cross-sectional area.
NOTE: This effect increases the resistance and becomes more marked as the frequency rises.

skin friction (Fl Mech) The friction or shear stress between a surface and a moving fluid that retards the flow at the surface, creating a **boundary layer**. Also called: **drag**.

skin opening (Build) The space between a floor slab and the skin or curtain wall of a building.

skip (Mining) A large elevating or hauling container used to hoist ore out of a mine. The front is often shorter than the back to facilitate loading.
NOTE: Some skips have hinged bottoms for dumping; others are dumped

by tipping. Skips can also be used to raise and lower miners and materials.

sky condition (Meteor) The state of the sky expressed in tenths of cloud cover. Clear means less than 1/10 cloud cover; scattered clouds, 1/10 to 5/10; broken clouds, 5/10 to 9/10; overcast, >9/10. Ten sky conditions are defined and coded for synoptic charts:
  0   no clouds
  1   <1/10
  2   1/10
  3   2/10 to 3/10
  4   4/10 to 6/10
  5   7/10 to 8/10
  6   9/10
  7   >9/10 with openings
  8   10/10
  9   sky obscured by fog, dust, snow, etc.

skylight (Build) A structure with translucent or transparent panes set in a roof to admit daylight into the room below. Syn: lantern light (Brit).

slag (Metall) The dross of flux and impurities that rises to the surface of molten metal during melting and refining.
  NOTE: The fused byproducts of smelting consist primarily of calcium and aluminum silicates, formed by the action of low-melting material (flux) on impurities in the metal ore.

slaked lime Cf: quicklime

slash (For Serv) Debris, including branches, chips, and brush left after logging, thinning, or brush-cutting operations.

slash disposal (For Serv) Removal or treatment of slash to reduce fire hazard. Also called: brush disposal.

sleeper fire (For Serv) A fire that rekindles and breaks out anew sometime after a wildland fire has been apparently extinguished. Also called: holdover fire.

sleet (Meteor) Frozen or partly frozen rain. See: precipitation.

sleeve[1] (Fire Pr) A large-diameter hose, used for hooking up pumpers to hydrants. Syn: suction hose; hard suction; soft sleeve.

sleeve[2] (Mech Eng) Any tubular part that fits over another part.

slide escape (Fire Pr) An open, chute-like fire escape used for emergency evacuation of a building or an aircraft. Syn: chute.
  NOTE: Slides are generally open, whereas chutes are in the form of tubes or sleeves.

sliding damper A plate normally installed perpendicular to the flow of gas through a duct and arranged to slide across it to regulate the flow.

sliding pole (Fire Pr) A vertical metal pole, usu of brass, used by firefighters to descend from an upstairs dormitory to the apparatus floor when responding to an alarm.

slip (Transp) An area of open water between structures for berthing or mooring vessels. Cf: dock.

slippage See: pump slippage

slippery water (Fire Pr) Water containing small quantities of polyethylene oxide. Also called: Rapid Water (trade name).
  NOTE: Friction in hose lines is reduced considerably by the additive, so that a slippery water stream has greater range than an ordinary water stream from the same hose at the same pressure.

slope[1] (Soil Mech) The angle with the horizontal at which a particular earth material will stand indefinitely without movement. Syn: angle of repose; angle of slip. See: grade.

slope[2] (Civ Eng) A natural or artificial incline to the plane of the horizon, such as a hillside, terrace, or ramp.
  NOTE: Usu expressed as a ratio of the vertical rise to the horizontal distance traveled, or if expressed as a percent, the percent slope or percent grade. A 1-foot rise in a distance of 5 feet is a 20% slope.

slopover[1] = breakover

slopover[2] (Fire Pr) Overflow from a tank caused by frothing of burning oil when foam or water is applied to the hot oil. See also: bleve; boilover.

slot velocity (Techn) Air velocity through the openings in a slot-type hood.

sloughing (Med) Separation and falling away of dead tissue from living tissue.

slow burning (Comb) A characterization of burning rate; meaningful when referred to a particular test method.

slow-burning construction See: mill-type construction.

slow combustion See: smoldering

sludge[1] (Techn) Any muddy or slushy mass, such as mud from a drill hole in boring, muddy sediment in a steam boiler, or precipitated solid matter resulting from sewage treatment processes.

sludge[2] (Techn) An accumulation of heavy impurities, including water, in the bottom of an internal combustion engine crankcase or a petroleum product storage tank. Syn: bottom settlings.

slug (Nucl Phys) A fuel element for a nuclear reactor, a piece of fissionable material. The slugs in large reactors consist of uranium metal coated with aluminum to prevent corrosion.

slurry (Techn) 1. A semifluid mixture of insoluble matter with water. 2. Any semifluid suspension.

small line (Fire Pr) A 1-in. or 1-1/2-in. hose line.

small lot storage (Storage) A quantity of supplies, comprising less than one stack.

smell (Physiol; Psych) A chemical sense of the olfactory organs by which the presence of small quantities of substances in the air are detected. Syn: olfaction. See: sense organs.
NOTE: Odors have been classified as fruity, floral, resinous, spicy, foul, and burnt. Sensitivity to odors tends to decrease rapidly with continued exposure.

Smithels separator (Comb) A device for separating the inner and outer cones of a Bunsen flame.

smog (Meteor) A fog contaminated with city, industrial, and automobile pollution, containing soot, other smoke products, dust particles, excess CO_2, SO_2, hydrocarbons, nitrogen oxides, and complex eye irritants, such as aldehydes. The term is a blend of smoke and fog. See also: aerosol; fog; air pollution.
NOTE: The haze results from the effect of sunlight on certain pollutants in the air.

smoke (Comb) A visible, nonluminous, airborne suspension of particles, originating from combustion or sublimation. Combustion smoke is the visible fraction of airborne combustion products. The invisible fraction of the combustion products, which invariably accompanies smoke, consists of various gases and vapors. Smoke consists largely of soot particles ranging from 10 to 0.001 micron in size and is produced by incomplete combustion of carbonaceous materials such as wood, coal, or oil. The carbon particles readily absorb the gases and vapors present in the products of combustion and thus become carriers of substances that are often toxic or irritant. Smoke generally contains other kinds of dry particles such as ash, as well as liquid or tarry droplets.
NOTE: The composition and quantity of smoke depend on the material that is burning, the temperature of the combustion process, the availability of oxygen, and in special cases on pressure as well. Smoke color and odor may indicate the nature of the fuel. Dense black smoke may indicate large quantities of carbon particles, such as from burning rubber, oil, tar, etc.; brown smoke may indicate nitrous fumes; yellowish-gray smoke may be a precursor to a back draft. Paper and cloth yield gray smoke. Almost all smoke tends to appear white in cold weather. Smoke from rubber, feathers, and many other materials have strong characteristic odors.

smoke alarm (Env Prot; Techn) A sensing instrument that continuously

measures and records the density of smoke from the attenuation of a light beam passing through the smoke. When the smoke density exceeds some preset value, an alarm goes off. Syn: smoke eye.
NOTE: More properly a smoke density monitor than an alarm.

smoke candle (For Serv) A pyrotechnic device that produces smoke of a uniform color at standard rate, simulating the appearance of a small fire.

smoke control door (Build) A door capable of restricting the spread of smoke and reducing drafts that might otherwise spread fire rapidly. A smoke control door is not completely smoketight, and usu is not the equivalent of a fire door.

smoke damper (Build) A shutter arranged to seal off air flow automatically through a part of an air duct system and thus block the passage of smoke.

smoke density (Comb) The proportion of solid matter contained in smoke, measured on various arbitrary scales, often by relating the grayness of smoke to an established standard.
NOTE: The optical density of smoke is defined as the logarithm to the base 10 of the incident light intensity divided by the intensity of the light transmitted through the smoke.

smoke detector (Fire Pr) A device, usu of the ionization type or photoelectric, that triggers an alarm when a sufficient concentration of smoke upsets a balanced electronic circuit or interrupts or otherwise affects a light beam. Cf: ionization detector.
NOTE: According to the Southern Building Code Congress, a smoke detector is a device that senses both visible and invisible products of combustion.

smoke development rating (Build; Testing) A relative numerical classification of the quantity of smoke produced by a building material.

smoke eater (Fire Pr) A firefighter known for his ability to work in heavy smoke and heat.

smoke ejector (Fire Pr) A power-driven fan used to remove smoke from burning buildings or to blow fresh air into a building to expel smoke and heat. Syn: smoke extractor; forced air blower.

smoke explosion (Fire Pr) An explosive combustion of hot smoke and gas. Also called: hot air explosion (erroneous). See: back draft.
NOTE: A confined fire can deplete the oxygen supply and generate considerable quantities of combustible gases. When air is admitted, the gases can burst explosively into flame.

smoke extractor = smoke ejector

smoke eye (Techn) A device consisting of a light source and a photoelectric cell that measures the degree to which smoke in a flue gas obscures light. Also called: smoke alarm.

smoke haze (For Serv) A haze caused by smoke alone.

smoke here = smoke point

smoke inhalation (Fire Pr; Med) Breathing in of smoke particles and combustion fumes.
NOTE: Medically, smoke is viewed as consisting of a particulate fraction and a gaseous fraction. Most of the particulate fraction is trapped on the mucous membranes of the mouth, nose, and upper airways of the respiratory tract. The gaseous fraction penetrates deeper into the lungs, where irritants can cause damage or be absorbed and lead to more extensive injury in the body.

smoke jumper (For Serv) A firefighter who is transported to a fire by air and is dropped by parachute. See also: helijumper.

smoke pall (Fire Pr; For Serv) An extensive, thick blanket of smoke spreading more or less horizontally from a fire.

smoke point (For Serv) A posted location where smoking is permitted in an area otherwise closed to smoking. Also called: fag station; smoke here.

smoke poisoning (Fire Pr; Med) A debilitated condition caused by exposure to gases and smoke from fire.

smokeproof stairway (Build) A stairway isolated from the rest of a building by self-closing doors. Some stairways have positive pressure ventilation to prevent smoke from seeping in and to remove any smoke that may enter. Also called: **smoke tower**; **fire tower** (erroneous); **smokeproof tower stairway**.

smokeproof tower (Build) 1. A continuous fire-resistive enclosure protecting a stairway from fire or smoke in the building served, with communication between the building and the tower by means of open balconies. Also called: **smoke tower**; erroneously, **fire tower**. 2. A similar structure for the same purpose, jutting out from or built into an outer wall of a building but adequately separated from the building by appropriate construction and fire-resistant doors. 3. = (Brit) **enclosed staircase**.

smoker fire (For Serv) A fire caused by a smoker's match or burning tobacco.

smoke scare (Fire Pr) Panic caused by the appearance of smoke or steam or other vapor mistaken for smoke.

smoke shaft (Build) An unobstructed shaft, with a fan at the top, extending the full height of a building, and with openings at each level. In case of fire, dampers open on the fire floor and the fan vents the combustion products.

smokestop door See: **smoke control door**

smoke toxicology (Comput) The study of the harmful effects on biological systems of chemical and physical agents in smoke. Cf: **combustion toxicology**.

smoke vent (Build) An opening, either normally open or one that opens automatically in case of fire to discharge smoke from a burning building. See: **vent**.

Smokey Bear (For Serv) A US and Canadian Forest Service symbol used to publicize forest-fire prevention programs.
NOTE: The original Smokey was found in 1950 as a cub, one of the few animals surviving an extensive forest-fire in New Mexico. Prior to that he was simply a cartoon character. He was flown to Washington DC and was officially designated as the living symbol of forest-fire prevention by the US Congress. Smokey died in 1975 at the age of 26. His successor now lives at the National Zoological Park in Washington, DC, where he has his own mailing zip code: 20252. In Mexico he is called Simon El Oso.

smolder (Comb) To burn without flame.

smoldering (Comb) Combusting without flame, usu with incandescence and moderate smoke. Cf: **glowing combustion**. See also: **creeping fire**.
NOTE: Smoldering can be initiated by small sources of ignition, esp in dust, and may persist for an extended period of time, after which a flame may kindle.

smooth-bore tip (Fire Pr) A straight-stream nozzle tip.

smother (Fire Pr) To extinguish a fire by blocking the oxygen supply or limiting it to a point below that required for combustion.
NOTE: Certain substances, such as cellulose nitrate, cannot be smothered because they contain sufficient oxygen to sustain self-combustion.

smothering blanket (Fire Pr; Brit) = **fire blanket**

smoulder = **smolder**

SMRE = **Safety in Mines Research Establishment** (Brit)

snag (For Serv) A dead tree from which the leaves and at least some of the branches have fallen. If less than 20 feet tall, it is called a **stub**.

snap coupling (Fire Pr) A type of coupling that clamps together instead of being held together by screw threads.

snorkel (Fire Pr) 1. A tube used to supply air to swimmers or engines underwater. 2. An aerial platform apparatus.

snotter becket (Fire Pr) A short piece of line with an eye spliced at one end, used on fireboats to lift or secure hose lines. Cf: hose becket.

snow[1] (Meteor) Precipitation in the form of ice crystals that have frozen from the vapor phase in the atmosphere into hexagonal crystals. See precipitation.
NOTE: Snow is variously described as wet, dry, or powdery. When partly melted it is called slush.

snow[2] (Chem) A substance that has the appearance of snow, such as solidified carbon dioxide.

snuff 1. To extinguish by pinching, such as a candlewick. To smother. 2. The charred portion of a candlewick.

soaking pit (Metall) A heated upright chamber used in steel manufacturing to hold ingots with still molten interiors until their solidification is complete.

soap (Chem) Ordinarily a metal salt of a fatty acid made by alkaline hydrolysis (saponification) of a triglyceride (fat).
NOTE: Soaps are used primarily as cleansing agents but are also used in lubricants, rust inhibitors, and jelled fuels. Synthetic soaps are called detergents or syndets.

social science (Soc Sci) A branch of science concerned with human society, its institutions, and relationships between and among individuals, including economics, history, political science, psychology, sociology, and the like.

Society of Fire Protection Engineers (Abb: SFPE) A professional organization of fire protection engineers, founded in 1950 and affiliated with the National Fire Protection Association. This formal relationship was discontinued in 1971.
NOTE: The objectives of the Society, in part, are "...to promote the art and science of fire protection engineering and its allied fields, to maintain a high professional standing among its members and to foster fire protection engineering education."

sociogram (Psych) A graphic representation of the interactions between members of a group.

soda-acid extinguisher (Fire Pr) 1. A common, hand-operated fire extinguisher charged with a water solution of bicarbonate of soda and a bottle of sulfuric acid. To use, the extinguisher is inverted, causing the acid to mix with the soda solution and produce carbon dioxide gas. The gas pressurizes the extinguisher and the contents are ejected through a short hose and nozzle. 2. = (Brit) tip-up extinguisher.

sodium (Chem; Symb: Na) An element belonging to the group of alkali metals, atomic number 11, at. wt 22.98977, mp 97.83°C, bp 882.9°C. Used as a heat exchange medium because of its excellent thermal conductivity.
NOTE: Reacts vigorously with water, yielding hydrogen, which may ignite from the heat of the reaction.

sodium bicarbonate (Chem) A dry chemical, $NaHCO_3$, used as a secondary extinguishing agent in combination with foam.

sodium carboxymethylcellulose (Chem; Fire Pr) An additive used to increase the viscosity of water and thus to reduce runoff in fire-fighting applications. See: viscous water.

sodium hydroxide (Chem) A white, solid, soluble compound, NaOH, mp 318°C. A strong alkali used extensively in the manufacture of soap, rayon, paper, rubber, textiles, and other products. Syn: caustic soda; soda lye; or broadly, lye.
NOTE: Sodium hydroxide is an extremely corrosive substance. It can generate sufficient heat spontaneously to ignite nearby combustibles.

soffit (Build) The underside of a structural overhang, such as an eave.

soft sleeve (Fire Pr) A short, large-diameter hose used to supply water to pumpers from hydrants. Compared with hard-suction hose, soft sleeves are easier to handle and hydrant connections can be made quicker.
NOTE: The term soft suction is

frequently used erroneously for soft sleeve. Soft sleeves are used only when water is supplied under pressure from a hydrant. If a soft sleeve were used to draft water, atmospheric pressure would cause it to collapse.

soft suction = soft sleeve

soil classification system (Geol; Soil Mech) Grouping of soils by characteristics and properties.

soil values (Transp) Parameters such as sinkage, slip, rolling resistance, and thrust used to describe soil properties in equations for predicting performance of vehicles on soil.

sol (Chem) A colloidal suspension of a solid in a liquid. If the continuous phase is water, the system is called a hydrosol; if air, an aerosol. See also: colloid; slurry.

solar constant See: solar radiation

solar radiation (Meteor) Sunlight or sunshine. The energy, ranging from long infrared to short ultraviolet, reaching the earth from the sun. The atmosphere strongly absorbs the energy at the extremes of this range.
 NOTE: The solar constant is defined as the quantity of energy that falls in a unit time on a unit area normal to the sun's rays at the edge of the earth's atmosphere. The mean value is $1.36 \cdot 10^6$ ergs/cm^2-sec, 1.938 cal/cm^2-min, or 1.8 hp/m^2. A third to a half of this energy is reflected back into space or scattered by the atmosphere.

solder (Metall) An alloy that is fused to join metal surfaces together.
 NOTE: The most commonly used solder is one containing lead and tin. Silver solder may contain cadmium. Zinc chloride and fluorides are commonly used as fluxes to clean the surfaces to be joined. Soldering takes place at temperatures up to 800°F; above this temperature the process is called brazing. See also: joint; welding.

solder-type head (Fire Pr) A fusible-link sprinkler system valve and spray device. See: sprinkler system.

solenoid (Electr) A multiturn coil of wire wound in one or more uniform layers on a hollow cylindrical form.

solid¹ (Phys) One of the four states of matter (gas, liquid, solid, and plasma) characterized by rigidity, resistance to deformation, and fixed shape.

solid² (Geom) A space bounded by a continuous surface in three dimensions.

solid angle (Geom) An angle subtended by a solid cone.

solid-state (Phys) A branch of physics dealing with solids, such as metals or crystals, and their properties with regard to heat, light, electricity, atomic arrangement, electronic movement, etc.

solid-state laser (Phys) A type of laser in which the active substance is a solid crystal such as ruby or glass. See: laser.

solid stream (Fire Pr) A jet of water with almost parallel sides used to obtain the longest range possible at the given pressure and nozzle size. Solid streams diverge and break up over distance because of friction with the air that causes turbulence at the surface of the water jet. Also called: straight stream.

solid waste (Env Prot) Useless, unwanted, or discarded material with insufficient liquid content to be free-flowing.
 NOTE: Among the more common wastes are Agricultural: resulting from the rearing and slaughtering of animals and the processing of animal products and orchard and field crops. Commercial: solid waste generated by stores, offices, and other activities that do not actually turn out a product. Industrial: resulting from industrial processes and manufacturing. Institutional: originating from educational, health care, and research facilities. Municipal: normally, residential and commercial solid waste generated within a community. Pesticide: the residue resulting from the manufacturing, handling, or use of chemicals for killing plant and animal pests. Residential: originating in a

residential environment, sometimes called <u>domestic solid waste</u>.

<u>sollar</u> (Mining) A sloping roof of planks on stulls used to protect miners working below from falling material.

<u>solubility</u> (Chem) The degree to which a solid, liquid, or gas dissolves in a solvent, which can be a liquid, solid, or gas, but most commonly is water.

<u>soluble</u> (Chem) Capable of being dissolved or of passing into solution.

<u>solute</u> (Chem) The substance dissolved in a <u>solvent</u>.

<u>solution</u> (Chem; Phys) A homogeneous mixture in which the components are uniformly dispersed. See also: <u>saturation</u>; <u>solubility</u>.
NOTE: All solutions are composed of a solvent (water or other fluid) and the substance dissolved, called the solute. Air is a solution of oxygen and nitrogen. Salts, acids, and bases (electrolytes) dissociate in solutions into positive and negative ions, and the solutions conduct electricity.

<u>solvent</u> (Chem) A material, generally liquid, capable of dissolving another material (solute) to form a homogenous, single-phase solution. The solvent is usu present in larger quantities than the solute.
NOTE: A <u>chemical solvent</u> is one that reacts chemically with a solute. A <u>physical solvent</u> is one that does not react chemically. A <u>polar solvent</u>, such as water, consists of molecules that exert local electrical forces, causing acids, bases, and salts to dissociate into ions, forming electrically conducting solutions called electrolytes. A <u>nonpolar solvent</u>, usu a hydrocarbon, consists of molecules that do not have dipole moments and is therefore unaffected by an electrical field.

<u>solvent cement joint</u> (Build) A joint made in thermoplastic piping by the use of a solvent or solvent cement, which forms a continuous bond between the mating parts.

<u>somatic</u> (Med) Pertaining to the body, as opposed to mental or psychological.

<u>somatic radiation effects</u> (Dosim) Effects of radiation limited to the exposed individual, as distinguished from genetic effects, which can be transmitted to offspring.
NOTE: Large radiation doses can be fatal. Smaller doses may make the individual noticeably ill, may merely produce temporary changes in blood-cell levels detectable only in the laboratory, or may produce no detectable effects whatever. Also called: <u>physiological effects of radiation</u>.

<u>sone</u> (Acoust) A unit of loudness. A 1000 Hz tone 40 dB above a listener's threshold produces a loudness of 1 sone. See: <u>phon</u>.
NOTE: The loudness of any sound that is judged by a listener to be n times that of the 1-sone tone is n sones.

<u>soot</u> (Comb) Carbonaceous particles resulting from incomplete combustion, usu carrying tarry products and found deposited on exposed surfaces. When suspended in the air, soot is a major constituent of <u>smoke</u>. See also: <u>smog</u>; <u>air pollution</u>.

<u>sorbent</u> (Techn) A material that removes toxic gases and vapors from air inhaled through a canister or cartridge.

<u>Soret effect</u> = <u>inverse thermal diffusion</u>. See: <u>transport processes</u>.

<u>sorption</u> (Chem; Phys) The general term for the process of one material being taken up by another, including both <u>absorption</u> and <u>adsorption</u>.
NOTE: In adsorption, molecules of a gas or liquid stick to the surface of an adsorbing material. In the case of a gas mask, the process is selective: molecules of toxic gas are adsorbed by the activated charcoal in the canister, while air molecules pass through. Absorption, on the other hand, is largely a capillary action, such as that of a sponge when it soaks up a liquid. Capillarity plays a part in both absorption and adsorption. Sorption is the general term for both phenomena.

sound¹ (Acoust) An alternation of pressure or particle displacement propagating in an elastic medium.
  NOTE: Sound waves from 1 to 16 Hz are called infrasonic; from 16 to 20,000 Hz, sonic; above 20,000 Hz, ultrasonic.

sound² (Psych) An auditory sensation perceived by the ear. See: hearing.

sound absorption (Acoust) The change of sound energy into some other form, usu heat, in passing through a medium or on striking a surface.

sound analyzer (Acoust; Audiom) A device for measuring the band-pressure level or pressure-spectrum level of a sound as a function of frequency.

sound intensity level (Acoust) The sound-energy flux density level, measured in decibels of sound. In decibels of sound, intensity level equals 10 times the logarithm to the base 10 of the ratio of the intensity of this sound to the reference intensity. See: decibel.

sound level meter (Audiom) An instrument for measuring sound pressure levels in decibels referenced to 0.0002 microbar. Readings can also be made in specific octave bands, usu beginning at 75 Hz and continuing through 10,000 Hz.

source (Nucl Phys) Any substance that emits radiation. Usually a quantity of radioactive material conveniently packaged for scientific or industrial use.

Southern Standard Building Code = Standard Building Code

space charge (Electr) A cloud of free electrons or ions between electrodes or near charged surfaces.

space heater An electrical, gas, or oil heat-generating device used to heat a single room or other limited area. Sometimes called a room heater.

spaghetti¹ (Electr) Flexible hollow sleeves that slip over wires to insulate them.

spaghetti² (Fire Pr) A snarl of charged hose lines from several fire companies at the scene of a fire.

spalling (Techn) Destruction of a surface by frost, heat, corrosion, or mechanical causes.
  NOTE: Concrete exposed to intense heat is capable of spalling explosively.

span (Build) The horizontal distance between the supports of a beam.

spandrel (Build) 1. The triangular space between the curve of an arch and an enclosing right angle. 2. The triangular space beneath the string of a stair. 3. Loosely, the area between the top of one window and the bottom of a window directly above.

spanner wrench (Fire Pr) A tool used to tighten or loosen hose couplings. Sometimes also used as a lever for prying and other tasks.

span of control (Man Sci) The number of subordinates directly supervised by one person.

spar (Fire Pr) 1. The side rail or beam of a ladder. 2. = (Brit) string.

spark¹ (Comb) A small, incandescent particle, usu related to burning wood, but also including glowing particles produced in a metal grinding operation.
  NOTE: A mechanical spark may be produced by impact between two hard objects.

spark² (Electr) A localized, instantaneous electrical discharge, accompanied by heat and light, occurring between bodies or points at different electrical potentials.
  NOTE: Electrical sparks are commonly associated with live electrical circuits, motor brushes, and the like, but they can also be produced by static electricity.

spark³ (Fire Pr) A fire buff. An individual interested in fires, fire equipment, and firefighting activity. Sometimes called: ringer.

spark arrester (Techn) A screen installed in a chimney or flue to stop sparks and burning fragments from rising up the chimney. Also called: spark guard.

spark chamber (Phys) An instrument for detecting and measuring the paths of elementary particles.
  NOTE: The chamber, analogous to the cloud chamber and bubble chamber, consists of numerous electrically charged metal plates mounted in a parallel array, the spaces between the plates being filled with an inert gas. Any ionizing event causes sparks to jump between the plates along the radiation path through the chamber.

spark guard = spark arrester

sparkproof (Techn) Incapable of producing sparks. Characterizing special copper alloy tools that do not give off sparks when striking hard objects.
  NOTE: Such tools are used in potentially flammable environments.

spark suppresser (Electr) A device, usu consisting of capacitance and inductance, used to protect electronic and communications equipment from line surges and noise produced by sparks.

sparky (Fire Pr) 1. A firefighter who enjoys his profession and fights fire agressively and with daring. 2. A fire department mascot or a special friend of the fire service. Syn: buff.

Sparky (Fire Pr) The name of the Dalmatian fire dog used as a fire prevention symbol by the National Fire Protection Association.

Sparky's Fire Department (Fire Pr) A fire prevention program sponsored by the National Fire Protection Association for school-aged children.

spasm (Med) Involuntary tightening or contraction of muscles. See: laryngeal spasm.

special attendance (Fire Pr; Brit) A response to an incident other than a fire.

special call (Fire Pr) A call for additional equipment or special units. Generally because of unusual circumstances or for an alarm greater than provided for on the running card.

special hazard See: hazard

special order (Fire Pr) An official order issued by a fire department for cases not covered by the official rule books. Special orders are usu temporary, in contrast to general orders, which are more or less permanent.

special-purpose ladder A portable ladder that is a modification or a combination of design or construction features that adapt it to special or specific uses.

special services (Fire Pr; Brit) Responses by fire department units to emergencies other than fire; e.g., road accidents, cliff rescues, etc.

specialized tissue (Anat) Differentiated tissue that performs a specific body function, such as muscles, nerves, connective tissues, bone, teeth, blood, eye, liver, and skin.

species (Chem; Phys) A particular kind of atomic nucleus, atom, molecule, or ion; a nuclide.

specific (Med) A drug or treatment that has a special action in the cure of a given disease or disorder.

specific activity (Phys) The radioactivity of a radioisotope of an element per unit weight of the element in a sample. The activity per unit mass of a pure radionuclide. The activity per unit weight of any sample of radioactive material.

specific gravity (Phys; Chem) The ratio of the mass of a sample to the mass of an equal volume of a standard material at the same temperature. The common standard for liquids and solids is water, and air or hydrogen at the same temperature and pressure for gases. Syn: relative density.

specific heat (Phys; Chem; Symb: c) The ratio of the heat required to raise one unit of mass of a substance one unit in temperature, compared to that required to raise an equal mass of water the same amount.
  NOTE: The specific heat of water is

1.0; ice, 0.49; dry air, 0.24; dry soil, 0.2. A gas has two specific heats: one at constant volume and one at constant pressure.

specific volume (Phys; Chem) The volume of a substance per unit mass; the reciprocal of density.

spectral emissivity (Spectr) The spectral emissivity of a radiator at any given wavelength is the ratio of its radiant flux density to that of a blackbody at the same temperature and under similar circumstances. Except for luminescent materials, the emissivity can never be greater than one. See: emissivity.

spectral sensitivity (Opt; Spectr) The response of a detector to one or more spectral bands or to a given wavelength of radiant energy.

spectrograph (Spectr) An instrument used to record a spectrum.

spectrophotometer (Spectr) An instrument for measuring transmission or emission at a particular wavelength.

spectroscopy (Abb: Spectr) The analysis of electromagnetic radiation spectra for the purpose of studying the nature of phenomena associated with radiating sources.
NOTE: Spectra are measured with instruments called spectroscopes, spectrographs, spectrometers, and spectrophotometers that separate radiant energy into arrays of colors, wavelengths, or frequencies by diffraction or refraction.

spectrum[1] (Opt) Electromagnetic radiation separated into its constituents by a prism, grating, or other means and displayed as an array of frequencies or wavelengths.

spectrum[2] (Spectr) A visual display, a photographic record, or a plot of the distribution of the intensity of a given type of radiation as a function of its wave length, energy, frequency, momentum, mass, or any related quantity.

speculation building (Build) A building constructed for investment without clear foreknowledge of occupancy. Upon completion of the shell, floors or parts of floors are leased and then partitioned and finished according to the requirements of the individual renters.
NOTE: Sometimes floorspace is sold, in which case the building becomes a condominium.

speech interference level (Audiom; Abb: SIL) The speech interference level of a noise is the average, in decibels, of the sound pressure levels of the noise in the three octave bands of frequency 600-1200, 1200-2400, and 2400-4800 Hz.

speech perception test (Audiom; Psych) A measurement of hearing acuity by the administration of a carefully controlled list of words. The identification of correct responses is evaluated in terms of norms established by the average performance of normal listeners.

speech reading Understanding spoken words by observing lip movements. Also called: lip reading; visual hearing.

speed (Phys) The magnitude of the velocity vector, expressed as units of distance divided by time. See: motion.

speed of attack (For Serv) The time elapsed from the start of a fire to the time of arrival of the first fire-suppression forces.

sphygmomanometer (Med) An apparatus for measuring blood pressure.

spike[1] (Fire Pr) The pointed metal pin protruding from the lower end of a tormentor pole, used to fix the pole in position during ladder climbing operations.

spike[2] (Graph) A sharp peak on a plotted curve.

spike camp = side camp

spiling (Mining) Shingled roof sections advanced into mine drifts that are being driven in loose or running ground. Also called: forepoling.

spill[1] (Safety) The accidental release of radioactive or other hazardous material, such as combustible or flammable liquid.

spill[2] (Fire Pr) a. To string out a line of hose from a moving fire truck. b. To intentionally drop a hose load on the ground.

spill fire (Fire Pr) Combustion of flammable liquid spilled on the ground. See also: ground spill

spillage (Safety) 1. Accidental dumping of material on the floor, ground, or water surface. 2. Material accidentally spilled.

spinneret (Techn) A small disk with very small holes through which viscose or other spinning solution is forced, emerging as thin filaments into a coagulating bath.

spit (Mining) To light a fuse.

splice (Techn) 1. a. To join, for example, by interweaving the strands of two ropes to form a single continuous line. b. To double back and interweave the strands of a rope end to prevent the end from raveling, such as a crown splice. c. To interweave the end strands of a rope a short distance from the end to form an eye splice. 2. To make such a joint. See: rope; line; thimble.

splint (First Aid) Any device used to immobolize a body part.

spontaneous combustion (Comb) Combustion resulting from spontaneous heating and spontaneous ignition. A fire started by spontaneous ignition.

spontaneous heating (Comb) Natural, chemical, or biological heating of material, for example, by atmospheric oxidation and bacterial fermentation. A precursor of spontaneous ignition. See: heat.

spontaneous ignition (Comb) Initiation of combustion of a material by an internal chemical or biological reaction that has produced sufficient heat to ignite the material. Cf: autoignition.

spontaneous ignition temperature = autoignition temperature

spot burning (For Serv) Slash burning confined to large accumulations of slash. A modified form of broadcast burning.

spot fire (For Serv) 1. A fire ignited at a distance from the edge of the main fire by sparks and embers carried by convection currents or wind. 2. = (Brit) secondary fire.

spot welding (Techn) One form of electrical resistance welding in which the current and pressure are restricted to the spots of metal surfaces directly in contact.

spotting[1] (Fire Pr) The advantageous placement of fire-fighting equipment and forces for an effective attack on a fire.

spotting[2] (For Serv) The ignition of new fires at a distance from the main body of a forest fire by flying firebrands. Spotting may also occur when burning materials roll down steep slopes.

spotting fire (For Serv) A fire that produces flying sparks and fire brands that start new fires at some distance from the main fire edge.

sprain (First Aid; Med) An injury in which the ligaments are torn, accompanied by pain, discoloration, and swelling.

spray (Fire Pr) Water in finely divided droplets produced by special nozzles used to absorb heat, act as a heat shield for firefighters, and to generate steam to smother burning materials. Spray is also used to wet down exposures to protect them from radiant heat. Often called: fog.

spray booth (Techn) A power-ventilated exhaust structure provided to enclose or accommodate a spraying operation to control the escape of spray, vapor, and residue.

spray-finishing (Techn) Application of materials in dispersed form on surfaces to be coated, painted, treated, or cleaned. Such methods may involve automatic, manual, or electrostatic deposition but not metal spraying or metallizing, dipping, flow coating, roller coating, tumbling, centrifuging, or spray washing and degreasing as conducted in self-contained washing and degreasing machines or systems.

spraying area (Techn) Any area in which dangerous quantities of flammable vapors or mists, or combustible residues, dusts, or deposits may be present because of spraying processes.

spray nozzle (Fire Pr) A nozzle that is designed to break up a water stream into fine droplets. Sometimes called a **fog nozzle**.

spray sprinkler (Fire Pr) A standard automatic sprinkler head installed across ceilings and designed to spray water horizontally to absorb heat rising from a fire. See: **sprinkler system**.

spread (Fire Pr) To propagate, travel, or grow in extent.

spreader[1] (Fire Pr) A wedge-type emergency tool, usu operated by hydraulic pressure, used to force structural members apart and to lift or move heavy objects in rescue work.

spreader[2] (Mining) A board fastened underneath a cap and butted against the posts of a timber set. Also called: **scab**.

spreader tip (Fire Pr) A special nozzle tip that spreads water into a fan-shaped discharge.

spring weight (Transp) The total weight of all of the vehicle components that are supported by the vehicle spring system. This includes such major components as frame, body, power plant, transmission, clutch, cargo, etc. It does not include such items as wheels, axles, road wheels, etc.

sprinkler alarm (Fire Pr) An audible local alarm that sounds when one or more automatic sprinkler heads are actuated. See: **sprinkler system**.
 NOTE: The alarm may also be transmitted to a local fire station.

sprinkler block (Fire Pr) A salvage device used to stop the flow of water from a sprinkler head. See: **sprinkler wedge**.

sprinkler connection (Fire Pr) A connection used by a fire department to boost the pressure of a sprinkler system. Often a sprinkler siamese with two pumper inlet connections.

sprinklered (Fire Pr) Equipped with an automatic sprinkler system.

sprinkler head (Fire Pr) A sprayer consisting of a threaded nipple that connects the head to a water pipe, a fusible link (or other releasing mechanism) held in place by a yoke (called a strut), and a deflector that breaks up the water jet into fine droplets.

sprinkler spacing (Fire Pr) The distribution of sprinkler heads to provide adequate protection for the given hazard.

sprinkler stopper = **sprinkler tongs**

sprinkler system (Build; Fire Pr) A grid system of water pipes and spaced discharge heads installed in a building to control and extinguish fires.
 NOTE: There are four basic types of sprinkler systems: Wet, dry, deluge, and preaction. **Wet pipe**: All piping is filled with pressurized water, which is released when a fusible link melts in a sprinkler head. **Dry pipe**: All piping contains pressurized air. When a sprinkler opens, air pressure is released, and the dry pipe valve opens to admit water into the system and to any open sprinkler head. **Deluge**: All sprinkler heads are open and water is held back at the main (deluge) valve. When this valve is actuated, water is delivered to all heads simultaneously. **Preaction**: The main water control valve is opened by an actuating device which permits water to flow to individual sprinkler heads. Sprinkler systems include a suitable water supply, such as a gravity tank, fire pump, reservoir, pressure tank, or connections to city mains. In addition there is usu a controlling valve and devices for signalling alarms when the system is actuated.

sprinkler tongs (Fire Pr) A tool used to stop the flow of water from sprinkler heads. Syn: **sprinkler stopper**.

sprinkler wedge (Fire Pr) A wedge-shaped wooden block used to stop the flow of water from a sprinkler head. See: **sprinkler block**.

spud See: grouser

spur (Fire Pr) A metal spike at the lower end of a ladder pole, used to hold the pole in position on the ground. Syn: spear; spike.

spurt test See: overload test

squad (Fire Pr) A fire department team, detail, or company that is specially trained and equipped to perform special tasks or carry out particular assignments, such as rescue, or first aid.

squad fire apparatus (Fire Pr) A fire department vehicle resembling a pumper but designed primarily to carry personnel. It is often the first-in fire apparatus. Syn: squad truck.
  NOTE: The vehicle is usu equipped with a tank, booster hose, and ladders. Depending on the area of use and purpose, a squad truck may also carry medical first-aid equipment, breathing apparatus, and a variety of rescue tools and other equipment.

squadrol (Police) A police department squad patrol vehicle.

squall (Meteor) A sudden onset of strong winds or gusts, sometimes accompanied by rain or snow showers. Squalls are usu of short duration, with wind-gust speeds 30 to as much as 100 mph. See: wind.

squall line (Meteor) A long, narrow line of violent thunder, rain, or snow storms developing from strong updrafts which produce cumulus clouds and severe atmospheric turbulence. A typical squall line is formed by a cold front moving under and lifting warm moist air ahead of it.

square knot (Techn) A common knot used to join the ends of two ropes. Called a granny knot when improperly tied. See: knot.

square of the diameter (Fire Pr) A factor obtained by squaring the diameter of a nozzle and used to estimate the flow from any circular orifice by comparison with the flow from a 1-in. nozzle tip.

square set (Mining) A supporting timber structure in a mine consisting of four posts, four caps, and four girts.

squeegee (Fire Pr) A rubber blade mounted on a handle, used to remove water from floors.

squib (Comb) 1. Any of various small-size pyrotechnic or explosive devices. 2. A small explosive device, similar in appearance to a detonator, but loaded with low explosive, so that its output is primarily heat (flash). Usu electrically initiated to ignite pyrotechnic devices and rocket propellants.

squirrel-tail section (Fire Pr) A long length of hard suction hose permanently connected to a pumper inlet and carried in a half-loop around the front of the apparatus. A squirrel tail permits rapid connection of a pumper to a hydrant.

SSBC = Southern Standard Building Code. Now called the Standard Building Code.

stability[1] (Phys) The tendency to remain in a given state or condition.
  NOTE: A stable dynamic system possesses restoring forces that return it to its equilibrium state when it is disturbed by an outside force.

stability[2] (Chem) The resistance of a chemical compound to decomposition.

stability[3] (Build; Brit) The capacity to resist failure when exposed to shock, stress, or other severe condition. A measure characterizing the lasting quality of building construction elements, and the like, under fire conditions.
  NOTE: In a fire-resistance test stability is measured to the time of collapse for nonbearing constructions; to the time of deflection exceeding specified limits for horizontal constructions; and to 80% of the time to collapse for load-bearing constructions, or to 80% of the heating time if collapse occurs in the reload test.

stabilizer[1] = foaming agent. See: foam concentrate.

stabilizer² = ground jack. See: jack.

stabilizer³ (Chem) A substance added to a plastic to protect it against degradation by heat, ultraviolet light, and oxidation.

stack¹ (Build) A vertical passage through which products of combustion are conducted to the atmosphere. A tall chimney.

stack² (Storage) A quantity of supplies stored vertically, occupying approx one pallet space on the floor and braced to assure its stability.

stack effect (Fire Pr) Descriptive of the normal movement of air in a building caused by the temperature difference between the interior and exterior. As the heated air inside rises, cool outside air is drawn in at the bottom. Syn: chimney effect; flue effect.
NOTE: The effect is responsible for much of the smoke migration in multistory buildings during fires.

staff function (Man Sci) An auxiliary service that provides technical or specialized support of operations and line management.
NOTE: Typically, staff functions include planning, costing or budgeting, inventory control, engineering, quality control, purchasing, personnel, training, etc.

staging (Mining) A working platform made of heavy timbers.

staging area (Fire Pr) An area on a floor below the fire floor in a highrise building where equipment is assembled and to which firefighters report to receive their operational assignments. Cf: location.

stagnation temperature (Comb) The temperature obtained on a surface that obstructs a gas flow and brings it to rest. For low-velocity flows the difference between local gas temperature and stagnation temperature is small, typically less than 0.01°C, but for supersonic flows it can be very large, even as much as 10,000°C for mach numbers exceeding 20. The difference between local gas temperature and stagnation temperature is the kinetic energy of flow of the gas, which increases quadratically with gas velocity.

stair platform (Build; Safety) An extended step or landing interrupting a continuous run of stairs. Syn: landing.

stair pressurization (Build) Maintenance of pressure slightly above atmospheric in a stairway to keep out smoke and combustion gases.

stair railing (Build) A vertical barrier erected along the exposed sides of a stairway to prevent falls.

stairs (Build) A series of steps leading from one level or floor to another, or leading to platforms, pits, boiler rooms, crossovers, or around machinery, tanks, and other equipment.
NOTE: A continuous run of steps between two floors, or between a floor and a landing is called a flight.

stair shaft (Build) A vertical, tower-like structure containing a stairway in a building.

stall¹ (Fire Pr) A space or berth assigned to an apparatus in a fire station.

stall² (Techn) To slow down and stop, for example an engine, because of overload.

stampede See: panic

stanchion (Build) An upright column, bar, post, or other support.

stand (For Serv) A growth of plants or trees in a given area.

standard¹ Normal, usual, in common usage, or easily obtainable.

standard² (Techn) a. An official, detailed statement of specifications and requirements, such as for equipment or testing. Cf: regulation. b. An object that serves as a basis for comparison or acceptance.

standard atmosphere¹ (Phys) A unit of pressure equal to 760 mm Hg at sea level. See: air pressure.

standard atmosphere² (Meteor) An atmosphere with a sea-level pressure of 1,013.25 millibars, temperature of 15°C, and lapse rate of 6.5°C up to 11 km (equal to 29.92 in. Hg, 59°F, and 3.5°F/1000 ft, respectively).

standard atmosphere[3] (Med) Any gas mixture used for respiration.

Standard Building Code (Build) A building code that is maintained and published by the Southern Building Code Congress. Formerly called the Southern Standard Building Code.

standard conditions (Phys; Chem) A temperature of 0°C and a pressure of one standard atmosphere. See: standard temperature and pressure.

standard coupling (Fire Pr) A fire hose coupling that has American National Standard fire hose threads.

standard deviation (Statist; Symb: s) The most common measure of dispersion in a frequency distribution; the variability of a set of measurements, numbers, or data points, obtained by taking the square root of the variance, q.v. Syn: standard error.
NOTE: In a normal distribution 68.2, 95.4, and 99.7% of the cases fall, respectively, between ±1, ±2, and ±3 standard deviations from the mean.

standard equipment (Techn) 1. Equipment in common use. 2. Equipment conforming to official specifications and requirements.

standard man (Dosim) A theoretical physically fit man of average height, weight, dimensions, and other parameters, used in studies of how heat or ionizing radiation affect humans.

standard score (Statist) Any of a variety of transformed scores, in terms of which raw scores may be expressed for reasons of convenience, comparability, ease of interpretation, etc.

standard temperature and pressure (Phys; Abb: S.T.P.) The conditions to which measurements of gases are referred: 760 mm Hg pressure at 0°C temperature (273°K). Also called: normal temperature and pressure (Abb: N.T.P.).

standard thread (Techn) US National Standard hose threads.

Nominal Hose Size	Threads per in.	Outside Dia. Male Coupling
4-1/2 in.	4	5-3/4 in.
2-1/2 in.	7-1/2	3-1/16 in.
1-1/2 in.	9	2 in.
1 in.	8	1-3/8 in.
Garden	11-1/2	1-1/16 in.

standard time-temperature curve (Build) A standardized program for test specimens of furnace temperature as a function of time, used in fire tests to simulate fire exposure.

standardization (Techn) The process of establishing engineering criteria for processes, equipment, and parts to achieve uniformity and interchangeability in engineering practices and manufactured items.

stand by (Fire Pr) 1. To be in immediate reserve. 2. An order to remain immediately available for an assignment, if the need arises. 3. A state of alert of neighboring units to cover stripped territories or to be ready for a threatening emergency.

standby (Techn) 1. A reliable substitute. 2. A reserve or backup item or piece of equipment.

standby crew (For Serv) A specially organized and trained team that can be dispatched for quick suppression work on fires.

standby equipment (Fire Pr) Equipment assigned and ready to protect specific hazards at which a fire or other emergency is anticipated, such as a standby detail of personnel and equipment at a circus grounds.

standby redundancy (Techn) A duplication of components or units in a system that are available to perform the function of a primary component or unit that fails.

standing by (Fire Pr) The presence and readiness of firefighters and equipment at a fire but not engaged in fire fighting.

standing rope (Techn) A supporting rope that maintains a constant

standing wave (Phys) A periodic wave having a fixed distribution in space, characterized by the existence of nodes or partial nodes and antinodes.

standpipe[1] (Build; Fire Pr) A vertical water pipe used to supply fire-hose outlets in tall buildings.
  NOTE: In contrast, a <u>riser</u> is a vertical water pipe that supplies a <u>sprinkler system</u>. Standpipes are either wet or dry, depending on whether they are continuously charged with water or empty when not in use.

standpipe[2] (Fire Pr) A short, permanent pipe with one or more hose outlets, or a pipe that is attached to a hydrant, esp to one that is recessed, to elevate the outlet above ground level.

standpipe[3] (Hydraul) An above-ground reservoir used to maintain uniform pressure in a water supply system.

standpipe hose = <u>house line</u>

Stanton number See: <u>dimensionless numbers</u>

staphylococcus (Med) A virulent, penicillin-resistant, gram-negative, spherically-shaped pathogen (bacterium) that forms pus and frequently infects burn injuries.

starboard (Navig) The right side of a ship or vehicle when viewed looking forward.

starting mix (Comb) An easily ignited mixture that transmits flame from an initiating device to a less readily ignitible composition.

starvation (Fire Pr) 1. Oxygen deprivation. 2. Fire extinguishment accomplished by limiting the fuel or oxidizer.

static[1] (Phys) At rest or in equilibrium.

static[2] (Mechanics) Acting through weight only. Pressure exerted by a motionless body or mass.

static[3] (Commun) a. Radio interference. b. Popping, hissing, or clicking noises heard over a loudspeaker. See: <u>sferics</u>.

static electricity (Electr) A charge of electricity built up on nonconducting materials, usu by friction.
  NOTE: Sufficiently large potentials can build up to cause sparks which can ignite flammable gases and vapors.

static head (Hydraul) The pressure exerted by a stationary column of water. The magnitude of the pressure is directly proportional to the height of the column.

static pressure (Hydraul) The pressure exerted in all directions at a point in a fluid at rest. For a fluid in motion, it is measured in a direction normal (at right angles) to the direction of flow. When added to velocity pressure, it gives total pressure.

static-steering torque (Transp) The torque required to turn the wheels of a stationary vehicle.

static water supply (Hydrol) Any body of water not under pressure, for example, a reservoir, pond, lake, or cistern.

statics (Mech) The branch of mechanics that deals with particles or bodies in equilibrium under the action of forces or torques, and esp with the properties of centers of mass and moments of inertia.

station[1] (Fire Pr) A fire department building housing on-duty firefighters and fire equipment. Also called: <u>fire station</u>; <u>fire hall</u>; or <u>fire house</u>.

station[2] (Fire Pr) A fire alarm box location.

station[3] (Transp) a. Any of several longitudinal reference points on the plans of a vehicle. b. The transverse section plane at one of these reference points.

statistical fire (For Serv) A fire suppressed entirely by or with the assistance of US Forest Service employees. Cf: <u>nonstatistical fire</u>.

statistics (Math) A mathematical science concerned with numerical data, its collection, and evaluation for the purpose of drawing rigorous conclusions from variable data in terms of well-defined probability.
  NOTE: An important application of statistics is the derivation of information about large populations by observing relatively small samples drawn randomly.

statute mile (Metrol) A distance of 5280 feet. See: mile. Cf: nautical mile.

statutory (Law) Legal; provided for or covered by law.

staypole (Fire Pr) A pole used to brace and stabilize a fire ladder. Syn: tormentor pole.

steam (Phys; Techn) Water vapor or gaseous water above the boiling temperature; or loosely, the mist formed by partial condensation of water vapor.
  NOTE: Steam under pressure can be superheated.

steamer (Fire Pr) An obsolete, coal-burning, steam-driven pumping engine, generally drawn by horses. A second-class steamer was capable of delivering 600 gpm; a first-class, 900 gpm; and an extra-first-class, 1100 gpm at 100 psi.

steamer connection (Fire Pr) A hose connection on a pumper of 4-1/2 inches diameter or larger, comparable to the steamer outlet on a hydrant.

steamer outlet (Fire Pr) The large connection on a hydrant, originally provided for steam fire engines.

steel fire door (Build) A door with interlocking steel slats or plate steel construction.

steering system (Transp) The assembly of linkages and components that enables a driver to control the direction of a vehicle.

Stefan-Boltzmann law (Thermodyn) An equation stating that the rate at which heat is emitted by a perfect radiator (black body) is proportional to the fourth power of its absolute temperature.

Steiner tunnel (Comb) An 18- by 12-inch rectangular duct in which a 25-foot sample of structural material can be tested to determine its flame-spread characteristics. See: flame-spread rating.

St Elmo's fire = coronal discharge

stemming (Civ Eng) A suitable inert noncombustible material or device used to confine or separate explosives in a drill hole or to cover explosives in mud-capping.

stench (Mining) A strong odor released into the ventilation system of a mine to warn miners of danger.
  NOTE: The odor is produced by a chemical, such as ethanethiol or ethyl mercaptan.

stenosis (Med) Narrowing of the lumen in a blood vessel.

stepladder A self-supporting portable ladder, nonadjustable in length, with flat steps and a hinge at the top. A common name for a trestle ladder.

step-up method = progressive method

sterile (Med) Free of living organisms.

sterility (Med) Inability, either temporary or permanent, to reproduce.

sterilization (Med) The destruction or removal of all forms of life from an object by heat, irradiation, filtration, or chemical action.

steroid (Pharm; Med) A chemical substance, containing the sterol carbon ring, present in bile acid, certain hormones, and glycerides.
  NOTE: An injection of corticosteroid is effective against pneumococci and is used in treating patients with severe burns. Steroid injections are usu given only once, because repeated use is harmful to the red blood cells.

stethoscope (Med) An instrument used in physical examinations to detect and listen to internal body sounds. See: auscultation.

steward An individual elected by a local group or unit of a labor union to represent the members in labor matters.

stiffleg (Mat Hand) A rigid member supporting a mast at its head.

still alarm (Fire Pr) An alarm sent only to the units designated to respond, usu the companies first due on the assignment card. Although no alarm box or audible signal is sounded, a still alarm may be transmitted by voice communication. Syn: **silent alarm**.

stink damp (Mining) Hydrogen sulfide. See: **damp**.

stirred reactor (Comb) A device for volumetric reaction in which flow conditions result in such complete mixing that the composition is approx uniform despite rapid inflow and exit.
  NOTE: The device has found use in characterizing maximum reaction rates for combustor designs.

stirrup (Fire Pr) 1. A wedge-shaped metal attachment on the heel of a ladder, provided to keep the ladder from slipping after it has been raised. Syn: **heel plate**. 2. A foot bracket on a stirrup pump.

stirrup pump (Fire Pr) A portable, hand-operated piston pump with a short hose and a stirrup-shaped foot bracket for holding the pump on the floor or ground. The pump is used as a first aid for small fires and in cramped quarters. The pump is supplied by water from buckets and is a common US Forest Service pump for back-pack operations.

stochastic (Math) Dependent upon chance occurrences; random.

stock company (Insur) An insurance company owned by stockholders, who share in the profits. Cf: **mutual company**.

stock cover = **salvage cover**

stoichiometry (Chem) The application of quantitative relations derived from the equations of chemical reaction and the law of definite proportions.
  NOTE: In stoichiometric combustion, the fuel and oxidizer are balanced.

stope (Mining) 1. An underground excavation from which the ore has been removed. 2. To mine ore in successive layers or slices.

stop message (Fire Pr; Brit) A fire service message indicating that no more help is required at the fireground, except for possible relief.

stopping[1] (Mining) A barrier across a mine tunnel or a seal that blocks off a mined-out working.

stopping[2] See: **firestopping**

stopping distance (Transp) The distance in which a vehicle comes to rest after the driver discovers he must stop. Includes driver reaction time, brake reaction time, and braking time.

stopping power (Nucl Phys) A measure of the ability of a substance to hinder the passage of a charged particle.

stopple-type extinguisher (Fire Pr) An older soda-acid type of extinguisher in which the acid is contained in a bottle with a loosely-fitted lead stopple that falls out when the extinguisher is inverted and releases acid that mixes with the soda solution.

stops (Fire Pr) Limiting devices on ladders to prevent fly sections from overextending when elevated and from retracting too far when nested.

storage[1] (Build) A building, structure, or area used primarily for sheltering goods, merchandise, products, vehicles, or animals. A warehouse, silo, or any of a variety of structures used for this purpose.

storage[2] (Storage) a. The interim safekeeping of materials after manufacturing and prior to disposal. b. The keeping or placing of property in a warehouse, shed, or open area. c. The state of being stored.

storage[3] (Comput) A device into which data can be introduced, held, and then extracted.

storage battery See: **battery**

storage life (Storage; Techn) The period of time during which a packaged item can be stored under specific temperature conditions and remain suitable for use. Sometimes called shelf life.

storm (Meteor) An atmospheric disturbance caused by pressure, temperature, and humidity differences, usu characterized by high winds and precipitation. Storms include cyclones, blizzards, tornadoes, waterspouts, squalls, and hurricanes, as well as sand, dust, and thunderstorms. See also: cloud; wind.

storm door (Build) An outer door, usu with glass or plastic panes, installed primarily to stop drafts from passing through the inner door.

story (Build) 1. The portion of a building between two adjacent floors. See: floor. 2. = (Brit) storey.

STP = standard temperature and pressure

straight ladder (Fire Pr) A single-section ladder, with parallel solid or trussed spars. Also called: wall ladder.

straight lay = forward lay. See: hose lay.

straight stream (Fire Pr) A stream of water ejected straight from a nozzle. Also called: solid stream.

straight third (Fire Pr) A third alarm given without sounding the second alarm, bringing both second and third alarm response.
 NOTE: A straight fourth alarm can skip both second and third alarms, or just the third if the second has been given.

strain[1] (Mech) The change in the linear dimensions or other deformation of a body placed under an applied force called stress.

strain[2] (Bacter) A pure culture of bacteria composed of the descendants of a single isolation.

strain[3] (Med) An injury to a muscle caused by overexertion.

strainer 1. (Salv) Wire mesh or other sieve-like device used to keep debris from clogging water drains or scuppers. 2. (Fire Pr) A screen used in pump and suction-hose intakes to keep dirt and other solid matter from clogging or damaging waterways and pumping equipment. Self-cleaning strainers have a valve provided for reverse flushing with water.

strainer basket (Fire Pr) A cup-shaped guard placed over a suction strainer to prevent the entry of dirt, leaves, etc. The canvas part of the basket is called a skirt. See also: conical strainer.

strategy (Fire Pr) The overall plan for attacking a fire, using available men and equipment to the best advantage and taking into account the extent and behavior of the fire and all factors and conditions affecting the endangered property.

stratosphere (Meteor) A relatively stable layer in the atmosphere extending from the tropopause to approx 15 miles altitude (sometimes placed as high as 30 miles).

straw boss (For Serv) A working crew leader who directs the work of from 3 to 8 firefighters.

strays = sferics

streak (Mining) An exceptionally rich portion of a vein.

stream See: fire stream

streamform branch (Fire Pr; Brit) A type of nozzle with a coaxial open tube supported in the center of the nozzle by three vanes. This design is said to produce a more solid jet.

street (Transp) A throughfare, such as an avenue, boulevard, or similar way, that is accessible to the public.

street box (Fire Pr) 1. A fire alarm signal box mounted on a post on a public street. Also called: call box; street alarm. 2. = (Brit) pillar.

street floor (Build) The floor or story of a multistory building that is accessible directly from street level. The main floor of a building.

Syn: first floor; ground floor.

strength of attack (Fire Pr) The number of firefighters and quantity of equipment deployed to attack a fire.

streptococcus (Bact) One of an important class of disease-causing bacteria.

streptomycin (Pharm) A colorless antibiotic effective in treating wounds and skin infections, and particularly pseudomonas aeruginosa, which frequently infects burn victims.

stress[1] (Mech Eng) a. A force applied to a body by an external load, measured in standard units, as pounds per square inch, etc. b. The intensity of internal, distributed forces resisting deformation.

stress[2] (Psych) A physical, chemical, or emotional factor that causes mental tension, erratic behavior, and may be a factor in disease causation or fatigue. Stress can arise from temperature extremes, threatening situations, injury, anoxia, exposure to toxic agents, and other adverse influences that call for defensive responses. If prolonged, the stress situation leads to physical exhaustion.

stress corrosion (Techn) Chemical corrosion that is accelerated by mechanical stress.

stretch (Fire Pr) 1. A command to lay out hose. 2. The length of a laid out hose. 3. To lay hose.

stretcher (First Aid) A collapsable rack, often of canvas stretched between two poles, used to carry injuried or disabled persons. A litter.

strike (Mining) 1. The line formed by the intersection of a horizontal plane with a tilted geological stratum. 2. The compass direction of a vein.

strike a box (Fire Pr) To actuate a fire alarm box.

strike appliance (Fire Pr; Brit) A fast, maneuverable, fire apparatus, usu smaller than ordinary, capable of reaching a fire with sufficient firefighters and equipment to engage in effective first attack against a fire.
NOTE: Comparable to the minipumper concept in the US.

string (Build) The side member supporting the treads and risers of a stair.

strike out (Fire Pr) A signal used to indicate that a fire is out or under control. See: all out.

stringer[1] (For Serv) A narrow finger or band of fuel that connects two or more patches or areas of wildland fuel.

stringer[2] (Civ Eng) The horizontal member of a shoring system whose sides bear against the uprights or earth. Also called: wale.

strip burning (For Serv) 1. Burning of a narrow strip of fuel along a control line and then successively burning wider adjacent strips as the preceding strip burns out. Syn: strip firing. 2. Burning one or more narrow strips of slash through a cutting unit and leaving the remainder.

strip firing = strip burning

stripped (Fire Pr) 1. Having no fire companies in an area that are in quarters or available to answer alarms. 2. A pumper or hose truck with all hose removed. 3. A ladder truck with all ladders removed, usu for use at a fire.

stripping (Chem Eng; Techn) Removing or separating products from a liquid by evaporation, distillation, or by passing a gas through it.

stroke (Med) A sudden, severe attack of illness, particularly apoplexy.

strong acid (Chem) An acid that can be completely ionized in a water solution.

strontium-90 (Chem) An isotope of strontium having a mass number of 90.
NOTE: An important fission product and constituent of fallout, it has a half-life of 25 years.

structural analysis (Build; Mech Eng) Study of structures and calculation of stresses and strains produced by various loads for the purpose of determining structural stability and safety.

structural integrity (Build; Techn) The quality or state of being complete and unimpaired.

strut (Fire Pr) A sprinkler head yoke that holds the fusible link in place and carries the water jet deflector on its top.

stub (For Serv) A snag less than about 20 feet tall.

stud (Build) An upright member, commonly of 2x4-in. size, used in the framing of a building wall. The element to which sheathing, paneling, and lath are fastened.

stull (Mining) A post or timber used as a wall support in a mine. A prop wedged between the sides of a mine stope and used to support a platform.
  NOTE: A *false stull* is one that is wedged at an angle below a main stull to give it added support. A *saddle-back stull* is one in the shape of an inverted V. A *reinforced stull* consists of a saddle-back stull with a horizontal timber placed across its apex.

styrene (Chem) A colorless, highly flammable, reactive liquid, $C_6H_5CH{:}CH_2$, bp 145.2°C, fp -33°C, used extensively in the production of synthetic rubber, plastics, resins, paints, and adhesives.
  NOTE: Styrene vapor causes respiratory, eye, and nose irritation. Liquid styrene irritates the skin.

sub-basement (Build) A basement below the first basement; a second story underground.

subacute (Med) An illness or condition that is not quite as serious or as dangerous as the acute phase but may become so if not properly attended.

subassembly (Techn) Two or more parts, each of which are individually replaceable, or are replaceable as a unit, and which form a portion of an assembly or a component.

subatomic particle (Nucl Phys) Particles smaller than an atom: an electron, neutron, proton, meson, etc.

subcutaneous (Med) Beneath the skin; usu below the dermis.

subjective (Psych) Arising from or affected by the mind or a particular state of mind. Resulting from personal opinion rather than being based on measureable (objective) external phenomena.

sublimation (Chem; Phys) 1. A change of state in which a material, such as iodine, passes directly from a solid into a gaseous state. 2. Purification of a substance by sublimation and recondensation of solid crystals (called a sublimate).
  NOTE: The heat involved in sublimation is the sum of the *heat of fusion* and the *heat of vaporization*. Frost and snow are familiar sublimates.

subpleural hemorrhage (Med) Bleeding beneath the covering envelope of the lung.

subroutine (Comput) A portion of a total program, stored in a computer and available upon call to accomplish a particular operation, usu a mathematical calculation. At its conclusion, control reverts to the master routine.

subsidance[1] (Meteor) The gradual sinking of an *air mass* over a relatively large area, generally accompanied by increasing pressure, decreasing relative humidity, and clear or partially clouded skies.

subsidence[2] (Mining) Gradual sinking or settling of soil, esp into an excavation.

substance (Chem) Matter having definite atomic or molecular composition.

substitute (Fire Pr) 1. A part-time paid firefighter employed to fill-in for full-time firemen on sick leave or vacation. 2. Sometimes refers to a probationary fireman waiting permanent assignment.

subsurface injection (Fire Pr) A method of fighting fuel tank fires in which foam is fed into the tank at a point below the burning fuel surface. The foam rises to the surface, cooling the fuel, and blanketing the fuel vapor at the surface. Also called: **base injection**.

suburban On the outskirts of a city; usu referring to a residential area in an outlying part of a city or town.

suction (Fire Pr) Drawing water from a static source below the level of the pump. Syn: **drafting**.

suction basin (Fire Pr) An enclosed cistern used to test the draft of fire department pumpers.

suction booster (Fire Pr) A type of jet siphon booster usu supplied from a tank and used to draft water over a large distance and to a higher elevation than is possible with a suction pump that depends solely on atmospheric pressure.

suction crutch (Fire Pr; Brit) A device for supporting the weight of a suction hose used with collapsible dams.

suction engine (Fire Pr) A hand engine or pumper capable of drafting static water.

suction eye (Hydraul) The opening in a centrifugal pump through which water enters and is accelerated by the impeller.

suction hose (Fire Pr) Hose designed to withstand both internal water pressure and external atmospheric pressure, used between a nonpressurized water supply and a pump. Also called: **hard suction**; **suction sleeve**. See also: **soft sleeve**.

suction lift (Hydraul) The vertical distance from the surface of a water supply to the center of the pump impeller.

suction pipe (Fire Pr) A permanent pipe with pumper connections installed at static water sources, such as lakes, or on piers, to supply water for fire.

suction strainer (Fire Pr) A mesh attached to the lower end of a suction hose to prevent solid debris from entering the hose when drafting from an open, static water source.

suction wrench (Fire Pr) A tool about 9 inches long with a semicircular jaw on one end and a handle on the other, used to tighten or loosen suction hose couplings.

suffocation See: **asphyxia**

suicide (Law) Intentional self-destruction. The purposeful taking of one's own life.

sulfadiazine See: **silver sulfadiazine**

sulfamylon (Pharm) A topical medication used to prevent sepsis of burn injuries.

sulfur dioxide (Chem) A noxious non-flammable gas, SO_2, highly toxic in concentrations of 400 ppm. Extremely irritating to the eyes and respiratory tract.

sulfur oxides (Chem) Pungent, colorless gases formed by the combustion of sulfur, primarily from fossil fuels, and considered major air pollutants. Sulfur oxides may damage the respiratory tract, as well as vegetation.

sulfuric acid (Chem) A corrosive, colorless, odorless mineral acid, H_2SO_4, fp 10.5°C, bp 330°C in undiluted form. Water poured into sulfuric acid produces a violent steam explosion. In reaction with most metals the acid evolves combustible hydrogen gas. The acid is extremely destructive to skin and eye tissue on contact. Also called: **oil of vitriol**; **vitriolic acid**.
  NOTE: Not flammable itself, the acid reacts with many compounds to produce highly flammable or explosive products. Sulfuric acid fumes or mists are extremely irritating to the upper respiratory tract. High-strength fuming sulfuric acid, called **oleum**, is yellow and has a sharp odor.

sump (Mining) An enlarged chamber at the bottom of a shaft for collecting drainage water.

sunstroke (Med) Physical distress characterized by headache, dry skin, rapid pulse, dizziness or coma, fever, and redness of the skin caused by overexposure to the sun.

supercooling (Phys) Decreasing the temperature of a liquid below the freezing point without freezing. Supercooled raindrops freeze almost instantly when disturbed.

superheating (Techn) The heating of a vapor, particularly saturated (wet) steam, to a temperature higher than that at which it would condense.
NOTE: This is done in power plants to reduce condensation in the turbines and to improve efficiency.

superimposed load (Build) Weights or forces applied to a specimen during a fire test to develop as nearly as practicable the working load expected for the piece in service.

superoxide (Chem) A compound that contains an O_2^- ion. Potassium superoxide, KO_2, for example, is used in self-contained breathing apparatus in fire fighting and spacecraft applications to remove exhaled carbon dioxide and provide oxygen. Sodium superoxide, NaO^2, is better for this purpose but is considerably more expensive.

supersaturated solution (Chem) A solution in which more solute is dissolved than in a saturated solution of the same substances at the same temperature and pressure.
NOTE: If crystallization occurs, a saturated solution will form that has solution and crystals in equilibrium.

supervisory management (Man Sci) The group directly responsible to the middle management group for implementation of policies and execution of directives by rank-and-file employees.

suppuration (Med) The formation of pus, commonly by a coccus infection.

supplied-air suit A one- or two-piece suit that is impermeable to most particulate and gaseous contaminants and is provided with an adequate supply of respirable air.

supply line (Fire Pr) A large-diameter hose that feeds water to two or more working lines.

supply officer (For Serv) A staff officer, reporting to the service chief, responsible for issuing supplies and equipment required to suppress a fire and for servicing and repairing tools and simple equipment.

suppression (Fire Pr) The sum of all the work done to extinguish a fire from the time of its discovery. Fire extinguishment.
NOTE: A distinction can be made between direct and indirect suppression. Direct suppression includes such actions as cutting off the fresh air supply, flooding with inert gas, and applying water or other extinguishing agents. Indirect suppression includes smoke management, heat venting, and in the case of wildland fires, building of fire control lines ahead of the advancing fire front.

suppression crew (For Serv) Two or more firefighters stationed at a strategic location, either permanently or during a temporary emergency, for the purpose of taking quick initial action on fires. Syn: suppression squad.

suppression foreman (For Serv) The supervisor of a suppression crew.

suppression squad = suppression crew

surety (Insur) The corporation or individual guaranteeing performance or faithfulness under a bond.

surety bond (Insur) An agreement providing for compensation on the failure of one party to perform certain acts for another within a specified time period.

surface-active agent (Phys; Chem) A substance that in small quantity alters the surface properties of gas, liquid, and solid interfaces in any combination, except perhaps gas-gas. Surface activity is usu associated with emulsification, wetting, adhesion, surface tension, and friction. Also called: surfactant.

surface burning See: flame spread

surface coating (Techn) Paint, lacquer, varnish, and other chemical compositions used to protect or decorate surfaces. See also: protective coating.

surface contamination (Safety) The deposition and attachment of radioactive materials to a surface. See: radioactive contamination.

surface drag (Meteor) Resistance to wind at or near ground level.

surface fire[1] (Fire Pr) Fire propagating on the surface of a combustible material as contrasted to a deep-seated fire.

surface fire[2] (For Serv) A fire burning through small vegetation and surface litter on the forest floor.

surface flame spread See: flame spread

surface spread of flame test (Build; Brit) A test for measuring the rate and extent of flame spread over the surface of a material when subjected to a radiant heat flux.

surface tension (Chem; Phys) The contractive force of a liquid surface, as if it were an elastic membrane.
NOTE: Surface tension plays an important role in the formation and lifetime of foam bubbles and emulsions, in capillarity, and curvature of free liquid surfaces, as well as in the action of detergents and antifrothing agents.

surface water (Hydrol) Any water collected in pools, lakes, marshes, glaciers, reservoirs, or other basins, or flowing in streams or rivers on the surface of the earth, as contrasted to ground water. See also: aquifer.

surface zero (Milit) The point on a water surface directly above or below a nuclear explosion. Equivalent to ground zero on land.

surfactant[1] (Fire Pr) = surface-active agent

surfactant[2] (Med) An agent that affects surface tension, esp of the mucous membranes of body airways.

surge (Hydraul) An abrupt jump in the delivery pressure of a fluid. See: water hammer.

surge impedance See: characteristic impedance

surgery (Chem; Abb: Surg) The field of medicine that treats diseases and injuries of the body by operative procedures. The general field is divided into the specialties of general surgery, thoracic, plastic, orthopedic, neurosurgery, ophthalmology, otolaryngology, proctology, and urology.

surgical benefits (Insur) Plans that provide employees, and in many cases their dependents, with specified surgical care or a cash allowance toward the cost of such care, usu in accordance with a schedule of surgeon's fees. Generally part of a health and insurance program.

surveillance (Safety) Observation, inspection, and classification of pyrotechnic and explosive items in movement, storage, and use with respect to degree of serviceability and rate of deterioration.

survey (Soc Sci) A sampling of data or opinion taken and used to derive an understanding of a given problem.

survey map (Fire Pr; Cartog) A detailed scaled map of an occupancy or area, prepared in the field during a prefire planning survey by a fire department to show existing conditions, relative sizes and locations of structures, life safety features, water supplies, private fire protection equipment, and the like.

survey meter (Dosim) 1. An instrument used to detect and measure gamma radiation. Also called: dose rate meter. 2. Any portable radiation detection instrument esp adapted for surveying or inspecting an area to establish the existence and amount of radioactivity present. Cf: counter; monitor.

survival curve (Dosim) A curve obtained by plotting the number or percentage of organisms surviving at a given time against the dose of radiation or the number surviving at different intervals after a particular dose of radiation. See: lethal dose.

survivors' benefits (Insur) Payments to dependents of employees who die prior to retirement, financed in whole or in part by an employer. Also called: transition benefits; bridge benefits; widow's allowance.
  NOTE: Benefits may be in the form of payments for a fixed period (e.g., 24 months) supplementing regular life insurance benefits, a benefit for life out of a pension program, a lump-sum payment, etc.

suspended ceiling (Geom) A decorative ceiling hanging on rods below the structural ceiling, usu consisting of metal gridwork holding acoustical tiles. Also called: hanging ceiling.

suspension[1] (Chem) An intimate mixture of particles and a fluid, such as smoke or fog in the air, in which the particles are dispersed but not dissolved.
  NOTE: Suspensions usu consist of particles larger than colloidal size (>0.2 micron) and tend to settle unless the fluid is in motion. Emulsions, by contrast, are liquid suspensions of liquid particles of colloidal dimensions and are stable for long periods. A gel is a suspension of a solid in a liquid phase. Foam is a colloid in which a continuous liquid medium surrounds gas bubbles. The film is so thin as to be colloidal, while the enclosed bubble may be either colloidal or macroscopic. A sol is a suspension of a solid in a liquid phase.

suspension[2] (Mech Eng) a. A device by which something is suspended, such as a gimbal. b. A body that is suspended, such as a bridge.

suspension[3] (Fire Pr) Relief from duty, generally as a punishment or during the period of an investigation of alleged misconduct.

suspension system (Mech Eng) The mechanical linkages and the elastic members that provide a flexible support, such as for a vehicle.

suspicious fire (Fire Pr) A fire of undetermined cause, which, because of unusual circumstances, is thought to be of incendiary origin. See: arson.

swamper (For Serv) An axman who cuts brush, small trees, tree limbs, and other fuel so that it can be cleared away more easily.

swamper burning = progressive burning

swash partition (Fire Pr; Transp) A baffle plate in a water tank carried on a pumper to impede water sloshing when the vehicle is accelerating, decelerating, or turning a corner.

swatter (For Serv) A tool similar in shape to a fly swatter and about as large as a snow shovel, with a large semiflexible flap attached to a handle, used to extinguish grass and ground fires by beating.

sweating[1] (Physiol) Loss of fluids through the skin.

sweating[2] (Techn) The process of uniting metal parts by heating and allowing molten solder to run between them.

sweat rate (Physiol) An index of heat stress on humans working in hot environments.

sweep (Fire Pr) To play a hose stream from side to side to wet down a large area. Syn: rake.

sweetdamp = carbon monoxide. See: damp.

sweetening (Chem Eng) The process by which petroleum products containing certain sulfur compounds are chemically deodorized.

swimming (Transp) The ability of a vehicle to negotiate a water obstacle by propelling itself across, without being in contact with the bottom.

swing (Electr) The variation in frequency or amplitude of an electrical quantity.

swing shift (Fire Pr) An irregular work period used to change one or more men from day to night duty or from one platoon to another.

swivel[1] (Mech Eng) A fastening device that rotates freely.

swivel[2] (Fire Pr) a. A freely rotating female coupling ring for connecting a hose line to a male nipple. b. The fastener that attaches a tormentor pole to a ladder.

symbols (Info Sci) Special characters, abbreviations, and notation conventions, such as Greek letters, mathematical signs, and the like, used in chemistry, physics, mathematics, engineering, and other sciences and technologies as shorthand for expressing quantitative and qualitative relationships.

symptom (Med) Subjective evidence of disease or physical disorder.

synchronous (Phys) Happening at the same time; having the same period and phase.

syncope (Med) Fainting spell. Unconsciousness caused by inadequate blood flow to the brain.

syndrome (Med) A group of symptoms or signs typical of a given disease condition or illness. A sign is an objective indication, whereas a symptom is a subjective indication.

synectics The joining together of different and apparently unrelated elements.

syneresis (Phys) The separation of a liquid from a gel by contraction.
NOTE: Syneresis causes foam to deteriorate and to lose its fire-extinguishing capacity.

synergism Cooperative action of substances whose total effect is greater than the sum of their separate effects.
NOTE: A kerosine and pyrethrum insecticide becomes ten times more toxic to houseflies if a small quantity of sesame oil is added, whereas kerosine and sesame alone have no toxic effect.

synthesis (Chem) The reaction or process by which a compound is obtained from other, simpler compounds or elements.

synthetic (Chem) 1. A man-made material that approximates a natural product. 2. Artificial, as opposed to natural.

synthetic detergent (Chem Eng) A chemical cleaning agent soluble in water or other solvents.
NOTE: Originally developed as a soap substitute.

synthetic rubber (Chem) A man-made polymer with rubber-like properties.
NOTE: Major types are designated as S-type, butyl, neoprene (chloroprene polymers), and N-type. Several synthetics duplicate the chemical structure of natural rubber.

synthetic trainer (Educ) A device that provides a substitute for practice under actual operational conditions.

syphon = **siphon**

system[1] (Techn) A set or arrangement of components so related or connected as to form a unity or organic whole.

system[2] (Man Sci) An orderly arrangement of interdependent activities and related procedures which implements and facilitates the performance of a major activity in an organization.

systemic (Med) 1. Internal; within the body. 2. Not localized to one part of the body.

systemic chemotherapy (Med) Treatment internally with drugs.

system safety (Safety) The detection of deficiencies in system components that have an accident potential. The system safety engineer analyzes task performance and conducts a systematic safety study of equipment and procedures.

systems analysis (Man Sci) An investigation of long-range problems of complex man-machine systems for the purpose of defining objectives, the means of attaining them, and formulating alternatives; comparing costs, effectiveness, and risks associated with the alternatives; and integrating the results of the analysis for effective decision-making.

systems engineering (Ind Eng) The practice of design, construction, prediction of performance, and testing of complex groups of interrelated components or subsystems, emphasizing optimum performance under varying conditions.

systole (Physiol) A period of heart contraction during a cardiac cycle, in which the blood is pumped into the arteries. See: diastole.

T

T (Phys) Abbreviation for temperature, time, ton.

tabular (Mining) Descriptive of a flat mineral deposit or other geological formation.

tachometer (Mech Eng) An instrument that measures the rotational speed of a shaft, engine, or the like, usu in rpm. On a pumping engine it is used to maintain safe speed and correct gear ratio for pumping.

tachycardia (Med) 1. Rapid heartbeat; heartbeat above 120 beats per minute. 2. A rapid pulse.

tachypnea (Med) Hyperventilation; breathing at a rate of over 25 breaths per minute.

tackle (Mat Hand) An assembly of ropes and sheaves arranged for hoisting and pulling.

tactical channel (Commun; Fire Pr) A radio channel reserved for communications between the fireground commander and the operational units on the fireground.

tactical squad (Fire Pr) A team of specialists that responds to large-scale, multiple-alarm fires to provide special skills, expert knowledge or particular services on the fireground. Cf: flying squad.

tactics (Fire Pr) The art and skill of deploying and maneuvering men and equipment on the fireground and using them effectively in extinguishing the fire. Cf: strategy.

tactile (Physiol) Pertaining to touch.

tactile organs (Anat) The sensory organs of touch and feeling located at the surface of the body and stimulated by pressure. See: sense organs; touch.

tag See: tracer

tag cup test (Comb; Testing) A test method for measuring the flash and fire point temperatures of flammable liquids. Both open and closed cup tests have been standardized by the ASTM.

NOTE: Named after its developer, Tagliabue.

Tagliabue tester (Testing) An apparatus for measuring the flash points of liquids. Also called: Tag tester.

tags (Safety) Temporary signs, usu attached to a piece of equipment or part of a structure, to warn of existing or immediate hazards.

tail (Statist) Either extreme of a frequency distribution.

tailboard (Fire Pr) A platform on the back of a hose or pumper truck on which firefighters stand while en route to a fire. Syn: back step.
NOTE: Nozzles, hand extinguishers, and various other small items are carried on the tailboard, as well as donut rolls of wet hose on the way back to the station.

tailings (Metall; Mining) Ore residues from which all or most of the metal has been extracted.

take a hydrant (Fire Pr) 1. An order for a pumper to stand by a fire hydrant. 2. An order to connect a pumper to a hydrant. 3. An order to lay out hose from a designated hydrant.

take charge (Fire Pr) To assume command of a unit or fire-fighting operation.

talc (Techn) A hydrous magnesium silicate used in ceramics, cosmetics, paint, and pharmaceuticals, and as a filler in soap, putty, and plaster.
NOTE: Used as a control agent for magnesium fires.

tally-in (Transp) 1. Itemized list of supplies received. 2. Process of recording the number of containers or quantity of material received.

tally-out (Transp) 1. Itemized list of supplies included in an issue or shipment. 2. Process of recording the number of containers or quantity of material issued or shipped.

tandem[1] Two or more entities or objects one behind another; in line, as contrasted to abreast.

tandem² (Fire Pr) A configuration in which one pumper supplies water under pressure to another pumper, and the second increases the pressure, as in relay or series operation, in which water must be pumped over long distances or uphill. See: relay pumping.

tank apparatus = tanker

tanker¹ (Fire Pr) A water tank fire apparatus, usu carrying 1000 gallons or more of water, and equipped with small hose and a pump. Also called: tank apparatus; tank truck. See: apparatus; tanker-pumper.
  NOTE: Sometimes a petroleum transport truck is converted for use as a fire department water tanker.

tanker² (Transp) A ship, railroad car, or highway transport truck, usu of large capacity, designed to carry liquids, esp petroleum products.

tanker boss (For Serv) A supervisor, usu of three to five tanker units, responsible for supplying water to pumpers both in fire attack and in mopup work.

tanker call (Fire Pr) A preplanned assignment of tank trucks to supply water at locations where local water supplies are limited or unavailable.

tanker manager = tanker boss

tanker-pumper (Fire Pr) A tanker equipped with at least a 500-gpm pump.

tanker shuttle (Fire Pr) A water-supply operation provided by a fleet of tankers in which tank trucks are in continuous rotation; empty trucks en route to the water-supply point to refill while loaded trucks are returning to the fire scene to supply the pumpers.

tank farm (Storage) A group of bulk liquid fuel-storage tanks.

tank fire (Fire Pr) A fire involving a bulk liquid fuel storage tank. See: bleve; boilover fire.

tank gage (Fire Pr) A gage that indicates the quantity of water in a fire apparatus tank.

tank line (Fire Pr) A small hose line supplied by a water tank, or the pipe leading from a tank to a pump. Also called: booster line.

tank line valve (Fire Pr) A shut-off valve in a tank line.

tank overflow (Fire Pr) A pipe that vents air and discharges excess water from a tank during filling.

tank truck = tank apparatus; see: tanker¹.

tape (Fire Pr) 1. The paper strip of a fire station register on which codes are punched by the register to indicate numbers of fire alarm boxes and other signals. 2. Magnetic tape used to record radio and telephone messages and conversations at a communications center.

taper (Fire Pr) A paper tape device that records fire alarms received at a fire alarm headquarters. Syn: punch register.

tapper (Fire Pr) 1. A bell that rings when a box alarm is pulled. Syn: joker. 2. A bell used to signal companies in and out of service.

tar (Chem Eng) Wood, coal, or petroleum exudations. In general, a complex mixture of chemicals from the bottom fraction of distillation.

tare weight (Transp) The weight of an empty container.

target¹ (Nucl Phys) a. A material subjected to particle bombardment (as in an accelerator) or irradiation (as in a research reactor) in order to induce a nuclear reaction. b. A nuclide that has been bombarded or irradiated. See: cross section; x-ray.

target² (Opt) The material into which a laser beam is fired.

task (Man Sci) A portion of a job assigned to or required of a person.

taste (Physiol) One of the chemical senses residing principally in the taste receptors of the tongue but dependent also on the tactile and temperature receptors of the mouth, as well as on the smell receptors of the nose. See also: sense organs.

NOTE: Primary taste sensations are classified as sweet, sour, bitter, and salty.

taxpayer[1] (Fire Pr) a. A citizen of a community who supports the fire service by paying taxes. b. (Brit) = ratepayer.

taxpayer[2] (Build; Fire Pr; Colloq) A poorly and cheaply constructed building housing several businesses, esp a structure without adequate firestops and potentially a serious fire hazard. A firetrap; quick burner.

TEAP = Transportation Emergency Assistance Program, operated by the Candian Chemical Producers' Association, providing assistance and guidance in emergencies involving chemicals in transit. Cf: CHEMTREC.

technical data (Techn) Specifications, standards, engineering drawings, instructions, reports, manuals, tabular data, or test results used in the development, production, testing, use, maintenance, and disposal of items, equipment, and systems.

technical evaluation (Techn; Testing) Test and analysis required to determine whether a system, component, equipment, or material meets design specifications, is functioning in a technically acceptable manner in its operational environment, and is technically suitable for use.

technical information (Techn) Information or processed data, including scientific information, that relates to research, development, engineering, test, evaluation, production, operation, use, or maintenance.

technical report (Techn) Any document written to describe and to record results obtained from scientific and technical activities.
NOTE: A technical report is considered to be unpublished material, even though it may be printed in many copies and distributed widely.

telescopic boom (Fire Pr) An arm or mast that extends and retracts, consisting of telescoping tubular sections, used to raise elevating platforms and the like.

teletherapy (Med) Radiation treatment, usu by gamma rays, administered by using a radioisotope source that is at a distance from the body. Cf: brachytherapy. See: radiation therapy.

telltale (Safety) A device, usu mechanical, used to signal a hazardous condition, such as inadequate overhead clearance.

temper[1] (Metall) To relieve the internal stresses in metal or glass and to increase ductility by heating the material to a point below its critical temperature and cooling it slowly. See: anneal.

temper[2] (Psych) a. The manner of behavior. Syn: temperament. b. Display of anger. An instance of uncontrolled behavior.

tempered glass (Build) Glass that has been treated to improve its stability and resistance to heat, distortion, and impact.

temperature (Phys) The intensity of sensible heat of a body as measured by a thermometer. Temperature is related to the kinetic energy of molecular motion. See: temperature scale.
NOTE: The lowest temperature (absolute zero) is reached when all molecular motion ceases; no upper limit is known, but the hottest temperature is assumed to exist in the centers of the hottest stars, estimated at over one billion degrees Kelvin.

temperature-controlled area (Storage) An enclosed space in which the temperature can be controlled within specific limits.

temperature failure (Build) 1. Buckling, collapse, fracture, or other destruction or deformation of a building structural member (often metallic) because of exposure to high temperature. 2. One of the limiting end points of a fire endurance test.

temperature gradient (Thermodyn) The difference in temperature per unit distance.

temperature inversion (Meteor) An increase in temperature with height in the lower atmosphere caused by radiative cooling of the lower layers of air, by subsidence heating of an upper layer, or by advection of warm air over cooler air or of cooler air under warm air. Inversions interfere with vertical air movement, so that smoke, air pollution, and smog is trapped near the ground. See also: lapse rate.

temperature scale 1. A graduated measure of the intensity of heat. 2. A thermometer scale. See also: absolute zero.
NOTE: Four temperature scales are used extensively: the Celsius (centigrade), Kelvin, and Rankine scales in science and technology, and the Fahrenheit scale, used popularly in English-speaking countries. The designation "centigrade" was officially replaced by "Celsius" by the Ninth General Conference on Weights and Measures in 1948. The Reaumur scale, on which the freezing point of water is 0° and boiling point 80°, is now largely obsolete. Conversion equations for the principal scales are

$$C = (5/9)(F - 32)$$
$$C = K - 273.16$$
$$C = (5/9)(R - 491.69)$$

$$F = (9/5)C + 32$$
$$F = (9/5)(K - 273.16) + 32$$
$$F = R - 459.69$$

$$K = C + 273.16$$
$$K = (5/9)(F - 32) + 273.16$$
$$K = (5/9)(R - 491.69) + 273.16$$

$$R = (9/5)C + 491.69$$
$$R = F + 459.69$$
$$R = (9/5)(K - 273.16) + 491.69$$

where C is Celsius, F is Fahrenheit, K is Kelvin, and R is Rankine. Several representative temperatures in the four scales are compared in the following table:

Temp	C	K	F	R
bp water	100	373.16	212	671.69
human body	37	310	98.6	558.29
fp water	0	273.16	32	491.69
abs zero	0	−273.16	0	−459.69

temperature senses (Physiol) The sensitive elements in the skin and body organs that produce feelings of warmth and cold. See also: homeostasis; sense organs.
NOTE: Within limits, the body adapts to warmth and cold. A hand plunged into cold water feels colder at first than later. Eventually the feeling of cold may disappear entirely. Thermal pain and skin injury occur at about 52°C (126°F).

temperature spots (Physiol) Small areas on the skin that are particularly sensitive to heat and cold.

tempering (Metall) The process of changing the physical properties of a substance, generally a metal alloy, such as steel, by heat treatment or mechanical deformation. The process is used to harden an alloy or to increase its toughness.

template (Techn) A pattern used as a guide in cutting, shaping, etc.

temporary storage (Comput) Storage locations reserved for intermediate and partial results.

temporary threshold shift (Audiom; Med) A hearing loss suffered as the result of noise exposure, all or part of which is recovered during a period of time after one is removed from the noise.

tenacity[1] (Techn) The property of a solid to resist rupture.

tenacity[2] (Fire Pr) A desirable quality of firefighting foam, referring to its ability to cover and cling to horizontal and vertical surfaces.

ten and fourteen (Fire Pr) A duty schedule consisting of 10-hour day shifts and 14-hour night shifts.

tenant space (Build) An area rented and used by a tenant, as opposed to a service area, public facility, elevator or stair shaft, and the like.

ten code (Commun) A code used in radio communication for frequently occurring transmissions.

NOTE: A few examples from the police ten code are: 10-1, receiving poorly; 10-2, receiving well; 10-3, stop transmitting; 10-4, ok; 10-5, repeat.

tendency (Psych) A disposition to act or behave in a certain way.

tender (Fire Pr) A fire truck, often equipped with a turret nozzle, that carries extra hose to supplement the hose loads of pumping engines or fireboats. Also called: hose tender; hose wagon.

tendon¹ (Anat) Fibrous tissue that connects muscle to bone.

tendon² (Build) A plastic coated cable used to tension concrete.

tenosynovitis (Med) Inflammation of the connective tissue sheath of a tendon.

tensile strength (Techn) 1. The resistance of a solid to elongation when a stretching load is applied. 2. The maximum load in tension that a material will withstand without fracture.
NOTE: Ductile materials stretch considerably and decrease in cross section before breaking. Maximum stress occurs just prior to necking down of the specimen. Tensile strength in this case is calculated from the maximum applied load divided by the original cross-sectional area of the test sample.

tension (Psych) A condition of anxiety and stress.

tent collapse See: floor collapse.

tenure (Fire Pr) Job security afforded to fire department officers and firefighters by law. It provides that dismissal can occur only for sufficient cause after an impartial hearing.

terminal (Comput) A keyboard device that prints on paper or displays characters on a cathode ray tube screen, used as an input-output device for a computer.
NOTE: Some terminals have light pens, function keys, and other means for inputting and editing data in a computer file.

ternary compound (Chem) A compound containing three elements.

territory (Fire Pr) An area covered by a given fire protection unit or organization.

test¹ (Educ) A set of questions given to an individual for the purpose of measuring aptitude or achievement in a given subject. Also called: examination.

test² (Statist) A procedure used to verify the significance or validity of a statistical result.

test fire (For Serv) A controlled fire set to evaluate firefighter performance, fire behavior, fire-detection devices and techniques, control measures, and other factors.

testimony (Law) Evidence presented by a witness, under oath, in a court of law.

testing (Techn) A procedure in inspection in which the properties, elements, or performance of materials or components are determined by established scientific principles.

test item (Educ) An element or question in a test.

test-wise (Educ) Characteristic of an individual who is experienced in taking tests and therefore may have an advantage over an inexperienced test-taker.

tetanic spasm (First Aid; Med) Involuntary contraction of the muscles, such as caused by an alternating current shock that "freezes" the victim to the conductor.

tetanus (Med) An infectious disease of sudden onset caused by the poison of the germ called Clostridium tetani. It is characterized by muscle spasms or prolonged contraction of a muscle. Also called: lockjaw.

tetany (Med) A physical condition marked by muscular twitching, tremors, spasms, cramps, or convulsions.
NOTE: The condition could be caused by calcium insufficiency.

tetraethyl lead (Chem; Abb: TEL) A colorless, oily, organometallic compound, $Pb(C_2H_5)_4$, that is added to gasoline to increase its <u>octane number</u> and reduce engine knock.
NOTE: A rather strong poison that is absorbed by the skin.

tetrahedron (Geom) A three-dimensional geometric figure having four triangular faces.

textile A material woven from natural or synthetic fibers or yarns, used chiefly for the manufacture of clothing, upholstery, curtains, wall coverings, etc. See also: <u>fabric</u>.

TGS = <u>Taguchi gas sensor</u>

theodolite (Surv) A precision surveying instrument. See: <u>transit</u>.

theorem (Math) A proposition derived from other propositions stating a demonstrable mathematical truth. Cf: <u>axiom</u>; <u>postulate</u>.

theoretical air (Comb) The quantity of air, calculated from the chemical composition of a substance, that is required to burn the substance completely. Also called: <u>stoichiometric air</u>; <u>theoretical combustion air</u>.

theory 1. A body of facts and principles used to explain a phenomenon. 2. One or more plausible or logical assumptions used as the basis for a policy or procedure.

therapeutic (Med) Pertaining to medical treatment; curative.

therapy (Med) Treatment administered to cure a pathological condition.

therm (Metrol; Brit) A unit of heat equal to 10^5 Btu.

thermal¹ (Thermodyn) Pertaining to heat.

thermal² (Meteor) A vertically rising column of warm air. See also: <u>thermal column</u>.

thermal burn (Med) A burn of the skin or other organic material due to radiant heat. See: <u>flash burn</u>; <u>radiation burn</u>.

thermal column (Fire Pr) The column of smoke and hot gases rising above a fire. Also called: <u>convection column</u>; <u>thermal updraft</u>.

thermal conduction (Phys) The transport of energy in a system due to a temperature gradient. See: <u>transport process</u>.

thermal conductivity (Phys; Symb: K) The capacity of a material to transmit heat, measured as the quantity of heat passing through a unit cube, the opposite faces of which are maintained at a unit temperature difference. The resultant value is called the <u>thermal conductivity coefficient</u>.
NOTE: Thermal conductivity is the measure of the ease or difficulty with which heat travels though a given substance. In general, dense, solid materials, such as metals, are better heat conductors than light or loose substances, such as granular or fibrous materials. Good thermal conductors are usu also good electrical conductors.

thermal convection See: <u>convection</u>

thermal decomposition (Comb) The breakdown of complex substances into simpler compounds under the influence of heat. Syn: <u>thermal degradation</u>. See: <u>pyrolysis</u>.

thermal degradation = <u>thermal decomposition</u>

thermal detector = <u>heat detector</u>

thermal diffusion (Phys) A molecular transport process induced by heat. The differential transport of various species in a gaseous mixture under the influence of a temperature gradient. See: <u>transport process</u>.

thermal efficiency (Techn) The ratio of heat used for a desired purpose, such as work to total heat generated.

thermal expansion (Phys) The proportional increase in length, volume, or superficial area of a body with rise in temperature. The amount of this increase per degree temperature, called the <u>coefficient of thermal expansion</u>, is different for different substances.
NOTE: Gases at constant pressure and almost any temperature have a practically uniform coefficient of

expansion, equal to 1/273 of their volume per Celsius degree temperature rise. Liquids also generally expand as temperature rises. Water has an anomalous contraction between 0° and 4°C. Otherwise it behaves normally.

thermal expansion coefficient (Techn) The fractional increase in the length, area, or volume of a body per degree of temperature rise.

thermal ignition (Comb) The initiation of combustion of a substance by supplying heat until the ignition point is reached.

thermal injury = burn

thermal insulation (Techn) Poorly conducting material used to retard the passage of heat. See: insulation[2].

thermal pollution (Env Prot) Discharge of heat into bodies of water.
NOTE: The increased warmth can deplete oxygen and eventually harm fish and other organisms in the water.

thermal radiation[1] (Phys) Heat energy in the electromagnetic spectrum, ranging from ultraviolet, through visible light, to the far infrared frequencies, emitted by molecules of substances excited to thermal energies. See also: emissivity; heat transfer; radiation.

thermal radiation[2] (Nucl Phys) Electromagnetic radiation emitted from the fireball produced by a nuclear explosion. Thirty-five percent of the total energy of a nuclear explosion is emitted in the form of thermal radiation, as light, ultraviolet, and infrared radiation.

thermal shield (Nucl Phys) One or more layers of high-density material used to reduce radiation heating.

thermal shock (Techn) The effects manifested in a substance or material that is heated or cooled rapidly.
NOTE: Thermal shock is often destructive. Hot glass shatters easily when it is suddenly splashed with water. The process is used in heat treatment to change the crystalline structure and properties of metals.

thermal shock resistance (Techn) The ability of a material to withstand sudden heating or cooling or both without cracking or spalling.

thermal stress (Techn) Mechanical stress produced in a body or system that is heated unequally and is not free to expand or contract to compensate for differences in expansion.

thermal transmittance (Thermodyn) The time rate of heat flow per unit area under steady conditions from a fluid on the warm side of a barrier to a fluid on the cold side per unit temperature difference between the two fluids. Cf: thermal conductivity.

thermal turbulence (Fire Pr) An often violent atmospheric disturbance above and downwind of a major fire.
NOTE: The disturbance is caused by the thermal column, in which hot rising gases displace cooler air.

thermal updraft = thermal column

thermionic conversion (Electr) The conversion of heat into electricity by evaporating electrons from a hot metal surface and condensing them on a cooler surface. No moving parts are required. Cf: thermoelectric conversion.

thermistor (Electr) A temperature-responsive resistor; one in which the resistance element changes in value in a specific way to changes in temperature.

thermite (Techn) A mixture of powdered metallic oxides (usually iron) and powdered aluminum that burns with great heat but little flame or gas. Used in thermite welding and for cutting through refractory materials.

thermochemistry (Chem) The branch of chemistry concerned with the heat involved in changes of state and chemical reactions. Reactions that evolve heat are called exothermic; those that absorb heat are endothermic.

thermocouple (Techn) A temperature-measuring device consisting of a

junction of two dissimilar metals. When heated, an electrical current is generated that is proportional to the temperature.
NOTE: Typical thermocouples are made of platinum-rhodium, chromel-alumel, and iron- or copper-constantan. Several thermocouples in parallel or in series are called a thermopile.

thermodynamics (Thermodyn) The branch of physics that deals with the relations between heat and mechanical, chemical, and electromagnetic energy in systems and the conversion of one into the other. Thermodynamics is concerned with matter and energy in bulk, and systems in thermal equilibrium.
NOTE: The basic principles are expressed as: 1) heat and work are related quantitatively; 2) heat flows naturally from a hotter body to a cooler one; and 3) every substance has an entropy, a measure of the order in its system.

thermoelectric conversion (Phys) The conversion of heat into electricity by the use of thermocouples.

thermoforming (Chem Eng) The process of shaping plastics after softening them by heat.

thermometer (Metrol) An instrument for measuring temperature, generally consisting of a bulb containing mercury or other liquid and a capillary mounted on a scale marked in degrees. The height of the liquid in the capillary indicates the temperature. See: temperature scale.
NOTE: Other types of temperature-measuring instruments are based on bimetallic strips, Bourdon gages, electrical resistance, thermocouples, vapor pressure, etc.

thermonuclear reaction (Nucl Phys) A reaction in which very high temperatures bring about the fusion of two light nuclei to form the nucleus of a heavier atom, releasing a large amount of energy.
NOTE: In a hydrogen bomb, the high temperature required to initiate the thermonuclear reaction is produced by a preliminary fission reaction.

thermopile (Electr) A device for converting heat energy into electrical energy, usu consisting of a series of thermocouples.

thermoplastic (Chem; Phys) A solid material capable of being reversibly softened or fused by heat.
NOTE: With reference to plastics, the vinyls, acrylics, and polyethylenes can be softened or melted repeatedly with heat and hardened by cooling.

thermoscreen (For Serv) A protective structure for meteorological instruments. Syn: instrument shelter.

thermosetting (Chem; Phys) Capable of undergoing a chemical reaction that results in a change from a soft to a hard substance when heated.
NOTE: With reference to plastics, descriptive of those that are heat-set in final processing to a permanent state of hardness. Examples are the phenolics, ureas, and melamines.

thermosphere (Atm Phys) The top layer of the atmosphere, lying above the mesosphere, characterized by increasing temperature with height. See also: atmosphere; pause.
NOTE: Although temperature increases with altitude in the thermosphere, there is little sensible heat, because the air is extremely thin.

thermostable (Phys) Resistant to changes by heat.

thermostat (Techn) An automatic control device responsive to changes in temperature, used to control heating systems, flows of coolants in engines, etc.
NOTE: A graduated thermostat is one in which the control element moves in proportion to the change in temperature; a snap-action type closes and opens completely and instantaneously when the preset temperature is reached.

thesis (Educ) 1. A formal proposition offered for proof or one advanced as a statement to be proved. 2. A treatise based on original research prepared by a candidate for an advanced degree.

thick water See: viscous water

thickening agent See: water thickening agent

thimble (Fire Pr) A metal U-shaped liner fitted into an eye-splice of a rope to reduce wear.

thinner (Techn) A solvent added to paint, varnish, lacquer, and the like to make them brushable and to improve their spreading characteristics.

Thiokol (Chem) A trade name for a copolymer of ethylene dichloride and sodium tetrasulfide.

third-class structure (Fire Pr; Insur) A building of wooden frame construction.

third-degree burn See: **burn**

thirst (Physiol) A craving for liquids, esp water. A sensation of dryness in the mouth and throat, accompanied by a desire to drink.

thixotropy (Chem) The property displayed by a substance that sets into a gel when left standing but liquifies when stirred.
 NOTE: Thixotropic substances usu gel faster as the temperature is raised. Such gels are formed when a small quantity of electrolyte is added to certain metallic oxide sols.

thoracic (Anat; Med) Pertaining to the chest, or thorax.

thorax (Anat; Med) The chest; the part of the body between the neck and abdomen.

thorium (Chem; Nucl Phys; Symb: Th) A heavy, trivalent, metallic element; atomic number 90 and atomic weight 232.
 NOTE: When bombarded with neutrons, thorium changes into uranium, becoming fissionable, and thus a source of nuclear energy. Powdered thorium has a low ignition temperature.

thou = **mil**

thread (Techn) A helical groove cut into a cylindrical surface. See: **iron pipe thread**; **standard thread**.

thread cutter (Techn) A tool used to cut new threads on pipes, couplings, rods, and the like or to recut worn ones.

three-bagger (Fire Pr; Colloq) A three-alarm fire.

three-company box (Fire Pr) A fire alarm box or location, usu in a congested or light mercantile district, which calls for a normal first-alarm response of three engine or pumper companies, ladder truck, rescue squad, chief officers, and other units assigned to the area.

three-door house (Fire Pr) A fire station with three large apparatus doors. Such a fire house usu has a pumper, ladder truck, and a squad truck or chief's car. Also called: **three-bay station**.

three-horse hitch (Fire Pr; Archaic) A team of three horses formerly used to pull steam fire engines and other wagons.

three-nurse carry (Med; First Aid) A method by which three nurses can lift and carry a patient or casualty. See: **rescue carry**.

three-platoon organization (Fire Pr) A fire-fighting force or company divided into three groups, each with a duty tour of 56 hours per week.

three-stage ignition (Comb) A separated three-stage flame process in which the products of the first stage (cool flame) goven the initiation of the second stage (blue flame), the products of which, in turn, govern the initiation of the third stage (normal or hot flame).

three-way radio (Commun) A radio communications system in which the base station transmits on one frequency and the mobile units transmit to base on another (two-frequency simplex), while the mobile units transmit to each other on a third frequency (duplex), thereby permitting conversation in both directions, between mobile unit and base station as well as between mobile units. Cf: **two-way radio**.

three-way wye See: **wye**

three-wire system (Electr) A system for distributing electrical energy through a three-wire cable conductor. Except for a "delta" system, two of the wires carry current; the third wire is neutral and is grounded. In a 115-volt system, the potential difference between the outer wires is 230 volts, and between either wire and neutral is 115 volts.

threshold (Audiom) The point at which a person just begins to notice that a sound is audible.

threshold dose (Dosim) The minimum dose of radiation that will produce a detectable biological effect. See: absorbed dose; biological dose.

threshold limit value (Toxicol; Abb: TLV) Concentrations of airborne substances to which it is believed that nearly all workers may be repeatedly exposed without adverse effect. These values may be breathed continually for 8 hours per day without harm.
NOTE: Threshold limits should be used as guides in the control of health hazards and should not be regarded as fine lines between safe and dangerous concentrations.

thrombosis (Med) The formation of a blood clot.

thrombus (Med) A blood clot that forms as a result of injury to the wall of a blood vessel, lodges in the vessel, and obstructs circulation. See: phlebitis.
NOTE: If the clot breaks loose and is carried away by the blood stream, it is an embolus, which may eventually plug a smaller blood vessel and cause an infarction. The formation of a blood clot is called thrombosis.

throttling (Fl Mech) An adiabatic expansion of a fluid through a constriction across which there is a pressure difference. In general, to throttle is to decrease the flow of a fluid by partly closing a valve or other restricting device. Also called: choking.

through ladder (Safety) A ladder that a man must step through in order to get off at the top end.

through the roof (Fire Pr) The state in which a fire has gained sufficient headway to vent itself by burning through a roof. See: vented fire.

throw¹ (Fire Pr) To raise or erect quickly, for example, a ladder.

throw² (Fire Pr) a. To deliver water rapidly from one or more hose streams. b. Referring to the rate of water discharge or the length (range) of a fire stream.

throw³ (Salv) To spread quickly, such as salvage covers at a fire.

throwaway (Techn) An assembly, subassembly, module, or component that is discarded when it fails; no repairs are made.

throwing covers (Salv) Spreading of salvage covers at a fire.

thunder (Meteor) A loud, explosive sound wave produced by an electrical discharge in the atmosphere. See: lightning.

thundercloud (Meteor) A cloud with vertical development reaching up to 13 miles. In cold air the top portion usu consists of ice crystals, and if so, it is called a thunderhead. A flat top is a sign of a fully mature thunderstorm. See: cloud.

thunderhead See: thundercloud

thunderstorm (Meteor) A violent weather phenomenon developing from strong updrafts producing cumulonimbus clouds with anvil-shaped tops, thunder, lightning, and showers of torrential rain and sometimes hail. Often accompanied by strong, gusty surface winds. See: wind.

tidal volume (Med) The quantity of air that is inhaled or exhaled during normal quiet breathing.

tie beam (Build) A beam across the span of a roof tying the feet of the rafters.

tied up (Fire Pr) 1. The state of a fire company already engaged and not available to cover its district. Also called: out of service. 2. The state of a pumper connected to a hydrant. 3. The state of a fire apparatus delayed by traffic or weather conditions.

tie-in¹ (Fire Pr) a. Securement to a ladder or other object with a rope, belt, or leg lock. b. The result of tying in.

tie-in² (Fire Pr) a. A connection made into a hose line, esp into one that has already been laid. b. The connection so made.

tier (Fire Pr) A layer of accordion-folded hose.

tie rod (Fire Pr) A metal rod that runs along a ladder rung and holds the beams together. See: ladder.

tiller (Fire Pr) A steering wheel at the rear of a large, tractor-drawn ladder trailer used for steering the rear wheels of the rig around corners.

tillerman (Fire Pr) The driver who operates the tiller wheel of a tractor-trailer type ladder truck.

timber (Build; For Serv; Mining) 1. Wood in its natural state. 2. A piece of cut lumber of large dimensions. 3. A massive wooden structural support member used in mines, such as a post, cap, or stull.

timber set See: square set

timbre (Acoust) The quality given to a sound by its overtones; the tone distinctive of a singing voice or a musical instrument.

time (Phys) The dimension of duration measured by a clock, usu by comparison with a known frequency.
   NOTE: The standard unit of time is the mean solar day, and the practical unit is the mean solar second (1/86,400-th of a mean solar day).

time and motion study (Ind Eng; Man Sci) An industrial engineering function involving a study of the time required and the motions involved in the performance of a job.
   NOTE: The purpose may be to establish standards of performance, the best way of doing a job, or to determine incentive wage rates.

time charge (Safety) A standard measure of disability in which death, loss of use of any body member, organ, or function is converted to a number of days for the purpose of computing accident severity rates.

time constant (Electr) A measure of the response time of an electrical or electronic circuit or other device.

timed test (Educ) A test for which a time limit has been set, obliging the taker to stop, whether he has finished or not.

time lag (For Serv; Abb: TL) The time interval during which a piece of fuel loses about 63% of the difference between its initial and equilibrium moisture contents.

time of arrival (Fire Pr) The time that a responding officer or fire department unit reaches the fireground.

time-of-flight spectrometer (Phys; Spectr) A device for separating and sorting neutrons (or other particles) into categories of similar energy, measured by the time it takes the particles to travel a known distance. Cf: mass spectrometer.

time sense (Psych) The ability to judge intervals of time. Also called: time perception.

time-sharing¹ (Comput) Descriptive of a computing system in which numerous terminal devices use a central computer concurrently for input, processing, and output functions.

time-sharing² (Man Sci) a. The apportionment of time-availability intervals of various items of equipment to complete several tasks by alternating and interlacing them. b. The use of a device for two or more purposes during the same overall time interval, accomplished by interspersing the separate actions in time.

time-temperature curve (Fire Pr) A graph that shows the increase in temperature of a fire as a function of time, beginning with ignition and ending with burnout or extinguishment. See: standard time-temperature curve.
   NOTE: A time-temperature curve is an important variable in fire-resis-

tance rests of materials and structures.

tin-clad door (Build) A door of two- or three-ply wood core construction, covered with galvanized steel or terne plate. See: fire door.

tincture (Pharm) An alcohol solution of a drug or chemical substance, for example, the antiseptic tincture of iodine.

tinder (Fire Pr) Dry, flammable material that can be ignited easily with a match or a spark.

tinning (Techn) Any work with tin, such as tin roofing, but in particular in soldering, the primary coating with solder of the two surfaces to be united.

tinnitus (Audiom; Med) A ringing sound in the ears.

tip[1] (Fire Pr) The upper end of a ladder, as opposed to the heel. Also called: top, or fly.

tip[2] (Fire Pr) A replaceable fitting that can be attached to a playpipe to change the size or pattern of a hose stream. Often synonymous with nozzle.

tip[3] (Techn) The nozzle of a burning torch or the heated point of a soldering iron.

tip-up extinguisher (Fire Pr; Brit) = soda-acid extinguisher

tissue (Anat) An aggregation of similar or related cells that perform a specific function in the body.
NOTE: Tissues are divided into four classes: 1) epithelial, surface or lining tissues (commonly called skin); 2) connective, including blood, bone, and cartilage; 3) muscular; and 4) nervous. Sometimes vascular tissue (blood, lymph, and related body fluids) is separated from connective tissue and distinguished as a fifth class.

titanium (Metall; Symb: Ti) A light, silvery-white metallic element, atomic number 22, atomic weight 47.9, mp 1668°C, bp 3260°C. Used mainly in military hardware and aircraft structures.
NOTE: In particulate form, titanium ignites readily and burns with intense heat. The metal is capable of igniting spontaneously in oxygen-rich atmospheres.

titles See: ranks and titles

titration (Chem) A gradual measured addition of one solution of known concentration to another until the end point of the reaction between them is reached.
NOTE: The process is used to determine the concentration of an unknown solution.

TLFM = time lag fuel moisture

TLV = threshold limit value

TMB = trimethoxyboroxine

TNT = trinitrotoluene

TNT equivalent (Milit) A measure of the energy released in the detonation of a nuclear explosive, expressed in terms of the weight of TNT (the chemical explosive trinitrotoluene) that would release the same amount of energy when exploded. It is usu expressed in kilotons or megatons.
NOTE: The TNT equivalence relationship is based on the fact that 1 ton of TNT releases one billion (10^9) calories of energy.

toe (Mining) 1. The bottom of a drill hole. 2. The lower end of a stull.

toe piece (Fire Pr; Brit) A metal projection on the underside of ladder spars that holds the ladder away from a building and thus allows room on the rungs for the toes in climbing.

toeboard (Safety) A low vertical barrier at floor level erected along exposed edges of a floor opening, wall opening, platform, runway, or ramp to prevent tools or material from falling off.

toggle[1] (Fire Pr) A device at the end of a tormentor pole that attaches the pole to a ladder to assist in raising operations. Also called: swivel.

toggle[2] (Electr) A flip-type electrical switch.

tolerance¹ (Mech Eng) The limits of permissible inaccuracy in the fabrication of an article above and below its design specifications.

tolerance² (Med; Psych) a. The ability of an organism to endure and adjust to hardship, stressful conditions, or unpleasant stimuli, such as noise, pain, heat, and the like. b. The ability to endure the effects of a drug or poisonous substance without showing adverse signs.

toluene (Chem) A volatile, flammable hydrocarbon, $C_6H_5CH_3$, derived mainly from petroleum, but also from coal. The basis of TNT, lacquers, saccharin, and many other chemicals.
NOTE: Toluene is similar to benzene, but is less flammable, volatile, and toxic.

ton (Metrol) A unit of weight equal to 2,000 pounds (short ton) or 2,240 pounds (long ton).
NOTE: The common short ton is equal to 907 kg, or 0.907 metric ton.

ton of refrigeration (Techn) A cooling effect equal to 12,000 Btu per hour.

tongs See: sprinkler tongs

tonne = metric ton

tonometer (Audiom) A device used to measure pitch or to produce tones of a given pitch.

tonus (Physiol) The normal state of partial contraction of a muscle.

topical (Med) Local, or on the surface, such as an antiseptic applied to a wound, contrasted with a medicine taken internally.

topless match (Techn) A self-extinguishing match.

topography (Geophys) 1. The natural and artificial features of the earth's surface, including elevations, waterways, communications and transportation systems, forests, fields, man-made buildings and structures, and the like. 2. The art of charting the surface features and elevations of the earth on maps. Cf: cartography.

topology (Math) A branch of mathematics that deals with transformations of geometric surfaces in space, esp those that retain one-to-one correspondence after stretching or twisting.

torch¹ (Fire Pr) a. A person hired to set a fire for fraudulent or criminal purposes. Syn: arsonist. b. To start a fire.

torch² (Fire Pr; Archaic) A flaming brand once carried to fires to provide illumination for fire-fighting operations.

torch³ (Comb) A kind of pyrotechnic or fireworks device.

torch⁴ (Metall) A device that produces a hot flame jet that can be used to make solder joints, to cut metals, etc.

torch⁵ See: drip torch

torch⁶ (Fire Pr; Brit) a. A flashlight. b. A signal flare.

torch job (Fire Pr) An arson or incendiary fire.

torchman (For Serv) A firefighter who is responsible for setting back fires.

tormentor pole (Fire Pr) A wooden, metal, or fiberglass pole attached by a swivel or toggle to the top of the main section of an extension ladder and used to guide and steady the ladder during raising and to prop it firmly while in use. Also called: pole; staypole. Cf: crotch pole.

tornado (Meteor) A violent, destructive whirlwind characterized by a low-pressure, funnel-shaped column extending downward from a cloud and touching the earth. The funnel ranges in thickness from 10 to 100 meters and circular wind speeds reach 200 to 400 mph. A tornado is usu associated with a squall line.

torque¹ (Mech Eng; Symb: L) The measure of a force acting to set a body in rotation and defined as L = r • F, where r is the position vector with respect to the origin and F is the applied force.

torque² (Mech Eng) The torsional moment (force times lever arm) that twists a fixed, rigid body, such as a shaft, about its axis of rotation.

torr (Phys) A unit of pressure equal to 1 mm Hg at 0°C or 1.33322 millibars, named after Torricelli. One atmosphere is equal to 760 torr.

Torricelli theorem (Fl Mech) The principle that a vertical jet of liquid will rise approx to the height of the free liquid in the vessel from which the jet is supplied. Also called: **Torricelli law**, which states that the velocity of flow from an orifice in a container of liquid is equal to the velocity of a body falling freely from rest over a distance equal to the depth of the orifice below the free surface of the liquid.

torsion (Mech) The twisting of a body about an axis. Within limits, a restoring force occurs.

torsion-bar spring (Mech Eng) A straight bar spring, usu cylindrical, serving as the elastic member in vehicular suspension. One end of the bar is secured to the vehicle frame while the other end is supported by and free to rotate in a bearing. Torsional loads are applied to the bar by means of an arm fastened to the free end and rotated in a plane perpendicular to the longitudinal axis of the torsion bar.

total emissivity (Thermodyn) The ratio of the radiant flux density of a radiating body to that of a **black body** at the same temperature. See also: **emissivity**; **spectral emissivity**.

total loss (Insur) 1. Complete loss of an insured property with no salvage value remaining. 2. A loss involving the maximum amount for which a policy is written.

total pressure (Fl Mech) The sum of the velocity pressure and the static pressure of a fluid.

total risk (For Serv) The sum of natural (e.g., lightning) fire and human (e.g., smoking) risks.

touch (Physiol) The special cutaneous sense of the **tactile organs** that responds to physical contacts and applied pressures. See: **sense organs**.
NOTE: Few touch sensations have been named, although tickle, vibration, deep pressure, and lively contact are sometimes used. As with heat and cold sensations, steady pressure is most strongly felt when it is first applied, after which it gradually fades and may disappear.

touch-off (Fire Pr) A fire believed to be of incendiary origin. See: **arson**.

tour¹ (Fire Pr) A work period of a platoon or group of firefighters.

tour² An inspection round. A complete patrol circuit by a security guard. See: **round**.³

tournament (Fire Pr) A gathering of fire department units, usu volunteer fire companies, at which competitions are held in ladder raising, laying hose, hooking up, replacing burst couplings, and the like. Cf: **muster**.

tourniquet (First Aid) A device used to check the flow of blood, as a bandage twisted tightly around a limb with the aid of a stick to squeeze the blood vessels shut.

tower¹ (Build) A structure typically several times taller than its diameter. See also: **smokeproof tower**.

tower² (Fire Pr) a. A structure used for training and drill in firefighting skills. Syn: **drill tower**; **training tower**. b. A structure used to hang hose for drying.

tower³ (Fire Pr) A fire department **water tower** truck.

tower⁴ (For Serv) An elevated forest-fire lookout platform, Also called: **lookout tower**.

towered (Build) Descriptive of a building in which upper floors have smaller areas than the base of the building, either because the exterior walls are slanted inward or are set back.

407

tower ladder (Fire Pr) An aerial platform apparatus with a ladder mounted on the boom.

towerman (For Serv) A lookout man stationed in a tower.

towing pad (Transp) An eye on the front or rear of a vehicle to permit attachment of a towing line.

toxemia (Med) A state in which the blood contains toxins, usu of bacterial origin, resulting in damage to tissues and organs, edema, and bleeding, with symptoms of fever, fatigue, pain, or other distress. Cf: septicemia.

toxic (Med; Toxicol) Poisonous; destructive to body tissues and organs or interfering with body functions.

toxic dust (Ind Hyg; Toxicol) Dust that may be harmful to the respiratory system or to other parts of the body when material passes from the respiratory tract into the blood stream.

toxicity (Med; Build) The state or degree of being poisonous. A relative property of a chemical agent, referring to a harmful effect on some biological mechanism and the condition under which this effect occurs.

toxicology (Abb: Toxicol) The study of poisons and their mechanisms, effects, and methods of treatment or counteraction. See: poison.

toxin (Med; Pharm) A noxious or poisonous substance produced by a microorganism.
NOTE: Exotoxins are secreted by bacteria into the surrounding fluids, whereas endotoxins are retained within the organism and released when the organism disintegrates. Toxins are often specific in their effects. Some attack certain organs, others damage blood cells, while still others produce dysentery, typhoid, diphtheria, etc.

tracer (Med; Techn) Usu an isotope of an element, a small amount of which may be incorporated into a sample of material (the carrier) in order to follow (trace) the course of that element through a chemical, biological, or physical process, and thus also to follow the larger sample. Syn: label; tag.
NOTE: If the tracer is radioactive, observations are made by measuring the radioactivity. If the tracer is stable, mass spectrometers, density measurement, or neutron activation analysis may be used to determine isotopic composition.

trachea (Anat) The windpipe; the tube extending from the bronchial tree of the lungs to the larynx.

tracheotomy (Med) An operation in which an opening is made in the trachea for the purpose of inserting a tube through which oxygen is pumped into the lungs. Used as an emergency measure when the airway in the throat is constricted or blocked. Also called: tracheostomy.
NOTE: The technique was once used freely to alleviate breathing difficulties of burn victims with upper respiratory involvement, but is now used cautiously because of the critical danger of ulceration and infection.

track (Fire Pr; Archaic) A berth for a fire apparatus in a fire station, the term deriving from the time steel rails were used to guide horse-drawn wagons into the fire house. See: rails.
NOTE: A company is said to be "on the track" when its apparatus is in its berth.

track-laying vehicle (Transp) A motorized vehicle that travels on endless belts or tracks that distribute its weight to achieve more uniform ground pressure for improved traction and mobility on adverse soils.

track link (Transp) A rigid unit that is flexibly connected to other links to form a jointed-type track.

traction[1] (Phys) The friction between a powered surface, such as a wheel, and another surface, such as a roadway.

traction[2] (Transp) The act of pulling a load over a surface.

traction[3] (Med) Application of a pulling force on a body part, as on a fractured limb, to maintain proper length during healing.

tractor¹ (Transp) A wheeled, self-propelled vehicle for hauling other vehicles, trailers, or equipment; used extensively in agriculture.

tractor² (Transp) The power and drive unit of a highway trailer truck. Also called: truck-tractor.
 NOTE: The tractor and trailer together are called a rig or a semi.

tractor boss (For Serv) An individual responsible for supervising the work of the tractors assigned to him for fireline construction. Syn: cat boss.

traffic accident (Transp) Any accident involving one or more motor vehicles in motion on a roadway.

traffic engineering (Civ Eng) The branch of engineering concerned with the determination of capacities and layouts of highways and streets for vehicular movement.

trafficability (Civ Eng) The capacity of a soil to support moving vehicles.

trailer¹ (Fire Pr) The rear portion of a tractor-drawn aerial ladder truck. The trailer carries the ladder and other fire-fighting equipment.
 NOTE: The larger ones require a tillerman to steer the rear wheels when turning corners and moving into position.

trailer² (Fire Pr) A long, fuse-like trail of fast-burning material used by arsonists to spread fire rapidly through a structure. See arson. See also: plant; set.

trailer³ See: mobile home.

trailer court (Build) A rental parking area for mobile homes, supplying utilities and other services. Syn: trailer park.

trailer pump (Fire Pr) A self-contained pumping unit on a two-wheeled chassis that is towed behind a fire truck or hauled into position manually.

train¹ (Educ) To teach a skill by instruction, demonstration, and drill.

train² (Mech Eng) A series of interconnected mechanical parts, such as gears, for transmitting motion.

train³ (Fire Pr) To aim or direct something toward a given point, such as a hose stream or aerial ladder.

training (Educ) A program of instruction, exercises, practice, review, testing, and the like, designed to develop proficiency and skill in a job or special activity.

training academy (Fire Pr) A fire department school equipped with classrooms, drill tower, structures in which fires can be set, water basin for pumper drafting, pits and tanks for flammable-liquid fire demonstrations, etc. Also called: drill school.

training manual (Fire Pr) A textbook used for training firefighters.

training officer (Fire Pr) A fire department officer responsible for organizing and conducting the training program for the department. Cf: drillmaster.

training tower (Fire Pr) A structure 3 to 6 stories high, equipped to train firefighters in basic hose, ladder, and rescue techniques. Syn: drill tower; or simply, tower.
 NOTE: Training towers are also used sometimes as hose towers to drip-dry wet hose.

trait (Biol; Psych) 1. A biologically inherited characteristic. 2. A consistent pattern or peculiarity of behavior.

trajectory (Phys) The path followed by a projectile.
 NOTE: By extension, the path described by a hose stream.

transceiver (Commun) A combination radio transmitter and receiver.

transducer (Electr) 1. A device that converts input mechanical, electromagnetic, or acoustical energy into an output energy of another form, as for example, a microphone that converts sound into electrical pulses. 2. A device, such as a piezoelectric crystal or photoelectric cell, that outputs a signal having a specific relationship to the input without necessarily converting the form of the energy.

transfer (Fire Pr) 1. Reassignment of a firefighter to a different firefighting unit. 2. Reassignment, usu temporary, of fire companies to cover territories in which firefighting forces are engaged or to strengthen forces in a given territory to meet emergency situations. Also called: move-up, q.v.

transfer valve (Fire Pr) A control valve used to switch a multistage pump between volume and pressure operation.

transformer (Electr) A device composed of two or more coils, linked by magnetic lines of force, used to transfer energy from one circuit to another. Variations in current in the primary coil induce corresponding variations in currents and voltage in the secondary.

transfusion (Med) The administration of whole blood, glucose, or plasma expanders to a patient to restore the blood level in the body in cases of hemorrhage, severe burns, certain forms of shock, and a variety of blood disorders. See also: blood type.

transient (Electr) A temporary variation in voltage or current, commonly caused by a sudden change in load.
 NOTE: Transients occur in power grids from lightning strokes during thunderstorms, sometimes causing power failure.

transistor (Electr) A semiconductor device with three or more electrodes for controlling and regulating the flow of electrons. Performs much the same function as a vacuum tube.

transit (Surv) An instrument, consisting of a telescope, various scales, and leveling devices, used for measuring horizontal and vertical angles in surveying. A precision transit is called a theodolite.

transit shed (Transp) A storage building normally located adjacent to a cargo ship berth and esp adapted for handling material received or shipped by water.

transition point (Phys) 1. A point at which a system changes phase, including changes from one solid phase into another. 2. Broadly, the temperature at which a liquid becomes a vapor (boiling point), a solid becomes a gas (sublimation point), or a solid becomes a liquid (melting point).

translation[1] (Math; Phys) Motion without rotation.

translation[2] (Med) Motion of a body, such as one hurled by a blast.

transmission (Transp) The gear box between the engine and differential of a vehicle, which allows a choice of gear ratios for efficient driving.

transmission line (Electr) A conductor carrying electrical power or a signal from one point to another, for example, from a power-generating station to a substation or from a telephone exchange to a subscriber.

transmission loss (Acoust; Build) The ratio, expressed in decibels, of sound energy incident on a structure to the sound energy that is transmitted.
 NOTE: Applicable both to building structures (walls, floors, etc.) and to air passages (muffler, ducts, etc.).

transmit[1] (Commun) To send a message or signal by electronic means.

transmit[2] (Fire Pr) To send an alarm from a central station over the alarm or voice circuits.

transmitter[1] (Commun) A device that converts sound waves into electromagnetic pulses and relays or broadcasts them to a remote receiver, for example, a telephone or radio.

transmitter[2] (Fire Pr) A device over which coded alarm signals are sent.

transmutation (Nucl Phys) The transformation of one element into another by a nuclear reaction or series of reactions. For example: the transmutation of uranium^{238} into plutonium^{239} by absorption of a neutron.

transpiration (Biol) The passage of water vapor from living tissue, for example, from plant leaves, to the atmosphere.

transport^1 (Transp) The movement of goods and passengers by land, sea, or air.

transport^2 (Phys) The movement of matter or physical properties from one point to another.

transportable unit (Build) A facility generally designed to be brought to a given location for semipermanent installation.

transport process (Phys) A molecular process by which energy, matter, or momentum is conveyed by temperature, concentration, velocity, or pressure gradients.
NOTE: The common transport processes are diffusion, the transport of a chemical species due to its concentration gradient; thermal conduction, the transport of energy due to a temperature gradient; thermal diffusion, the transport of a species due to a temperature gradient; viscosity, the transport of momentum due to a velocity gradient; pressure diffusion, the differential transport of a species due to a pressure gradient; and the inverse themal diffusion, or Soret effect, the transport of energy due to a concentration gradient.

transport vehicle (Transp) A vehicle primarily intended for carrying personnel and cargo.

transport velocity (Phys) The minimum air velocity required to move suspended particles in an air stream.

transverse (Geom) Perpendicular to the fore-and-aft or longitudinal axis.

trapezoid (Geom) A four-sided figure with two opposite sides parallel.
NOTE: In Britain, a trapezoid is called a trapezium. A trapezium in the US is any general four-sided figure.

trauma (Med; Psych) A wound or injury to tissue by physical or chemical means, including burns, cuts, abrasions, bone fractures, sprains, and asphyxia. Often accompanied by shock and hemorrhage.
NOTE: Psychic trauma is the physiological result of an emotional shock.

travel^1 (Mat Hand) The distance between the centers of the top and bottom pulleys of a block and tackle.

travel^2 (Civ Eng) The movement of a machine from one location to another on a jobsite.

travel distance (Build) The distance occupants of a building have to travel to reach a place of safety.

travel time (Fire Pr; For Serv) The elapsed time between the departure of firefighters to a fire and their arrival at the scene. See: response time.

travel-time map (For Serv) A map that shows the time required to reach various parts of a protected area by a firefighter or crew from given starting points. Cf: fire protection map.

traveling cable (Build) A cable made up of electrical or communication conductors or both, and providing electrical connection between a working platform and a roof car or other fixed point.

travel trailer See: mobile home

traverse (Surv) A series of connected straight lines on a diagram outlining a plot of land, marking the center of a roadway, or the like, as determined by a survey.
NOTE: A traverse map indicates distances, bearings, and angles between adjacent lines.

tray burning (Comb) An experimental pool fire contained in a shallow tray.

tread^1 (Build) The horizontal part of a stair between the risers.

tread^2 (Techn) The grooved peripheral surface of an automotive tread.

tread width (Build) The horizontal distance from front to back of a stair tread, including nosing. Also called: **tread run**.

tree farm (For Serv) A tract of land planted with trees that are cultivated as a crop.
  NOTE: Tree farms produce commercial lumber, pulpwood, Christmas trees, and trees for landscaping.

tremor (Med) Involuntary shaking, trembling, or quivering.

trench = **fireline**

trench jack (Civ Eng) A screw or hydraulic type of jack used as a cross brace in a trench shoring system.

trench shield (Civ Eng) A shoring system composed of steel plates and bracing, welded or bolted together, which support the walls of a trench from the ground level to the trench bottom and which can be moved along as work progresses.

trestle[1] (Civ Eng) A braced framework of timbers, piles, or steelwork for carrying a road or railroad over a depression.

trestle[2] (Fire Pr; Brit) A frame, usu of steel tubing, used to carry a portable motor pump.

trestle ladder (Techn) A self-supporting, portable ladder, nonadjustable in length, consisting of two sections hinged at the top to form equal angles with the base. The size is designated by the length of the side rails measured along the front edge. Commonly called a **stepladder**.

triage (Med) The sorting of casualties according to the nature and seriousness of their injuries. The process of determining which of many casualties need urgent treatment, which are well enough to go untreated temporarily, and which are beyond hope of benefit from treatment.
  NOTE: Used in medical aspects of war, civil defense, catastrophes, etc.

triangle (Geom) A geometric three-sided, plane figure, the sides of which meet by pairs at three points called vertexes. The sum of the three interior angles is 180 degrees.
  NOTE: A triangle is called equilateral if all sides are equal, isosceles if two sides are equal, and scalene if no two sides are equal. Triangles are called acute if the largest angle is less than 90 degrees, right if one angle is 90 degrees, and obtuse if one angle is greater than 90 degrees. Trigonometry is the study of the relations between the angles and sides of a triangle.

triaxial shear test (Testing) A shear test in which a cylindrical specimen is subjected to a confining restraint and/or pressure and then loaded axially to failure.

tribasic acid (Chem) An acid containing three replaceable hydrogen atoms per molecule. See: **acid**.

tribology (Techn) The science of friction.

trichloroethylene (Chem) A stable, low-boiling solvent, $CHCl:CCl_2$, with a chloroform-like odor.
  NOTE: Toxic; should be used only with adequate ventilation or in closed systems.

trickle charge (Electr) A small charging current maintained through a storage battery when not in use to keep it in fully charged condition.

trigonometry (Math) The study of triangles, their sides, angles, and relationships between them.

trim (Build) Light woodwork used in finishing a building.

trimethoxyboroxine (Fire Pr; Abb: TMB) A colorless liquid compound used to extinguish burning metals. Esp effective for controlling burning magnesium, titanium, and zirconium.

trinitrotoluene (Chem; Abb: TNT) A flammable derivative of toluene, consisting of colorless to yellow crystals, $C_7H_5N_3O_6$. Emits toxic fumes of oxides of nitrogen when heated to decomposition. A highly poisonous explosive.

triple combination (Fire Pr) A pumping engine with hose, water tank, and pump.

triple hydrant (Fire Pr) A hydrant with a steamer or pumper outlet and two smaller outlets.

triple point (Phys) The temperature and pressure conditions under which the liquid, solid, and gas phases of a substance coexist in equilibrium. Also called: three-phase equilibrium.
NOTE: Triple points can also exist with three solid phases, two solid and one liquid, or two solid and one gas phase. The triple point of water is at 0.072°C and 4.6 mm Hg.

trip the horses (Fire Pr; Archaic) To turn out a fire company in response to an alarm.
NOTE: The term arose when horses were automatically released from their stalls when a fire alarm was received.

trip the lights (Fire Pr) To turn on the fire station lights at night when an alarm is received.

tritium (Nucl Phys; Symb: T) A radioactive isotope of hydrogen with two neutrons and one proton in the nucleus, which is called a triton. It is man-made and is heavier than deuterium (heavy hydrogen). Also called: hydrogen-three. Cf: deuterium. See: hydrogen.
NOTE: Tritium is used in industrial thickness gauges and as a label in experiments in chemistry and biology.

trolley ladder (Techn) A semifixed ladder, nonadjustable in length, supported by attachments to an overhead track.

tropopause (Atm Phys) A transition zone in the atmosphere between the troposphere and the stratosphere, at approx 10 to 15 miles above the earth. See also: atmosphere; pause.
NOTE: The tropopause usu marks a temperature minimum and the upper limit of convective activity.

troposphere (Atm Phys) The lowest main layer of the atmosphere from ground level up to approx 11 km, characterized by a steep, uniform lapse rate (6.5°K/km), a low degree of hydrostatic stability, and, as a result, subject to frequent overturnings.

trouble signal (Fire Pr) 1. A signal indicating a malfunction, such as an interruption or short circuit, in a protective signaling system. 2. = (Brit) fault warning.

truck¹ (Transp) A wheeled, self-propelled vehicle with a platform, usu enclosed, for transporting goods and materials in quantity.

truck² (Fire Pr) A fire department ladder truck carrying ladders, tools, and other fire-fighting and emergency equipment. Cf: fire truck.

truck company (Fire Pr) A fire department company operating a ladder truck. A ladder company.

truck crane (Mat Hand) A rotating superstructure with power plant, operating machinery, and boom, mounted on an automotive truck equipped with a power plant for travel, used to hoist and swing loads.

truck-tractor (Transp) A short-wheel-base, wheeled vehicle designed to tow and partially support a semitrailer through a fifth-wheel coupling device. See: tractor².

truckman (Fire Pr) A firefighter assigned to a ladder company. Also called: ladderman; trucker.

trumpet (Fire Pr) A symbol of rank in the fire service, dating from the time speaking trumpets were used to give voice commands at fires. Also called: bugle. See also: ranks and titles.
NOTE: Senior officers wear gold insignia with crossed trumpets, junior officers wear silver insignia with trumpets not crossed. Common usage is as follows:
Five gold - chief of department
Four gold - deputy or assistant chief (whichever is higher)
Three gold - deputy or assistant chief
Two gold - battalion or district chief
Two silver - captain
One silver - lieutenant

truss¹ (Build) A webbed or framed metallic or wooden structural member consisting of a series of interconnecting triangles, used as a beam to support a ceiling, roof, bridge span, and the like.

truss² (Fire Pr) The side member of a trussed ladder.

truss block (Fire Pr) A structural piece used to separate and hold the beams of a truss beam ladder.
NOTE: If the block also holds the rung it is called a rung block.

trussed ladder (Fire Pr) A ladder with side members built with open construction instead of solid spars.
NOTE: Trussed construction provides greater strength for a given weight as compared to a solid spar ladder.

tuberculosis (Med) A contagious disease caused by infection with the germ Nycobacterium tuberculosis. It usu affects the lung, but bone, lymph glands, and other tissues may be affected.

tuck pointing (Build) Removal by grinding of cement, mortar, or other nonmetallic jointing material.

tumbling (Techn) An industrial process, such as in founding, in which small castings are cleaned by friction in a revolving drum (tumbling mill, tumbling barrel) which may contain an abrasive.

tumor (Med) A swelling or a growth of useless cells and tissues anywhere in the body.

tunnel (Mining) 1. An underground passageway, both ends of which open to the surface. 2. A mine passageway, one end of which opens to daylight. An adit. 3. Any underground passageway.

tunnel test (Testing) A popular name for a standard surface burning test of building materials.
NOTE: Developed by the Underwriters' Laboratories, Inc. and standardized by ASTM as procedure E-84.

tunnel vision (Med) Inability to see toward the sides. A narrow field of vision.

turbid 1. Cloudy; roiled with mud or sediment. 2. Heavy with smoke or mist.

turbulence (Fl Mech) Irregular fluid motion caused by eddies (local variations in pressure and velocity) superimposed on the overall flow.

turbulence scale (Comb) The size or distribution of eddies in turbulent flow.

turbulent diffusion See: eddy diffusivity

turbulent flame (Comb) A flame propagating through a turbulent stream. Propagation is governed by eddy diffusivity. Turbulent flames can be either premixed or nonpremixed. A typical example is a jet-engine flame. See: flame; turbulence.

turbulent-flame velocity (Comb) The velocity of propagation of a turbulent flame. See: burning velocity; flame velocity.

turnaround (Comput) Referring to the usual time elapsed between submission of a job and receipt of the output from a computer.

turn out (Fire Pr) To alert a fire company to answer a fire alarm.

turnout (Fire Pr) a. The activity preparatory to departure to a fire in response to an alarm. b. Assembly of firefighters in response to an alarm.

turnout area (Fire Pr) The territory or geographic area protected by a given fire service unit.

turnout board (Fire Pr) A dispatcher's display board showing units that are in service and those out of quarters or out of service.

turnout coat (Fire Pr) A waterproof fire coat that can be put on quickly when turning out for a fire alarm. Syn: fire coat.

turnouts (Fire Pr) Fire clothes worn when turning out for a fire alarm. Also called: bunker clothes.

turnout time (Fire Pr) The time between the moment a fire suppression unit is notified of a fire and the time it leaves the fire station.

turnstile (Build) A type of gate with a post and arms pivoted at its top that allows only one person at a time to pass through. Used at points where fares or admission tickets are collected.

turntable (Fire Pr) The rotating platform on which an aerial ladder is mounted.

turntable ladder (Fire Pr) A multisection ladder mounted on a self-propelled chassis. The ladder can be raised either mechanically or hydraulically, extended up to about 100 feet, and rotated through a complete circle. A monitor nozzle is often provided at the head of the ladder.

turret lathe (Techn) A power-driven, metal-working machine in which the stock is held and rotated about a horizontal axis to be machined successively by a number of cutting tools.

turret nozzle (Fire Pr) A heavy stream nozzle (usually for foam) mounted on the roof of a self-sufficient fire apparatus used for fighting crashed aircraft fires.

turret pipe (Fire Pr) A heavy stream nozzle mounted on a pumper or fireboat and connected directly to a pump. Also called: gun; deck pipe.

turret wagon = monitor wagon

twinned agent unit = combined agent unit

two-hand controls (Safety) Tripping devices for a machine which require simultaneous application of both hands to operate so that the operator's hands are kept out of the point-of-operation area while the machine is operating.

two-man carry (Fire Pr; First Aid) 1. A method by which two persons can conveniently lift and carry a third who has become incapacitated. See: rescue carry. 2. A method for carrying a ladder by two firefighters.

two-man fold (Fire Pr) A method of folding salvage covers that requires two persons to unfold and spread.

two-piece company (Fire Pr) A company that regularly responds to alarms with two pieces of fire apparatus.
  NOTE: Commonly referring to two pumpers: one lays supply lines to the fire and the other pumps from the water source.

two-platoon company (Fire Pr) A company consisting of two groups, each averaging 84 hours of on-duty time per week.

two-position attack (Fire Pr) A firefighting tactic in which a fire is attacked from two separate vantage points.

two-way radio (Commun; Fire Pr) A radio communications system in which the central station transmitter and mobile transmitters can talk to each other. Cf: three-way radio.
  NOTE: In duplex they use the same frequency. In dual-frequency simplex, the base station uses one frequency and the mobiles use another.

tying in¹ (Fire Pr) a. Securing oneself with a rope or life belt to free both hands for work. b. Securing oneself to a ladder with a leg lock.

tying in² (Fire Pr) a. The act of making a connection or a securement. b. Securing a ladder to a building or other structure with rope tools or the like. See: dog; dog chain. c. Connecting one hose to another, esp to one that has already been laid. d. The act of making a connection.

tympanic cavity (Anat) The chamber of the middle ear.

typhoon (Meteor) A violent hurricane in the China Sea or the Pacific Ocean area. See: wind.

415

U

U (Chem) Symbol for uranium.

UL = Underwriters' Laboratories, Incorporated

UL label (Testing) An identification authorized by the Underwriters' Laboratories, Inc. and affixed to a building material, component, or product, indicating 1) that the labeled article is subject to the follow-up inspection service of UL, or 2) that it is from a production lot found by follow-up inspection to be made from materials and by processes essentially identical to those of representative articles subjected to appropriate fire or other tests.

ulcer (Med) An inflammatory lesion resulting from loss of superficial tissue, occurring most often in the gastrointestinal tract with or without infection or pain.
NOTE: Gastric ulcers, called Curling's ulcers, are often seen after severe body burns. Ulcers can also occur on the skin and the mucous tissues of the mouth and tongue. Thermal ulcers caused by flame or electrical burns are treated by removing the dead tissue, ensuring asepsis, and applying skin grafts, if large areas are involved.

ullage (Techn) 1. The unfilled portion of a barrel or tank. See: heel.
2. Sometimes, erroneously, the contents of a tank.

ultimate analysis (Chem; Comb) A report on the chemical composition of a substance giving the chemical elements present and their proportions by weight.

ultrasonic (Acoust) Referring to sound frequency beyond the range of human hearing, usu above 20,000 Hz.

ultrasonic testing (Techn) A nondestructive method for testing materials for flaws by inducing ultrasonic waves into the material and monitoring the effects the material has on the waves being propagated.

ultrasonics (Acoust) The branch of physics that deals with the study of vibrational waves above the audible range of the human ear. Ultrasonic techniques are used in sonar, flaw detection in metal castings, measurement of thickness, and mixing or emulsification through induced cavitation. Syn: supersonics.

ultraviolet (Opt) Wavelengths of the electromagnetic spectrum that are shorter than those of visible light and longer than x-rays, 10^{-5} cm to 10^{-6} cm.

ultraviolet detector (Fire Pr) A device that detects fires by sensing the electromagnetic waves emitted in the ultraviolet range by the fire.

ultraviolet radiation (Opt; Abb: uv) Electromagnetic radiation in the wavelength range from 400 to 4000 A, between visible light and long-wavelength X-rays. Also called: black light.
NOTE: Ultraviolet radiation causes sunburn, excites fluorescence and phosphorescence, is destructive to plastics, and produces other photoeffects. Phototubes and other detectors are sensitive to uv radiation.

umbilical cord (Anat) The cord that joins the fetus at the navel to the placenta.

umbrella liability (Insur) Insurance protection against losses in excess of amounts covered by other liability insurance.

underbreathing = hypoventilation

under control (Fire Pr) A stage reached in fire fighting in which the fire has been contained and extinguished to the extent that fire authorities are confident of its complete extinguishment, and in certain cases overhauling can begin. Syn: tapped out.

undercut (For Serv) A notch cut in a tree to guide the tree in felling.

undercut line (For Serv) A forest-fire control line constructed below a fire on a slope. Also called: underslung line. See: gutter trench.

underhand stoping (Mining) Breaking ore in horizontal slices in descending order; miners stand on the ore and mine downward. Cf: overhand stoping.

underlie angle (Mining) The angle between the normal to a hanging wall and the longitudinal axis of a stull.

understory (For Serv) A layer of foliage spreading beneath and shaded by the main canopy of a forest.

underwire raise (Fire Pr) A method of raising a ladder under overhead telephone and electric power lines. See: ladder raise.

underwriter (Insur) An insurer who assumes the reponsibility for a loss up to a specified amount in return for payment of an agreed premium.

Underwriter's Laboratories, Incorporated (Testing; Abb: UL) An independent, nonprofit research and testing organization established to study life and property hazards associated with industrial and consumer products.
NOTE: Founded by William Merrill in Chicago in 1894. By test the UL determines if products and materials meet appropriate safety standards in the fields of electricity; fire protection; burglary protection and signaling; heating, air-conditioning, and refrigeration; consumer product and chemical hazards; and marine transportation.

underwriter thread (Fire Pr) An obsolete hose thread with 12 threads per inch, used to attach tips to underwriter playpipes and other nozzles. See: standard thread.

unequal liability = accident liability

unexposed surface (Build; Testing) The side of a structural assembly opposite to the one exposed to a fire.

unexposed surface temperature (Build; Testing) The temperature, single-point or average, of the surface of a floor, roof, wall, or partition opposite to the surface exposed to fire.

unexposed surface temperature rise (Build; Testing) A performance criterion applicable to floors, walls, roofs, or partitions that have been subjected to appropriate fire tests. A measure of the thermal conductivity of a building assembly, important in characterizing its fire-separating capacity.

unfriendly fire (Fire Pr; Insur) A heating or cooking fire that has spread to materials not intended as fuel. Syn: hostile fire; unwanted fire. Ant: friendly fire.

uniform (Fire Pr) The official fire department dress clothing, as distinguished from turnout clothes.

uniform atmosphere (Meteor) A region of the atmosphere in which the density is taken, for computational purposes, to be independent of height.

Uniform Building Code (Build) A building code that is maintained and published by the International Conference of Building Officials.

uniform continuity See: fuel continuity.

uniformed force (Fire Pr) The uniformed members of a fire department, as contrasted with civilian personnel.

unit of exit width (Build; Safety) A measure relating the dimension of doors, stairways, or other means of egress to the rate at which people can pass through them in an emergency.

unit vent (Build) An opening ranging from 4 by 4 feet to 10 by 10 feet, installed in a roof. Generally equipped with a lightweight metal frame and housing and with a hinged damper that may be opened manually or automatically in the event of fire.

unit weight (Transp) Weight per unit volume.

units system (Metrol) Related sets of units (based on primary units of mass, length, and time) used to express measurements of physical quantities.

universal coupling (Fire Pr) A device for connecting mismatched couplings.

universal gas constant = gas constant

universal joint (Mech Eng) A mechani-

417

cal linkage for transmitting rotational motion between shafts that are at an angle to each other.

unlined fire hose (Fire Pr) A fire hose made of cotton, linen (flax), or synthetic fiber and lacking a rubber lining. See also: hose; linen (flax) hose.

unpaired electron (Chem; Phys) A single electron available for chemical bonding.

unprotected openings (Build) Openings through floors, walls, and other constructions that lack the means for blocking the passage of smoke, heat, and flame.
 NOTE: Vertical openings include utility ducts, stairwells, elevator shafts, and the like; horizontal openings include air and heating ducts, vents, windows, doorless passageways, and various concealed spaces in walls, floors, and ceilings. Cf: interstitial space; concealed space; vertical opening.

unprotected steel (Build) Structural steel members that are not covered and thus may be exposed to the direct heat of a fire.

unsafe act (Safety) A departure from an accepted, normal, or correct procedure or practice. An act that produces injury or property damage or has the potential to do so. An unnecessary exposure to a hazard; or conduct reducing the degree of safety.
 NOTE: Not every unsafe act produces an injury or loss but, by definition, all unsafe acts have the potential for producing them. An unsafe act may be doing something that is unsafe or failing to do something that should have been done.

unsafe personal factor (Safety) A mental or physical characteristic that causes or permits an unsafe act to occur.

unsatisfactory operation (Fire Pr) Failure of fire protection equipment to meet performance standards.

unsprung weight (Transp) The total weight of vehicle components that are not supported by the vehicle spring system, including such items as tracks, axles, road wheels, etc.

unstable liquid (Chem) A liquid which, in the pure state or as commercially produced or transported, will vigorously polymerize, decompose, condense, or will become self-reactive under conditions of shock, pressure, or temperature.

unstable soil (Soil Mech) Earth that cannot be depended upon to remain in place without extra support, such as by shoring.

unwanted fire (Fire Pr) Any undesired fire, usu destructive to property or life, regardless of origin. Syn: hostile fire; unfriendly fire.

upper (Mining) Referring to a hole drilled steeply upward. Also called: dry hole.

upper atmosphere[1] (Atm Phys) The upper reaches of the envelope of air surrounding the earth.

upper atmosphere[2] (Fire Pr) The space above a fire where a large portion of the heat generated by the fire collects. More properly called the upper atmospheric area.

upper explosive limit (Comb) The maximum proportion of vapor or gas in air above which propagation of flame does not occur. The upper limit of the flammable or explosive range. See: explosive limits.

upper limit of flammability (Comb) The highest concentration of flammable vapor or gas mixed with air that will burn with a flame. See: flammability limits.

upright sprinkler (Fire Pr) A sprinkler head that points vertically upward from the water-supply grid; the direct opposite of a pendant sprinkler head, which hangs downward. See: sprinkler system.

up-time (Man Sci) The percent of total working time in which a system or unit of equipment is operating satisfactorily.

uranium (Chem; Symb: U) A naturally radioactive element that emits alpha and beta particles. Metallic uranium is a lustrous, silvery solid, mp 1132°C, bp 3818°C.

NOTE: The metal combines directly with oxygen, reacts with water, and is combustible even in solid form. In powdered form, uranium can ignite spontaneously at room temperature. Highly toxic.

urban (Fire Pr) Referring to geographic areas that have 2500 or more inhabitants and towns, townships, and other areas classified as urban by the US Census Bureau.

urea formaldehyde (Chem) A resin produced by condensation of urea and formaldehyde, often used with alpha cellulose filler in molded products. The resin is also used as an adhesive, surface coating, textile surface treatment, and for improving the wet-strength of paper.

urticaria (Med) Hives.

USDA = United States Department of Agriculture

uterus (Nucl Safety) A muscular organ in the female body in which the fetus develops. Syn: womb.

utility fire (Fire Pr; Brit) = friendly fire

utility gas (Techn) Natural gas, manufactured gas, liquefied petroleum gas, or a mixture of them.

utility hole (Build) A hole through a concrete floor or through a wall or ceiling, made to pass wiring, pipes, and the like. A pokethrough.

V

v (Phys) Abbreviation for vector, velocity.

V (Chem; Phys) Symbol for volt, vanadium.

vacant net (Storage) Usable space that is not occupied by material in storage bins or racks.

vaccine (Med) A suspension of disease-producing microorganisms modified by killing or attenuation. Upon injection, it stimulates the formation of antibodies and thereby immunity to the disease.

vacuum (Phys) A space ideally completely void of matter. In practice, pressures below atmospheric are referred to as vacuum, or more accurately, as a partial vacuum.

vacuum gage (Techn) An instrument for measuring subatmospheric pressures, usu expressed in inches of water or millimeters of mercury.
  NOTE: Pressures above 1 mm Hg are measured by liquid-column manometers, diaphragms, bellows, and Bourdon gages. Lower pressures are best determined indirectly by measuring thermal conductivity or ionization.

vacuum packaging (Packag) Packaging in containers, either rigid or flexible, from which most of the air has been removed prior to final sealing.

vacuum pump (Hydraul) A pump that extracts air or other gas from an enclosed space or chamber.

vacuum test¹ (Techn) A test for detecting leaks in a closed system.

vacuum test² (Fire Pr) A measure of the efficiency of the priming system of a pump.

vacuum-tube voltmeter (Electr; Abb: VTVM) A device that uses the amplifier characteristic or the rectifier characteristic of a vacuum tube or both to measure d-c or a-c voltages.

vagina (Anat) The birth canal; a body passage in the female leading from the vulva to the cervix.

valence (Chem) a. A number characterizing the combining power of one element with another, expressed as the number of atoms of hydrogen (or their equivalent) that will combine with or be replaced by the atom in question. For example, two hydrogen atoms combine with one oxygen atom to form water. Each hydrogen atom has a valence of 1 and the oxygen atom has a valence of 2. Nitrogen has a valence of 2, 3, 4, and 5, respectively, in NO, NH^3, NO^2, and N^2O^5. b. **Polar valence** is the excess of positive or negative charges on an atom or radical. c. **Nonpolar valence** is the number of electron pairs shared with other atoms.

validity (Metrol) The adequacy of an instrument in measuring what it is designed to measure. The validity of a measuring instrument always depends on the specific purpose for which the instrument is used.

value engineering (Ind Eng) An organized effort to appraise the elements of design, manufacture, inspection, operation, and maintenance of an item to keep components, maintainability and reliability at minimum cost.

valve (Techn) Any device for restricting, regulating, or stopping the flow of a fluid.
  NOTE: A **pressure valve** opens and closes a relief vent depending on whether the pressure is above or below a given value. **Temperature valves** include fusible link types that melt at a set temperature; manual or automatic resetting types; and vacuum types that respond to pressures below atmospheric. **Gate valves** provide throttling action; **check valves** allow flow in one direction only; **pressure relief valves** operate automatically when a predetermined critical pressure is exceeded; and **needle valves** allow fine regulation of fluid flows. Many other special types of valve exist, such as plug, butterfly, poppet, slide, feather, etc.

vamp (Fire Pr; Colloq) 1. A volunteer fireman. 2. An inactive fireman who maintains social contacts with the fire service.

vandalism (Law) Malicious damage to

property. Also called <u>malicious mischief</u>.

<u>van der Waals forces</u> (Chem) Relatively weak forces of attraction between atoms and/or molecules other than those of chemical bonding. Cf: <u>Boyle's law</u>.

<u>vapor</u> (Phys; Chem) The gas phase component of a substance; particularly of those that are normally liquids or solids at ordinary temperatures.

<u>vapor area</u> (Safety) Any area containing dangerous quantities of flammable vapors.

<u>vapor density</u> (Chem; Phys) The ratio of the weight of a given volume of gas or vapor to the weight of an equal volume of a standard gas, usu hydrogen or air, at the same temperature and pressure.
NOTE: A gas or vapor heavier than the standard has a vapor density greater than 1. Gasoline vapor, for example, has a vapor density of approx 3.5 with respect to air.

<u>vaporization</u> (Chem; Phys) The process of formation of a vapor from a liquid or solid phase.
NOTE: The readiness with which vaporization takes place is called <u>volatility</u>.

<u>vaporizing liquid</u> (Fire Pr) A liquid extinguishing agent that vaporizes and decomposes from the heat of the fire, chemically extinguishing it. Used esp for flammable liquids and electrical fires. See: <u>Halon</u>.

<u>vapor lock</u> (Transp) Interruption of the flow of fuel to an engine by the formation of vapor in the fuel lines at a rate faster than the fuel pump can draw it off.

<u>vapor openings</u> (Techn) Vents or other openings in a tank above the surface of a stored liquid provided for tank breathing, tank gauging, fire fighting, or operating purposes.

<u>vapor plume</u> (Comb; Techn) Flue gas emerging from a stack, containing visible condensed water droplets or mist.

<u>vapor pressure</u> (Chem; Phys) The equilibrium pressure exerted by the vapor of a liquid (or solid) at a given temperature, reached when the number of molecules escaping from the liquid (or solid) equals the number returning.

<u>vapor proof</u> (Mat Hand; Packag) Impervious to vapor. Syn: <u>vapor tight</u>.

<u>vapor tension</u> = <u>vapor pressure</u>

<u>vapor volume</u> (Techn) The number of cubic feet of solvent vapor formed by the evaporation of one gallon of liquid.

<u>variable</u> (Math) A quantity that can take many values. Ant: <u>constant</u>.

<u>variable danger</u> (For Serv) The sum of all fire danger factors that change from day to day, such as weather, fuel moisture, plant growth, and varying hazards related to human activities and the like. See: <u>constant danger</u>.

<u>variance</u>[1] (Build) A deviation from a building code or standard, usu granted on the basis of evidence or assurance that the deviation does not materially conflict with the sense or intent of the code or standard.

<u>variance</u>[2] (Env Prot) A license, granted by government, permitting an exception to some restriction of use, such as pollution beyond the acceptable limit, in return for a promise to curb pollution at a later time.

<u>variance</u>[3] (Statist) The square of the standard deviation. A measure of the variability of a set of data points calculated by squaring the differences between the individual values and their mean and dividing the sum of these squares by the number of values minus 1. Also called: <u>second moment about the mean</u>. See: <u>standard deviation</u>.

<u>vasoconstriction</u> (Med) Narrowing of the blood vessels.

<u>vasodilation</u> (Med) Enlargement of the blood vessels.

<u>vascular system</u> See: <u>blood vessels</u>

<u>vault</u> (Storage) A space in a specially constructed burglarproof and

vector (Math; Phys) A quantity that has both a magnitude and a direction, in contrast to a scalar quantity, that has only magnitude. A force is an example of a vector quantity, since it depends on direction; mass is an example of a scalar quantity, since it is independent of direction.
  NOTE: A vector is depicted graphically as an arrow, called a stroke. Its length indicates the scalar magnitude and its direction the direction of the vector. The tail of the stroke is the origin and the head the terminus.

veering wind (Meteor) A wind that changes direction clockwise. Ant: backing wind. See: wind.

vegetation (For Serv) Plant cover of an area.

vegetation stage See: condition of vegetation

vein¹ (Anat) A blood vessel in the body that returns blood to the heart. See: circulatory system.

vein² (Mining) a. A tabular, nonsedimentary bed of ore or a layer of mineral deposits. b. A fissure in rock that has been filled with mineral matter.

velocity (Phys; Symb: v) The time rate of motion of a body. Linear velocity is expressed in units of distance per unit of time. Angular speed is usu expressed in radians per second or as rotations per minute or per second. A point on a rotating body has a tangential velocity. See also: acceleration. Cf: speed.
  NOTE: Velocity is a vector quantity. Speed is a scalar measure of motion through distance per unit of time without consideration of direction.

velocity pressure (Fl Mech; Abb: vp) The kinetic pressure in the direction of flow necessary to cause a fluid at rest to flow at a given velocity. It is usu expressed in inches of water. Syn: velocity head.

velocity shock (Mat Hand) A particular type of shock motion characterized by a sudden change in velocity.

venipuncture (Med) Penetration of a vein by a needle or catheter.

vent¹ (Build) An opening for the free passage of or dissipation of fluids, such as gases, fumes, smoke, and the like.
  NOTE: An explosion vent is often a deliberately weakened portion of a confining chamber, building, compartment, etc., or a duct that fails mechanically so as to release overpressure and thus prevent structural damage to the enclosure.

vent² (Fire Pr) To release pent-up smoke and heat from an enclosure or other structure.

vented fire (Fire Pr) A fire that has penetrated through windows or has burned through the walls or roof of a burning structure, releasing heat and smoke and allowing the outside air to enter.

ventilate (Fire Pr) To open up a burning structure to release pent-up smoke and heat. Syn: open up, q.v.

ventilation¹ (Build) Circulation of air in any space by natural wind or convection, or by fans blowing air into or exhausting air out of a building.

ventilation² (Fire Pr) Opening of windows and doors or making holes in the roof of a burning building to allow smoke and heat to escape.

ventilation³ (Med; Physiol) Breathing. The process of respiration.

ventilation⁴ (Safety) Adequate circulation of air to prevent accumulation of flammable, explosive, or toxic concentrations.

ventilatory assistance (Med) Aid to breathing, such as an oxygen mask or tracheotomy tube supplying positive-pressure oxygen.

venting = ventilation

ventral (Anat) Pertaining to the front of the body.

ventricle (Anat) A small cavity, esp one in the heart or brain. Either of the two lower chambers of the heart.
NOTE: The ventricles of the heart receive blood from the auricles and force it into the arteries.

ventricular fibrillation See: fibrillation

ventricular tachycardia See: tachycardia

venturi tube (Techn) A shaped pipe with a short, narrow, straight throat between two tapered sections. When a fluid is flowing through the tube, its velocity increases and pressure decreases in the throat section. Thus the tube can be used to measure flow velocities or to meter fuel or other liquids, such as foam compounds into a water stream.

verify (Comput) To check, sometimes by automatic means, one set of data against another, in order to miminize the number of human errors in the data transcription.

vermiculite (Techn) An expanded mica (hydrated magnesium-aluminum iron silicate) used in light-weight aggregates, insulation, fertilizer, and soil conditioners, as a filler in rubber and paints, and as a catalyst carrier.

vernier 1. An auxiliary scale next to the main scale of an instrument that provides accurate fractional readings of the finest division of the main scale. 2. A control for fine adjustment or tuning of a radio receiver or transmitter.

verruca = wart

vertebra (Anat) Any one of the small bones in the spinal column.

vertical opening (Build; Fire Pr) Any aperture through floors in buildings, such as stairways, elevators, conveyers, and ducts for heating, ventilation, and utility lines. See also: balloon construction; fire door; unprotected opening.
NOTE: Such openings can act as channels for the vertical spread of smoke and fire.

vertical reach (Fire Pr) The maximum height of a fire stream.

vertical spread (Fire Pr) The upward propagation of fire.

vertical tower An aerial device designed to elevate a platform in an essentially vertical direction.

vertigo (Med) Dizziness.

vesicant (Toxicol) A substance that attacks the skin and produces blisters. Ant: nonvesicant.

vesicle (Med) A small blister on the skin.

vessel (Navig; Transp) Any watercraft or other contrivance used as a means of transportation on water, including special-purpose floating structures not primarily designed for transportation on water.

vestibule (Build) A small lobby, reception room, or enclosed space between the outer doors and the interior of a building. Sometimes located between two parts of the same building.

viable (Med) Capable of living.

vibration (Mech Eng) A periodic oscillatory displacement of a body with respect to a reference point, line, or plane. Vibrations range from simple harmonic motions, such as the swing of a pendulum, to extremely complex motions, such as a truck on a rough road. The magnitude of excursion by the body is called the amplitude of vibration. See also: degrees of freedom.

vibration isolation (Mech Eng) The separation of a body from its support by one or more resilient members that damp the vibrations passing between the body and its support.

vicinity box = adjacent box

viral invasion (Med) Infection by viruses.

virulence (Med) The capacity of microorganisms to infect and injure or kill a susceptible host. Some microorganisms are virulent with respect to one host and harmless to another. Virulence depends on the invasiveness of the microorganism and its ability to produce tissue-damaging toxins. Ant: a<u>virulence</u>.

<u>virus</u> (Bact; Med) An ultramicroscopic infectious agent consisting of nucleic acid in a protein shell, capable of infecting animals, plants, and even bacteria. Viruses reproduce only within appropriate host tissue cells.

<u>viscera</u> (Anat) The internal organs of the body, particularly those of the trunk (intestines, liver, etc.).

<u>viscose</u> (Chem) A viscous liquid composed of cellulose xanthate. Used in making rayon.

<u>viscosity</u> (Fl Mech) 1. The resistance of a fluid to flow. 2. The internal friction of a fluid. 3. The transport of momentum in a system due to a velocity gradient. See also: <u>boundary layer</u>.
NOTE: Viscosity is defined as the force per unit area resisting the differential flow of two parallel layers of fluid a unit distance apart. The viscous force, in dynes, $f = e \cdot A(dv/dx)$, where e is the absolute viscosity coefficient in <u>poise</u>, A is the area in cm², and dv/dx is the velocity gradient in cm-sec⁻¹. For water at 20°C, e = 10.087 millipoise. For fluids in general, the coefficient of viscosity decreases with increase in temperature.

<u>viscosity coefficient</u> See: <u>viscosity</u>

<u>viscosity index</u> (Techn) A number, usu in the range of ±100, given to a lubricating oil to indicate its performance, particularly as to change of viscosity with temperature variation, as compared with the average of two groups of test oils.

<u>viscous damping</u> (Techn) The dissipation of energy that occurs when a particle in a vibrating system is resisted by a force that has a velocity-dependent component.

<u>viscous water</u> (For Serv) Water that contains a thickening agent, such as sodium carboxymethylcellulose, to reduce surface runoff. Also called: <u>thick water</u>.
NOTE: Thickened water tends to cling to burning fuels and to spread in layers that are several times thicker than plain water.

<u>visibility</u>¹ (Opt) The relative clearness with which objects stand out from their surroundings under good seeing conditions. Also called: <u>visual range</u>.

<u>visibility</u>² (Meteor) The distance at which it is just possible to distinguish a dark object against the horizon.

<u>visibility distance</u> (For Serv) The maximum distance at which a standard column of smoke of a specified size and density can be seen and recognized as smoke by a person with normal vision.

<u>visibility meter</u> = <u>haze meter</u>

<u>visible area</u> (For Serv) The expanse of ground or vegetation that can be observed directly from a given lookout point under favorable atmospheric conditions. Syn: <u>seen area</u>. Ant: <u>blind area</u>.

<u>visible area map</u> (For Serv) A map that shows the territory that can be observed directly from a given lookout point. Syn: <u>seen area map</u>.

<u>vision</u> (Physiol) The sense of sight by which the shape, color, size, movement, and distance of objects are perceived when light rays reflected from the objects enter the eyes and form images on the retina. Human vision is most sensitive to light in the wavelength range of 380 to 720 millimicrons. See also: <u>blindness</u>; <u>sense organs</u>.

<u>Vistanex</u> (Chem) A trade name for polyisobutalene, a compound that increases the viscosity of a liquid without affecting its other properties. See: <u>slippery water</u>.

<u>visual range</u> = <u>visibility</u>

<u>vital capacity</u> (Med) The volume of air that can be forcefully expelled from the lungs following full inspiration.

<u>vital signs</u> (Med) Pulse, heart beat, blood pressure, temperature, and respiration.
NOTE: Vital signs are the fundamental indicators of the physiological

state of the body; being readily observable, they provide the first and most easily measured diagnostic indications of the condition of a patient.

vital statistics (Demog) Data concerning births and deaths.

vitrification (Chem Eng) A high-temperature process used to produce permanent chemical and physical changes in a ceramic body, most of which is transformed into glass.

void[1] (Phys; Techn) An empty space; a vacuum.

void[2] (Law) Having no legal force or effect. Referring to a legal instrument that is no longer valid.

void ratio (Soil Mech) Ratio of the volume of void space to the volume of solid mineral matter in a soil mass.

volatile corrosion inhibitor (Mat Hand) A chemical that slowly gives off vapor that reduces or inhibits corrosion. Usu applied to chemically treated paper used in packaging ferrous metal products. Also called: vapor-phase inhibitor.

volatility[1] (Phys; Chem) The readiness with which a substance vaporizes. See also: sublimation; vaporization.

volatility[2] (Comput) The likelihood of information loss by a computer memory unit in the event of power interruption.

volt (Electr; Abb: v) A unit of electrical potential. A potential difference of one volt causes a current of one ampere to flow through a resistance of one ohm.

voltage divider (Electr) An impedance connected across a voltage source. The load is connected across a fraction of this impedance so that the load voltage is essentially in proportion to this fraction.

volume[1] (Geom) The three-dimensional space occupied by a body.

volume[2] (Fire Pr) The rate of water delivery.

volume[3] (Acoust) The loudness or intensity of a sound.

volume operation (Fire Pr) The operating mode of a centrifugal pump that produces maximum discharge. Syn: capacity operation Cf: parallel operation.

volunteer fire department See: fire department

volunteer fireman (Fire Pr) A regularly enrolled firefighter in a community fire service who contributes his time to fire protection and suppression activities without pay, except for workman's compensation or other injury or death benefits.

volute (Techn) The spiral, divergent chamber of a centrifugal pump in which the kinetic energy imparted to water by the impeller blades is converted into pressure.

volute spring (Transp) A form of conical compression spring, usu made of flat spring stock and wound in a spiral helix with the successive coils telescoping into each other. It is characterized by its compactness, variable spring rate, and high friction damping. It is used as the spring element in certain bogie suspensions of tracked vehicles and as bottoming springs on vehicles with soft suspensions.

von Neumann spike (Comb) The front of a shock wave that produces very rapid increases in pressure, density, and temperature. The pressure peak preceding a detonation wave prior to the establishment of a steady-state condition.

vortex (Fl Mech) A rotational motion in a fluid.
 NOTE: Typical vortexes in nature are tornadoes, water spouts, dust and fire devils, smoke rings, and whirlwinds. See: Karman vortex street; vorticity.

vorticity (Fl Mech) A measure of the local rate of rotation, usu two-dimensional, of a fluid.

V-type collapse See: floor collapse

vulcanization (Techn) The process of combining rubber (natural or synthetic) with sulfur and accelerators in the presence of zinc oxide under heat and usu pressure in order to change the material permanently from a thermoplastic to a thermosetting composition, or from a plastic to an elastic condition. Strength, elasticity, and abrasion resistance also are improved.

vulva (Anat) The external female organs of reproduction.

W

W (Chem; Phys) Symbol for energy, tungsten, work

WAB = Western Actuarial Bureau

waders (Fire Pr) A one-piece, rubberized garment consisting of boots and overalls or boots and trousers.

wading suit (Fire Pr) A rubberized, water-repellent coverall worn with a self-contained gas mask to protect firefighters from injurious or noxious gases, esp vesicants, or toxic substances that can be absorbed through the skin. Syn: ammonia suit. See also: protective clothing.

wagon (Fire Pr; Archaic) 1. A fire department hose truck, usu equipped with a water tank and small hose lines. 2. Originally a horse-drawn fire wagon. 3. Loosely, a ladder wagon, fuel wagon, or pumper serving as a hose wagon.

wagon battery (Fire Pr) A heavy-stream nozzle mounted on a hose truck or pumper and usu connected to the pump by internal piping. Also called: monitor; truck pipe.

wagon pipe = wagon battery

wagon steering (Transp) Steering of a vehicle consisting of one or more units by a single pivot system with the pivot point located over the front axle.

waiting period (Insur) The time interval stipulated in certain loss and disability policies between the loss or disabling incident and payment of benefits.

walkie-talkie (Commun) A two-way portable radio set used for short-range communication. See: Handie-Talkie.

wall hole (Build) An opening, other than a doorway or window, in any wall or partition, such as a ventilation hole or drainage scupper.

wall hydrant (Fire Pr) A hose connection from a stationary fire pump brought out through the wall of a building and used in fire fighting in the same manner as a street hydrant.

NOTE: Wall hydrants at fire stations are used for testing pumpers.

wall ladder (Fire Pr) A single-section ladder, up to 35 feet in length, without hooks, used for scaling purposes by leaning it against a wall. Also called: straight ladder. See: ladder.

wall nozzle (Fire Pr) A straight-stream nozzle with a small tip, inserted into walls to extinguish hidden flames.

wall opening[1] (Build; Safety) An opening in any wall or partition through which persons may fall, such as a yardarm doorway or chute opening. See: wall hole.

wall opening[2] (Build; Fire Pr) An aperture in a wall, such as a door leading to a room or corridor, passageway, through a partition, and the like. See also vertical opening; unprotected opening.
NOTE: Classes of wall openings are: Class A, through fire walls; Class B, into vertical enclosures, such as elevator shafts, stairwells, etc.; Class C, into corridors or through room partitions; Class D, through exterior walls subject to severe fire exposure; Class E, doors through exterior walls subject to moderate fire exposure. Each class has a recommended fire resistance rating.

wall sprinkler (Fire Pr) An automatic sprinkler head designed for installation near a wall and set to spray in the direction away from the wall. Also called: side wall sprinkler. See: sprinkler system.

warden See: fire warden

warehouse (Storage) Any building, structure, or area within a building used principally for storage purposes.

warehousing (Storage) The physical activity involved in receiving, storing, and issuing goods and materials at a storage facility.

warp thread (Fire Pr) The set of threads running lengthwise in a fabric or woven hose. Cf: weft

thread.

wart (Med) A localized growth on the skin, generally regarded as a virus infection. Syn: **verruca**.

wash down (Fire Pr) 1. To flush away spilled combustible materials, esp flammable liquids. 2. To soak debris with water after a fire has been controlled but not completely extinguished. Syn: **wet down**.

washer (Techn) A round, usu metallic disk with a hole used with nuts and bolts to protect the material being fastened and to spread the load.
NOTE: A split ring, called a spring lock washer, is used to prevent nuts and bolts from loosening. An antiturn (toothed lock washer) is used to prevent the fastener from loosening in service from vibration. Some nuts and bolts come with permanent locking washers built in.

washing down (Fire Pr) 1. Drenching a fire with several heavy streams. 2. Flushing away spilled material, such as a flammable liquid, with water.

waste[1] (Env Prot; Techn) Discarded material.
NOTE: **Bulky waste**: Items whose large size precludes or complicates their handling by normal collection, processing, or disposal methods. **Construction waste; demolition waste**: Building materials and rubble resulting from construction, remodeling, repair, and demolition operations. **Hazardous waste**: Those wastes that require special handling to avoid illness or injury to persons or damage to property. **Wood-pulp waste**: Wood or paper fiber residue resulting from a manufacturing process. **Special waste**: Wastes that require extraordinary management. **Yard waste**: Plant clippings, prunings, and other discarded material from yards and gardens. Also called: **yard rubbish**.

waste[2] (Mining) Worthless minerals or rocks.

waste processing (Env Prot; Techn) Any operation such as shredding, compaction, composting, and incineration, in which the physical or chemical properties of wastes are changed.

watch (Fire Pr) The tour of duty at the watch desk, including the responsibility for keeping the company journal, receiving visitors, receiving alarms, and turning out the company when required.

watch desk (Fire Pr) The control point at a fire station.

watch detail (Fire Pr) 1. A team of firefighters assigned to remain on a fireground after a fire to guard against rekindling. 2. = (Brit) **firewatch**.

watchdog (Safety; Colloq) A safety device, such as an electrical fuse or circuit breaker, that guards against malfunctions and dangerous conditions.

watch line (Fire Pr) A charged line of hose left at the scene of a large fire with a detail of firefighters to guard against rekindling of the fire.

watchman An individual hired to patrol the premises of a plant or property to guard against intruders, fire, or other incidents while the plant is not in operation or the property is otherwise not attended.

watch room (Police) A guard post or central office of a guard force.

water (Chem) A liquid compound, H_2O, fp 0°C (32°F), bp 100°C (212°F), triple point 0.07°C.
NOTE: The solid phase is called **ice**, the vapor phase **steam**. Fresh water weighs 62.35 pounds per cubic foot; salt water, 64 pounds.

water bag (For Serv) A container of water carried by an aircraft to be used to bomb forest fires with the water.

water bombing (For Serv) Fighting wildland fire by dropping water or fire-retardant chemicals on it from aircraft. Also called: **water drop**.

water chute (Fire Pr; Salv) A construction of salvage covers or other materials used to channel excess water out of a building.

water curtain[1] (Techn) A flow of water running down a wall into which excess paint from spray painting is

drawn or blown by fans so as to carry the paint downward to a collecting point.
  NOTE: An enclosure containing a water curtain is called a <u>waterfall booth</u>.

<u>water curtain</u>² (Fire Pr) a. A screen of water thrown between a fire and an exposed surface to prevent it from igniting. b. A row of outside sprinklers provided to protect a building from exposure fires.

<u>water damage</u> (Fire Pr; Salv) Destruction or injury to goods and materials caused by water used to fight a fire.
  NOTE: In certain cases careless application of hose streams may cause as much or even more damage than the fire that was put out.

<u>water drop</u> = <u>water bombing</u>

<u>waterfall booth</u> See: <u>water curtain</u>

<u>water fog</u> (Fire Pr) Finely divided water droplets produced by a <u>fog nozzle</u>. Properly, a spray.

<u>water gas</u> (Techn) A combustible fuel gas, consisting mainly of hydrogen and carbon monoxide, produced by treating coal with steam.

<u>water gel</u> (Civ Eng; Techn) One of a wide variety of materials used for blasting, containing substantial proportions of water and high proportions of ammonium nitrate, some of which is in solution in the water.
  NOTE: Two broad classes of water gels are 1) those containing an explosive, such as TNT or smokeless powder, and 2) those not containing an explosive. These are sensitized with metals, such as powdered aluminum. Water gels may be premixed or mixed at the site immediately before delivery. Also called: <u>slurry explosives</u>.

<u>water glass</u> See: <u>adhesive</u>

<u>water hammer</u> (Fl Mech) A pressure wave traveling along a column of liquid when flow is suddenly halted or the velocity of the fluid is changed abruptly.
  NOTE: The pressure surge travels from the origin to the point of relief and back. The velocity of the surge wave is sonic, and depends on the diameter, thickness, and elasticity of the conduit and on the specific gravity and compressibility of the liquid. With fast-flowing streams, abrupt shut-off creates pressure waves sufficiently strong to damage hoses and pumps.

<u>water head</u> (Fl Mech) The pressure created by the weight of a column of water. More properly, a hydraulic or static pressure head.
  NOTE: A column of water 10 feet high exerts a pressure of 4.33 pounds per square inch.

<u>water hole</u> (Fire Pr) A rural cistern or pond used for fire protection as well as for watering farm stock and other purposes.

<u>water horsepower</u> (Mech; Abb: whp) A unit of work equal to 33,000 foot-pounds, or one horsepower-minute.
  NOTE: One whp is required to lift 33,000 lb (3958.61 gal of water) a distance of 1 foot or to provide a 100-ft pressure head to about 40 gallons of water. Approx 2.5 whp are required to provide a stream of 100 gpm at a pressure equal to 100 feet of head.

<u>water main</u> (Fire Pr) A large conduit that distributes water to smaller branches.

<u>water motor</u> (Fire Pr) A turbine in a water motor alarm that is driven by water flow and in turn drives the water motor gong.

<u>water motor alarm</u> (Fire Pr) A device in a sprinkler system that consists of a water motor and a water motor gong that gives an audible alarm when water flows from opened sprinkler heads or damaged piping.

<u>water motor gong</u> (Fire Pr) A gong driven by a water motor in a water motor alarm to indicate that water is flowing in a sprinkler system.

<u>water motor proportioner</u> (Fire Pr) A device for continuously proportioning foam concentrate into a water flow, independent of flow rate. Two positive-displacement rotary pumps operate on a common shaft in such a manner that one pump acts as a metering device with respect to water flow volume, and the other

injects the foam concentrate.

water pressure See: **hydraulic pressure**

waterproofing agent (Techn) A substance applied to a material to shed water or to prevent water from soaking in.
NOTE: Three types of formulations are common: 1) coating materials; 2) solvents; and 3) plasticizers. Among such materials are cellulose esters and ethers, polyvinylchloride resins or acetates, and variations of vinyl chloride-vinylidine chloride polymers.

watershed (For Serv) A geographic region that drains to a common water course.

water spout (Meteor) A rapidly whirling, funnel-shaped **vortex** over the sea or inland lakes, extending from several hundred to several thousand feet from a cumulus-type cloud to the water surface.
NOTE: The spout can be several feet to several hundred feet thick. The funnel consists of atmospheric moisture and spray drawn up from the water surface. The interior of the funnel is below atmospheric pressure due to the rotation, which can induce wind speeds up to 600 mph. See: **wind**.

water supply (Fire Pr) The quantity of water available for fire fighting.

water supply map (Fire Pr) A map showing the network of water mains, their sizes, locations of hydrants, water flow rates, and static water sources.

water supply point (Fire Pr) A location at which water can be drawn in sufficient quantities for fire-fighting purposes.

water table (Hydrol) The level to which the ground is saturated with subsurface water.

water tanker See: **tanker**

water tender (Fire Pr; Brit; Abb: WrT) 1. A self-propelled apparatus with a 400 to 500 gpm pump, carrying water, hose reel, and extension ladder or wheeled escape. 2. = (US) **pumper-ladder**.

water-thickening agent (Fire Pr) A chemical compound, such as sodium carboxymethylcellulose, that increases the viscosity of water and improves its fire-extinguishing effectiveness.

water thief (Fire Pr) A dividing connector for hoses, usu with one female inlet and two to five male outlets, each with a shut-off valve. Used to branch one or more small lines from one large line. Cf: **siamese**; **wye**.

watertight (Techn) Impervious to water.

water tower[1] (Fire Pr) A fire apparatus with an extendable mast having one or more heavy stream nozzles. The mast can be raised to heights of 60 feet or more. Also called: **water tower truck**.
NOTE: Water towers are not widely used because the same result can be obtained with ladder pipes on aerial ladders.

water tower[2] (Civ Eng) An elevated tank used to store water.

water utility A water supply company. Syn: **waterworks**.

water vacuum (Fire Pr) A suction appliance used to pick up water from floors after a fire.

water vapor (Phys) Water in a gaseous state.
NOTE: The water-vapor content of air may be expressed in terms of dewpoint, but it may also be stated in parts per million by volume (ppm/v), parts per million by weight (ppm/w), pounds of water per pound of dry air, relative humidity, etc.

waterway (Hydraul; Techn) The internal channel for water in a hose, pump, or other equipment.

waterworks = **water utility**

watt (Electr) A unit of power equal to 1 joule/sec or 10^7 erg/sec.
NOTE: A watt of electrical power is developed by one ampere at one volt potential. One horsepower equals 746 watts. When applied to heat measurements, 1 watt = 0.239 cal/sec =

0.000948 Btu/sec.

watt/cm² (Techn) A unit of power density used in measuring the amount of power per area of absorbing surface.

waveform (Electr) The shape of the wave obtained when instantaneous values of an alternating current are plotted against time in rectangular coordinates.

wave height (Navig) The average height from crest to trough of the 1/3 highest waves. Also called: significant wave height.

wavelength (Phys; Symb: l) The distance between two consecutive corresponding points on a wave train.
NOTE: Wavelength is equal to the velocity of the wave divided by its frequency.

wave motion (Phys) The propagation of a disturbance or energy from one point to another without a corresponding transport of the propagating medium.

wave train (Phys) A continuous series of waves.

WBGT = wet bulb globe temperature

weak acid (Chem) An acid that ionizes only slightly in water solutions.

weather (Meteor) The state of the atmosphere as affected by temperature, pressure, wind, humidity, topography, and other factors. See also: fire weather; clouds; wind.

weathering (Civ Eng; Ind Eng) The combined effect of chemical and physical action of wind, rain, freezing water, changes in temperature, plants, and bacteria, that deteriorates exposed surfaces.

weatherproof (Mech Eng) Protected from the weather and environment by design, coatings, coverings, special structural materials, watertightness, and so forth.

weed control (For Serv) A program for minimizing fire hazard by eliminating or reducing weeds.

weft thread (Techn) The set of threads running crosswise in fabrics or woven hose. Syn: woof. See also: warp thread.

weight (Phys) The force with which a gravitational field attracts a body. This force is proportional to the mass of the body and the position of the body in the field. On earth, a body weighs the most at the poles and least at the equator because of centrifugal force. Weight also decreases with distance from the surface of the earth.
NOTE: Weight and mass are easily confused. Mass refers to the quantity of matter in a body (essentially constant) and weight to the force of gravity on a body (variable). The pound (16 ounces) is properly a unit of force equal to the gravitational attraction at the earth's surface on 453.6 grams (mass units).

welding (Techn) A process of joining two metallic pieces together by heat and pressure or by depositing a fillet or spot of fused metal between them. Cf: brazing; soldering. See also: joint.
NOTE: The American Welding Society lists 37 different commercially significant welding processes, grouped under arc welding; gas welding; resistance welding; brazing; thermite welding; induction welding; forge welding; and flow welding.

welding and cutting permit (Fire Pr; Techn) Authorization for the use of open flame and spark-producing devices in areas of combustible or hazardous materials, where their use is normally prohibited.
NOTE: Frequently extended to the use of open flame devices or high-heat producing devices.

welding rod (Techn) A rod or heavy wire that is melted and fused into metals in arc welding.

well (Build; Safety) A permanent enclosure around a fixed ladder which is attached to the walls of the well. Proper clearance for a well will give the person who must climb the ladder the same protection as a cage.

well-head fire (Fire Pr) A fire burning at an oil well.

well-hole car Transp) A flatcar with a depression or opening in the center to allow the load to extend below the normal floor level when it could not otherwise come within the overhead clearance limits.

Western Actuarial Bureau (Insur; Abb: WAB) A research and service bureau headquartered in Chicago that provides technical assistance to the fire rating organizations of several midwestern states. The WAB is a cosponsor of the Fire Department Instructors Conference.

wet adiabatic (Meteor) An adiabatic process involving the latent heat given up in moisture condensation. See: adiabatic.

wet-bulb globe temperature index (Ind Hyg; Abb: WBGT) An index of the heat stress in humans when moderate or heavy work is being performed in a hot environment.

wet-bulb temperature (Meteor) Temperature determined by the wet bulb thermometer of a standard sling psychrometer or its equivalent.
 NOTE: This temperature is influenced by the evaporation rate of the water, which in turn depends on the humidity (quantity of water vapor) in the air.

wet down (Fire Pr) 1. To cool a surface exposed to heat by applying water. 2. To thoroughly soak the remains of a fire. Syn: wash down.

wet pipe system (Fire Pr) An automatic sprinkler system that is charged with water at all times. See: sprinkler system.

wet riser (Build; Fire Pr) A vertical pipe, filled with water, supplying a sprinkler system. Cf: wet standpipe.

wet standpipe (Build; Fire Pr) A charged water pipe installed vertically in a building, connected to a water main, and equipped with hose connections on the individual floors of the building.
 NOTE: In contrast to a dry standpipe, a wet standpipe is charged with water at all times.

wet tank (Techn) A tank designated for the storage of liquids.

wetting agent (Fire Pr) A chemical additive that reduces the surface tension of water or other liquid to make it spread and penetrate more easily. The effect is similar to that of detergents that make water wetter. Also called: penetrant. See: surface-active agent.

wet water (Fire Pr) Water to which a wetting agent has been added to increase its spreading and penetrating characteristics.

wharf (Transp) A platform built parallel to a waterway that permits vessels to moor alongside to load and unload cargo.

Wheatstone bridge (Electr) A circuit used to measure electrical resistance under zero-current conditions.

wheelbase (Transp) The distance between centers of front and rear axles.
 NOTE: For a multiple-axle assembly, the axle center for wheelbase measurement is taken as the midpoint of the assembly. In vehicles equipped with bogie axles, it is the distance from the center of the front wheels to a vertical line that is equidistant from the centers of the two bogie axles.

wheel blocks (Fire Pr) Metal, rubber, or wooden wedges used to prop the wheels of an apparatus to prevent it from rolling. Syn: chocks.

wheel dance (Transp) The vertical vibration of the upsprung mass of a suspension system occurring at the natural frequency of the spring-mass system, consisting of the primary spring elements, the upsprung mass, and the spring characteristics of the tires. Also called: wheel hop.

wheel wobble See: shimmy

whip (Fire Pr) 1. To churn, stir up, or drive, such as flames by a strong wind. 2. To thrash about, as a wild line.

whipline (Mat Hand) A separate hoist rope system of lighter load capacity and higher speed than provided by the main hoist. Also called: auxiliary hoist.

whirlwind (Meteor) A localized cyclonic motion of air, usu only a few hundred feet high and several tens of feet in diameter, originating in a shallow ground layer of unstable air, most often in the afternoon. See: vortex; wind. Cf: dust devil; water spout.

whistle (Fire Pr) 1. A steam whistle once used to sound coded fire alarms and other emergency signals. See: horn. 2. Signals sounded by fire-alarm air horns. 3. Formerly, a whistle on a steam engine used to signal for more fuel.

whitedamp (Mining) Carbon monoxide gas. Also called: sweetdamp. See: damp.

white hat (Fire Pr) A chief officer, from the white helmet worn at the scene of a fire or the white top on the summer uniform cap.

white light (Phys) Polychromatic light.

white noise (Acoust; Electr) A noise whose spectrum density (or spectrum level) is essentially independent of frequency over a specified range.

white spirit (Techn; Brit) 1. A solvent used as a paint thinner, for dry cleaning, etc. 2. = (US) petroleum spirit.

whole body counter (Dosim) A device used to identify and measure the radiation in the body (body burden) of human beings and animals; it uses heavy shielding to keep out background radiation and ultrasensitive scintillation detectors and electronic equipment. Cf: scanner. See: body burden.

whole body irradiation (Dosim) Exposure of the entire body to incident electromagnetic energy or the case in which the cross section of the body is smaller than the cross section of the incident radiation beam.

whole-time (Fire Pr; Brit) Referring to paid, full-time professional firefighters.

whp = water horsepower

wide band (Electr) Characterizing an electromagnetic filter that passes a broad range of frequencies, with neither the critical nor cutoff frequencies being zero or infinite.

widow maker (For Serv; Colloq) An overhanging limb or section of a tree, usu diseased or dead, that could break off and drop to the ground and possibly strike someone on the head.

wildfire (For Serv) 1. = wildland fire. 2. A freely-burning wildland fire that is not affected by fire-suppression measures.

wildland (For Serv) Forests and fields with natural cover, such as trees, brush, and grass, excluding fallow and cultivated land.

wildland fire (For Serv) An unintended forest, brush, grass, or field fire involving land not under cultivation, as contrasted to a prescribed burning fire. Cf: forest fire.

wild line (Fire Pr) An operating hose line not under control or one that has broken loose and thrashes about from the force of the water flowing out of the butt or nozzle.

willful misconduct (Law) Deliberate failure to comply with statutory regulations.

winch (Mat Hand) A hand-powered or machine-powered hoisting machine consisting of a rope drum driven through reduction gearing by a crank handle or power mechanism.

wind (Meteor) A movement of air with respect to the earth's surface due to pressure, centrifugal force, temperature differences, and other factors. Wind is a vector quantity, having both speed and direction. See: Beaufort wind scale; geostrophic wind; gradient wind.
NOTE: Wind is generally one of three types: 1) a worldwide permanent system such as the prevailing and trade winds, 2) a seasonal wind, such as a monsoon, or 3) a local wind.

wind-chill index (Meteor; Physiol) A number that indicates the severity of the effect produced by the combination of low temperature and wind on exposed skin.

NOTE: At a temperature of 20°F, the effect of a 10-mph wind produces an effect subjectively equal to 2°F on the skin; a 20-mph wind, -9°F; 30-mph, -18°F; and in a 40-mph wind exposed flesh will freeze (equal to -22 to -25°F). Wind speeds above 40 mph do not appreciably increase the wind-chill effect. The wind-chill factor is obtained by adding the Fahrenheit temperature to the wind-speed; e.g., -20°F temperature and 15 mph wind speed = wind-chill factor of 35.

wind drag (Aerodyn) Resistance to the motion of a body caused by the wind.

winder (Build) A pie-shaped step of a winding stairway; a step that is wider at one end than at the other.

windfall (For Serv) Trees or parts of trees broken and thrown to the ground by wind.

NOTE: An area in which many trees have been broken by the wind is called a blowdown.

window indicator (Fire Pr) A colorimetric indicator for gas mask canisters that denotes the service life for a particular gas.

wind rose (Cartog) A statistical diagram showing wind direction and speed at a given location.

NOTE: A standard wind rose consists of two concentric circles, representing 10% and 20% frequency, divided radially into eight segments (octants) by lines representing the cardinal and collateral compass directions. Strong, moderate, and light winds are represented by thick, medium, and fine lines whose length and direction on the rose represent wind frequency and direction, respectively.

wind speed (For Serv; Meteor) The speed of the wind, in miles per hour, measured 20 feet above the ground and averaged over a period of at least 10 minutes.

windthrow (For Serv) a. Uprooting of trees by strong winds. b. A tree uprooted by wind. Cf: windfall.

wing curtain (Mining) A baffle used in mines to deflect or channel ventilating air currents.

winze (Mining) A steep downward passageway connecting one working place in a mine with a lower one. A small shaft sunk from an interior point in a mine. The reverse of a raise.

wire cutters (Fire Pr) A tool with insulated handles used to cut electrical power lines, fence wire, and the like. See: forcible-entry tools.

wire rope (Mat Hand; Techn) Steel rope made by twisting or laying a number of strands over a central core, the strands themselves being formed by twisting together steel wires.

wired glass (Build) Window glass with a wire mesh embedded within it to improve its stability under moderate fire exposure and give it shatter-resistant qualities. Sometimes called wire glass.

witness (Law) An individual who is able to testify in a court of law by reason of special knowledge, having been present when an event took place, or the like.

wood (Build; For Serv) A forest product, one of the most widely used structural materials, consisting of about 50% cellulose, 20% hemicellulose, and 30% lignin. Wood ignites above 500°F.

NOTE: Some commonly-used structural woods are fir, spruce, hemlock, pine, oak, and maple; floors and furniture woods include oak, maple, beach, birch, and cypress; furniture and interior finishing woods include walnut, mahogany, teak, gumwood, basswood, and various fruitwoods, such as cherry; shingles and siding are usu of cedar or redwood. Certain woods, such as yellow pine, have a high resin content and therefore burn hotter than other species. Veneer is usu more flammable than solid wood because of its small thickness and the fact that the sheets may tend to delaminate when exposed to heat.

wood alcohol = methyl alcohol

wood cylinder = fuel-moisture indicator stick

wood flour (Techn) Wood reduced to fine particles.

wood frame construction (Build) Construction in which exterior walls, bearing walls and partitions, floor and roof constructions and their supports are of wood or other combustible material, but the construction does not qualify as heavy timber construction or ordinary construction. See: building construction.

wood hook (For Serv) A tool with a point used to assist in the manual handling of bolts of wood. See: pickaroon.

wood irregularities (Build; For Serv) Natural characteristics in or on wood that may lower its durability, strength, or utility.

woodland (For Serv) A tract of land covered with woody vegetation.

wood roof (Build) A roof made of wooden shingles or shakes. Such roofs of untreated shingles are easily ignited by firebrands and present a major hazard in spreading fire.

woods fire (Fire Pr; For Serv) A fire consuming woodland fuels, including small trees and shrubs in wooded areas. Syn: forest fire; wildland fire.

work (Phys; Symb W) A scalar quantity defined as the product of force displacing a body and the distance through which it acts. Specifically, the work (W) done by a force (f) can be found by the equation $W = f \times s \cos\phi$, where s is the displacement of the body and ϕ is the angle between the force vector and the direction of displacement.
NOTE: Common work units are the foot pound, foot poundal, erg, and joule.

worker = working fire

work hardening (Metall) The property of metal to become harder and more brittle on being worked, that is, bent repeatedly or drawn.

working (Mining) An excavation in a mine or an area being excavated.

working chamber (Civ Eng; Techn) A pressurized space or compartment in which work is being done.

working face (Mining) The surface of ore, stone, or coal exposed by mining or quarrying operation.

working fire (Fire Pr) A fire that requires fire-fighting efforts from all fire department personnel assigned to the alarm.

working level month (Dosim) The exposure received by a worker breathing air at one working level concentration for 4-1/3 weeks of 40 hours each.

working line (Fire Pr) An operating line of fire hose.

working load (Techn; Transp) The load imposed by firefighters, materials, and equipment on a platform, vehicle, or hoist.

working platform (Build) A suspended structure arranged for vertical travel that provides access to the exterior of a building or structure.

working rules = shop rules

working surface (Safety) Any surface or plane on which an employee walks or works.

work injury See: occupational injury

workload (Man Sci) Amount of work to be performed by an employee, or output expected in a given period of time.

workmen's compensation (Insur) A system of insurance required by state law and financed by employers that provides payment for medical costs and loss of income to employees or their families for occupational illness, on-the-job injuries, or fatalities, regardless of the cause or blame for the accident.

works fire brigade (Fire Pr; Brit) An on-site, private fire-suppression organization that provides fire protection to a factory or industrial plant. Syn: industrial fire brigade.

work study (Ind Eng) Methods and study of work measurement. The investigation of factors affecting the efficiency, economy, and safety of the operations being examined.

wound (Med) An abrasion, cut, laceration, or puncture of the skin.

woven-jacket hose (Fire Pr) Conventional fire hose made of woven cotton or synthetic fibers, usu lined inside with a layer of rubber. See: hose.
 NOTE: Most fire department hoses have a double jacket. The outer jacket protects the inner one from wear, abrasion, and other damage.

wraparound fire (Fire Pr) A fire that has encircled the central core of a building.

wrecking bar (Fire Pr) A short, heavy-duty steel or iron crowbar with a chisel edge on one end and a curved nail puller on the other. Used as a lever for prying.

writ (Law) An official order requiring an individual to carry out a specified act or to refrain from engaging in some activity.

WrT (Brit) = water tender

wye (Fire Pr) A hose fitting with one female inlet and two male outlets, used to supply two small lines from one large hose line. Wyes are available with or without shut-off valves. Syn: (Brit) dividing breeching. See: breeching. See also: water thief; siamese; gated wye.
 NOTE: A similar coupling with three outlets is called a three-way wye. A similar fitting that unites two lines into one is called a siamese. A hose line divided into two or more branches is called a wyed line.

X

x-ray (Phys) A highly penetrating radiation similar to the gamma ray. Produced by high-energy electron bombardment of a metal target. Syn: Roentgen ray.
 NOTE: X-rays are differentially absorbed by substances of different molecular weights. X-rays are widely used in medical diagnostics and therapy, chemical and physical analyses, spectrometry, etc.

xylonite (Brit) = celluloid

Y

Y See: wye

yard¹ (Metrol) A distance of three feet, or 36 inches; equal to 91.44 centimeters.

yard² (Build) An open space adjacent to a building, usu toward the back of the building; esp a residence. A space not enclosed on two or more sides by the exterior walls of a building. See court.
  NOTE: In reference to institutions, yard may mean a completely enclosed area for physical exercise.

yard hydrant (Fire Pr) An industrial type of hydrant with independent gates for each outlet. The outlets sometimes have fire department pumper connections.

yarding See: skid

yard tractor (Mat Hand) A small semitractor used exclusively for maneuvering transfer trailers into and out of loading position.

yaw (Navig) Deviation about the vertical axis of a body to one side or the other from the desired direction of motion.

yelp (Fire Pr) A whooping sound made by a certain type of fire siren.

yield (Nucl Phys) The total energy released in a nuclear explosion. It is usu expressed in equivalent tons of TNT (the quantity of TNT required to produce a corresponding amount of energy). Cf: fission yield. See: TNT equivalent.
  NOTE: Low yield is generally considered to be less than 20 kilotons; intermediate yield from 200 kilotons to 1 megaton. There is no standardized term to cover yields from 1 megaton upward.

yield point (Mech Eng) a. The stress at which a metal undergoes plastic deformation without increase in load. b. Stress at which a material exhibits a permanent set of 0.2 percent.

yield ratio (Metall) The ratio of yield strength to ultimate tensile strength.

yield strength (Metall) The stress at which a metal exhibits a specified limit of permanent strain.

yield temperature (Metall) The temperature at which a fusible metal or alloy will yield when tested.

Young's modulus (Symb: E) The ratio of the stress (force applied) to the strain (elongation) of an elastic material, usu in reference to a structural metal. Syn: modulus of elasticity.

Z

Z (Chem) The symbol for atomic number.

zenith (Geophys) The point directly overhead in the sky. Ant: nadir.

zephyr (Meteor) A light wind. See: wind.

zero (Math) 1. A mathematical symbol indicating the absence of quantity (the empty set) or magnitude (the number zero). 2. The value of the independent variable of a function for which the function value is zero. 3. A mathematical marker designating an empty position in a positional system of number notation. See also: absolute zero; number system.
  NOTE: As a number, zero has special properties: $a \pm 0 = a$, $a0 = 0$, where a is any number. Division by zero is not a permissible operation in mathematics.

zinc chloride (Chem) A salt, $ZnCl_2$, used in fire-retardant treatment of wood.

zinc oxide fume (Chem; Toxicol) Amorphous white or yellow powder. The powder is essentially nontoxic, but freshly generated fume may cause metal fume fever.

zinced fastening (Techn) A wire rope attachment in which the splayed or fanned wire ends are held in a tapered socket by means of poured molten zinc.

zirconium (Metall; Symb: Zr) A gray, ductile metallic element, at. no. 40, at. wt 91.22, mp 1856°C, bp 4377°C. Used in thermonuclear reactors.
  NOTE: Finely-divided zirconium can ignite spontaneously in air. Once ignited, zirconium burns more violently under water than in air.

zone (Build; Brit) An area protected by a group of detectors. A signal transmitted by any detector in the group also identifies the area. Syn: sector.

zoned system (Build) An air-handling system in which service to different areas in the building is controlled separately, in contrast to a single system that serves the entire building as a unit.

zone of aeration (Hydrol) The area above a water table where the interstices (pores) are not completely filled with water.

zone of capillarity (Hydrol) The area above a water table where some or all of the interstices (pores) are filled with water that is held by capillarity.